Modern Introductory Physics

Charles H. Holbrow • James N. Lloyd
Joseph C. Amato • Enrique Galvez
M. Elizabeth Parks

Modern Introductory Physics

Second Edition

 Springer

Charles H. Holbrow
Charles A. Dana Professor of Physics,
Emeritus
231 Pearl St.
Cambridge, Massachusetts 02139
USA
cholbrow@colgate.edu

James N. Lloyd
Professor of Physics, Emeritus
Colgate University
Department of Physics & Astronomy
37 University Ave.
Hamilton, New York 13346
USA
jlloyd@colgate.edu

Joseph C. Amato
William R. Kenan, Jr. Professor of
Physics, Emeritus
Colgate University
Department of Physics & Astronomy
13 Oak Drive
Hamilton, New York 13346
USA
jamato@colgate.edu

Enrique Galvez
Professor of Physics
Colgate University
Department of Physics & Astronomy
13 Oak Drive
Hamilton, New York 13346
USA
egalvez@colgate.edu

M. Elizabeth Parks
Associate Professor of Physics
Colgate University
Department of Physics & Astronomy
13 Oak Drive
Hamilton, New York 13346
USA
meparks@colgate.edu

ISBN 978-1-4939-3707-3   ISBN 978-0-387-79080-0 (eBook)
DOI 10.1007/978-0-387-79080-0
Springer New York Dordrecht Heidelberg London

Cover: ATM images of O_2 molecules and two O atoms courtesy of Dr. Wilson Ho, Donald
Bren Professor of Physics and Chemistry & Astronomy, Department of Physics and Astronomy,
University of California, Irvine.

Printed on acid-free paper

Springer is part of Springer Science+Business Media (www.springer.com)

... all things are made of atoms—little particles that move around in perpetual motion, attracting each other when they are a little distance apart, but repelling upon being squeezed into one another.

In that one sentence, you will see, there is an *enormous* amount of information about the world, if just a little imagination and thinking are applied.

— Richard P. Feynman

Preface

This book grew out of an ongoing effort to modernize Colgate University's three-term, introductory, calculus-level physics course. The book is for the first term of this course and is intended to help first-year college students make a good transition from high-school physics to university physics.

The book concentrates on the physics that explains why we believe that atoms exist and have the properties we ascribe to them. This story line, which motivates much of our professional research, has helped us limit the material presented to a more humane and more realistic amount than is presented in many beginning university physics courses. The theme of atoms also supports the presentation of more non-Newtonian topics and ideas than is customary in the first term of calculus-level physics. We think it is important and desirable to introduce students sooner than usual to some of the major ideas that shape contemporary physicists' views of the nature and behavior of matter. Here in the second decade of the twenty-first century such a goal seems particularly appropriate.

The quantum nature of atoms and light and the mysteries associated with quantum behavior clearly interest our students. By adding and emphasizing more modern content, we seek not only to present some of the physics that engages contemporary physicists but also to attract students to take more physics. Only a few of our beginning physics students come to us sharply focused on physics or astronomy. Nearly all of them, however, have taken physics in high school and found it interesting. Because we love physics and believe that its study will open students' minds to an extraordinary view of the world and the universe and also prepare them well for an enormous range of roles—citizen, manager, Wall-Street broker, lawyer, physician, engineer, professional scientist, teachers of all kinds—we want them all to choose undergraduate physics as a major.

We think the theme and content of this book help us to missionize more effectively by stimulating student interest. This approach also makes our weekly physics colloquia somewhat accessible to students before the end of their first year.[1]

In parallel with presenting more twentieth-century physics earlier than is usual in beginning physics, this book also emphasizes the exercise and development of skills of quantitative reasoning and analysis. Many of our students come fairly well prepared in both physics and math—an appreciable number have had some calculus—but they are often rusty in basic quantitative skills. Many quite capable students lack facility in working with powers-of-ten notation, performing simple algebraic manipulation, making and understanding scaling arguments, and applying the rudiments of trigonometry. The frustrations that result when such students are exposed to what we would like to think is "normal discourse" in a physics lecture or recitation clearly drive many of them out of physics. Therefore, in this first term of calculus-level physics we use very little calculus but strongly emphasize problems, order-of-magnitude calculations, and descriptions of physics that exercise students in basic quantitative skills.

To reduce the amount of confusing detail in the book, we often omit interesting (to the authors) facts that are not immediately pertinent to the topic under consideration. We also limit the precision with which we treat topics. If we think that a less precise presentation will give the student a better intuitive grasp of the physics, we use that approach. For example, for the physical quantities mass, length, time, and charge, we stress definitions more directly connected to perceivable experience and pay little attention to the detailed, technically correct SI definitions. This same emphasis on physical understanding guides us in our use of the history of physics. Many physical concepts and their interrelations require a historical framework if they are to be understood well. Often history illustrates how physics works by showing *how* we come to new knowledge. But if we think that the historical framework will hinder understanding, we take other approaches. This means that although we have tried diligently to

---

[1] These and other aspects of the approach of this book are discussed in more detail in C.H. Holbrow, J.C. Amato, E.J. Galvez, and J.N. Lloyd, "Modernizing introductory physics," *Am. J. Phys.* **63**, 1078–1090 (1995); J.C. Amato, E.J. Galvez, H. Helm, C.H. Holbrow, D.F. Holcomb, J.N. Lloyd and V.N. Mansfield, "Modern introductory physics at Colgate," pp. 153–157, *Conference on the Introductory Physics Course on the Occasion of the Retirement of Robert Resnick*, edited by Jack Wilson, John Wiley & Sons, Inc., New York, 1997; C.H. Holbrow and J.C. Amato, "Inward bound/outward bound: modern introductory physics at Colgate," in *The Changing Role of Physics Departments in Modern Universities*, pp. 615–622, Proceedings of International Conference on Undergraduate Physics Education, College Park, Maryland, August 1996, edited by E.F. Redish and J.S. Rigden, AIP Conference Proceedings **399**, Woodbury, New York, 1997.

avoid saying things that are flat out historically wrong, we do subordinate history to our pedagogical goals.

We believe that it is important for students to see how the ideas of physics are inferred from data and how data are acquired. Clarity and concision put limits on how much of this messy process beginning students should be exposed to, but we have attempted to introduce them to the realities of experimentation by including diagrams of apparatus and tables of data from actual experiments. Inference from tables and graphs of data is as important a quantitative skill as the others mentioned above.

Asking students to interpret data as physicists have (or might have) published them fits well with having beginning physics students use computer spreadsheets to analyze data and make graphical displays. Because computer spreadsheets are relatively easy to learn and are widely used outside of physics, knowledge of them is likely to be useful to our students whether they go on in physics or not. Therefore, we are willing to have our students take a little time from learning physics to learn to use a spreadsheet package. Some spreadsheet exercises are included as problems in this book.

The examination of significant experiments and their data is all very well, but nothing substitutes for actual experiences of observation and measurement. The ten or so laboratory experiments that we have developed to go along with this course are very important to its aims. This is particularly so, since we observe that increasingly our students come to us with little experience with actual physical phenomena and objects. We think it is critically important for students themselves to produce beams of electrons and bend them in magnetic fields, to create and measure interference patterns, to observe and measure electrolysis, etc. Therefore, although we believe our book will be useful without an accompanying laboratory, it is our heartfelt recommendation that there be one.

Although our book has been developed for the first of three terms of introductory physics taken by reasonably well-prepared and well-motivated students, it can be useful in other circumstances. The book is particularly suitable for students whose high-school physics has left them with a desire to know more physics, but not much more. For them a course based on this book can stand alone as an introduction to modern physics. The book can also work with less well prepared students if the material is spread out over two terms. Then the teacher can supplement the coverage of the material of the first several chapters and build a solid foundation for the last half of the book.

The format and techniques in which physics is presented strongly affect student learning. In teaching from this book we have used many innovative pedagogical ideas and techniques of the sort vigorously presented over many years by well-known physics pedagogues such as Arnold Arons, Lillian McDermott, Priscilla Laws, Eric Mazur, David Hestenes, and Alan van Heuvelen. In one form or another they emphasize actively engaging the students and shaping instruction in such a way as to force students to confront, recognize, and correct their misconceptions. To apply these ideas we teach the course as two lectures and two small-group recitations each week. In the lectures we use Mazur-style questions; in the recitations we have students work in-class exercises together; we spend considerable effort to make exams and special exercises reach deeper than simple numerical substitution.

Drawing on more than ten years of experience teaching from *Modern Introductory Physics*, we have significantly revised it. Our revisions correct errors in the 1999 edition and provide clearer language and more complete presentation of important concepts. We have also reordered the chapters on the discovery of the nucleus, the Bohr model of the atom, and the Heisenberg uncertainty principle to better tell the story of the ongoing discovery of the atom.

Our boldest innovation is the addition of two chapters on basic features of quantum mechanics. In the context of real experiments, these chapters introduce students to some of the profoundly puzzling consequences of quantum theory. Chapter 19 introduces superposition using Richard Feynman's approach; Chap. 20 discusses quantum entanglement, the violation of Bells inequalities, and experiments that vindicate quantum mechanics. Superposition, entanglement, non-locality, and Bell's inequalities are part of the remarkable success story of quantum mechanics. We want acquaintance with these important ideas to alert students to themes and technologies of twenty-first century physics. We want our book, which unfolds the ideas and discoveries that led to the quantum revolution, to end by opening for students a window into a future shaped by themes and emerging technologies that rely fundamentally on quantum mechanics.

Many colleagues helped us make this a more effective book with their useful critiques, problems, exercises, insights, or encouragement. For these we are grateful to Victor Mansfield (1941–2008), Hugh Helm (1931–2007), Shimon Malin, Stephen FitzGerald, Scott Lacey, Prabasaj Paul, Kurt Andresen, Pat Crotty, Jonathan Levine, Jeff Buboltz, and Ken Segall.

Deciding what specific subject matter should go into beginning physics has been a relatively small part of the past 30 years' lively discussions of pedagogical innovation in introductory physics. We hope our book will help to move this important concern further up the agenda of physics teachers. We think the content and subject emphases of introductory physics are a central responsibility of physics teachers and of great importance to the long-term health of the physics community. This book represents our idea of a significant step toward making introductory physics better represent what physics is. Whether or not we have succeeded, we hope this book will stimulate discussion about, encourage experimentation with, and draw more attention to the content of undergraduate introductory physics.

Charles H. Holbrow
James N. Lloyd
Joseph C. Amato
Enrique Galvez
M. Elizabeth Parks
Colgate University
August, 2010

Contents

C H A P T E R

1

What's Going On Here?

1.1 WHAT IS PHYSICS?

From earliest times humans have speculated about the nature of matter. The Greeks with their characteristic genius developed a highly systematic set of ideas about matter. They called these ideas "physics," but physics in the modern sense of the word comes into being only in the seventeenth century.

In 1638 Galileo Galilei published *Dialogues Concerning Two New Sciences*[1] which summarized a lifetime's work that created the description of motion that we use today. A generation later Isaac Newton made a grand synthesis with his laws of motion and his famous law of universal gravitation.[2]

These two great physicists introduced two exceptionally important ideas that characterize physics still. First, physics is a *mathematical* description of natural phenomena, a description of underlying simple relationships from which the complicated and various behavior of observed matter can be inferred. Second, the predictions or inferences must be

---

[1]This is a wonderful book available from Dover Publications in a paperback edition. It describes basic features of the science of the strength of materials; and it presents the first mathematical account of that part of physics that we call "mechanics." The mathematics is plane geometry and accessible to anyone with a good high-school education. Here Galileo presents his arguments, both theoretical and experimental, for the law of falling bodies and the resulting possible motions.

[2]These ideas were published in *Principia Mathematica*. This book was written in Latin but is available in English translation. (R.T. Jones, the eminent NASA engineer who played an important role in developing the delta-wing aircraft once said that he learned physics by reading the *Principia*. That is a very strong endorsement.)

C.H. Holbrow et al., *Modern Introductory Physics, Second Edition*,
DOI 10.1007/978-0-387-79080-0_1, © Springer Science+Business Media, LLC 1999, 2010

checked by measurements and observations. Physicists create quantitative descriptions of the behavior of matter and then examine the consistency and accuracy of these descriptions by philosophical, mathematical, and experimental study.

Thus, when you say that bodies fall, you are not really doing physics. But when you say all bodies fall with constant acceleration, you are propounding a generalization in mathematical form, and you have begun to do some physics. When from that statement you deduce logically that trajectories are parabolas and that the maximum range occurs when a body is launched at 45°, you are doing physics. When you devise arguments and instruments to measure and show that near the surface of the Earth all bodies fall with a constant acceleration $g = 9.8\,\mathrm{m\,s^{-2}}$ and that actual bodies do move almost as you predict, then you are doing more physics. And when you are able to explain quantitatively that observed deviations from your predictions are due to variations in the distance from the surface of the earth and the effects of air resistance, you are doing deep physics. And when you create new concepts in order to construct a quantitative explanation of why falling bodies have constant acceleration in the first place, you are doing physics at a deeper level yet.

Physicists are students of the behavior and structure of matter. This phrase covers a multitude of activities. The Physics and Astronomy Classification Scheme[3] or PACS, as it is also called, lists approximately 4000 short phrases describing the different things physicists are busy at—from "communication, education, history, and philosophy" through "exotic atoms and molecules (containing mesons, muons and other unusual particles)" to "stellar systems; interstellar medium; galactic and extragalactic objects and systems; the Universe." The variety is astonishing.

▪ EXERCISES

---

1. In the PACS listing, find the "General Physics" category, and then find the subcategory that features "Instruments, apparatus, and components common to several branches of physics and astronomy." List four subject headings pertaining to techniques of producing and measuring vacuum.

---

[3]Use a Web browser and go to http://www.aip.org/pacs/. A look at the myriad of categories and subcategories and sub-subcategories reveals a wonderland of strange words and jargon. If you like language, you might like to peruse the PACS.

> **2.** In the "Condensed Matter: Structure, Mechanical, and Thermal Properties" category, find four listings having to do with defects in crystals.
>
> **3.** If you were looking for papers that discussed the decay of pi mesons (an elementary particle), under what listing might you search?
>
> **4.** Find four listings that discuss different kinds of galaxies.

1.2 WHAT IS INTRODUCTORY PHYSICS ABOUT?

You can see that physics can include almost everything. What then is going to be in this book? Well, it is a physics book intended for people with a serious interest in science, but it is different from the usual introductory physics textbook.

Most introductions to physics begin with the mathematical description of motion. They talk about forces, momentum, energy, rotational motion, oscillations. They discuss heat and temperature and the laws of thermodynamics, and they treat electricity and magnetism plus some optics. There is a notoriously numbing quality about this approach. That may be unavoidable, since a goal of the course is to change the structure of your brain, which is full of deeply ingrained misconceptions. The misconceptions have to be straightened out. Also, you need to overcome your resistance to the sharpness and lack of ambiguity that are part of quantitative thought, and you need to be strengthened against blanking out during the long chains of inference by which physicists connect the basic ideas of physics to the observable world. Restructuring your thinking is uncomfortable, and many people are not able to accept the very real "present pain" for the prospective "future pleasure" of greatly enhanced powers of understanding the natural world. Our official recommendation to you is: "Be strong, be brave, be persistent. Hang in there."[4]

---

[4]Perhaps Winston Churchill's words say it more firmly: "Never give in. Never give in! Never, never, never, never. Never give in except to convictions of honor and good sense."

1.3   WHAT WE'RE UP TO

This book is based on some different ideas about how to start physics. They are the basis of a significant change in the teaching of the introductory course. Rather than start with seventeenth-century physics and work our way through to the nineteenth century, we are going to emphasize some ideas that have dominated physics in the twentieth and twenty-first centuries. Modern physics is quite different from physics of past centuries. It seems to us that introductory physics should introduce you to what we physicists actually do.

Isn't this dangerous? The ideas of physics are cumulative. To talk meaningfully about what is going on deep in an atomic nucleus, you must understand velocity as Galileo used the idea; you need to know about potential energy—an idea developed in the eighteenth century; you need to know about electric charge, about momentum, about kinetic energy. The usual theory of teaching physics is to introduce these ideas in terms of simple, more directly observable phenomena, and then apply the ideas in increasingly complicated ways. Build the foundation first, then put up the building. By starting with the physics of this century isn't there a danger that we will erect a superstructure with no foundation?

We don't think so. For one thing, you all have a bit of foundation. You know what velocity is, you have heard about acceleration. You have talked about energy and momentum in your high-school physics course. For another, we are not going to be dogmatic about sticking to the twentieth century. If we need to spend some time reviewing or introducing some basic ideas, we will. Furthermore, we are not going to do the hardest parts of modern physics. We introduce enough quantum mechanics to explore some deep questions, but we do this without advanced mathematics. We present Einstein's special theory of relativity in a very "nuts and bolts" fashion. You will have to wait for more advanced courses to see the powerful and elegant mathematical treatments of these two cornerstones of modern physics.

But there is a more important reason why our approach should work. An enormous amount of twentieth-century physics is done with simple ideas and mathematics no more complicated than algebra. Do not think that because ideas are simple, they are trivial. Simple ideas are often used with elegant subtlety to do physics. You can learn enough about waves, particles, energy, momentum, uncertainty, scattering, and mass to make a remarkably comprehensive and consistent picture of the nature of matter without having to know all the underlying connections among the ideas. The more complete elucidation of the connections can wait until later courses.

After all, you would not familiarize yourself with a skyscraper by first studying all its plumbing diagrams and then its wiring diagrams, and then its ductwork, and the arrangement of its girders, and so on until you are familiar with all the parts, and only then assemble them in your mind to create the skyscraper. You must do that if you are building a new building, but if the building is already there, you need to know first where the main doors are, where the express elevators are, and on what floors are the important offices, and how some of the suites are connected. You can visit what seems to be of particular interest without knowing the details of the building's construction. Of course, to operate and really appreciate the building you will eventually need to know and understand the details. But not right away.

Physics is a skyscraper of imposing dimensions. This course will show you some of its rooms and some of the furniture in those rooms. You should learn enough so that you can rearrange the furniture in interesting ways as well as get from one room to another. Later courses will go back to the seventeenth century and look at the foundations of physics; then you will go down into the utility rooms of our edifice and see what's there. In this book we will stay upstairs where the view is better (Fig. 1.1). Once you know how to get to the windows in the skyscraper of physics, you can look out over the entire panorama of nature laid out in the PACS, from subnucleonic quarks and leptons to the ends of the visible universe.

1.4   THIS COURSE TELLS A STORY

The Short Story of the Atom

Physics helps us to understand the physical world. It extends our perceptions beyond our immediate senses and opens new vistas of comprehension. We think you can understand physics better if the physics you learn tells a story. There are many stories to tell with physics, so we had to choose one. We chose what we think to be the most significant story of the past two centuries. It is the story of the atom and its nucleus. We want you to know both **what** physics teaches us about atoms and their remarkable properties and **why** we believe atoms and nuclei exist and have the properties we ascribe to them.

The story is a good one. It starts in the early nineteenth century with hard, featureless atoms. They become more complicated as more is learned about them. They explain many observations by chemists and many of the observed properties of gases. By the middle of the nineteenth century, the kinetic theory of hard-sphere atoms makes it possible to know that the diameters of atoms are of the order of 10^{-9} m, some nine or ten orders of

FIGURE 1.1 You don't need to know how to build a skyscraper to appreciate the view from it. *Photo courtesy of Mary Holbrow.*

magnitude smaller than familiar everyday objects. Then their electrical nature is discovered, and by the end of the century atoms are known to be made of positive and negative charges. The negative charges are found to be tiny elementary particles that are named "electrons." Their mass and charge are determined.

At the threshold of the twentieth century, radioactivity reveals new complexity of the atom. The compact core, or "nucleus," of the atom is discovered. It is 10^{-5} the size of the atom and contains 99.97% of its mass. It signals the existence of new elementary particles, the proton and the neutron, and the existence of a previously unknown fundamental force of nature more than 100 times stronger than the familiar electrical forces, and 10^{40} times stronger than gravity. This new force is the agent by which nuclei store the extraordinarily large amounts of energy that can be released by nuclear fission and fusion. The search for political and social controls of these energies remains a major preoccupation of our twenty-first century world.

Special Relativity and Quantum Mechanics

At the heart of our story lie two strange new ideas that revolutionized our view of the physical world. The first idea is that there is a limiting velocity in the universe; nothing can travel in a vacuum faster than the speed of light. This idea and the idea that the laws of physics must not depend on the frame of reference in which they are studied are central to Einstein's special theory of relativity. The consequences of this theory are necessary to understand the behavior of atoms or their components at high energies. This behavior is surprising and unfamiliar to beings whose experience with the physical world is limited to velocities much less than that of light.

■ EXERCISES

> **5.** What beings might these be?

Stranger still are the ideas of quantum mechanics. The behavior of atoms and their components can only be understood if, unlike the familiar particles of our world—marbles, raindrops, BBs, baseballs, planets, sand grains, bacteria, etc.—they do not have well-defined locations in space, but are spread out in some fashion like water waves or sound waves. In fact, in some sense they must be in more than one place at the same time. To describe atoms and the details of their behavior we must use these peculiar ideas mixed together with a fundamental randomness that physicists schooled in the ideas of Newton have found difficult to accept.

Physics Is Not a Spectator Sport

In this course the "why" of your understanding is extremely important. After all, you already believe in atoms. You don't need convincing. You accept their existence as matters of faith, and you will probably believe most things we tell you about them. Of course, you will need to know many things of and about physics, but it is also important that you learn to make arguments like physicists. We want you to learn what convinces physicists and what does not. In the end, we want you not so much to know the story as to be able to convince yourself and others that it is true. We want you to learn to follow and use quantitative arguments and to be able to describe the posing of questions of physics as experiments.

Of course, physicists, like everyone else, teach and learn as much by authority as by proof. Because there is not time or will, we will often just tell you that something is so in order to pass on to larger issues. Nevertheless, this introductory physics course lays more stress on argument than the traditional course. There are reasons you may not like this. It requires thinking, and thinking is uncomfortable, muddy, difficult, ambiguous, and inefficient. It requires you to participate actively rather than passively. It means that your textbook—this very book—must be different from the traditional text.

Most introductory physics courses greatly emphasize the working of problems. Homework and quizzes and exams require you to work problems that illustrate the topics of the book. Most students respond by reading the assigned problems and leafing backwards through the chapter until they find the equations that produce an answer. The text becomes a reference manual for solving problems, and it is not read for any broader comprehension.

We have tried to create a text that has to be read for broader comprehension, a book that does not serve merely as a user's manual for solving assigned problems. We want you to read the book and think about it as you go along. This does not mean that problems are unimportant; they are very important. They are how you test your understanding. Trying to work a problem is the quickest way to show the emptiness of the understanding you thought you gained when you passed your eyes over pages of print without repeated pauses to think. As you read physics, you should be asking yourself questions. To show you how this works we have put questions in among the paragraphs of the text where you should be asking them. In general, you should work exercises as you go along; if they aren't provided, you should make them up yourself.

For starters and to establish our basically kindly nature as authors, we have provided some questions for you. For instance, when you were reading above that a new extremely strong force was discovered, did you ask:

◼ EXERCISES

> **6.** How much stronger is the electromagnetic force than the gravitational force?
>
> Or you might have wondered:

> **7.** What does it mean for one force to be stronger than another?
>
> And because we think it might help you with your thinking about that question, we might ask:
>
> **8.** What do we mean when we say, "Lead is heavier than air"? Which weighs more, a pound of air or a pound of lead?
>
> And, of course, you might wonder:
>
> **9.** If the new force is as strong as we say, why wasn't it discovered much earlier in time?

If you did not already know, you can see that reading physics is slow work. Ten pages an hour is quite fast; five pages an hour is not unreasonable. And for the new and very strange, a page a day or a week is not inconceivable. Reading and working problems go hand in hand.

1.5 WHY THIS STORY?

An Important Idea

We have chosen to make atoms the central theme of introductory physics for three main reasons. First, the idea of the atom is extremely important. Our ideas about atoms color our understanding of all of nature and of all other sciences. One of the greatest physicists of the twentieth century, Richard Feynman, has written[5]

> If, in some cataclysm, all of scientific knowledge were to be destroyed, and only one sentence passed on to the next generations of creatures, what statement would contain the most information in the fewest words? I believe it is the *atomic hypothesis* (or the atomic *fact*, or whatever you wish to call it) that *all things are made of atoms—little particles that move around in perpetual motion, attracting each other when they are a little distance apart, but repelling upon being squeezed into one another.* In that one sentence, you will see, there is an *enormous* amount of information about the world, if just a little imagination and thinking are applied.

---

[5] Richard P. Feynman, Robert B. Leighton, and Matthew Sands, *The Feynman Lectures on Physics*, pp. 1–2, vol. I (Addison-Wesley, Reading MA, 1963).

Tools for Quantitative Thought

Second, the arguments and evidence we use to infer the existence and properties of atoms are in many ways easier to understand than the arguments of traditional Newtonian physics. Some of the ideas are stranger than Newton's because they are unfamiliar, but, up to a point, the mathematics underlying them is simpler. We can learn a great deal by rough, order-of-magnitude, numerical calculations, and by using proportionality, plane geometry, some trigonometry, and how the sizes of simple functions scale as their variables are changed. These tools of rational argument are basic in all the branches of physics, in all sciences, and in any kind of practical work you may do—from making dinner to running a large corporation.[6] A major aim of this course is to have you become skillful with these simple mathematical tools.

An Introduction to Physics

Third, an introductory physics course built around the theme of atoms will give you a better sense of what physics is and what physicists do than a traditional course would. Most physicists today study atoms or their components and how they interact and behave under different conditions. Many of the deep unanswered questions of physics center on aspects of the behavior of atoms or their parts.

1.6   JUST DO IT!

This book will teach you the basic physics you need to know in order to understand why we believe in atoms and their properties. This will require learning much traditional physics, but it will be applied in a different context than is usual in beginning physics. We think that the physics you learn this way will make more sense to you, that the larger context will help you perceive that physics is not a disconnected set of formulas used to solve disconnected sets of problems. We also want you to learn what physicists think they know, what they think they don't know, and how they go about learning new physics. The "how" is very important, because as you learn "how" physicists do physics and persuade each other of the truth of what they do, you will be learning how to teach yourself physics. Learning to teach yourself is a goal for the long term. For most people it takes years, but then real mastery becomes possible.

---

[6]Some people argue that these tools are so basic to constructive thought and practical action and physics is such a good place to learn them that every college student should take physics. The same sort of argument could be made for lifting weights.

PROBLEMS

1. Check the website for your school's physics department. Find the listing of seminars. Look at the upcoming or recent seminar titles, and find a few that look related to the topics you'll learn this semester.

2. Look at the faculty research interests for your department. Which ones, if any, look like they might be related to the subject of this course? You'll want to look again at the end of the semester; it's likely that you'll find you've learned a lot that is related to current research areas.

C H A P T E R

2.

Some Physics You Need to Know

2.1 INTRODUCTION

In this chapter you get reviews of length, mass, time, velocity, acceleration, momentum, and energy along with accounts of their characteristic physical dimensions and the units used to describe them quantitatively. You get a review of angle measure with particular emphasis on the radian and its use. And you get two important tips: how to check the consistency of physics equations, and how to work efficiently with SI prefixes.

2.2 LENGTH, MASS, TIME: FUNDAMENTAL PHYSICAL PROPERTIES

Length, mass, and time are fundamental physics ideas important in all the sciences. From them physicists build up more elaborate physics concepts such as velocity, acceleration, momentum, and energy. To work comfortably with these concepts you need some intuitive physical feeling about each; you need to know about their physical dimensions (to be explained); and you need to be able to describe each of them quantitatively in a consistent set of units.

Most of the units used in this book are part of the internationally agreed-upon Système International (SI).[1] Although the basic units of length, mass, and time have been very precisely defined by an

---

[1] The US National Institute of Science and Technology (NIST) provides a complete presentation of SI units at http://physics.nist.gov/cuu/Units/units.html.

C.H. Holbrow et al., *Modern Introductory Physics, Second Edition*,
DOI 10.1007/978-0-387-79080-0_2, © Springer Science+Business Media, LLC 1999, 2010

international committee, there are approximate definitions of these units that are more useful for you to know:

- one meter is about the length between your nose and the end of your fingers when your arm is stretched out to the side;

- one kilogram is the mass of a liter of water, a little more than a quart;

- One second is about the time between beats of your heart when you are at rest (as when reading a textbook?).

All SI units can be scaled by powers of ten by means of standard prefixes such as "micro," "kilo," and "mega," and corresponding standard abbreviations like μ, k, and M. Many of these are introduced in this chapter. Watch for them and make a special effort to learn them. To work with units you have to be able to manipulate these standard multipliers and convert from one to another efficiently. There is a summary of SI prefixes on p. 633.

Length

You already have a good intuitive idea of length. However, there are so many different units for measuring lengths—barley corns, furlongs, chains, rods, yards, feet, miles, and light-years to name a few—that an international effort has defined the meter and made it the standard unit of length: all other units of length are now defined in terms of the meter. In the 1790s, at the time of the French Revolution, the French Academy of Sciences set up a consistent set of units of length, mass, and time in terms of standards existing in nature and so, at least in principle, accessible to any observer anywhere. The meter was then defined to be one ten-millionth (10^{-7}) of one-fourth of the circumference of Earth, (see Fig. 2.1), and two marks this distance apart were put on a particular bar of metal that

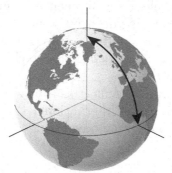

FIGURE 2.1 The meter was originally defined as one ten-millionth (10^{-7}) of one-fourth of Earth's circumference, the distance from the Equator to the North Pole.

became the international standard. Today, the official definition of the meter used by all scientists depends on the speed of light and how we measure time. For now, however, ignore the official definition, and use the historical definition: To sufficient accuracy for our purposes the meter is 10^{-7} of one quadrant of Earth's circumference.

By knowing the historical definition of the meter, you know that Earth's circumference is very nearly 40 million meters, that is, 40×10^6 m. This system of units—the metric system—also introduced the idea of using prefixes to scale units up or down by factors of 10. Thus distances on Earth are commonly measured in multiples of a thousand meters, i.e., kilometers, where, as suggested above, the prefix "kilo" stands for 1000 (10^3). Thus it is common to say that the circumference of Earth is 40 000 km (where "k" and "m" are the standard abbreviations respectively for "kilo" and "meter"). Indeed "kilo" means 1000 wherever it is used in scientific work and can be used to scale any unit: A kiloanything (ka?) is 1000 anythings.

You could also say the circumference is 40 megameters, i.e., 40 Mm, where capital "M" is the abbreviation for "mega" which stands for 10^6, one million. You could even say that Earth's circumference is 40×10^9 mm, where "mm" means millimeters because the first lower case "m" stands for "milli" or 10^{-3} or 1 one-thousandth. But why would you?

▼ EXAMPLES

1. Knowing Earth's circumference, you can calculate its radius or diameter. The circumference of a circle of radius R is $2\pi R$. Therefore, $R_{\text{Earth}} = (40 \times 10^6 \text{ m})/(2\pi) = 6.37 \times 10^6 \text{ m}$, or 6370 km.

Mass

You experience mass when you push an object. It takes great effort to get an automobile rolling on a level surface, and most of the car's resistance is due to its mass. When you pull a quart of milk from the refrigerator, you sense its mass of almost one kilogram (10^3 grams).

Using the standard abbreviation kg for kilogram, you would write or say that a quart of milk[2] has a mass of 0.979 kg. In handling the quart of milk you also resist its weight due to the pull of gravity, but we won't be overly concerned about the difference between weight and mass and will follow the custom of most of the people of the world who measure the

---

[2] A quart of water has mass 0.946 kg, but milk is slightly more dense than water.

TABLE 2.1   Masses of some familiar objects

Object	Mass (kg)	Object	Mass (kg)
Golf ball	0.050	Basketball	0.600
Tennis ball	0.057	1 liter of water	1.000
Baseball	0.149	Bowling ball	7.0
Hockey puck	0.160	4 × 4 SUV	1500.

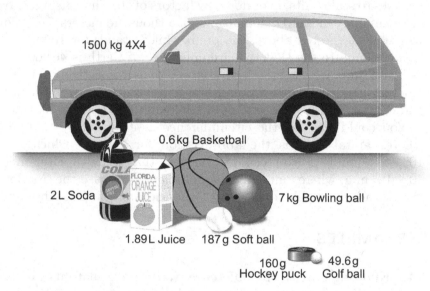

FIGURE 2.2   Some familiar objects with masses you may have experienced.

weight of flour, potatoes, and the like in "kilos." Masses of some objects with which you may be familiar are listed in Table 2.1, and some of these objects are shown in Fig. 2.2.

The unit of mass, the kilogram, was originally chosen to be the mass of one liter (abbreviated L) of water at a particular temperature and pressure. Cola and other important beverages are sold in 2 L bottles. From the definition of the kilogram, you can quickly estimate the mass of the liquid in a bottle of soda. (What do you assume when you do this?)

Because a liter is defined to be 1000 cubic centimeters and one kg is 1000 g, the mass of $1\,\mathrm{mL} = 1\,\mathrm{cm}^3$ of water is $1\,\mathrm{g}$. The metric system was constructed to have these interrelationships, and it is very convenient to know them. You can see that the kilogram's size was defined to make the density of water equal to $1\,\mathrm{kg/L}$ or $1\,\mathrm{g\,cm}^{-3}$—a useful fact and, thanks to these units, easy to remember.

Density

Notice that density is measured in units composed from other units—those of mass and volume. The SI unit of volume, the cubic meter, is also composite, i.e., m^3. Most physical properties are measured in composite units; Sect. 2.3 will discuss how such units obey the rules of algebra.

The concept of density was devised to compare the relative masses of different materials independently of their volumes. In everyday language you might say iron is "heavier" than water. Of course this does not mean that any iron object is heavier than any other amount of water. You certainly can have a small piece of iron that is lighter than a bucket of water. Yet we are quite clear when we see a stone sink in a lake or a cork float on water that the first is heavier than water and the second is lighter. The idea of density comes to the rescue here: The important thing is whether for equal volumes one object has more or less mass than the other. For making such comparisons it is convenient to use a unit volume and observe that 1 L of iron has a mass of 7.9 kg, i.e., a density 7.9 kg/L, compared to water's density of 1 kg/L. Equivalently, 1 cubic centimeter ($1 \, cm^3$) of iron has a mass of 7.9 g, and $1 \, cm^3$ of water has a mass of 1 g.

■ EXERCISES

1. A cubic foot of water has a mass of 62.4 pounds. What is the ratio of the mass of a cubic foot of iron to the mass of a cubic foot of water? Hint: Reread the last sentence of the preceding paragraph.

With these simple ideas you can do a lot of science.

■ EXERCISES

2. For example, what is your density? You know that when you are swimming you can float but only just barely. Therefore your density must be a trifle less than that of water. From that fact what might you guess to be your principal chemical ingredient? Estimate your volume in liters and in cubic centimeters.

The above exercise is not really as obvious as it might seem. As an experienced student, you may have arrived at the answer more by what the instructor seems to want than by thoughtful analysis. But if you were aware that you assumed an average density for the human body, then you are beginning to do real science.

Is that assumption realistic? You might imagine continuing your scientific analysis by testing the consequences of the assumption. Get a steak from the store and determine its density, then dry it thoroughly and measure the mass of the dried remains. You would indeed conclude that the meat was mostly water. But the bones in the meat are clearly much more dense than water. How, then, can the average density of the body be so close to that of water? How is it that an incorrect assumption led to the right answer? You would have to conclude that there were compensating volumes of density less than that of water, such as the lungs and head cavities.

But now an interesting question comes up. Why is the average density of the human body so close to that of water? You are getting into some profound evolutionary questions at this point and straying from our main topic. But you see how simple questions can lead to much deeper ones.

▼ EXAMPLES

2. Suppose you decide to become fabulously wealthy by running the "guess-your-mass concession" at your favorite carnival. If someone 1.8 m tall with a waist size of 0.8 m approaches, what would you estimate his mass to be? Try modeling him as a cylinder. His waist size is his circumference $2\pi r$, where r is his radius. The volume V of a cylinder with a height $h = 1.8$ m and a radius $r = (0.8 \text{ m})/(2\pi)$ is $V = \pi r^2 h = 0.092 \text{ m}^3$. Notice that we were careful to do this problem in consistent units, so the answer comes out in cubic meters. Since $1 \text{ m} = 10^2 \text{ cm}$, it follows that $1 \text{ m}^3 = 10^6 \text{ cm}^3$. Therefore, the volume V is $92.0 \times 10^3 \text{ cm}^3$, which is 92 L, or about 92 kg.

We have been doing some physics here. First, we developed quantitative concepts of volume, mass, and density; then we made a mathematical model of our subject and applied the concepts. But that's not enough; one always needs to verify a model with an experiment. So we weighed a 1.8 m tall, 0.8 m circumference author and found his actual mass to be 82 kg.

▮ EXERCISES

3. Our prediction was more than 10% higher than our experimental result. Is that bad? How might we do better? Why is it off by so much? Why by 10%? Answering these questions will take you into yet another round of doing physics.

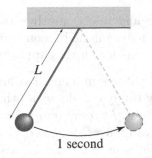

1 second

FIGURE 2.3 On the surface of Earth a pendulum of length $L = 0.993 \approx 1\,\text{m}$ will take 1 s to swing through a small angle from one side to the other.

Time

Time is measured by repetitive behavior—the swinging of a pendulum, the annual cycle of the seasons, the heartbeat's regular thud. The SI unit of time is the second. It is roughly the duration between heartbeats of a person sitting at rest. It is very nearly the time that it takes a mass near Earth's surface at the end of a 1 m long string to swing through a small angle from one side to the other (see Fig. 2.3).

Although the basic units of time (as also of length and mass) are human in scale, much of physics deals with other scales. For example, in the realm of atoms and nuclei, 10^{-9} s can be a very long time; some nuclear phenomena happen in 10^{-21} s. At the other end of the time scale, we think the Universe has existed for longer than 10^{17} s. Some examples show how knowing the scale of a phenomenon can help to understand it.

▼ **EXAMPLES**

3. If in a laboratory you are working with the transmission of light between objects a few tens of centimeters apart, then an important time scale is 30 cm divided by the speed of light. Light travels $3 \times 10^8\,\text{m s}^{-1}$, so the time to go 30 cm is

$$30\,\text{cm} \times \frac{1\,\text{m}}{100\,\text{cm}} \times \frac{1\,\text{s}}{3 \times 10^8\,\text{m}} = 10^{-9}\,\text{s}, \tag{1}$$

i.e., 1 nanosecond (where "nano" is the standard prefix for 10^{-9}—a billionth—of anything), usually written 1 ns. This result means that it will take a nanosecond or so for light from one part of your lab apparatus to reach and affect another part.[3] You can also see that the SI prefixes like nano might be convenient for talking about times involving light transmission in this apparatus.

4. When you study atoms in a small volume so thoroughly evacuated that the atoms very rarely collide with other atoms, collisions with the walls may be important. At room temperature, the nitrogen and oxygen molecules in air have an average speed of about $500\,\mathrm{m\,s^{-1}}$. Thus, in a cylinder $2\,\mathrm{cm}$ in diameter the time between collisions with the walls will be roughly $2\,\mathrm{cm}/50\,000\,\mathrm{cm\,s^{-1}} = 4 \times 10^{-5}\,\mathrm{s}$. Customarily we would call this $40\,\mu s$ and write it as $40\,\mu s$ where ($1\,\mu s = 10^{-6}\,\mathrm{s}$ and μ, the lowercase Greek letter mu, is the SI prefix used to denote "micro," or millionth). Thus, interactions of atoms with the walls of this cell occur on a scale of millionths of a second. This kind of simple information is often useful. For example, if while studying these atoms you find something that happens in nanoseconds, you know that it has nothing to do with the walls. On the other hand, if the time scale of whatever you are observing is microseconds or longer, you may be seeing some effect of the walls.

Some Important Masses, Lengths, and Times

Table 2.2 lists some important masses, lengths, and times and gives the SI units in which they are measured. The table illustrates that these quantities are used over ranges from the human scale down to the very small and up to the very large. Every physical system has a characteristic time, a characteristic length, and a characteristic mass. The table gives examples of some of these. When doing physics, you need to have in mind concrete examples of physical objects or systems and to know their characteristic time, length, and mass scales.

2.3   UNITS AND DIMENSIONS

SI units are a consistent set of units defined in terms of standards accepted everywhere in the world.[4] SI units offer you the hope that, when evaluating equations, everything will come out all right if you make sure all the

---

[3]It is useful to know that light travels just about 1 foot in 1 ns—one of the rare occasions when English units produce a convenient number.

[4]Even culturally backward countries that do not use SI units in everyday life have redefined their historically quaint units in terms of metric standards. For example, in America the inch is defined to be exactly 2.54 centimeters long.

TABLE 2.2 Basic quantities of physics

Name of Quantity	SI Unit of Measure	Abbreviation	Examples
mass	kilogram	kg	1 L of water has a mass of 1 kg. The mass of a proton is 1.67×10^{-27} kg. Earth's mass is 5.98×10^{24} kg. Masses of typical American adults range from 50 kg to 90 kg.
length	meter	m	A long stride is about 1 m. A range of typical heights of American adults is from 1.6 m to 1.9 m. Earth is 40×10^6 m in circumference. An atom's diameter is 0.2×10^{-9} m.
time	second	s	Your heart probably beats a little faster than once a second. Light travels 30 cm in 10^{-9} s. There are 3.15×10^7 s in a year. The age of the Universe is $\sim 4 \times 10^{17}$ s.

quantities you use in the equations are measured in appropriate combinations of kg, m, and s. If you do this, the units of every term on both sides of the equal sign will be the same. You always want this to be the case.

Three complications undermine this hope. First, the SI assigns certain groups of units distinctive names, e.g., the group of units $\mathrm{kg\,m\,s^{-2}}$ is called a newton; a $\mathrm{m^3}$ is called a "stere" (pronounced steer). There are many of these names, and you need to know them to be able to check if your units are consistent. You will learn several of them in this chapter.

Second, values of quantities are often given using the SI prefixes. Thus, a length might be given in centimeters (cm = 10^{-2} m) or millimeters (mm = 10^{-3} m) or micrometers (μm = 10^{-6}) or nanometers (nm = 10^{-9} m). You have to know these prefixes and convert the values of the quantities in your equations to meters before evaluating the terms.

Finally, there are situations in physics where physicists do not use pure SI units.[5] Other units may be more appropriate to the scale of the phenomena; other units may be better than SI units for revealing important connections or simplicities. Physicists measure energies of atoms in electron volts, a non-SI unit of energy; continents drift a few centimeters per year; the mass of an oxygen atom is about 16 hydrogen atom masses.

---

[5] *A foolish consistency is the hobgoblin of little minds....* Ralph Waldo Emerson.

Composite Units

For most physical quantities the units are more complicated than those for length, mass, and time. Most physical quantities require units that are composed from the basic units according to the rules of algebra. To see what this means, consider "volume" as a simple example.

What is the volume of Earth? To answer the question you use the fact that the volume V of a sphere of radius R is $V = \frac{4}{3}\pi R^3$, and the fact (noted above) that Earth's radius R is 6.37×10^6 m. Then

$$V = \frac{4}{3}\pi (6.37 \times 10^6 \text{ m})(6.37 \times 10^6 \text{ m})(6.37 \times 10^6 \text{ m}) = 1.08 \times 10^{21} \text{ m}^3,$$

where m^3 is the composite unit called a cubic meter.

Notice how the notation m^3 arises from the natural algebraic combination of the factors in the formula for the volume of a sphere: $\frac{4}{3}\pi R^3$. This is a general property; any formula for a volume must contain exactly three factors of length. Correspondingly, units of volume always have three factors of length in their definition.[6] Any time you multiply three lengths together you obtain a volume.

Table 2.3 shows formulas for calculating volumes of some different shapes. Notice that each formula multiplies three lengths together, so in each case the volume has SI units of m^3—cubic meters.

Using SI Multipliers

The units of the cubic meter are straightforward. But what if one dimension is in Mm (megameters) and another in cm? For example, suppose

TABLE 2.3 Volume formulas always contain three factors of length

Shape	Dimensions	Volume
Cube	edge ℓ	ℓ^3
Box	length ℓ width w height h	$\ell w h$
Cylinder	height h radius r	$\pi r^2 h$
Cone	height h radius r	$\frac{1}{3}\pi r^2 h$
Sphere	radius r	$\frac{4}{3}\pi r^3$

---

[6]The three factors may not always be apparent as, for example, when the volume of water is measured in quarts or liters or acre-feet, but they are always present if only implicitly.

topsoil is on average a layer $d = 10\,\text{cm}$ thick over 30% of Earth's surface. What is the volume V of the topsoil? The area of Earth is $4\pi R^2$ where $2\pi R = 40\,\text{Mm}$. So to evaluate

$$V = 0.3 \times 4\pi R^2\, d$$

you might want first to convert every dimension to meters. To do that use Tip 1.

- Tip 1: To manipulate SI prefixes efficiently, think of each prefix as a number ($\text{M} = 10^6$, $\text{c} = 10^{-2}$) multiplying the unit.
  You already know how this works with a "prefix" like "dozen." How many is 3 dozen eggs? $3 \times 12\,\text{eggs} = 36\,\text{eggs}$ where you replaced "dozen" with its numerical value. Similarly, the expression "3 cm" is "$3 \times 10^{-2}\,\text{m}$" when you replace c with its numerical value.

Here's how to use Tip 1 to find Earth's radius R. From the definition of the meter you know Earth's circumference is 40 Mm. Therefore,

$$R = \frac{2\pi R}{2\pi} = \frac{40\,\text{Mm}}{2\pi} = \frac{40 \times 10^6\,\text{m}}{2\pi} = 6.37 \times 10^6\,\text{m}.$$

Also use Tip 1 to convert d to meters:

$$d = 10\,\text{cm} = 10 \times 10^{-2}\,\text{m} = 0.10\,\text{m}.$$

Now substitute these into the equation to get an estimate of the volume of topsoil on Earth:

$$V = 0.3 \times 4\pi \times (6.37 \times 10^6)^2\,\text{m}^2 \times .1\,\text{m} = 1.53 \times 10^{13}\,\text{m}^3.$$

EXERCISES

4. If you had not converted the values of R and d to meters, what would have been the units of volume of your answer?

5. Suppose your answer had come out to be $1500\,(\text{Mm})^2\,(\text{cm})$. Is this a unit of volume? What factor converts the units to cubic meters?

The example showed how to convert megameters and centimeters to meters. What if you want to go in the other direction? Suppose you have a box that is 0.1 m wide, 0.2 m long, and 0.05 m deep. Its volume is $10 \times 10^{-4}\,\text{m}^3$, but it would be easier to visualize this volume if it were in cubic centimeters. You can convert meters to cm by multiplying the meter unit "m" by $1 = 10^2\,\text{c}$, i.e.,

$$0.1 \times 0.2 \times 0.05\,\text{m}^3 = 0.1 \times 0.2 \times 0.05\,(10^2\,\text{cm})^3.$$

Now pull the factor of 10^2 out from the parentheses, being sure to cube it because the contents of the parentheses are to the third power, and you get

$$10 \times 10^{-4} \times 10^6 \, \text{cm}^3 = 1000 \, \text{cm}^3.$$

■ EXERCISES

> **6.** What would be your answer in cubic micrometers? Keep in mind that when you ask how many little volumes go to make a large one, the number you get should be (much) larger than the one you started with.

Practice doing conversions of units as often as you can so that you become good at it. If you have learned a cumbersome process for doing conversions, replace it with the quicker more efficient way shown here.

SI multipliers are often used with named combinations of metric units their own names. The "liter" is an example; it's a volume of $10 \, \text{cm}$ cubed: $(10 \, \text{cm})^3 = 1 \, \text{L}$ where L is the abbreviation for liter. Notice that $1000 \, \text{cm}^3 = 1 \, \text{L} = 1000 \, \text{mL}$, showing you that $1 \, \text{mL} = 1 \, \text{cm}^3$, i.e., $1 \, \text{mL}$ equals 1 cubic centimeter.

▼ EXAMPLES

> **5.** What is one liter expressed in cubic meters? To answer this keep in mind that the notation cm^3 really means $(\text{cm})^3$ so the "c" is cubed along with the "m." Then
>
> $$1000 \, \text{cm}^3 = 10^3 \times (10^{-2})^3 \, \text{m}^3 = 10^{-3} \, \text{m}^3 \qquad (2)$$
>
> and your answer is $1 \, \text{L} = 10^{-3} \, \text{m}^3$, or $1000 \, \text{L} = 1 \, \text{m}^3$.

The later parts of this chapter review the important physics ideas of momentum, force, and energy. Each of these concepts has a quantitative measure in terms of composite units, and these are developed along with the concepts themselves. Be attentive to these groups of units and their names.

Consistency of Units

The equation for centripetal acceleration illustrates an important aspect of units: they must be the same on both sides of the equal sign. If v^2/R is acceleration, it must have units of acceleration. You can see that it does. The SI units of v are $\mathrm{m\,s^{-1}}$ and the SI units of R are m. It must always be the case that when you combine these according to the formula v^2/R you get the units of acceleration. Do the algebra of the units of centripetal acceleration, and you get

$$\frac{v^2}{R} = \frac{(\mathrm{ms}^{-1})^2}{\mathrm{m}} = \mathrm{m}^2\mathrm{s}^{-2}\,\mathrm{m}^{-1} = \mathrm{m\,s}^{-2}.$$

These are units of acceleration; there is consistency of units.

The units of equations must always be consistent. This is important! Every term on both sides of the equals sign must have the same units. Consider the following where a and g are accelerations, v is a velocity, ℓ is a length, and t is time:

$$\frac{1}{2}at^2 + vt = \sqrt{\frac{\ell}{g}}\,at$$

The first term has SI units of $(\mathrm{m\,s}^{-2})(\mathrm{s}^2) = \mathrm{m}$; the second term has units of $(\mathrm{m\,s}^{-1})(\mathrm{s}) = \mathrm{m}$. Each term on the left side has the same units—meters; they are consistent. The term on the right side has units of $\mathrm{m}^{\frac{1}{2}}\,(\mathrm{m\,s}^{-2})^{-\frac{1}{2}}\,\mathrm{m\,s}^{-2}\,\mathrm{s} = \mathrm{m}$, and the whole equation has consistent units.

Having consistent units does not guarantee an equation is correct, but having inconsistent units absolutely guarantees it is wrong. You need to know this not as a curious and interesting fact, but so you don't humiliate yourself. Checking for consistent units is the first thing a physicist does when reading equations. Presenting equations with inconsistent units is like walking around with a booger on your nose; you look foolish.

- Tip 2: To avoid looking foolish, always make sure the units in your equations are consistent.

Physical Dimensions

Instead of talking about "units," physicists often refer to the more general idea of physical dimensions. These are not the spatial dimensions of a line, a surface, or a volume; they are a way to talk about length, time, and mass independently of units. Although a length can be measured in any of many different units, the property of length-ness exists independently of the units in which it is measured. That property is called its "dimension" of length and denoted as "L" (not to be confused with the abbreviation for liter). Similarly the dimension of mass is denoted by "M" and the dimension of time by "T."

In this special sense every physical quantity must have dimensions composed of M, L, and T.[7] Thus, whether you measure area in square meters, square inches, or acres, the dimensions of the area are L^2. Similarly, a volume, whether measured in liters, quarts, pints, or steres, has dimensions of L^3. Another way to summarize the point of Table 2.3 (p. 22) is to say that every formula for a volume must have dimensions of L^3. Velocity has dimensions of length over time; that is, its dimensions are LT^{-1}. Acceleration has dimensions of LT^{-2}. Density has dimensions of ML^{-3}.

▨ EXERCISES

> **7.** What are the dimensions of an acre-foot, the measure for water used to irrigate fields?
>
> **8.** What are the dimensions of a hectare?
>
> **9.** You learned in high school that force F is mass m times acceleration a. What are the dimensions of force?

Like units, dimensions of physical quantities combine according to the rules of algebra. If a velocity has dimensions LT^{-1}, then the square of a velocity has dimensions L^2T^{-2}. The square of an acceleration has dimensions L^2T^{-4}.

- Tip 2 in terms of "dimensions": To avoid looking foolish, always make sure your equations are dimensionally consistent.

2.4 ANGLES AND ANGULAR MEASURE

A lot of reasoning in physics involves angles, so you need to understand how to measure them and how to talk about them. There are several different measures of angles: degrees, fractions of a circle, clock time, and *radians*. Radians will be the measure we use most, because they connect simply and directly to trigonometry and because they are often easy to measure.

Let's review some of the vocabulary and ideas associated with angles and their measure.

---

[7] In later chapters you will add electric charge and temperature to the set of basic quantities.

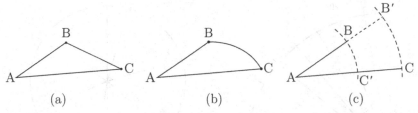

FIGURE 2.4 (a) An angle $\angle BAC$ with its vertex at A. The segment $\overline{BC}$ subtends the angle at A. (b) The arc $\overset{\frown}{BC}$ also subtends this angle at A. (c) Here the circular arcs $\overset{\frown}{B'C}$ and $\overset{\frown}{BC'}$ subtend the angle $\angle BAC$ at A.

Vertex and Rays

An angle is the figure formed by the spreading of two rays from a point. That point is called the "vertex" of the angle.

In Fig. 2.4 the three points B, A, and C define an angle with A as its vertex and $\overline{AB}$ and $\overline{AC}$ as its rays. It is usual to denote the angle as $\angle BAC$, where the middle letter is the vertex.

What Does "Subtend" Mean?

Imagine that some distance out from the vertex something stretches across between two rays to form a closed figure. For example, imagine two lines diverging from your eye straight to the edges of a white area on the blackboard. The thing that stretches across the diverging lines is said to "subtend" some angle "at the point," i.e., the vertex, from which the rays diverge—here, your eye. So we say that the white mark subtends an angle of some amount at your eye. The phrase tells you two things: what sits at the mouth of the angle (the white mark) and the location of the starting point of the rays that define the angle (your eye). In Fig. 2.4 the line $\overline{BC}$ and the arcs $\overset{\frown}{B'C}$ and $\overset{\frown}{BC'}$ each subtend the angle $\angle BAC$ at A.

■ EXERCISES

10. What angle does the hypotenuse of a right triangle subtend?

Degrees

Two principal measures of angles are used in physics: degrees and radians. Each of these expresses the angle in terms of segments of a circle. In effect,

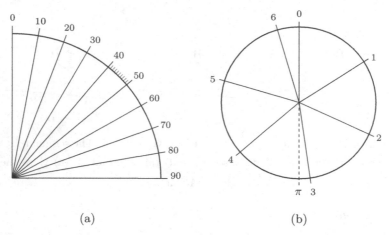

(a)   (b)

FIGURE 2.5   Angular measure: (a) degrees; (b) radians.

an angle is measured by specifying what fraction of a circle's circumference is subtended at the vertex by a circular arc. To see how this works consider how you define a "degree."

Given an angle, construct any circular arc centered on the vertex (as in Fig. 2.4c). One degree is defined as the angle formed by two rays from the vertex that intercept an arc that is 1/360 of the circle's circumference. In other words, we imagine a circle divided into 360 equal arc lengths, and each of these arcs connected by lines to the center of the circle. In Fig. 2.5 a quadrant of a circle has been divided into nine equal arc lengths; consequently the angles are each 10°. Expressed in algebraic terms, the angle θ in degrees is

$$\theta = \frac{\overset{\frown}{s}}{2\pi R} \times 360 \text{ degrees.} \tag{3}$$

The symbol $\overset{\frown}{s}$ stands for the length of the circular arc subtending θ at the circle's center. The angle's measure is independent of what circle you choose, because the *ratio* of the arc length to the radius will not change whether the circle is large or small.

It is common to use the symbol ° for degrees. Each degree is in turn divided into 60 equal parts called "minutes," and each minute is divided into 60 equal parts called "seconds." When it is necessary to distinguish between seconds of time and seconds of angular measure, the latter are called "arc seconds." You learned a version of this so-called "sexagesimal" system when you learned to tell time. It is a measure of the strength of cultural inertia that the sexagesimal system survives and is widely used in terrestrial and astronomical measurements of angle, even though the degree and its curious subunits are awkward for many calculations.

▽ EXAMPLES

6. Let's see what angle corresponds to a circular arc $\overset{\frown}{s} = 2\pi R/8$, one-eighth of the circumference. From Eq. 3,

$$\theta = \frac{2\pi R}{2\pi R} \frac{360°}{8} = 45°.$$

▪ EXERCISES

11. Show how Eq. 3 will correctly yield 90° for the angle subtended by a quarter of a circle's circumference.

As Eq. 3 shows, the definition of units of angle measure involves the ratio of two lengths, the arc length and the radius. This is why angles are dimensionless quantities. They have units of measure, such as degrees, but no physical dimensions.

Radians

The calculation in Eq. 3 is basic to the determination of the size of an angle. To get the angle in degrees, you take the length of the circular arc length subtending an angle and find what fraction it is of the circumference of the circle of which the arc is part. Then you multiply by 360°. As Eq. 3 shows, this means that an answer in degrees is always $360/(2\pi)$ times the ratio of the subtending arc length to the radius of the circle.

By choosing a different definition of the measure of angle, you can make this factor of $360/(2\pi)$ disappear. Instead of dividing the circle into 360 parts, so that you measure angle as arc length $\overset{\frown}{s}$ over circumference $2\pi R$ times 360°, why not divide the circle into 2π parts? Do you see that this will eliminate the constant factor? For a circle divided into 2π equal parts, the measure of an angle is just the ratio of the circular arc length subtending the angle at the center of the circle to the radius of the circle on which the arc length $\overset{\frown}{s}$ lies. This measure of angle is called the radian. We have in general

$$\theta = \frac{\overset{\frown}{s}}{R} \text{ rad}, \tag{4}$$

where "rad" is the usual abbreviation for "radians."

Of course this means that a full circle contains

$$\frac{2\pi R}{R} = 2\pi \text{ radians},$$

and this means that $360° = 2\pi$ rad or that 1 rad $= 57.3°$. This 57.3 is a useful number to remember.

Radians are more convenient than degrees for measuring angles and doing simple trigonometry. This especially so for angles that are small, where by small we mean situations where the length of the chord (the straight line) connecting two points is nearly the same as the length of the circular arc connecting the same two points. The following examples show that, for angles smaller than $10°$, using the chord in place of the arc length introduces negligible error (less than $\frac{1}{2}$%). The examples also show how to calculate angles in radians when the angles are small. It is important for you to understand how to do such calculations because the small angle situation occurs frequently in physics and astronomy.

EXAMPLES

7. Consider the angle subtended at your eye by a dime held an arm's length away. A dime has a diameter of 1.83 cm. For an arm's length of 60 cm, the angle subtended by the dime is the length of arc of radius 60 cm connecting two points 1.83 cm apart divided by 60. For a chord of exactly 1.83 cm, this arc length is 1.830071 cm. You make essentially no error when you calculate the angle as the chord length over the distance to your eye: $\theta = \frac{1.83}{60} = .0305$ rad.

8. Suppose you held the dime just 10 cm from your eye. What angle does it then subtend at your eye? You can figure out that the length of arc of radius 10 cm between two points 1.83 cm apart is 1.8326 cm, so the angle is 0.1833 rad. But you see that if you use the chord length of 1.83 cm instead of the arc length, you get an answer that is only about half a percent smaller.

9. The Moon subtends an angle of $\approx 0.5°$ from a point on the Earth. The Sun also subtends an angle of $\approx 0.5°$ from Earth. If the Sun is 1.5×10^{11} m from Earth, what is the Sun's diameter? This angle is about 8.7 mrad. Therefore the diameter of the Sun is $8.7 \times 10^{-3} \times 1.5 \times 10^{11}$ m $= 1.31 \times 10^6$ km.

More About the Small-Angle Approximation

Radians are also convenient for making useful approximations to trigonometric functions. Referred to the large right triangle in Fig. 2.6a, the

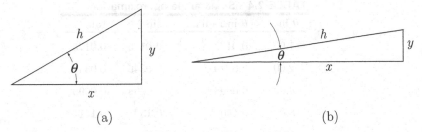

FIGURE 2.6 (a) A right triangle for defining the trigonometric functions. (b) For small angles $h \approx x$.

trigonometric functions—sine, cosine, and tangent—of θ, are respectively $\sin \theta = y/h$, $\cos \theta = x/h$, and $\tan \theta = y/x$. In Fig. 2.6b the very acute triangle shows the important fact that as θ gets small, the lengths of the hypotenuse h and the long leg x of the triangle become almost equal, and the right triangle more and more closely approximates an isosceles triangle. Therefore, as θ gets small, h and x better and better approximate radii of a circle, and the small leg y becomes a better and better approximation to the circular arc length $\overset{\frown}{s}$ connecting two radial legs. This is the basis for approximating the sine or tangent of a small angle by the angle itself in radian measure:

$$\sin \theta = \frac{y}{h} \approx \frac{\overset{\frown}{s}}{h} = \theta \approx \frac{y}{x} = \tan \theta.$$

▽ EXAMPLES

10. What is the sine of 5.7°? Since 5.7° is about 0.1 rad, and this is fairly small compared to unity, $\sin 5.7° \approx 0.100$.

If you do this with a calculator, you will get $\sin 5.7° = 0.0993$, showing that the approximation agrees with the exact value to better than 1%. The quality of the approximation is apparent from the entries in Table 2.4, which shows that even at angles as large as 15° the sine is only about 1% different and the tangent about 2% different from the radian measure of the angle.

The small-angle approximation makes many calculations easier.

TABLE 2.4   Small-angle approximation

θ in °	θ in radians	$\sin\theta$	$\tan\theta$
1.00	0.0175	0.0175	0.0175
2.00	0.0349	0.0349	0.0349
4.00	0.0698	0.0698	0.0699
8.00	0.140	0.139	0.141
10.0	0.175	0.174	0.176
15.0	0.262	0.259	0.268

▽ EXAMPLES

11. For example, given that the Moon is 60 Earth radii distant from Earth, what angle does Earth subtend at the Moon? Since Earth's diameter is $2\,R_E$, then the angle subtended at the Moon is $(2\,R_E)/(60\,R_E) = 1/30 = 0.0333\,\text{rad}$. If you're asked to find the angle in degrees, use the fact that $180° = \pi\,\text{rad}$, so

$$0.0333\,\text{rad} \times \frac{180°}{\pi\,\text{rad}} = 1.91°. \tag{5}$$

2.5   THINKING ABOUT NUMBERS

Although it is important to have quantitative values for such things as the volume of the Earth, it is not enough just to have the number. You need to think about it; you need to find ways to make it comprehensible. A number like 10^{21} m^3 does not spontaneously inform your imagination. Making very large and very small numbers meaningful is a recurring problem in physics. One way to understand them is by comparison.

▽ EXAMPLES

12. Now try some comparisons for a number that may have some more immediate interest than the volume of the Earth. The annual federal budget is on the order of four trillion dollars ($\$4 \times 10^{12}$). What is a trillion dollars? Try breaking the number into more manageable units, say 100 million dollars. It takes ten thousand sets of 100 million to make one trillion. One hundred million is still hard to imagine, but its scale is more tangible. For example, a medium-sized liberal arts college has a yearly budget on this order, or quite a nice hospital might be built for $100 million. Even so, forty thousand hospitals a year is hard to imagine. But when you discover that 4 trillion dollars could build more than 100 new, fully equipped, 40-bed hospitals every *day* for a year, you begin to get a sense of what $4 trillion means.

EXERCISES

12. Example 11 showed that Earth subtends at the Moon an angle of 0.0333 rad. Does this help you imagine what a full Earth looks like from the Moon? How can you make the number more meaningful?

13. The US federal budget runs annual deficits of about 5×10^{11}. How high would a stack of $100 bills be if it contained this much money? Notice here, as is often the case in such questions, that you need to make some reasonable estimate of some physical quantity important to your answer—in this case, the thickness of a $100 bill. Most bills you see daily are not very flat, but you'll probably decide that a bill at the bottom of this stack would be quite flat!

14. Write out all the figures in Example 12 numerically; write down the relevant relations; and confirm the statements made.

15. What is the largest object for which you have some sense of its size? Estimate its volume. What is Earth's volume in units of your object's volume?

The comparison you developed in Exercise 15 is probably still not very meaningful. To make the volume of Earth meaningful try comparing it with other similar objects—the Moon, the planets, the Sun. The following paragraphs show how to make such comparisons.

EXERCISES

16. The Moon is 3.8×10^8 m from Earth. You can just block out the Moon with a disk 9.28 mm in diameter held 1 m away from your eye. What is the diameter of the Moon? (See Fig. 2.12.) What is its volume?

17. If you like this sort of argument, apply it to the Sun. Curiously, the Sun subtends at Earth almost exactly the same angle as the Moon, although the Sun is more distant. As you saw in Example 9, knowing the Sun is about 1.5×10^{11} m from Earth, you can find its diameter. What is its volume? How does that volume compare to Earth's?

Physicists are always trying to find ways to make numbers meaningful. Here is an example of one approach. A large object often has a large mass, so it is not surprising that a planet with a volume of 1.08×10^{21} m^3 has a mass of 5.94×10^{24} kg, or that a star (our Sun) with a volume of 1.41×10^{27} m^3 has a mass of 1.99×10^{30} kg. But what do these numbers mean? A common trick is to use the large numbers to describe some property that is not itself a large number. For instance, consider how much mass there is in a unit volume, i.e., look at the density. Earth's density is $\rho = 5.52 \times 10^3$ kg m^{-3}. Another trick is to rescale the units. A cubic meter is a pretty large volume; let's look at the density in grams per cubic centimeter, i.e., g cm^{-3}. Then the average density of the Earth is 5.5 g cm^{-3}.

Now that's a number a person can deal with. You already know that water has a density of 1 g cm^{-3}. So Earth is about 5.5 times denser than water. Does that make sense? You could check by measuring the density of some other things. Iron (Fe) has a density of 7.85 g cm^{-3}. Mercury (Hg) has a density of 13.6 g cm^{-3}. More interesting, the granitic rocks of which Earth's crust is made have a density of about 2.8 g cm^{-3}.

▪ EXERCISES

18. How can this be? You just found that the density of Earth is 5.5 g cm^{-3}. How can Earth be denser than its crust? Make up a reasonable explanation for the discrepancy.

19. Now calculate the density of the Sun and compare it to Earth's. Are you surprised? We hope so. But in any case, do you see how useful it is to play with the numbers from different points of view?

Such rescaling and such comparisons are essential because so little of the universe is set to human scale. The visible universe continues out beyond 10^{26} m; subatomic particles are smaller than 10^{-15} m. Physicists deal casually with the unimaginably large and the inconceivably small. It is hard to know what is important about a mass of 3.35×10^{-27} kg.

It becomes more meaningful when you know that it is twice the mass of an atomic nucleus of hydrogen or half that of a helium nucleus. Part of thinking about physics is the search for meaningful comparisons among the numbers used to describe nature and the interactions of matter.

2.6 MOMENTUM, FORCE, AND CONSERVATION OF MOMENTUM

This section describes velocity, acceleration, momentum, and force; Sect. 2.7 surveys the idea of energy. The ideas of momentum and energy are especially important and are used repeatedly throughout this book. They are the basis of two of the most fundamental, universally applicable laws of physics: the conservation of momentum and the conservation of energy.

Velocity and Acceleration

Velocity describes how far (a length) a body travels in a unit time in some particular direction. For bodies in steady motion the idea is simple. If at the end of 1 s a car has moved 24.6 meters and then again at the end of 2 s another 24.6 m, we characterize the car as having a speed of $24.6\,\mathrm{m\,s^{-1}}$. Direction is an important part of velocity, but for now worry only about the numerical value, or magnitude, of velocity. This number is called the "speed."

For the case when the magnitude of the speed or its direction varies with time, calculus provides a precise definition of velocity. We'll explain this when necessary.

You also need a measure of the time rate-of-change of velocity; this is called "acceleration." You already know that near Earth's surface the speed of a body falling in a vacuum increases $9.8\,\mathrm{m\,s^{-1}}$ every second of its fall. Just as for the case of constant speed, the special case of steady (constant) rate-of-change of velocity is simple to visualize. If with every passing second the speed of an object increases by $9.8\,\mathrm{m\,s^{-1}}$, the rate-of-change is $9.8\,\mathrm{m\,s^{-1}}$ per second or, following the algebraic logic of the units $9.8\,(\mathrm{m\,s^{-1}})\,\mathrm{s^{-1}}$. It is customary to complete the algebra and write the acceleration as $9.8\,\mathrm{m\,s^{-2}}$. As velocity specifies the time rate-of-change of magnitude and direction of a distance traveled , so acceleration specifies the change of magnitude and direction of a velocity.

An object can accelerate without changing its speed. For example, when moving in a circular path at constant speed v, it is accelerating because it is changing direction even though v stays constant in magnitude. This kind of motion is called "uniform circular motion." It occurs when you swing a ball on a string in a horizontal circle or when Earth pulls a satellite around it in a circular orbit or when a uniform magnetic field bends a moving electrically charged atom in a circle. For uniform circular motion the acceleration is always toward the center of the circle, and, for this reason, it is called "centripetal" acceleration. Although we haven't proved it, keep in mind that an object moving with constant speed v around a circle of radius R, always has centripetal acceleration a_c where

$$a_\mathrm{c} = \frac{v^2}{R}. \tag{6}$$

■ EXERCISES

20. Show that the dimensions of $\frac{v^2}{R}$ are the dimensions of acceleration.

21. The Space Shuttle moves in a circular orbit about 300 km above Earth's surface. It takes 90.5 min to go once around Earth. What is its acceleration?

22. Why do you know that the Moon is accelerating toward Earth? What is the magnitude of that acceleration?

Momentum

You know from personal experience that when two bodies move at the same speed, the heavier one possesses more of something associated with its motion than the lighter one does. A baseball delivered into the catcher's mitt at $24.6\,\mathrm{m\,s^{-1}}$ is not especially intimidating. In the major leagues such a pitch would be so slow that it very likely would be hit before reaching the catcher. However, a bowling ball delivered at the same speed is quite another story, promising severe bodily damage, and an SUV at the same speed would probably kill you on impact.

Newton thought of moving bodies as possessing different amounts of motion, and he devised a useful measure of this "quantity of motion." It is the product of the mass and the velocity, i.e., mv. Today we use the word "momentum" instead of "quantity of motion," but it means exactly the same thing.

In cases where we are interested in momentum alone and are not calculating it as the product of mass with velocity, it is often convenient to give it a separate symbol, most commonly p,

$$\text{momentum} = \text{mass} \times \text{velocity},$$

$$p = mv.$$

Assuming that they move at $24.6\,\text{m}\,\text{s}^{-1}$, you see that the baseball, with its mass[8] of about $150\,\text{g}$, has a momentum of $3.69\,\text{kg}\,\text{m}\,\text{s}^{-1}$, while the bowling ball,[9] with a mass of $5.0\,\text{kg}$ has a momentum of $123.0\,\text{kg}\,\text{m}\,\text{s}^{-1}$, and the SUV,[10] with its mass of $1500\,\text{kg}$, has a momentum of $36\,900\,\text{kg}\,\text{m}\,\text{s}^{-1}$. The difference between being on the receiving end of $3.69\,\text{kg}\,\text{m}\,\text{s}^{-1}$ and $36\,900\,\text{kg}\,\text{m}\,\text{s}^{-1}$ is made evident daily in unpleasant ways.

■ EXERCISES

---

23. Estimate the momentum of you and your bicycle together when riding at a typical speed. Compare your answer to the momenta given above for the baseball, the bowling ball, and the automobile.

Force

Newton used his definition of momentum to specify a meaning for another word you use daily: force. *Anything that changes the momentum of a given body is a force.* This definition includes changes of the direction of momentum as well as of its amount.

The size, or magnitude, of a force depends upon how quickly the momentum changes. In fact, the magnitude of a force is just how much the momentum changes per unit time, i.e., the time rate-of-change of momentum. Suppose you start with some momentum p_0 at a time t_0. Suppose also that a little later, at time t_1, your momentum has changed to p_1. To find the average rate-of-change you divide the actual change by the number of units of time it took to make the change. In symbols this is

$$F = \frac{p_1 - p_0}{t_1 - t_0}.$$

A more compact notation uses the capital Greek letter delta, Δ, to denote a difference between the final and initial values of a quantity. Thus,

$$F = \frac{\Delta p}{\Delta t}. \tag{7}$$

---

[8]This kind of information can be found online, but often you can find it more directly. We measured the mass of a hardball on a triple-beam balance and got $151.6\,\text{g}$. The regulation American League and National League baseball is 5 to 5.25 ounces, i.e., between 142 and $149\,\text{g}$.

[9]Bowling balls range between 4.5 and $7.3\,\text{kg}$.

[10]Incidentally, $24.6\,\text{m}\,\text{s}^{-1}$ is about 55 mph.

In terms of mass and velocity, the expression reads

$$F = \frac{\Delta(mv)}{\Delta t},$$

where

$$\Delta(mv) = m_1 v_1 - m_0 v_0$$

and

$$\Delta t = t_1 - t_0.$$

EXAMPLES

13. If the baseball stops in the catcher's glove in 0.01 s, the average rate of change of its momentum is $-3.69\,\mathrm{kg\,m\,s^{-1}}/0.01\,\mathrm{s} = -369\,\mathrm{kg\,m\,s^{-2}}$. Similarly, if a 5-kg bowling ball moving at $8\,\mathrm{m\,s^{-1}}$ stops at the end of the lane in (perhaps) a time of 0.01 s, its rate-of-change of momentum is $-40/0.01 = -4000\,\mathrm{kg\,m\,s^{-2}}$. These two results mean that an average force of $369\,\mathrm{kg\,m\,s^{-2}}$ acted on the baseball and an average force of $4000\,\mathrm{kg\,m\,s^{-2}}$ acted on the bowling ball. Catching the baseball may sting a little; trying to catch the bowling ball would really hurt!

Composite Units Again

Here is a good place to learn about some more composite units and their special names. These names are handy for several reasons. For one thing, you get weary of writing "$\mathrm{kg\,m\,s^{-2}}$" all the time. For another, this group of units does not shout "force!" at the reader. Labels of physical quantities are more compact and more recognizable if you have standard names and abbreviations for groups of units.

In the SI the group "$\mathrm{kg\,m\,s^{-2}}$" is called the "newton." We say "the newton is the unit of force when using the meter, the kilogram, and the second as basic units of measurement." The newton is abbreviated "N." Thus to stop the baseball in 0.01 s requires a force of 369 N, while stopping the bowling ball in 0.01 s requires a force of 4000 N. Like velocity and momentum, force has direction as well as magnitude, but for now you can neglect this important aspect of its definition.

An important point: Once a group of units has a name (such as newton), you can use all the usual SI prefixes. Physicists talk of Meganewtons (MN), micronewtons (μN), millinewtons (mN), etc.

It's possible to make composites of composites. For example, once you have defined $\mathrm{kg\,m\,s^{-2}}$ to be a newton, you can express the units of momentum in terms of newtons too. Use the algebra of the units to write:

$$\frac{\text{kg m}}{\text{s}} \times \frac{\text{s}}{\text{s}} = \frac{\text{kg m}}{\text{s}^2} \times \text{s} = \text{N s}. \tag{8}$$

This means $368\,\text{kg m s}^{-1}$ is the same as 368 newton-seconds, N s, N·s, or N-s. Momentum is often described in units of N s.

Why Does $F = ma$?

Usually a body's momentum changes because its velocity changes. Then

$$\Delta p = m\Delta v,$$

and you can rewrite the force relation by substituting for Δp:

$$F = m\frac{\Delta v}{\Delta t}.$$

This says that time rate-of-change of momentum is the same thing as mass times the rate-of-change of velocity. But rate-of-change of velocity is the definition of acceleration. The average acceleration of a body is defined to be $a = \Delta v/\Delta t$.

Now you see where the famous relation $F = ma$ comes from. When the mass is constant, a force changes a body's momentum by producing acceleration, that is, the force changes the velocity in some time interval. $F = ma$ is just a particular way of writing that force causes a rate-of-change of momentum.

■ EXERCISES

24. Find the average accelerations that occurred while stopping the two objects in Example 13.

25. If the speedometer of your car reads 5 mph more than it did 2 s earlier, what was your average acceleration in mph per second? Such a mixture of units can be awkward, so convert them to m s^{-2}.

26. This and the next two exercises are to help you to see whether you understand that $F = \Delta p/\Delta t = ma$ means what it says.

A skydiver without her parachute open falls at a steady $70\,\text{m·s}^{-1}$. What is the total force acting on her?

27. After the skydiver opens her chute she falls at a steady speed of about $4\,\text{m s}^{-1}$. What is now the total force acting on her?

28. In the absence of air resistance, a body falling near Earth accelerates at $g = 9.8\,\mathrm{m\,s^{-2}}$. If that is the case, what is the force acting on a 10 kg weight as it falls in the absence of air? If the accelerating body has a mass of 27 kg, what is the magnitude of the acting force? What is the agent producing the force?

Conservation of Momentum

In the discussion of momentum, we asked you to imagine stopping a baseball and a bowling ball, because we wanted to evoke in you an intuitive sense of the quantity of motion (momentum) possessed by a moving object. The discussion was about the forces on the baseball and the bowling ball, not about the forces on the catcher or on the walls of the bowling alley where the ball stopped. There is, however, an intimate connection between the force exerted by an object A acting on an object B, and the force that B exerts on A. It's important for you to understand what it is.

Suppose you were seated on a very slippery surface and tried to stop a bowling ball coming at you. Assuming that you were successful in bringing it to rest relative to you without undue pain, what do you think might happen? You already have some idea that the ball would exert a force on you as you try to slow it down. With negligible friction to hold you in place, that force would have to impart some momentum to you. A remarkable thing happens. The momentum imparted to you is exactly equal in magnitude to the amount lost by the bowling ball. In fact, in all such interactions when there is no outside, i.e., external, force acting, the net change in momentum is zero. This is what we mean by *conservation* of momentum. The cartoon in Fig. 2.7 depicts the collision interaction.

If the initial momentum of the bowling ball is p_0 (the person's initial momentum is zero, since he is at rest), and the final momenta of the ball and person are p_1 and p_1', respectively, the conservation of momentum says that

$$p_0 = p_1 + p_1'.$$

▽ EXAMPLES

14. A $24.6\,\mathrm{m\,s^{-1}}$ bowling ball is a bit fearsome, so suppose that a person is on very slippery ice with a ball approaching at $4.0\,\mathrm{m\,s^{-1}}$ as

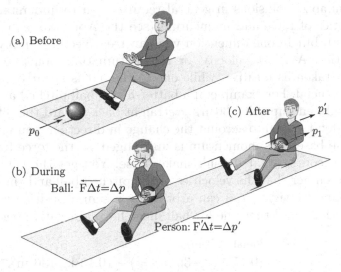

FIGURE 2.7 Collision between a bowling ball and a person on ice. (Don't try this at home).

suggested by Fig. 2.7. With what speed do the ball and person slide off together? Take the bowling ball's mass to be $5\,\text{kg}$ and the person's mass to be $70\,\text{kg}$. You can calculate the final momentum of each and their common velocity, v, by using conservation of momentum to find the velocity v:

$$p_0 = p_1 + p_1',$$
$$5\ \text{kg} \times 4\ \text{m s}^{-1} = (5\ \text{kg})(v) + (70\ \text{kg})(v) = (75\ \text{kg})(v),$$
$$v = \frac{20\ \text{kg m s}^{-1}}{75\ \text{kg}} = 0.27\,\text{m s}^{-1}.$$

Knowing v, the individual momenta are easily calculated.

■ EXERCISES

29. Calculate the individual momenta.

30. Why is it important that the person is on slippery ice? Why wouldn't momentum of the ball and person be conserved if he were seated on the ground?

To analyze collisions in general requires you to understand the vector properties of force and momentum (see the Appendix at the end of this chapter), but in one dimension you can treat vectors simply as algebraic quantities: A force, velocity, or momentum that points to the right is usually taken as positive, while one to the left is negative. This negative sign is crucial. For example, if a batter hits a ball pitched at $30\,\mathrm{m\,s^{-1}}$, the ball might end up going $30\,\mathrm{m\,s^{-1}}$ straight back toward the pitcher (ouch!). If you don't take into account the change in direction, you might conclude that the baseball's momentum is unchanged, so the force from the bat is zero. Of course, that doesn't make sense! You get the correct answer if you remember that the velocities of the pitched ball and the hit ball must have opposite signs. You can arbitrarily assign a positive velocity to the pitched ball, and then the hit ball must have a negative velocity. So now

$$\Delta p = p_{\text{final}} - p_{\text{initial}}$$
$$= (0.15\ \mathrm{kg})(-30\ \mathrm{m\,s^{-1}}) - (0.15\ \mathrm{kg})(30\ \mathrm{m\,s^{-1}})$$
$$= -9\ \mathrm{kg\,m\,s^{-1}}$$
$$\Delta p = -9\ \mathrm{N\,s}.$$

This negative sign in Δp is meaningful; it tells you that the force applied must be in the direction from home plate to the pitcher's mound, just as you would expect.

With the signs straightened out, go back to the brave (foolhardy?) catcher of bowling balls. A slight rearrangement of the equation defining force Eq. 7 allows you to calculate the change in momentum of the ball and the change of momentum of the person:

$$F\Delta t = \Delta p,$$

$$F'\Delta t = \Delta p',$$

where the primed quantities refer to the action on the person.

Conservation of momentum tells you that any momentum lost by the bowling ball must be gained by the person: $\Delta p = -\Delta p'$. Combining this equality with the two expressions above yields $F\Delta t = -F'\Delta t$, or, since the two Δt's are the same:

$$F = -F'. \tag{9}$$

That is, if you are exerting a force on a body, it will exert a force of the same magnitude and opposite direction back on you. You may already know this fact as Newton's third law of motion: When a body A acts on B with a force F_{AB}, body B acts on A with an exactly equal and opposite force $F_{\text{BA}} = -F_{\text{AB}}$. You have just seen this law is equivalent to the law of conservation of momentum.

EXERCISES

31. Suppose the person in Example 14 gave the ball an extra shove that sent it backwards at $2\,\mathrm{m\,s^{-1}}$ after the collision. Find the person's final momentum.

32. Earlier you found the average force required to stop the 5 kg bowling ball moving at $8\,\mathrm{m\,s^{-1}}$ in 0.01 s. Suppose the wall is very elastic and rebounds with the same speed, the time of collision being the same. What is the average force on the ball and on the wall for this new situation? In specifying directions, take the initial momentum to be positive.

33. When many collisions occur at a steady rate, one after the other, it is useful to describe the average force that they exert over time. We will use this idea of averaging when we study gases. Suppose 20 baseballs are thrown at a wall in 5 s and that each rebounds from the wall with the same $24.6\,\mathrm{m\,s^{-1}}$ speed that it came in with. Find the average force exerted on the wall during the 5 s.

Centripetal Forces

For an object to move in a circle it must have a force acting on it. As mentioned in Sect. 2.6, an object can only move with constant speed v in a circle of radius R by accelerating toward the center of the circle with an acceleration of $a_{\mathrm{c}} = \frac{v^2}{R}$ [Eq. 6, p. 36]. Because $F = ma$, an object of mass m moving with uniform circular motion must be experiencing a force of magnitude $m\frac{v^2}{R}$. Notice that this does not tell you what is exerting the force, only its magnitude. Because the acceleration is toward the center of the circle, the force is also toward the center. For this reason, whatever kind of force is producing the uniform circular motion—gravitational, electrical, frictional, magnetic, or string tension—it can be called "centripetal," a word that means "acting toward the center."

EXAMPLES

15. An electron is moving in a circle with radius 5 m. Its speed is 10^5 $\mathrm{m\,s^{-1}}$. Find the magnitude of the centripetal force that must be acting on it:

$$F = \frac{mv^2}{r}$$

$$= \frac{(9.1 \times 10^{-31} \text{ kg})(10^5 \text{ m s}^{-1})^2}{5 \text{ m}}$$
$$= 1.8 \times 10^{-21} \text{ kg m s}^{-2}$$
$$= 1.8 \times 10^{-21} \text{ N}$$

■ EXERCISES

34. If the speed of an electron moving in a circle is doubled, what force would keep the radius the same?

35. If the speed of the electron is doubled, but the force is unchanged, what is the new radius?

2.7   ENERGY

'Energy" is a word you use in daily speech. You talk about having enough energy to get up and do what has to be done. You hear reminiscences of "energy crises" and predictions of energy shortages to come. People talk about energy needs, energy efficiency, energy conservation, and the need for a national energy policy.

The idea of energy is fundamental to the story this book tells. Energy is useful for discussing remarkably different phenomena over a huge range of magnitudes—tiny particles, large planets, flowing electric charge, light waves, and colliding atoms or nuclei. Because of this general applicability and because the behavior and interactions of radiation and atomic and subatomic matter are more easily described in terms of energy than in terms of force, we use the idea over and over in this book.

For physicists energy can be a measure of a body's ability to do work. This, of course, tells you nothing until "work" is defined. That tells you nothing, of course, until "work" is defined. In the simple case of a constant force pushing parallel to the line of motion of the object on which it acts, and pushing on it over some distance d, the work W done is defined as the amount of force F times the distance d over which the force acts:

$W = Fd$.

EXAMPLES

16. Thus, a force of 2 N applied over a distance of 3 m does an amount of work $W = 2\,\text{N} \times 3\,\text{m} = 6\,\text{N}\,\text{m}$. A newton meter has its own name, "joule," so you can as well say that 2 N acting over 3 m does 6 joules of work. The abbreviation for joule is "J," and you should write that the work done was 6 J.

17. The rate at which energy is supplied over time is power. Power is measured in units of $\text{J}\,\text{s}^{-1}$, and this group of units is called a watt: $1\,\text{J}\,\text{s}^{-1} \equiv 1\,\text{watt}$ (abbreviated W). A watt-second is the same thing as a joule, i.e., $1\,\text{W}\,\text{s} \equiv 1\,\text{J}$.

All this semantic information has its uses, but it does not answer your central question: What *is* energy? One of the best answers to this question is an analogy, given by the renowned American physicist Richard Feynman, that conveys the essence of the idea.[11]

Feynman's Energy Analogy

Imagine a child, perhaps "Dennis the Menace," who has blocks which are absolutely indestructible, and cannot be divided into pieces. Each is the same as the other. Let us suppose that he has 28 blocks. His mother puts him with his 28 blocks into a room at the beginning of the day. At the end of the day, being curious, she counts the blocks very carefully, and discovers a phenomenal law—no matter what he does with the blocks, there are always 28 remaining! This continues for a number of days, until one day there are only 27 blocks, but a little investigating shows that there is one under the rug—she must look everywhere to be sure that the number of blocks has not changed. One day, however, the number appears to change—there are only 26 blocks. Careful investigation indicates that the window was open, and upon looking outside, the other two blocks are found. Another day, careful count indicates that there are 30 blocks! This causes considerable consternation, until it is realized that Bruce came to visit, bringing his blocks with him, and he left a few at Dennis' house. After she has disposed of the extra blocks, she closes the window, does not let Bruce in,

---

[11] Richard Feynman in volume I of *The Feynman Lectures on Physics* by Richard P. Feynman, Robert B. Leighton, Matthew Sands ©1963 by the California Institute of Technology. Published by Addison-Wesley Publishing Co. Inc., 1963. on pp. 4-1 to 4-2.

and then everything is going along all right, until one time she counts and finds only 25 blocks. However, there is a box in the room, a toy box, and the mother goes to open the toy box, but the boy says "No, do not open my toy box," and screams. Mother is not allowed to open the toy box. Being extremely curious, and somewhat ingenious, she invents a scheme! She knows that a block weighs three ounces, so she weighs the box at a time when she sees 28 blocks, and it weighs 16 ounces. The next time she wishes to check, she weighs the box again, subtracts sixteen ounces and divides by three. She discovers the following:

$$\begin{pmatrix} \text{number of} \\ \text{blocks} \\ \text{seen} \end{pmatrix} + \frac{(\text{weight of box}) - 16 \text{ ounces}}{3 \text{ ounces}} = \text{constant.} \quad (10)$$

There then appear to be some new deviations, but careful study indicates that the dirty water in the bathtub is changing its level. The child is throwing blocks into the water, and she cannot see them because it is so dirty, but she can find out how many blocks are in the water by adding another term to her formula. Since the original height of the water was 6 inches and each block raises the water a quarter of an inch, this new formula would be

$$\begin{pmatrix} \text{number of} \\ \text{blocks} \\ \text{seen} \end{pmatrix} + \frac{(\text{weight of box}) - 16 \text{ ounces}}{3 \text{ ounces}}$$

$$+ \frac{(\text{height of water}) - 6 \text{ inches}}{1/4 \text{ inch}} = \text{constant.}$$

$$(11)$$

In the gradual increase in the complexity of her world, she finds a whole series of terms representing ways of calculating how many blocks are in places where she is not allowed to look. As a result, she finds a complex formula, a quantity which *has to be computed*, which always stays the same in her situation.

What is the analogy of this to the conservation of energy? The most remarkable aspect that must be abstracted from this picture is that *there are no blocks*. Take away the first terms in Eqs. 10 and 11 and we find ourselves calculating more or less abstract things. The analogy has the following points. First, when we are calculating the energy, sometimes some of it leaves the system and goes away, or sometimes some comes in. In order to verify the conservation of energy, we must be careful that we have not put any in or taken any out. Second, the energy has a large number of *different forms*, and there is a formula for each one.

Energy Costs Money

Here is some more evidence that energy for all its abstraction is something real. It costs money. Whatever energy is, if you want some, you usually have to pay for it. You buy electrical energy to run your household appliances; you buy oil or natural gas to heat your house; you buy gasoline to run your automobile; you buy food to run your body.

A kilowatt-hour of electricity is the same thing as 3.6 MJ. In the U.S., the average cost of a kW-h is about $0.10. This is not very expensive, which is one of the reasons many Americans' lives are often pleasant. A representative price for all forms of energy is $25 for 10^9 joules. Scientists use the prefix "giga" to represent the factor 10^9 (an American billion), so 10^9 joules is called a gigajoule. Giga is abbreviated G, so you could write that energy costs roughly $25 \, \text{GJ}^{-1}$.

■ EXERCISES

36. A gallon of gasoline contains about 131 MJ. At $3 per gallon, how much are you paying for 1 GJ of gasoline?

37. At $0.10 a kW-h, how much are you paying for a GJ of electricity?

Actual prices vary a great deal depending on special features of the energy: is it easy to handle? Is it very concentrated? Is the energy accessible easily? Can you get out a lot of energy quickly? The concerns of this book with energy will usually be quite remote from practical considerations of cost and availability. Rather, energy will be a guide to studying atoms and their structure.

Conservation of Energy

As Feynman's analogy suggests, energy comes in many forms. There is heat energy, kinetic energy, gravitational potential energy, electrical potential energy, energies of electric and magnetic fields, nuclear energy. There is energy stored in the compression of a spring, in the compression of gas, in the arrangement of molecules and atoms. Although energy may change from one form to another, the sum of all the forms of energy in a system remains constant unless some agent moves energy into or out of the system. We say, therefore, that energy in a closed system is conserved, i.e., the total does not change. This property of conservation makes energy an exceptionally useful quantity. (Notice that physicists use

the word "conservation" differently from economists and environmental-
ists, who usually mean "use available forms of energy as efficiently as
possible.")

Kinetic Energy

One of the most familiar forms of energy is "kinetic energy," the energy
a body has by virtue of its motion. It is another kind of "quantity of
motion" that was found to be useful at about the same time that Newton
began to think in terms of momentum. A body of mass m moving with a
speed v substantially smaller than the speed of light has a kinetic energy
K given by the formula

$$K = \frac{1}{2}mv^2.$$

Knowing this, you can calculate the kinetic energy of the baseball, the
bowling ball, and the SUV described earlier.

▪ EXERCISES

38. The baseball's kinetic energy is $1/2 \times 0.15\,\mathrm{kg} \times (24.6\,\mathrm{m\,s^{-1}})^2 =$
$45.4\,\mathrm{J}$. The kinetic energy of the bowling ball is $1513\,\mathrm{J}$. That of the automobile is $0.454\,\mathrm{MJ}$ (where "M" is the usual abbreviation for the prefix
"mega," which stands for 10^6). Verify that these are correct numbers.
Notice that since each mass has the same speed, their kinetic energies
vary only by the ratios of their masses, i.e.,

$$K_{\text{bowling ball}} = \frac{5.0\text{ kg}}{0.15\text{ kg}} \times K_{\text{baseball}}.$$

(Insights like this are useful because they make calculating easier.)

39. A pitcher warms up by throwing a baseball to the catcher at 45
mph; this means it has a kinetic energy of about $30\,\mathrm{J}$. During the game
he throws a fastball at $90\,\mathrm{mph}$. What is its kinetic energy then?

Gravitational Potential Energy

Because you are familiar with it, gravity at the surface of Earth provides
a good way to introduce the concept of potential energy, an idea that we
will later apply to electrical properties of atoms and their internal parts.
When you lift a mass m to some height—call that height h, you add energy
to the mass. This energy is called "gravitational potential energy," or for
brevity just "potential energy." (This brevity is sloppy usage because

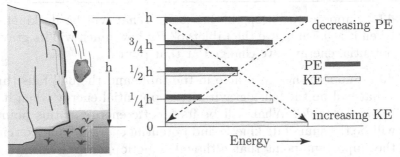

FIGURE 2.8 A rock falls off a cliff of height h. Its potential energy is converted into kinetic energy, but the total energy remains constant.

their are other kinds of potential energy than gravitational.) The amount of gravitational potential energy U you add by lifting the mass m a height h is

$$U = mgh,$$

where g is the acceleration due to gravity near Earth's surface, $9.80\,\mathrm{m\,s^{-2}}$.

Why is it called "potential"? Perhaps because it has the potential for becoming kinetic energy. If you lift the mass a height h and then release it, experience shows you that its speed increases as it falls. This means that its kinetic energy increases. Of course, as m falls, the height changes from h to some smaller height y, so the gravitational potential energy of m diminishes all the while its kinetic energy increases. What you are seeing here is the conversion of one form of energy—gravitational potential energy—into another—kinetic energy—as the mass falls. What makes this way of looking at the fall so useful is that the sum of the two forms of energy remains constant throughout the fall: an example of the conservation of energy:

$$mgh = mgy + \frac{1}{2}mv^2.$$

Figure 2.8 graphically illustrates this remarkable property of energy. As a stone falls off a cliff of height h its PE decreases steadily (dark bars) while its KE increases steadily (light bars). At any instant the sum of the dark and light bars is always the same.

▮ EXERCISES

40. What is the value of the sum of the two bars when $y = h/2$?

41. Suppose you lift a baseball ($m = 150\,\mathrm{g}$) $1\,\mathrm{m}$ above a table. By how much do you increase its gravitational potential energy?

---

42. Suppose you drop the baseball. What will be its kinetic energy when it is 0.5 m above the tabletop? By how much will its gravitational potential energy have changed at that point?

43. Suppose there is a hole in the table and the ball falls through it. What will be the ball's gravitational potential energy when it is 20 cm *below* the table? What will be its kinetic energy at this point? What will be the sum of its kinetic and potential energy? This exercise makes the important point that although kinetic energy is always positive, potential energy can be negative.

What happens after the ball hits the table and stops? Clearly, its kinetic energy becomes 0 J. Also, the ball has reached the point from which we chose to measure potential energy, and so its potential energy is 0 J. What has become of its 1.47 J? It has gone into heating up the point of impact, into the compression of the spot on which it is resting, and into acoustic energy—the sound of its impact. A fascinating aspect of energy is that so far in the history of physics there has always been an answer to the question: What has become of the initial energy? And very often the answer casts revealing light on the nature or behavior of matter.

It is an interesting and useful fact that gravitational potential energy depends only on *vertical* distance; sideways movements of a body do not change its gravitational potential energy. This means that no matter how a body falls from one height to another, the change in gravitational potential energy will be the same. Then as long as the only other form of energy can be kinetic, the change in kinetic energy will also be the same. Look at Fig. 2.9, where a flat object teeters at the top of two different inclines.

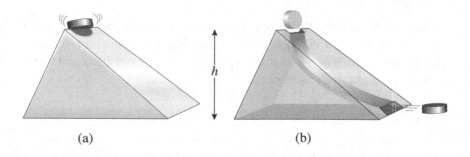

(a)   (b)

FIGURE 2.9 (a) A frictionless hockey puck teeters indecisively. Regardless of which way it slides, it will have the same kinetic energy when it reaches the bottom. (b) Now the puck may slide down through a tunnel drilled in the block, but its kinetic energy will still be the same at the bottom.

It can slide without friction down the left side or the right side, but whichever path it slides down, it will have the same kinetic energy at the bottom.

▽ EXAMPLES

18. Suppose that in Fig. 2.9 the puck has a mass of 160 g and that $h = 20$ cm. What will be its kinetic energy when it reaches the bottom of the left-hand incline? The bottom of the right-hand incline?

Relative to the bottom of the inclines, the puck has a gravitational potential energy of $mgh = 0.16\,\text{kg} \times 9.8\,\text{m s}^{-2} \times 0.2\,\text{m} = 0.314\,\text{J}$. As the puck slides without friction down either incline, this amount of gravitational potential energy is converted to kinetic energy. The kinetic energy of the puck is the same at the bottom of either side; it is 0.314 J.

▨ EXERCISES

44. Suppose a chute was drilled through the block, curving off to the side and arriving at the bottom right-hand corner as shown in Fig. 2.9b. If the puck fell down the chute, what would be its kinetic energy when it arrived at the bottom? How fast would it be moving?

Pendulums and Energy

A pendulum is a concentrated mass hanging by a tether from some pivot point. Its motion is familiar if you have ever swung on a swing or looked inside a grandfather clock. Now you can understand that what you observe when you watch a mass swing back and forth at the end of a string is the cyclical conversion of gravitational potential energy into kinetic energy. The pendulum's motion is begun by pulling its bob to one side; this has the effect of lifting it some vertical distance h, as shown in Fig. 2.10a, and it acquires gravitational potential energy. When released, the pendulum swings back to its lowest position, where it is moving its fastest because all its gravitational potential energy has been converted into kinetic energy. As the bob rises to the other side, it slows down because of the conversion of kinetic energy into gravitational potential energy. It reaches the highest point of its swing when all its kinetic energy has been converted to potential energy; then it moves back toward the lowest point, beginning another cycle of conversion.

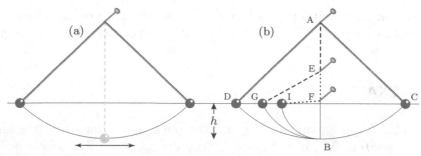

FIGURE 2.10 (a) A pendulum converts energy back and forth from gravitational potential energy to kinetic energy. (b) For any given amount of total energy, changing the pivot point will not change the height to which the pendulum can rise.

Figure 2.10b illustrates an argument made by Galileo that you can explain on the basis of the conservation of energy. He writes:

Imagine this page to represent a vertical wall, with a nail driven into it; and from the nail let there be suspended a lead bullet of one or two ounces by means of a fine vertical thread, AB, say from four to six feet long; on the wall draw a horizontal line DC, at right angles to the vertical thread AB, which hangs about two finger-breadths in front of the wall. Now bring the thread AB with the attached ball into the position AC and set it free; first, it will be observed to descend along the arc CB, to pass the point B, and to travel along the arc BD, till it almost reaches the horizontal CD, a slight shortage being caused by the resistance of the air and the string; from this we may rightly infer that the ball in its descent through the arc CB acquired a momentum [he means kinetic energy; the difference between momentum and kinetic energy was not clear until two hundred years after Galileo] on reaching B, which was just sufficient to carry it through a similar arc BD to the same height. Having repeated this experiment many times, now drive a nail into the wall close to the perpendicular AB, say at E or F, so that it projects out some five or six finger-breadths in order that the thread, again carrying the bullet through the arc CB, may strike upon the nail E when the bullet reaches B, and thus compel it to traverse the arc BG, described about E as center. From this we can see what can be done by the same momentum [kinetic energy] which previously starting at the same point B carried the same body through the arc BD to the horizontal CD. Now, gentlemen, you will observe with pleasure that the ball swings to the point G in the horizontal, and you would see the same thing happen if the obstacle were placed at some lower point, say at F, about which the ball would describe the arc BI, the rise of the ball always terminating exactly on the line CD. But when the nail is placed so low that the remainder of the thread below it will not reach to the height CD (which would happen if the nail were placed nearer B than to the intersection of AB with the horizontal CD) then the thread leaps over the nail and twists itself about it.[12]

■ EXERCISES

45. Using the conservation of energy, give your own explanation of the demonstration described here by Galileo.

Forces As Variations in Potential Energy

Figure 2.9 illustrates an important feature of potential energy. Wherever there is a spatial variation of potential energy, there is a force. Notice that as the puck slides down the right slope, its potential energy changes more gradually than it does when it slides down the left slope. The steeper the spatial change of potential energy, the greater is the force.

▼ EXAMPLES

19. Suppose the angle θ of the incline in Fig. 2.9 is 30°. Then as the puck slides along the right slope a distance $\Delta s = 1\,\text{cm}$, it drops a vertical distance of $\Delta z = 0.5\,\text{cm}$ because $\Delta z / \Delta s = \sin \theta$ and $\sin 30° = 0.5$. The force F produced by this change in potential energy is just

$$F = -\frac{\Delta U}{\Delta s} = -\sin\theta \frac{\Delta U}{\Delta z}, \tag{12}$$

which means that $F = -mg\sin\theta$.

You may have known this already, but the point here is not that the force down an incline is proportional to the sine of the angle of the incline. The point is that the force is equal to how much the potential energy varies over a small distance.

---

[12]In Galileo Galilei *Dialogues Concerning Two New Sciences*, Northwestern University, 1939 (Dover, New York), pp. 170–171.

This property of potential energy can be nicely expressed using calculus, but for the purposes of this book the important idea is that if you have two points A and B close together in space, and the potential energy of a body is higher at A than at B, then there is a force pushing the body from A toward B. Differences in potential energy tell you that forces are acting. Big differences of potential energy over small distances mean the forces are large.

2.8 SUMMARY AND EXHORTATIONS

This review of basic concepts and units of physics omits many interesting subtleties and ignores the vector nature of many of the quantities. Nevertheless, you should now have a better understanding of the ideas of mass, length, time, velocity, momentum, and energy, their units, the prefixes that give the powers of ten that become parts of the units, and the way they are used.

The ideas of length and time provide a basis for describing motion in terms of velocity—the direction and rate at which a body covers distance—and in terms of acceleration—the rate at which velocity changes.

The quantity of motion in a body is the product of its mass and its velocity, mv, and is called momentum. A body's momentum is changed only by a force; a force is anything that causes momentum to change. Force is measured as the time-rate-of-change of momentum $\Delta(mv)/\Delta t$. In a closed system momentum is conserved.

A body of mass m moving with a speed v has kinetic energy which at velocities of familiar objects is given by $\frac{1}{2}mv^2$. Energy comes in many forms. In a closed system energy may change from one form to another, but the total amount of energy does not change. In a closed system energy is conserved.

Connect Concepts to Physical Reality

These concepts and ideas are important, but so are the techniques for thinking about them. Practice assigning numbers to them and developing a sense of their physical scale and significance. You will need these techniques to make good use of the rest of this book.

Remember, there is more to reading equations than assuring that their units are consistent. You want to connect what they are telling you about the behavior of a physical system to personal experience or to other related phenomena. Often this means calculating numerical examples for the system under consideration and comparing them.

TABLE 2.5 Important quantities for describing the physics of anything

Concept	Compound units	Name	Abbrev- iation	Examples
velocity	$m\,s^{-1}$	None	None	A person walks at a rate of about $1.5\,m\,s^{-1}$. A snail goes at a few $mm\,s^{-1}$. The speed limit for U.S. autos on highways in urban areas is $24.6\,m\,s^{-1}$. The speed of light in a vacuum is $3 \times 10^8\,m\,s^{-1}$.
acceleration	$m\,s^{-2}$	None*		A body falling freely near the surface of Earth accelerates at $9.8\,m\,s^{-2}$
force	$kg\,m\,s^{-2}$	newton	N	Earth exerts a force of $98\,N$ on a $10\,kg$ mass near its surface. Earth's atmosphere exerts a force of $1.01 \times 10^5\,N$ on each square meter of Earth's surface.

*An acceleration of $1\,cm\,s^{-2}$ is sometimes called a gal (in honor of Galileo).

As you come across them, think about the numerical values of physical quantities and try to connect them to specific phenomena with which you are familiar. This will help you understand the physical significance of the concepts. Part of the great power of physics is that it works for very large-scale systems and very small-scale systems. When you are introduced to a new concept, you should try it out at several different scales. Table 2.5 offers some numerical values of real velocities, accelerations, and forces as concrete examples of these concepts for you to consider.

■ EXERCISES

46. Notice that Table 2.5 has no entries for energy or momentum. Make up appropriate entries for these two quantities.

Know the SI Prefixes

SI prefixes—micro, mega, kilo, nano, giga—have been introduced and used in several places in this chapter. You *must* know them, their abbreviations, and their numerical values well enough so that you can convert among them quickly and accurately. There is a list of all the SI prefixes at the end of the book. Maybe you can learn them by osmosis, but if you can't, then just memorize them. Do whatever it takes, but learn them.

APPENDIX: VECTORS

Some of the quantities discussed above have a special property that we will use occasionally. When the direction is of importance in the full description of the quantity, the quantity is a "vector." Examples of vectors we will use are change of position of an object, its velocity, and its momentum. Examples of quantities that are *not* vectors are mass, energy, and time.

Representing Vectors

An arrow is often used to represent a vector quantity. The arrow's direction shows the direction of the quantity and the arrow's length shows the magnitude (size) of the quantity. You can see how a vector represents a "displacement," i.e., a change of position, from point A to point B. The displacement vector is just a scaled down copy of the arrow connecting A to B. For velocity, acceleration, momentum, force, electric field, etc., the relation of the physical quantity and the representative arrow is not so obvious, but the mathematical correspondence is exact and very useful.

Figure 2.11 depicts a typical displacement vector, $\vec{R}$. We use the notation of a letter with an arrow over it to indicate a vector quantity, one with both magnitude and direction. The plain letter, such as R, just represents the magnitude, or length, of the vector. In this case it is the shortest distance an object could move while traveling from the origin to the point (X, Y).

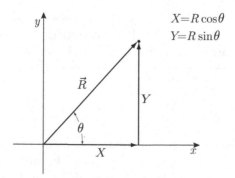

$$X = R\cos\theta$$
$$Y = R\sin\theta$$

FIGURE 2.11   Components $(X$ and $Y)$ of the vector $\vec{R}$.

Components

The diagram in Fig. 2.11 suggests another way of looking at the displacement. Arriving at the point (X, Y) could have been accomplished by a displacement in the x direction by an amount X and then in the y direction by an amount Y. If we make these two quantities themselves vectors, the vector $\vec{R}$ is entirely equivalent to adding the two displacements in the coordinate directions, $\vec{X}$ and $\vec{Y}$. The magnitudes of these two displacements are particularly handy quantities for dealing with vectors. They are called the "Cartesian components" of the vector. Figure 2.11 exhibits the trigonometry and geometry used to go back and forth between a vector and its components using the Pythagorean theorem and basic definitions of the sine and cosine:

$$X = R\cos\theta,$$
$$Y = R\sin\theta,$$
$$X^2 + Y^2 = R^2.$$

■ EXERCISES

> **47.** Prove that if $X = R\cos\theta$ and $Y = R\sin\theta$ then $R^2 = X^2 + Y^2$.

You can see from Fig. 2.11 that the magnitude of a vector is just the square root of the sum of the squares of its components.

What if the overall displacement involves a displacement Z in the third or z dimension? Then

$$R^2 = X^2 + Y^2 + Z^2.$$

Adding Vectors

Adding vectors can get somewhat more complicated in the general case. You have already seen that reconstituting a vector from its components is equivalent to adding two vectors at right angles. One other case is important to deal with at this time. Adding vectors that are all along one particular coordinate axis is a simple matter of addition and subtraction. Imagine walking ten paces west. If you reverse and come east for three paces, your second displacement undoes some of the first, so it is in effect negative. The systematic way of handling this is to take displacements in the positive coordinate direction to be positive and displacements in the negative coordinate direction to be negative. Then a simple algebraic sum of displacements gives the net displacement as long as you stick to the one dimension. Other vector quantities have the same kind of additive properties as does displacement, so you can follow exactly the same rules for adding them up, even if a geometric picture seems inappropriate.

PROBLEMS

1. What is the order of magnitude of your height? What meaning are you using for "order of magnitude"?

2. What is the circumference of Earth? How do you know this?

3. Estimate the volume of your body.

4. Without looking at a ruler or other measuring device, draw a line 1 cm long.

5. A large speck of dust has a mass $m = 0.00000412$ g (grams).
  a. Rewrite the mass in grams using scientific notation.
  b. Express m in (i) kg; (ii) mg (iii) μg.

6. Solid aluminum has density 2.7 g cm^{-3}. What is this density in units of kg m^{-3}? Use scientific notation and show your calculation.

7. Notice that one of your answers will be a number density and the other will be a mass density.
  a. A rectangular box has dimensions $L = 2$ cm, $W = 5$ cm, and $H = 10$ cm. How many of these boxes will fit into a cube of volume 1 m^3?
  b. If each box has a mass $m = 0.01$ kg, what is the density (mass per unit volume) of the assembly of boxes?

8. The speed of light is 3×10^{10} cm s^{-1}; the circumference of Earth is 40 Mm. How long would it take light to circle the Earth?

9. The electron volt is a unit of energy, abbreviated eV. What is the ratio of the energies of a 20 GeV electron and a 3 keV electron?

10. A plucked violin string gives off a sound with frequency $f = \frac{1}{2\ell}\sqrt{\frac{\tau}{\mu}}$, where ℓ is the length of the string, τ is the tension in the string, and μ is the mass per unit length of the string, i.e., μ has dimensions of $\mathrm{M\,L^{-1}}$.

    a. Verify that this equation is dimensionally correct. (Tension is a force, and the dimensions of frequency are s^{-1}.)

    b. At some point before the A string is tightened, its frequency is 220 Hz (low A). To bring it up to the correct frequency (440 Hz), by what factor must the tension be changed?

    c. Oh no! You're in the middle of your physics test, and you need to calculate how long it takes sound to travel 5 m. You think "All I need to do is multiply the distance by the speed of sound." Use dimensional analysis to show that this is wrong thinking.

11. A mass of 2 kg travels at $4\,\mathrm{m\,s}^{-1}$ towards a wall. It hits the wall and bounces directly back, now traveling at $2\,\mathrm{m\,s}^{-1}$ away from it. By how much does the momentum of the mass change in this collision?

12. When a paper airplane flies, sometimes the airlift makes it possible for the plane to travel horizontally (without falling) for a short distance. While the plane is traveling horizontally, what must be the lift due to air? (Hint: What quantity do you need to estimate in order to answer this question?)

13. Suppose an 80 kg mass falls off a table 30 inches high and moves toward the floor. Does its momentum remain constant? Why?

14. What will be the kinetic energy of the mass in the preceding question just before it hits the floor? Explain how you know.

15. An object of mass 2 kg, initially at rest, is dropped from a tower in order to determine its height. The object's velocity is $0.02\,\mu\mathrm{m}\,\mathrm{ns}^{-1}$ just before it hits the ground.

    a. Find the object's velocity in $\mathrm{m\,s}^{-1}$.

    b. What is its kinetic energy just before impact?

    c. What is the height of the tower?

16. A 50 g ball rolls up a short incline and then continues along level ground. The speed of the ball at the bottom of the incline is $2\,\mathrm{m\,s}^{-1}$, and the height of the incline is 10 cm. What will be the speed of the ball once it reaches the level ground?

17. If the ball in the previous problem had a mass of 100 g, and the same initial speed, what would be its final speed?

18. What is the maximum height of incline that either ball could climb?

19. If a 20 N force is applied to a 10 kg mass for 10 ms, by how much does the momentum of the mass change?

20. In icy weather, cars A (mass 1000 kg) and B (mass 2000 kg) collide head-on while traveling in opposite directions in the same lane of a highway.

  a. A and B each had speed 15 m s^{-1} just before the collision, and A was heading due east. Sketch vectors (arrows) representing the momenta of the cars before the collision, and calculate their total momentum.

  b. The cars stick together immediately after the collision. What is their speed immediately after the collision, and in what direction are they moving?

  c. The average force on a body is defined as the change in momentum divided by the time it takes for that change to occur. If the duration of the impact is 0.125 s, find the average force exerted on car A by car B.

21. A pole vaulter can clear a bar set at a height of 5 m.

1. By what factor should he increase the speed of his approach if he hopes to gain entry to the prestigious Six Meter Club?

2. In reality a vaulter who runs at 9.5 m s^{-1} can clear a 6 m bar.

  a. Compare his kinetic energy while running to his gravitational potential energy at 6 m height.

  b. Give several reasons why this experimental observation does not violate energy conservation.

22. A baseball and a bowling ball both have the same momentum. Which one (if either) has the greater kinetic energy?

23. Suppose you have a circle 1.2 m in diameter. Imagine that lines are drawn from the center of the circle to two points on the circumference 0.05 m apart. What is the (small) angle between the two lines? Give your answer both in degrees and in radians.

24. For the circle in Problem 23, what would be the angle if the length of the arc between the two points was 1.2 m? Find your answer both in degrees and in radians.

25. What diameter disk held at arm's length (call it 60 cm) just covers the full Moon (see Fig. 2.12)?

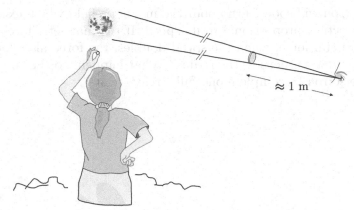

FIGURE 2.12   How a small disk can block out your view of the Moon.

26.   If a U.S. dime held at a distance of 2 m from your eye just covers the full Moon, what is the diameter D of the Moon?

27.   Take a common foodstuff—a candy bar, a bottle of soda, bread, breakfast cereal, or whatever—and from its label determine what a giga-joule of this foodstuff costs. Remember that Calories on labels are each 4180 J. Hint: When you buy a five pound bag of house-brand cane sugar you pay about $85 for a gigajoule.

28.   Complete the following table of standard prefixes for units

prefix	abbreviation	value
nano		
	M	
		10^{-2}
milli		
	p	
		10^9
kilo		
	μ	
	T	

29.   How many μg are there in 2 kg?

30.   List all the SI prefixes used in this chapter, giving their names, abbreviations, and numerical values. Use them with units to describe various physical phenomena.

31. You will soon learn about the magnetic field which exerts a force on the electron proportional to its speed. If the magnetic force provides the centripetal force, when the speed doubles, the force also doubles. In this case, what happens to the radius? What happens to the time required for the electron to complete one full circle?

3

The
Chemist's Atoms

3.1 INTRODUCTION

The idea of an atom has a long history. Around twenty-five hundred years ago, Greek philosophers argued that matter must be built up of small, hard, identical pieces. Because these pieces were thought to be irreducible they were called "atoms." The word "atom" is derived from the Greek for "un-cuttable." The pieces, or atoms, come in only a few kinds, said the Greek philosopher Thales, and the complexity we observe in nature arises from the variety of ways in which these building blocks go together and come apart.

This simplified picture of the Greek concept of atom contains the essentials of the modern idea. Why then did it take until the early 1800s for chemists and physicists to produce convincing physical evidence for the existence of atoms? The answer, in part, is that to obtain and interpret their evidence they needed both the modern concept of the chemical element and the development of quantitative techniques of measurement. These became precise enough to yield useful information at the end of the eighteenth century. Only then was the stage set for obtaining and understanding the physical evidence for atoms that is the subject of this chapter.

3.2 CHEMICAL ELEMENTS

A major step toward an atomic theory of matter was the recognition of the existence of the "chemical elements." Centuries of study of the qualitative properties of matter had made it possible to recognize and distinguish

C.H. Holbrow et al., *Modern Introductory Physics, Second Edition*,
DOI 10.1007/978-0-387-79080-0_3, © Springer Science+Business Media, LLC 1999, 2010

among a large variety of substances. Chemists, notably among them Lavoisier, recognized that while many chemical substances could be broken down into others, certain chemicals, like carbon, sulfur, and oxygen, could not be broken down by heat or grinding or other known chemical processes. These irreducible chemicals were viewed as "elements," basic species of which all other chemical substances were compounded. This is the origin of the idea of chemical "compounds" built up from chemical elements.

Although the idea of the chemical element was established by the end of the eighteenth century, notice that the idea does not necessarily imply that there are such things as atoms. All that has been established is that there seem to be a number of distinct types of chemicals from which others can be constructed. What then would be good evidence that elements are composed of atoms?

3.3   ATOMS AND INTEGERS

An essential feature of atomicity is countability. If discrete building blocks of matter exist, they should be countable like chairs, students, or money. Countability is closely associated with the integers. The number of chairs in your classroom is probably an integer. So is the number of students in your classroom, or the number of coins in your piggy bank. Experience suggests an important general idea: Countable sets are integer multiples of some basic, individual unit: one chair, an individual student, a coin. These units are indivisible at least in the sense that a set of these objects can only be subdivided so far and no farther without changing the nature of the elements of the set: to kindling, to a scene in a horror movie, to bits of metal.

Atomicity is closely related to integer countability. The observation of integer relationships in the formation of chemical compounds was early, important evidence that atoms exist as discrete building blocks of these compounds. By the beginning of the nineteenth century, techniques for weighing small masses had become precise enough to permit accurate quantitative measurements of the amounts of different chemicals before and after combination. By their ingenious analyses of such data, Proust, Dalton, Gay-Lussac, Avogadro, and other scientists at the beginning of the nineteenth century revealed that *elements combine in integer ratios that convincingly imply the existence of atoms.*

Proust's Evidence: The Law of Constant Proportions

The chemist Joseph-Louis Proust observed that when chemicals combine to make a particular substance, the reacting elements always make up the same percentage of the weight of the new substance. For example, stannous oxide always is made up of 88.1% tin (Sn) and 11.9% oxygen (O); and stannic oxide always is made up of 78.7% Sn and 21.3% O. This behavior suggests that a fixed amount of tin can combine only with one or another fixed amount of oxygen. This kind of constancy was verified by many careful measurements on many different substances. Table 3.1 shows the constant proportions of the weights measured by Sir Humphrey Davy for three different compounds of nitrogen (N) and oxygen. Because Proust was the first to notice that all chemicals are made up of fixed percentages of their constituents, this behavior is known as Proust's law of constant proportions.

Dalton's Evidence: The Law of Multiple Proportions

It was by looking at the data in Table 3.1 in a different way that John Dalton obtained the first convincing scientific evidence that substances are built up out of small, individual blocks of matter. Dalton's work was first published in 1808.[1]

Instead of dealing in percentages, Dalton first asked what weights of one element would combine with a fixed amount of another. Thus, he found that 13.5 g of oxygen would combine with 100 g of tin to make stannous oxide, while 27 g of oxygen would combine with 100 g of tin to make stannic oxide. (Nowadays, we would write that he was observing the formation of SnO and SnO_2.) Then he examined the ratio of the two different amounts of oxygen and observed that the ratio is of simple

TABLE 3.1   Davy's percentage of mass of nitrogen and oxygen in oxides of nitrogen

Compound gas (modern names)	% Mass of nitrogen	% Mass of oxygen
nitrous oxide	63.30	36.70
nitric oxide	44.05	55.95
nitrogen dioxide	29.50	70.50

---

[1]See *From Atomos to Atom* by Andrew G. Van Melsen, Harper & Brothers, New York, 1960. This book quotes extensively from a later edition of Dalton's work *A New System of Chemical Philosophy*, London 1842.

integers, 1:2. Examining many cases, he found that always when different amounts of one element combine with another to make different chemical compounds, the amounts are in simple integer ratios to each other.

Dalton realized that this law of multiple proportions, as he called it, was evidence that chemicals combined in integer multiples of some basic unit, or *atom*. His data could be explained if an integer number—one or two—of atoms of one element combined with an integer number—one, two, three, or more—of atoms of another element. More important, Dalton's way of comparing the masses of chemicals before and after combination was a quantitative method that could be used by other scientists and applied to many chemical reactions. On the basis of such analysis he proposed the following ideas.

Dalton's ideas about atoms:

• All elements consist of minute discrete particles called atoms.

• Atoms of a given element are alike and have the same mass.

• Atoms of different elements differ, each element having unique atoms.

• Chemical changes involve the union or separation of *undivided atoms* in fixed simple numerical ratios. Atoms are not created or destroyed when chemical change occurs.

▽ EXAMPLES

1. A striking example of the law of multiple proportions can be obtained from Davy's data on the combinations of nitrogen and oxygen (Table 3.1) if you look at them from Dalton's point of view. To do this, calculate how much oxygen reacts with a fixed amount of nitrogen, say 100 g. The first line of the table states that nitrous oxide is 63.3% nitrogen and 36.7% is oxygen. To scale the amount of nitrogen to 100 g, just multiply by 100/63.3; use this same scale factor to get the amount of oxygen:

$$\frac{100 \text{ g}}{63.3} \times 36.7 = 58 \text{ g}.$$

A similar procedure applied to the next two lines in the table implies that 127 g and 239 g of O combine with 100 g of N to form nitric oxide and nitrogen dioxide, respectively. It is the near-integer ratios of these different amounts of oxygen reacting with the fixed amount of nitrogen that supports the idea of atoms:

$$58 : 127 : 239 \approx 1 : 2 : 4.$$

Although Davy's data deviate somewhat from simple integer ratios, later more careful measurements confirm Dalton's law of multiple proportions very well. More complete studies show that 100 g of nitrogen will combine with 57 g or 113 g or 171 g or 229 g or 286 g of oxygen. These are respectively the compounds of nitrous oxide, nitric oxide, nitrous anhydride, nitrogen dioxide, and nitric anhydride.

■ EXERCISES

1. What are the "simple proportions" of these quantities? What would you guess to be the chemical formulas for these compounds? Hint: nitric oxide is NO.

2. Given that the density of nitrogen at one atmosphere of pressure and $0\,^{\circ}\mathrm{C}$ is $1.2506\,\mathrm{kg\,m^{-3}}$ and that of oxygen at the same temperature and pressure is $1.429\,\mathrm{kg\,m^{-3}}$, calculate the relative volumes of the two gases that combine to form the compounds named above. If you do this problem correctly, you will discover what Gay-Lussac discovered. These results are shown in Table 3.2.

Dalton's analysis shows how important it is not only to collect high quality data, but also to find the best way to analyze them. The information in Table 3.1 is not particularly compelling evidence for the existence of atoms, but the simple integer ratios found in the exercises above strongly suggest that elements are composed of fundamental, countable objects that we call atoms.

Gay-Lussac's Evidence: The Law of Combining Volumes

The appearance of simple integer relationships is even clearer in an interesting version of Dalton's law that was discovered by the chemist and physicist Joseph Louis Gay-Lussac (1778–1850). He studied the chemical combination of gases and observed that at the same temperature and

TABLE 3.2   Combining volumes of nitrogen and oxygen

Compound gas	Volume of nitrogen	Volume of oxygen	Final volume (approximate)
nitrous oxide	100	49.5	100
nitric oxide	100	108.9	200
nitrogen dioxide	100	204.7	200

pressure their *volumes* combined in ratios of small integers. For example, he found that $100\,cm^3$ of oxygen combined with $198.6\,cm^3$ of hydrogen. Within the uncertainty of his measurements, this is a ratio of 1:2. Using contemporary values of densities, he also converted Davy's data on the various compounds of nitrogen and oxygen (Table 3.1) to volumes. Table 3.2 shows the striking result, which is known as the "law of combining volumes."

Clearly the volumes of these gases combine in simple ratios of small integers like 2:1 and 1:1. Furthermore, for gases, the simplicity of the ratio shows up in a single reaction. There is no need to examine different compounds of the same elements as is required to exhibit Dalton's law of multiple proportions. The implication is plausible that the gases react by combining simple building blocks in whole-number, i.e., integer, amounts.

When Gay-Lussac performed experiments himself, he observed that not only did the gases combine in simple proportions, but the final volume of the reacted gas was simply related to the combining volumes. For example, when two volumes of hydrogen combined with one volume of oxygen, the resulting water vapor occupied two volumes. It is simple to interpret this result as two atoms of hydrogen combining with one atom of oxygen to make a water molecule. But if that were so, the end product should occupy only *one* volume of water. The appearance of *two* volumes of water was, therefore, puzzling.

A similar puzzle occurs with the combination of nitrogen and oxygen. If you combine $100\,cm^3$ of nitrogen gas with $100\,cm^3$ of oxygen you get $200\,cm^3$ of nitric oxide gas, as suggested in Table 3.2 after rounding off the oxygen volume. Apparently, the combination of one elemental unit of nitrogen gas with one elemental unit of oxygen produces two such units, or *molecules*, of nitric oxide gas. This is strange because one *atom* of nitrogen combining with one *atom* of oxygen should produce just one molecule of nitric oxide.[2]

---

[2]The oxides of nitrogen are confusing. Here is a list of their modern names and chemical formulas. Don't they make a nice application of Dalton's law of multiple proportions?

Nitrous oxide	N_2O
Nitric oxide	NO
Dinitrogen trioxide	N_2O_3
Nitrogen dioxide	NO_2
Dinitrogen tetroxide	N_2O_4
Dinitrogen pentoxide	N_2O_5

How can this be? In 1811 the Italian chemist Amadeo Avogadro suggested the solution to the puzzle.[3] He proposed that the volumes of hydrogen, nitrogen, and oxygen contained molecules made up of two identical atoms: H_2, N_2, O_2.[4] His idea helped us to understand that molecules are structures made up of atoms.

EXERCISES

3. Gay-Lussac's work made it possible to determine chemical formulas. To see how, use Avogadro's idea and Gay-Lussac's data (Table 3.2) to find the chemical formula for nitrogen dioxide, which he called "nitric acid." (You see that it is quite different from the HNO_3 that we today call nitric acid.)

4. What are the chemical formulas for the other two compounds in Table 3.2?

Avogadro's Principle

Avogadro and the Swedish chemist Jons Berzelius realized that if volumes of gases (at the same temperature and pressure) always combine in simple ratios, then equal volumes of gas contain equal numbers of molecules. To see why this is so, consider our earlier example of two volumes of hydrogen combining with one volume of oxygen to form two volumes of water. You can read the chemical equation

$$2H_2 + O_2 \rightarrow 2H_2O$$

as though it is a statement about the combination of volumes, but you can also read it as a statement that two molecules of hydrogen combine with one molecule of oxygen to form two molecules of water. *There is a one-to-one correspondence between volumes and numbers of molecules.* There must be twice as many molecules in two volumes as in one. Because this statement does not depend on the kind of molecule, we,

---

[3] A. Avogadro, "Essay on a Manner of Determining the Relative Masses of the Elementary Molecules of Bodies and the Proportions in Which They Enter into These Compounds" *Journal de Physique*, 1811, as excerpted in *Readings in the Literature of Science*, ed. W.C. Dampier and M. Dampier, Harper Torchbook, Harper & Brothers, New York, 1959. You might be interested to know that Avogadro's idea that hydrogen, nitrogen, and oxygen were molecules consisting of two atoms of the same kind was not accepted for several decades. His contemporaries, among them Dalton and Gay-Lussac, considered the idea too strange to be likely.

[4] Some other diatomic gases are F_2 and Cl_2 and, at high enough temperatures, Br_2 and I_2.

like Avogadro, conclude that equal volumes of gases (at the same pressure and temperature) contain equal numbers of molecules independent of their kind.

■ EXERCISES

5. Make an argument like the one above, only do it for combinations of nitrogen and oxygen instead of hydrogen and oxygen.

The full and precise statement of Avogadro's principle is this:

Equal volumes of all gases under the same conditions of temperature and pressure contain the same number of molecules.

You should remember this remarkable feature of gases. You should also think about it enough to realize that it is somewhat surprising. After all, you probably know that a hydrogen molecule is very small and light, while a chlorine molecule is much heavier. And, as you might guess, the volume of an atom of helium atom is only a small fraction of the volume of a molecule of methane (CH_4). Why then should the same number of molecules of gases of such different properties occupy the same volume? We will return to this question in the next chapter.

3.4   ATOMIC WEIGHTS

The law of combining volumes and Avogadro's principle made it possible to determine the relative masses of atoms and to establish a table of atomic weights that is very useful. Here is how this is done.

Because equal volumes hold the same number of gas molecules, the ratio of the masses of these volumes must equal the ratio of the masses of the individual molecules. Imagine a jar containing 1 liter of oxygen at a temperature of $20\,°C$ at atmospheric pressure. You can weigh the jar while it is full; then you can pump out the gas and weigh the empty jar. You will find that it contains about 1.33 g of oxygen. If you replace the oxygen gas with xenon gas and repeat the weighings, you will find that the jar contains 5.46 g of xenon. Because each volume contains the same number of molecules, it follows that the ratio of the mass of a xenon molecule to the mass of an oxygen molecule is $5.46/1.33 = 4.10$.

It takes additional experimentation to learn the atomic composition of a molecule. You have already seen how Avogadro concluded that gaseous oxygen is diatomic, O_2. Some gases are more complicated, e. g., carbon dioxide is CO_2, and some are simpler, e. g., when xenon was discovered at the end of the nineteenth century, it was quickly apparent that it is monatomic Xe.

▼ EXAMPLES

2. Knowing the atomic composition of the gas molecules, you can see that a xenon atom must have a mass that is $5.46/(1.33/2) = 8.21$ times the mass of an oxygen atom.

Using this and many other methods, chemists and physicists over time established a complete set of the relative masses of atoms of the chemical elements. For convenience they set up a standard scale of relative masses that by an international agreement assigns a mass of 12 units to carbon. Because relative masses are experimentally determined using various chemical combinations of atoms, carbon is a good choice for the standard mass; it combines chemically with many other atoms and is convenient to use. The choice of 12 for the mass of a carbon atom has the virtue that then the masses of all other atoms are close to being integers. One-twelfth of this mass is called the "atomic mass unit" and abbreviated as u.

As the standard for the scale of atomic masses, carbon has one complication. It turns out that carbon atoms come in several different masses. In a sample of naturally occurring carbon on Earth, 98.9% of the carbon atoms are of one mass and 1.1% are slightly heavier.[5] It is the more abundant species of carbon atom that is assigned a value of exactly 12.0000 u. In terms of this standard unit, the heavier carbon atom's mass is measured to be 13.003355 u.

▼ EXAMPLES

3. If you look at a periodic table of the elements, you will see that the atomic weight of carbon is given as 12.011 u. This is because chemistry is done with chemicals as they occur on Earth, and a natural sample of carbon is a mix of carbon atoms of two different weights:

$$m_C = .989 \times 12.0000 \text{ u} + .011 \times 13.003355 \text{ u} = 12.011 \text{ u}.$$

---

[5] Atoms of the same element but with different masses are called isotopes. Most elements have more than one isotope. You will learn more about isotopes in Chaps. 8.2 and 16.5.

In general the atomic weight listed in a periodic table for a chemical element is the average of the different masses of atoms of that element weighted by their fractional abundance. Some people reserve the words "atomic weight" for such an average and use the words "atomic mass" for the mass of a single kind of atom. For molecules, you can speak in the same way of molecular weights and molecular masses.

▉ EXERCISES

6. Look at the atomic weight of chlorine in Table 3.3. Why doesn't this value contradict the claim that the mass of any atom comes out close to an integer when the scale of atomic masses is set by assigning a mass of 12 u to carbon?

Table 3.3 lists some elements and the masses of their atoms. If you don't already know them, learn the nearest integer values of the atomic weights of H, C, N, and O.

Example 2 shows how to use Avogadro's principle and measured densities to find the ratio of the masses of two different molecules, Xe and O_2. Exercises 3 and 4 ask you to find the chemical composition of a molecule using the law of combining volumes. By putting these pieces of information together, you can determine an unknown atomic mass.

For example, to find the atomic mass of oxygen you could use the results of another of Gay-Lussac's experiments. Two volumes of CO were observed to combine with one volume of O_2 to form two volumes of another carbon–oxygen compound. The results imply that this compound is CO_2.

From Avogadro's principle and measured densities you can find the ratio of the molecular weights of CO_2 and O_2. The densities of these two gases are such that at room temperature (20 °C) and pressure (1 atm) a cubic meter of O_2 has a mass of 1.331 kg and a cubic meter of CO_2 has a mass of 1.830 kg. The ratio of these two numbers is the ratio of the molecular weights of O_2 and CO_2.

▼ EXAMPLES

4. Here is a direct way to find the atomic mass of oxygen from these data. Using the atomic idea greatly simplifies the analysis.

From Avogadro's principle you know that equal volumes contain equal numbers of molecules. In our example one liter of O_2 has a mass

TABLE 3.3 Some chemical atomic weights and gas densities

Element	Symbol	Chemical atomic weight[a]	Gas	Density at STP[b] (kg/m^3)
hydrogen	H	1.00794	Air	1.293
helium	He	4.002602	O_2	1.429
lithium	Li	6.941	N_2	1.251
beryllium	Be	9.01218	Cl_2	3.21
boron	B	10.811	H_2	0.0899
carbon	C	12.0107	CO_2	1.965
nitrogen	N	14.0067		·
oxygen	O	15.9994		·
fluorine	F	18.9984		·
neon	Ne	20.1797		·
chlorine	Cl	35.453		·
argon	Ar	39.948		·
krypton	Kr	83.80		·
xenon	Xe	131.293		·
radon	Rn	220.		·

[a] These data are taken from *The Chart of the Nuclides*, 13th ed., revised to July 1983, by F. William Walker, Dudley G. Miller, and Frank Feiner, distributed by General Electric Company, San Jose, CA. A complete table of data is at http://www.chem.qmul.ac.uk/iupac/AtWt/.

[b] STP means standard temperature and pressure. But there is more than one standard. This table uses the IUPAC standard temperature of 273.15 K, and standard pressure of 100 kPa. The NIST STP are 293.15 K and 101.3 kPa.

of 1.331 g, and one liter of CO_2 has a mass of 1.830 g. Imagine separating the CO_2 molecules into C atoms and O_2 molecules. Then the liter would contain 1.331 g of O_2 and $1.830 - 1.331 = 0.499$ g of C. The molecular weight of O_2 must then be

$$\frac{1.331\,g}{0.499\,g} \times 12.011\,u = 32.0\,u,$$

and because an oxygen molecule contains 2 atoms, it follows that the atomic weight of oxygen is 16.0.

■ EXERCISES

7. Using this atomic weight for oxygen and the data given earlier for the combining masses of gases of nitrogen and oxygen, find the atomic mass of nitrogen.

Notice that any mass of oxygen will contain the same number of atoms as a mass of carbon for which the two masses are in the ratio of 16 to 12.

3.5   NUMBERS OF ATOMS IN A SAMPLE

Atomic masses are not actual masses of individual atoms; they are ratios; they are a set of relative masses. Although they do not tell you the actual masses of individual atoms, atomic masses do allow you to compare the numbers of atoms in samples of different elements. An example with familiar objects shows how this works. The U.S. 5-cent coin, the nickel, has a mass exactly twice that of the U.S. 1-cent coin, the penny. Do you see that it follows that 200 kg of nickels will consist of the same number of coins as 100 kg of pennies? If you know that the ratio of the weights of a single nickel and a single penny is 2:1, then you know that any two masses of them that are in the ratio 2:1 contain equal numbers of coins.

The same argument applies to chemical elements. From their atomic masses you know that an atom of O has a mass 16 times the mass of an H atom; therefore the number of O atoms in a sample of 16 kg of oxygen must be the same as the number of H atoms in 1 kg of hydrogen. And the statement is true for 320 g of O and 20 g of H, or for 11.2 tonne of O and 0.7 tonne of H. As long as the ratio of the masses of the samples is 16:1, the numbers in the samples will be equal. Of course, the number of H atoms in 20 g is 1/50 times the number in 1 kg, and the number of H atoms in 0.7 tonne is 700 times the number in 1 kg, but in each case the corresponding amount of O has the same number of atoms.

In general, there are equal numbers of atoms in samples of any two elements that have masses in the ratio of their atomic weights. A simple way to be sure that samples have this ratio is to weigh them out with masses numerically equal to their atomic masses. Such samples contain equal numbers of atoms. Thus 131.3 g of xenon has the same number of atoms as 220 g of radon or 16 g of oxygen.

■ EXERCISES

8. What are masses of CO_2 and O_2 that would contain the same number of molecules? This question and the next require you to distinguish between molecules and atoms.

9. What are masses of CO_2 and O_2 that would contain the same number of atoms? There are an infinitude of possible answers. Explain the answer you chose to give.

The Mole

But how many atoms are in your sample? The answer depends on how large your sample is, and it is customary to describe sample sizes in terms of the number of grams equal to the atomic or molecular weight of a substance, the so-called "gram atomic" or "gram molecular" weight. Examples are 1 g of hydrogen, 4 g of helium, 12 g of carbon, 18 g of water, 44 g of CO_2, etc. A gram atomic weight of any element contains the same number of atoms as the gram atomic weight of any other element; the number of molecules in any gram molecular weight of any molecule is the same as in the gram molecular weight of any other molecule. The number of atoms (molecules) in a gram atomic (molecular) weight has its own name; it is called a "mole," abbreviated "mol" (why abbreviate four letters with three?).

The mole is a large number. Its formal SI definition says that 1 mole is the number of entities equal to the number of carbon atoms in 12 g of carbon if the carbon atoms are all of atomic mass 12.000 u. Then from Table 3.3 a sample of 16.0 g of oxygen also contains 1 mole of oxygen atoms, and 35.5 g of chlorine contains 1 mole of chlorine atoms. For any atom with atomic mass M there is 1 mole of these atoms in M grams of that substance.

Avogadro's Constant

But the number? *What* is the number? It is 6.022×10^{23}, and it is called "Avogadro's constant" and given its own symbol N_A. This means that just as the "c" in cm can be replaced by its value 10^{-2}, the "mol" in an equation can be replaced with its value of N_A. As "dozen" is the name for 12 of anything, and "score" is the name for 20 of anything, so "mole" is the name for 6.022×10^{23} of any objects.

EXAMPLES

5. One way to remember the mass of Earth is to think of it as 10 moles of kilograms.

6. The mass of one mole of CO_2 is 44 g. A mole of CO_2 molecules contains 1 mole of carbon atoms and 2 moles of oxygen atoms.

EXERCISES

10. Rewrite the opening line of the Gettysburg Address to use "moles" instead of "score."

Remember that the symbol n_M represents the number of moles. Don't confuse n_M with the n used to tell how many of something there are in a unit volume, i.e., the number in a unit volume or "number density." For example, equal volumes of any gases at the same temperature and pressure contain equal numbers of molecules. As you will see later, at $0\,°C$ and atmospheric pressure the number density is $n = 2.7 \times 10^{19}\,\mathrm{cm}^{-3}$.

EXERCISES

11. How many molecules of gas are there in 11.2 liters under the above conditions? $(1\,\mathrm{L} = 10^3\,\mathrm{cm}^3)$

12. What is the number of moles n_M of gas molecules in 11.2 liters of gas under the above conditions?

13. How many liters of gas does it take to hold a mole of molecules at the above temperature and pressure?

14. If you have $n_M = 3.2$ moles of gas molecules in 4 liters of volume, what is the number density of this gas?

15. If the molecular weight of the above gas is 28 u, what is the mass density of the gas in the previous problem?

Avogadro's number is fundamental because it connects the scale of atomic sizes to the macroscopic scale of our everyday world. Knowing it you can find the mass and the size of an individual atom or molecule.

For example, given that 18.998 g of fluorine contains 6.02×10^{23} atoms, you can see that fluorine atom has a mass of $18.998/(6.02 \times 10^{23}) = 3.16 \times 10^{-23}$ g. It took more than forty years from the time of the first scientific evidence for atoms before physicists were able to measure N_A, and it was nearly a hundred years before they could measure N_A to three significant figures.

■ EXERCISES

16. What is the mass of a hydrogen atom? A Be atom?

17. From the density of liquid water, estimate the diameter of a water molecule.

3.6 THE CHEMIST'S ATOM

Summary

In the space of a few years around 1810, atoms were established as an important concept in chemistry and physics. These atoms have some interesting properties. First, there are different kinds, one for each "chemical element." "Chemical element" is defined operationally: If the mass and chemical behavior of a given substance are not changed by heat, electricity, or other chemical reactions, it is said to be a chemical element. By this criterion chemists identified nearly sixty chemical elements before 1850. We now know of 117 elements, 23 of which do not occur naturally on Earth but can be made in the laboratory or in nuclear explosions.[6]

Second, atoms connect to one another. Atoms seem to have "hooks." For instance, a hydrogen atom has one hook, so two atoms of H can be hooked together to make H_2. Oxygen has two hooks, so we can hook a hydrogen atom on to each and make H_2O, or we can hook an H atom and an O atom onto one O, and another H onto the second O, and make hydrogen peroxide, H_2O_2. The somewhat whimsical diagram in Fig. 3.1 gives the idea. Chemists call the number of hooks the "valence" of the atom.

---

[6]These numbers change as new discoveries are made, and they depend on what scientists are willing to accept as a discovery. For example, there has been some argument about whether seeing 1 atom of element 118 in 2002 and then two more in 2005 is enough to be sure that the new element 118 has been observed.

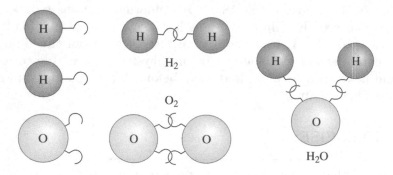

FIGURE 3.1   Ball-and-hook model of simple chemical bonds.

Questions

Much of chemistry in the first two-thirds of the nineteenth century was the sorting and describing of chemical compounds and reactions. The idea of atoms was used by many chemists to produce other useful ideas and good science, but it raised as many questions as it answered.

What are these "hooks" that connect one atom to another? Why do some elements have several different valences? For example, nitrogen seems to have valences of 1, 2, 3, 4, or 5 depending upon circumstances. What makes one element different from another? Hydrogen is very reactive; helium is not. Yet both are gases and very light in mass. What makes some elements surprisingly similar to others? Lithium, sodium, potassium, cesium, rubidium, francium all show similar chemical behavior.

These questions are related to more general, deeper questions: Do atoms have internal parts, i.e., do they have structure? If so, what are the insides of atoms like? How do the parts connect? What forces hold them together? How does the behavior of the parts explain the similarities and differences of atoms?

Answers

It took physicists over one hundred years to find satisfactory answers to these difficult questions. The answers required surprising elaboration of our concept of the atom itself and radical changes in our formulations of the principles that govern the behavior of matter. The central material of this book is the physics needed to begin answering these questions. But before looking inside the atom, we'll examine the very simple model of the atom that physicists used in the nineteenth century to answer two basic questions: How big is an atom, and what is its mass?

These questions were hard to address directly because until recently the extremely small size of atoms made it impossible to observe them individually. The first answers were obtained by studying large numbers of identical atoms or simple molecules in gases at low pressures, where they are separated by relatively large distances and therefore unaffected by each other. If the physical behavior of a gas depends only on the properties of the individual atoms or molecules of which it is composed, then by studying gases we can hope to learn something about the individual atoms and molecules themselves. The next two chapters trace the connection between atoms and the nature of gases.

PROBLEMS

1. The table below on the left shows hypothetical data obtained from a chemical reaction

$$X + Y \rightarrow A, B, C$$

in which three different compounds A, B, and C are formed from the reaction of elements X and Y.

Proust's law of constant proportions			Dalton's law of multiple proportions		
	X	Y		X	Y
A	57%	43%	A	100 g	
B	73%	27%	B	100 g	
C	47%	53%	C	100 g	

  a. To exhibit Dalton's law of multiple proportions find the mass of Y that interacts with 100 g of X and fill in the blanks in the table. Explain how these results illustrate Dalton's law.

  b. From your results write down possible expressions for the chemical compounds A, B, and C, e.g., X_2Y_5.

2. Explain Gay-Lussac's law of combining volumes. Why does this law support the existence of atoms?

3. Why when $100 \, cm^3$ of nitrogen gas is combined with $100 \, cm^3$ of oxygen gas do you get $200 \, cm^3$ of the compound gas at the same pressure and temperature?

4. How did Avogadro explain that when $200\,\mathrm{cm}^3$ of hydrogen at STP is combined with $100\,\mathrm{cm}^3$ of oxygen at STP they form $200\,\mathrm{cm}^3$ of water vapor at the same temperature and pressure?

5. Dalton noticed that $100\,\mathrm{g}$ of tin combined with either exactly $13.5\,\mathrm{g}$ or exactly $27\,\mathrm{g}$ of oxygen. He said that this result was evidence for the existence of atoms.

   a. Why? Explain his reasons.

   b. Take the atomic weight of oxygen to be 16. From Dalton's data calculate the atomic weight of tin. State what you assume in order to get an answer.

6. Observer A notes that water always consists of 11.1% hydrogen and 88.9% oxygen by weight, while hydrogen peroxide always consists of 5.88% hydrogen and 94.1% oxygen by weight. Observer B notes that in the formation of water $800\,\mathrm{g}$ of oxygen always combines with $100\,\mathrm{g}$ of hydrogen, while in the formation of hydrogen peroxide $1600\,\mathrm{g}$ of oxygen combines with $100\,\mathrm{g}$ of hydrogen.

   a. Whose observations illustrate the law of

     i. combining volumes?

     ii. constant proportions?

     iii. multiple proportions?

   b. Show how to deduce what B sees from A's observations.

   c. Explain in what way B's observations suggest that atoms exist.

7. Stearic acid is described by the chemical formula $C_{17}H_{35}COOH$.

   a. Find the mass of 1 mole of stearic acid in grams.

   b. Use the fact that there are 2.1×10^{21} molecules in a gram of stearic acid and your answer to (a) to find a value of Avogadro's constant (N_A).

   c. If the density of stearic acid is $8.52 \times 10^2\,\mathrm{kg\,m}^{-3}$, how many atoms are in $1\,\mathrm{cm}^3$?

8. Obeying Proust's law of constant proportions, stannous oxide is always 88.1% tin and 11.9% oxygen by weight; stannic oxide is 78.7% tin and 21.3% oxygen. Show that these numbers imply that in forming stannous oxide $100\,\mathrm{g}$ of tin combines with $13.5\,\mathrm{g}$ of oxygen.

9. Certain volumes of nitrogen gas and hydrogen gas react to form 4 L of ammonia gas, according to the reaction shown below.

$$N_2 + 3H_2 \rightarrow 2NH_3.$$

The atomic masses of nitrogen *atoms* and hydrogen *atoms* are 14.0 u and 1.0 u, respectively. The mass of NH_3 formed in this reaction is 3.4 g.

    a. How many moles of NH_3 were formed?

    b. How many hydrogen *atoms* took part in the reaction?

    c. Determine the initial volumes of N_2 and H_2.

10. What is a "mole" as the term is used in physics and chemistry? Give both qualitative and quantitative answers.

11. You wish to determine the identity of element X in the gaseous chemical reaction

$$X_2 + 3\,H_2 \rightarrow 2XH_3.$$

The densities of H_2 and XH_3 are measured to be $8.4 \times 10^{-5}\,\mathrm{g\,cm^{-3}}$ and $7.14 \times 10^{-4}\,\mathrm{g\,cm^{-3}}$, respectively.

    a. Find the mass of H_2 in a 30 L volume of gas.

    b. If you react enough X_2 to combine with 30 L of H_2, what volume of XH_3 is generated?

    c. From the information given, determine the atomic weight of element X and identify the element.

12. The chemical reaction for making nitrogen dioxide is given by

$$2\,O_2 + N_2 \rightarrow 2\,NO_2.$$

Suppose we combine nitrogen and oxygen to make 23 g of NO_2.

    a. How many moles of NO_2 does that correspond to?

    b. How many moles of N_2 are required?

    c. How many molecules of O_2 are required? Express your answer in terms of N_A.

    d. If the reaction produces 10 L of NO_2, what are the volumes of O_2 and N_2? Explain your reasoning.

Gas Laws

4.1 INTRODUCTION

The early chemical evidence for the existence of atoms led physicists to ask: What are the basic properties of atoms? How are atoms alike? How are they different? What properties must they have to produce the observable properties of matter?

Studies of gases helped to answer these questions because gases are physically simple compared to solids and liquids. Gases can be compressed and expanded easily by applying or relaxing pressures, and small changes in temperature will produce large changes in the volume or pressure of a gas. This chapter explains the ideas of temperature, pressure, and thermal expansion, and shows you how to describe these properties quantitatively. Using these ideas you can then describe the experimentally observed properties of many gases in terms of simple mathematical functions—the gas laws. These in turn give rise to the concept of an ideal gas.

The next chapter will show you how an atomic model of the ideal gas explains its properties and—of great importance—provides evidence that heat energy is the random motions of atoms.

4.2 PRESSURE

The Idea of Pressure

Imagine a tiny, empty hole in a liquid. (Not a bubble filled with air, but a true vacuum.) Your intuition correctly tells you that on Earth the weight of the surrounding fluid will cause it to squeeze in and make the hole

C.H. Holbrow et al., *Modern Introductory Physics, Second Edition,*
DOI 10.1007/978-0-387-79080-0_4, © Springer Science+Business Media, LLC 1999, 2010

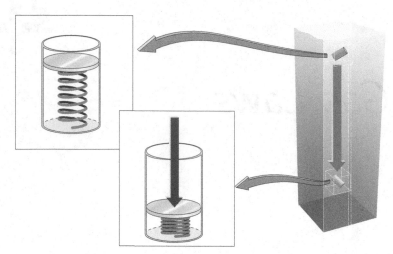

FIGURE 4.1 The spring inside an evacuated can is compressed further when the can is submerged more deeply.

disappear. This squeeze, which will be from all directions, is what we call pressure. It is what you feel in your ears when you swim under water.

Now imagine that instead of the hole in the fluid, there is a tiny, completely empty can with a lid that slides in and out like the piston in the cylinder of an automobile engine.

If there is really nothing in the can, the weight of the column of liquid and atmosphere above the can will push the lid to the bottom of the can. But suppose, as in Fig. 4.1 there is a tiny spring between the lid and the bottom of the can. Then as the weight of the liquid and atmosphere pushes the lid into the can, the spring compresses and pushes back. The lid will slide in only to the point where the force of the spring equals the weight of the liquid and atmosphere.

If you move the can deeper into the water, the spring compresses more. If you raise it to a shallower depth, the weight of water on the lid diminishes and the lid moves away from the bottom of the can. If you hold the can at any given depth and turn it around in different directions, the compression of the spring does not change because, as noted above, the pressure (squeeze) of the fluid is the same in all directions.

Are you surprised that the spring compression is the same regardless of which way the can is turned? Since gravity acts vertically, you might reasonably expect the compression to be much greater when the can is vertical. But if the forces were not the same in all directions, the liquid would flow until the forces did balance. The forces on a liquid in a container rearrange the liquid until the forces are balanced and the pressure is the same in all directions.

If you put the same spring in a smaller can with a lid of smaller area, then at any given depth the spring under the smaller lid will be compressed less than the spring under the larger lid. This is because a larger lid has more water pressing on it than a smaller one. The force due to water pressure is proportional to the area on which it acts.

Definition of Pressure

As always in physics, given a concept like "pressure," you want a way to measure it and assign a number to it. You could use a device like the little can: The amount the spring is compressed is a direct measure of how much pressure there is. But that is not so good, because, as you just saw, the observed result depends on how big the can is. A better measure is the ratio of the force to the area of the can's lid. This will work because even though the force on the can and the amount of compression of the spring vary with the size of the lid, the ratio of force to area remains constant.[1] This ratio is both conceptually and practically useful, and therefore we assign it the name "pressure."

$$P = \frac{F}{A}.$$ (1)

In SI units pressure is measured in $N\,m^{-2}$.

Discovery of Vacuum and the Atmosphere

Around the year 1640 Evangelista Torricelli produced the first vacuum, i.e., a volume from which matter—including air—has been excluded, and showed the existence of the pressure of the atmosphere. His technique was very direct. He poured mercury into a glass tube closed at one end, put his thumb over the open end (don't try this at home!), inverted the full tube, put the open end into a bowl of mercury, and removed his thumb. Without flowing entirely out of the tube, the mercury fell away from the tube's closed end and left an empty space—a vacuum. As shown in Fig. 4.2, a column of mercury about 760 mm high remained standing in the tube above the level of the mercury in the bowl.

Torricelli realized that the empty space above the mercury column was a vacuum; he also realized that the weight of the column of mercury was being balanced by a force arising from the weight of the atmosphere pushing on the surface of the mercury in the bowl, i.e., atmospheric pressure. This pressure, transmitted through the mercury, produces an upward force

---

[1]If different parts of the lid are at different depths then the pressure will vary over its surface. This is why pressure at any position is defined as the limiting value of the ratio as the area approaches zero.

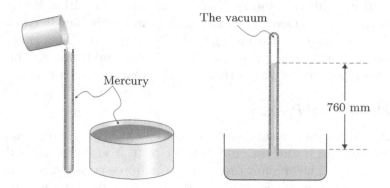

FIGURE 4.2 Apparatus for producing a Torricellian vacuum.

at the base of the column just equal to the weight of the mercury in the tube. Torricelli realized that he was observing the balance between the weight of a column of mercury and the weight of a column of air. This meant, as he put it, that we all live at the bottom of an ocean of air.

It is remarkable that the height h of the column for which this balance occurs does not depend on the column's cross-sectional area A. Some algebra shows you why this is true. Call the height of the column h, the column's cross-sectional area A, the density of mercury ρ, the acceleration due to gravity g, and the pressure of the air P. The mass of the mercury is its volume hA times its density ρ, so gravity exerts a downward force F_g on the mercury column of $F_g = mg = \rho hAg$. Now, the force exerted by the air is $F = PA$, but since F balances the weight of the mercury, it must be true that $F = hA\rho g = PA$. From this follows the important fact that

$$P = \rho gh \qquad \text{barometer equation} \qquad (2)$$

because A divides out from both sides. This result means that regardless of the size of the mouth of the tube, atmospheric pressure will support the same height h of mercury. This height h is a direct measure of the magnitude of atmospheric pressure, and Eq. 2 is often called "the barometer equation."

EXAMPLES

1. From your measurement of h you can directly calculate the magnitude of the atmospheric pressure. Suppose $h = 760\,\mathrm{mm}$, then

$$P = \left(13.6 \times 10^3\,\mathrm{kg\,m^{-3}}\right)\left(9.80\,\mathrm{m\,s^{-2}}\right)(0.76\,\mathrm{m}) = 101 \times 10^3\,\mathrm{N\,m^{-2}},$$

where, as you expect, the units are of force divided by area.

2. You can calculate h the height of the mercury column above the surface of the mercury in the bowl, i.e., the reservoir of mercury, from the barometer equation, Eq. 2: $h = P/(\rho g)$, but it is good practice to understand the rise of the Hg column as a balance between the force of gravity and the force exerted by the pressure of the atmosphere acting over the area of the cross section of the tube.

Suppose the area of the tube's cross section is $10\,\mathrm{cm}^2$. Then the force produced by 1 atm of pressure P acting on this area is $PA = 101\,\mathrm{N}$. The weight of mercury this force can support is the mass of the mercury in the tube times the acceleration due to gravity, g. The mass of the mercury is its volume $V = hA$, (that is, its height h times its cross-sectional area, $A = 10^{-3}\,\mathrm{m}^2$) times its density, $\rho = 13.6 \times 10^3\,\mathrm{kg\,m^{-3}}$. The pressure P at the surface of the reservoir is 1 atmosphere so the pressure inside the tube at this same level must also be 1 atmosphere. Pushing on the tube's cross sectional area A, P supplies a force $PA = 101\,\mathrm{N}$ that balances the weight $mg = \rho hAg$ so

$$h = \frac{PA}{\rho g A} = \frac{101\,\mathrm{N}}{13.6 \times 10^3\,\mathrm{kg\,m^{-3}} \times 10^{-3}\,\mathrm{m}^2 \times 9.80\,\mathrm{m\,s^{-2}}} = 0.76\,\mathrm{m}$$

Now you can see what happens to h if you make the cross section of the tube 10 times bigger, i.e., $100\,\mathrm{cm}^2$. The volume of mercury in the tube to be supported by atmospheric pressure becomes 10 times larger, but because the cross sectional area A is bigger, the force exerted across that surface by the pressure of the atmosphere is bigger by the same factor of 10, and the same height of column will be balanced.

EXERCISES

1. Suppose on a day when the atmospheric pressure is $101\,\mathrm{kN\,m^{-2}}$ you inflate a spherical balloon to a diameter of $20\,\mathrm{cm}$. Assuming that the elastic force of the balloon rubber is small enough to neglect, what will be the size of the force pushing out on $1\,\mathrm{cm}^2$ of the balloon? What will be the size of the force pushing in on $1\,\mathrm{cm}^2$?

Several different kinds of units are used to measure pressure.

$\mathbf{N\,m^{-2}}$ This is the most straightforward unit because it explicitly shows the relationship $P = F/A$.

pascal The combination of units $N\,m^{-2}$ is given its own special name, pascal, abbreviated Pa. Thus, you can say that atmospheric pressure is 101.3 kPa, where as usual the prefix "k" means "kilo" or 10^3.

atm Pressures are often measured in "atmospheres," which is a convenient unit because, as long as you don't stray too far from sea level, the atmospheric pressure is likely to be reasonably close to the same value all over the world. By convention, a "standard atmosphere" is defined to be 101 325 Pa.

bar There is another unit of pressure developed by meteorologists called the "bar." It is *almost* equal to the pressure of one atmosphere: 1 bar = 100 kPa exactly. All the sub- and supermultiples are used: mbar, kbar, Mbar, etc.

mmHg Often U.S. physicists measure pressure in terms of the height of the column of mercury that it would balance. The units are given as millimeters of mercury, or mmHg. Thus we say that atmospheric pressure is 760 mmHg. To convert this to more conventional units, multiply by the density of mercury and the acceleration due to gravity g. Or remember that standard pressure, 101 325 Pa, is very nearly equal to 760 mmHg and use this as a conversion factor.

in Hg U.S. weather reporters usually give barometric pressure in inches of mercury. To convert mmHg to in Hg divide by 25.4 mm per inch. Then 760 mmHg becomes 29.9 in Hg, a number you often hear from your TV weather reporter.

Torr The mmHg is also given its own name, the torricelli[2], abbreviated as Torr. This unit is frequently used for measuring low pressures inside vacuums. It is rather easy to obtain a vacuum of 10^{-3} Torr; a good vacuum of 10^{-6} Torr usually requires more than one kind of pump; an ultra-high vacuum of 10^{-10}–10^{-12} Torr is routine but expensive and tedious to achieve, especially in large volumes.

psi In American engineering practice pressures are measured in pounds per square inch, or psi, or p.s.i. Atmospheric pressure in these units is about 14.7 psi. In physics we sometimes write this as $14.7\,lb/in^2$.

gauge pressure Quite often, high pressures in tanks are measured as the *difference* between the internal tank pressure and the external atmospheric pressure. These are called "gauge" pressures. Sometimes

---

[2]One torr is now defined to be exactly(101325/760) Pa, so 1 Torr equals 1 mmHg only to about 1 part in 10^6. This is good enough for most situations.

the pressure-measuring device will tell you that the pressure being measured is gauge pressure by giving the units as psig or p.s.i.g., but frequently it won't.

Gas Pressure

Gases in closed containers exert pressure on the containers' walls. A simple experiment shows that this is so. Put an air-filled balloon inside a glass jar, pump air from the jar, and the balloon will expand; let air back into the jar, and the balloon contracts. The gas in the balloon is always pushing outward, but, as you saw in Exercise 1, normally that outward force is exactly balanced by the inward force of the atmospheric gas. Only when the effect of the atmosphere is removed by putting the balloon in a vacuum does the unbalanced gas pressure become apparent. The balloon expands until the internal pressure is balanced by the force from the balloon's latex fabric.

What causes gas pressure? It arises from frequent and repeated collisions of gas atoms with the walls of their container. We draw this conclusion because a simple model of atoms in motion quantitatively accounts for the experimentally observed dependence of the pressure of a gas on its temperature and the volume of its container. To understand this argument, you need to know about the gas laws—Boyle's Law, Gay-Lussac's Law, and the Ideal Gas Law.

4.3   BOYLE'S LAW: THE SPRINGINESS OF GASES

A modified form of Torricelli's barometer was used by the English physicist Sir Robert Boyle to perform what he called "Two New Experiments Touching the Measure of the Force of the Spring of Air Compressed and Dilated."[3]

Boyle's Experiment

Boyle made a J-shaped tube sealed at one end and open at the other. He pasted a scale along both arms of the tube (only one scale is shown in Fig. 4.3). He poured mercury into the open end and trapped a small volume of air under the closed end. (This is different from Torricelli's

---

[3]Reprinted in *A Treasury of World Science*, ed. Dagobert Runes, Philosophical Library, New York, 1962.

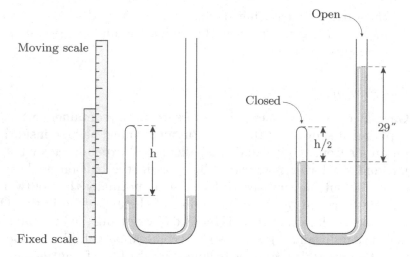

FIGURE 4.3 Apparatus for establishing Boyle's law.

experiment, in which the volume at the closed end was a vacuum.) He measured the volume of the trapped air by measuring the height of the column of air between the mercury and the end of the sealed tube, and he measured the pressure of the trapped air by measuring the height of the column of mercury in the open end. He varied the pressure by pouring in more mercury to compress the trapped gas. He also did a separate experiment in which he pumped out some air from the open end of the tube and saw the trapped volume of air expand. In his own words:

> ... we took care, by frequently inclining the tube, so that the air might freely pass from one leg into the other by the sides of the mercury (we took, I say, care) that the air at last included in the shorter cylinder should be of the same laxity with the rest of the air about it. This done we began to pour quicksilver into the longer leg of the siphon, which by its weight pressing up that in the shorter leg, did by degrees straighten the included air: and continuing this pouring in of quicksilver till the air in the shorter leg was by condensation reduced to take up by half the space it possessed ... before; we cast our eyes upon the longer leg of the glass, on which was likewise pasted a list of paper carefully divided into inches and parts, and we observed, not without delight and satisfaction, that the quicksilver in that longer part of the tube was 29 inches higher than the other .... For this being considered, it will appear to agree rarely well with the hypothesis, that as according to it the air in that degree of density and correspondent measure of resistance, to which the weight of the incumbent atmosphere had brought it,

was able to counterbalance and resist the pressure of a mercurial cylinder of about 29 inches, as we are taught by the Torricellian experiment; so here the same air being brought to a degree of density about twice as great as that it had before, obtains a spring twice as strong as formerly. As may appear by its being able to sustain or resist a cylinder of 29 inches in the longer tube, together with the weight of the atmospherical cylinder, that leaned upon those 29 inches of mercury; and, as we just now inferred from the Torricellian experiment, was equivalent to them.

This experiment is illustrated by Fig. 4.3. The data that Boyle published are given in Table 4.1. These data support the conclusion that the pressure P and volume V were inversely proportional, i.e., $P \propto 1/V$, or, as written in the form called "Boyle's law,"

$$PV = \text{constant} \tag{3}$$

at a constant temperature.

Notice that in the note for column E in Table 4.1 Boyle states his law clearly. How well do his data agree with his hypothesis? It appears that Boyle assumed that his first data point V_0, P_0 was exactly correct and then used his subsequent measured values of V_m to calculate values of P_c. These are given in column E of the table for comparison with the measured values P_m in column D.

◼ EXERCISES

2. Check Boyle's work by calculating the value of P_c at three or four rows spread throughout Table 4.1. (Notice that Boyle did not use decimals. When did decimals come into use?)

Another technique for seeing whether your data agree with your theory is to put the theory in a form such that a plot of your data comes out a straight line. Your eye is quite a good judge of the straightness of a line, and so of the quality of agreement between theory and experiment.

◼ EXERCISES

3. A good test of Boyle's law is to plot the pressure P against the reciprocal of the volume $1/V$. (Why might this be a useful thing to do?) Do this for the data in Table 4.1. Use reasonable scales and a nice layout of your graph. Graphs are quantitative tools, not just pretty pictures, but they need to be easy to read.

TABLE 4.1   A table of the condensation of the air

Volume	Excess pressure	Atmospheric pressure	Total pressure	$P_c = P_0 V_0/V_m$
V_m		P_0	P_m	P_c
A	B	C	D	E
(1/4 inches)	(inches)	(inches)	(inches)	(inches)
48	00	29 2/16	29 2/16	29 2/16
46	01 7/16	"	30 9/16	30 6/16
44	02 13/16	"	31 15/16	31 12/16
42	04 6/16	"	33 8/16	33 1/7
40	06 3/16	"	35 5/16	35
38	07 14/16	"	37	36 15/19
36	10 2/16	"	39 5/16	38 7/8
34	12 8/16	"	41 10/16	41 2/17
32	15 1/16	"	44 3/16	43 11/16
30	17 15/16	"	47 1/16	46 3/5
28	21 3/16	"	50 5/16	50
26	25 3/16	"	54 5/16	53 10/13
24	29 11/16	"	58 13/16	58 2/8
23	32 3/16	"	61 5/16	60 18/23
22	34 15/16	"	64 1/16	63 6/11
21	37 15/16	"	67 1/16	66 4/7
20	41 9/16	"	70 11/16	70
19	45	"	74 2/16	73 11/19
18	48 12/16	"	77 14/16	77 2/3
17	53 11/16	"	82 12/16	82 4/17

Column A: The number of equal spaces in the shorter leg, that contained the same parcel of air diversely extended [and hence proportional to the volume].

Column B: The height of the mercurial cylinder in the longer leg that compressed the air into those dimensions.

Column C: The height of the mercurial cylinder that counterbalanced the pressure of the atmosphere. (This was measured to be 29 1/8 inches.)

Column D: The aggregate of the two last columns, B and C, exhibiting the pressure sustained by the included air.

Column E: What the pressure should be according to the hypothesis, that supposes the pressures and expansions to be in reciprocal proportion.

TABLE 4.1   (continued)

Volume	Excess pressure	Atmospheric pressure	Total pressure	$P_c = P_0\, V_0/V_m$
V_m	P_0		P_m	P_c
A	B	C	D	E
(1/4 inches)	(inches)	(inches)	(inches)	(inches)
16	58 12/16	"	87 14/16	87 3/8
15	63 15/16	"	93 1/16	93 1/5
14	71 5/16	"	100 7/16	99 6/7
13	78 11/16	"	107 13/16	107 7/13
12	88 7/16	"	117 9/16	116 4/8

---

4. Boyle, writing in 1660, presented his data in a format that we now call a "spreadsheet." Today you can use computer programs to enter and manipulate data in the form of a spreadsheet. Enter Boyle's data into a spreadsheet program and then perform his calculations and make a graph to show how the data conform to Boyle's Law. Hand in a printout of your spreadsheet and a printout of your graph.

Now you see why you can believe Boyle's Law Eq. 3, i.e., at any given constant temperature and for many gases under conditions that are not too drastic PV = constant. It is important to understand that this law is an idealization based on measurement. Because measurements are never exact, you can never be sure that the law will continue to hold when you make more and more precise measurements or when you change the conditions under which the measurements are made. The incomplete verification of the law is evident just from examining the data. There are a couple of disagreements between theory and experiment that are close to 1%. However, Boyle's experimental uncertainties prevented him from attributing such discrepancies to inaccuracies in the law. We would say that Boyle's law is valid within his experimental uncertainty.

■ EXERCISES

---

5. What is the largest percent variation between theory and experiment in the data of Table 4.1?

4.4   TEMPERATURE, GASES, AND IDEAL GASES

Take a glass of cold milk and set it on the table. It will warm up because the motions of its water and protein molecules speed up. Take a hot bagel from the toaster and set it on a plate. It will cool down because the motions of its water, starch and other molecules will become less vigorous. In general an object gets hotter when the random motions of its atoms and molecules become more vigorous; the object gets cooler if these motions diminish. This association of hot and cold with atomic and molecular motions is an extremely important fundamental idea of physics, first suggested by studies of the properties of gases. This section examines some of these properties; the next chapter shows how a simple atomic model explains important properties of gases and relates atomic motion to temperature.

Asked "What is temperature?," you might answer "High temperature means hot, and low temperature means cold." For what we do in this book, that answer is good enough. However, the idea is more subtle. Temperature is an indicator of the direction of the flow of heat energy. The milk warms and the bagel cools by exchanging heat energy with the surroundings. In each case, heat energy flows in both directions, but for the cold milk more heat energy flows in from the surroundings than flows out, while for the hot bagel more flows out than in. As a result, in each case the difference between the inward and outward flows of heat energy diminishes until the flows become equal. When the rate of heat energy flow into the milk equals the rate of flow out, we say that the milk has come to thermal equilibrium with its environment; the same for the bagel. Two objects in thermal equilibrium with each other have the same temperature; objects with the same temperature will be in thermal equilibrium with each other if they are brought together.

To measure temperature quantitatively we want a device with a physical property that changes in an easily observable way when the device gets hotter or colder. Thermal expansion is one such property, and the mercury thermometer and gas thermometer are such devices. The next sections describe thermal expansion, the mercury thermometer, the gas thermometer, and the Kelvin temperature scale.

Thermal Expansion

At most temperatures a given volume of a solid, liquid, or gas at constant pressure expands when heated and shrinks when cooled.[4] When heated,

---

[4]Over small temperature ranges there are important exceptions. For example, a volume of water expands when cooled from $4\,^{\circ}C$ to its freezing point at $0\,^{\circ}C$. This is why ice forms on the surface of a pond rather than at its bottom.

a rod of metal gets longer (and also fatter), a pane of glass grows in area (and also in thickness), a block of iron swells in volume. If you heat a glass tube containing mercury, both the tube and the mercury increase in volume, but for the same amount of warming the mercury expands so much more than the glass that the expanding mercury moves up the tube.

The Swedish astronomer Anders Celsius used these properties to devise a thermometer and a scale of measurement. He put mercury, which is liquid from well below the freezing point of water to well above the boiling point of water, into a small diameter tube with a bulb at one end. He put the device into a freezing mixture of ice and water and made a mark on the tube where the mercury reached after it was cold; then he put the device into the vapor of boiling water and marked on the tube where the mercury reached when heated by the boiling vapor. He divided the length between the two marks into 100 intervals that we now call degrees Celsius. We call the freezing point $0\,°C$ and the boiling point $100\,°C$.[5]

Over temperature ranges that are not too large, the fractional change in volume of many substances is proportional to the change in temperature. For example, for a $1\,°C$ rise in temperature a volume of mercury (Hg) will expand by 0.0182% while for a $10\,°C$ rise it will expand by 0.182%. For a given temperature rise, say $10\,°C$, the fractional change is constant, but the actual change of volume depends on how much you start with. Thus $1\,cm^3$ of Hg expands by $0.00182\,cm^3$; $2\,L$ of Hg expands by $3.64\,cm^3$; and $1/2\,cm^3$ expands by $0.00091\,cm^3$. For a $20\,°C$ rise all these changes of volume would double.

Expressed in algebra the fractional change is $\frac{\Delta V}{V_0}$ where V_0 is the volume at the initial temperature and ΔV is the change in volume that occurs when the temperature changes by an amount Δt. Thermal expansion is described by

$$\frac{\Delta V}{V_0} = k_V \Delta t. \tag{4}$$

In Eq. 4 the constant of proportionality k_V is called the volume coefficient of thermal expansion. You can see that the units of k_V are $(°C)^{-1}$.

Equation 4 tells you what you need to do to determine k_V: Measure the volume and temperature of a chunk of stuff; heat it a little and measure its new volume and new temperature; then calculate k_V using Eq. 4. Values of k_V for aluminum, glass, and mercury at room temperature ($20\,°C$) are shown in Table 4.2; k_V values for some gases are also shown. The values in the table illustrate the facts that every state of matter—solid, liquid, or gas—exhibits thermal expansion and that k_V is small for a solid, larger for a liquid, and largest for gases.

---

[5]Celsius called boiling "0" and freezing "100"!

TABLE 4.2   Coefficients of volume thermal expansion at temperature t and Pressure 101 kPa

Substance[a]	k_V $(10^{-6}$ $(°C)^{-1})$	t (°C)
aluminum	69	20
mercury	182	20
pyrex glass	10	20
hydrogen	3664	0
nitrogen	3672	0
helium	3659	0
chlorine	3883	0
carbon dioxide	3724	0
	3725	0
ideal gas	$3661 \times 10^{-6} = 1/273.15$	0

[a] The values of the volume thermal expansivity of the gases are taken from J. R. Partington, *An Advanced Treatise on Physical Chemistry*, Longmans, Green, and Co., London, 1949. p. 547.

▩ EXERCISES

6. By what percentage will the volume of 2 cubic inches of Hg change when heated from 20 °C to 30 °C?

7. A cube of Al, 2 cm on each side, is heated from 21 °C to 37 °C. By how much does its volume change?

8. Suppose you have a glass sphere 100 cm³ in volume. When you heat it up from 20 °C to 80 °C, its volume increases by 0.024 cm³. What is the fractional change in its volume?

If k_V were exactly constant, then it would be exactly true to write

$$V = V_0(1 + k_V \, \Delta t) \tag{5}$$

where ΔV has been replaced by $V - V_0$, where V is the volume after the temperature has been changed by an amount Δt. But k_V is only approximately constant. For any given substance precise measurements of $\Delta V/V_0$ at different temperatures or over different ranges of temperature will usually yield different values of k_V; thermal expansion is non-linear. That is why Table 4.2 includes with each value of k_V the temperature at which it was measured. Nevertheless, for many materials the dependence of k_V on temperature is small enough to ignore, and Eq. 5 is accurate enough for practical purposes.

Notice how the non-linear thermal expansion of matter leads to a logical problem. If you measure the coefficient of thermal expansion k_V using the thermal expansion of liquid mercury for your thermometer, how do you know whether variation of measured values of k_V is in the measured substance or in your thermometer? Given the way Celsius defined his temperature scale, you can not answer this question. There is circular reasoning here.

One way to escape the circle is to create a hypothetical model substance—an ideal gas—for which Eq. 5 is exactly true and then find a real substance that behaves like the model. An important clue for constructing the model comes from studies of the thermal expansion of gases. When in 1802 Gay-Lussac measured the thermal expansion of several different gases heated from $0\,^\circ$C to $100\,^\circ$C, he found that they all had the same coefficient of thermal expansion. His result suggests there is some basic property that is the same for all gases, a property that our model gas will need to have.

His experiment was direct.[6] After filling a 205 mL flask with gas from which all water vapor had been carefully removed, he placed the flask in an iron frame and submerged the apparatus first in a tank of boiling water and then in a tank containing a mixture of water and melting ice. For all his measurements $\Delta t = 100\,^\circ$C.

When his apparatus was at the temperature of boiling water ($100\,^\circ$C), he used the cords attached to the ends of the bar LL to open the valve (see Fig. 4.4) to allow gas to escape through tube ID and bring the pressure in the flask to atmospheric pressure P_a. Then he closed LL, cooled the apparatus enough so that he could handle it and detach the tube ID. With the tube removed, he put the apparatus in a bath of ice and water. As the gas cooled, the pressure in the flask decreased, so now when he opened LL, water moved out of the bath and up into the flask partially filling it. Then he closed LL, carefully dried off the outside of the flask, and weighed the flask containing the water that had risen into it. From this he subtracted the weight of the empty flask to find ΔV, the decrease in the volume of the gas when cooled by $100\,^\circ$C. In a similar way he found the volume V_{100} of the empty flask. He could then calculate V_0 the volume of gas at $0\,^\circ$C: $V_0 = V_{100} - \Delta V$.

---

[6]Gay-Lussac's measurements were simple in conception, but it was several decades after his early work before refinement of apparatus and close attention to detail yielded values of k_V accurate to $\pm 0.1\%$. See the articles by Gay-Lussac, by Regnault, and by Chappuis (all in English) in *The Expansion of Gases by Heat* edited by W. W. Randall, The American Book Co., New York, 1902. This book is available online as a free download.

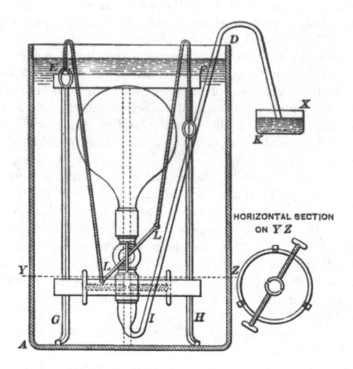

FIGURE 4.4 Gay-Lussac's 1802 apparatus for measuring the coefficient of thermal expansion of gases.

He found that $k_V = 0.00375$ $(°C)^{-1}$ was the same for hydrogen, nitrogen, oxygen, carbon dioxide, and the other gases that he studied. Later, other scientists made more precise measurements[7] and found that

$$k_V = \frac{\Delta V}{V_0}\frac{1}{\Delta t} = 0.003667 \ (°C)^{-1} = \frac{1}{273} \ (°C)^{-1}.$$

▪ EXERCISES

9. Explain how Gay-Lussac found the volume of the empty flask.

10. From the value of k_V Show that a given volume of gas at $0\,°C$ and constant pressure will expand by $\sim 37\%$ when its temperature is increased by $100\,°C$.

---

[7]Actually Gay-Lussac's measurements were quite precise, but because his apparatus was submerged in a tank of water, the final pressure was not quite the same as the initial pressure. See C. H. Holbrow and J. C. Amato, "What Gay-Lussac didn't tell us," *Am. J. Phys.*, 2010

11. How do you know from Fig. 4.4 that after Gay-Lussac let water into the flask at $0\,°C$, the pressure in the flask would have been more than the pressure at $100\,°C$?

▼ EXAMPLES

3. Suppose the flask weighed $100\,g$ when empty, $305\,g$ when full, and $156\,g$ after passing through the steps of his experiment. What is the value of $(V_{100} - V_0)/V_0$?

If the flask weighs $305\,g$ full of water and $100\,g$ when empty, then it must contain $305 - 100 = 205\,g$ of water. Because the density of water is $1.0\,g\,cm^{-3}$, the volume of the flask is $205\,cm^3$. By the same reasoning the decrease in volume of gas upon cooling is $56\,cm^3$. Consequently, $(V_{100} - V_0)/V_0 = 56/149 = 0.376$. To be exactly correct V_0 should be the volume at P_a, but it is not because the flask is under a few centimeters of water.

Of course the glass of the flask also expands when heated, so the observed change in the volume of the gas is not entirely from the expansion of the gas. Although Gay-Lussac did not correct for this effect, later experimenters did.

Imagining an Ideal Gas

Gay-Lussac's results provide a starting point for an argument that connects the idea of temperature to a property common to all gases: they expand when heated and contract when cooled. This connection is made quantitative by imagining a gas that exactly obeys Eq. 5. This model gas provides a temperature scale that does not depend on how mercury expands and contracts when heated or cooled.

Contrary to what Gay-Lussac thought, the modern values of k_V (see Table 4.2) are slightly different for different gases. In Table 4.2 the variation between k_V for nitrogen and helium is 0.2%. This small difference seems to be real because two experimentally measured values for the coefficient of thermal expansion of CO_2 differ by only 0.08%. If this is the case, what makes different gases have different values of k_V?

To answer such a question you need to do experiments to find what makes the differences between k_V of different gases larger or smaller. From such experiments you infer what properties of gases make them more or less similar. After identifying the properties that make one gas like

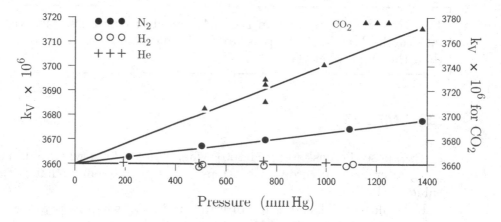

FIGURE 4.5 Coefficient of thermal expansion vs. pressure for four gases.

another, you can then imagine a gas that has only those properties and none of the ones that produce differences. This limiting case or simplified model is called an "ideal gas."

For example when you measure k_V for different gases at pressures other than 760 mmHg (101 kPa) with enough precision, you get values different from each other and from those in Table 4.2. But Fig. 4.5 shows an interesting result emerges when you graph the value of k_V versus the gas pressure at which the measurement was made. As the pressure goes down, all the values of k_V tend toward the same limit:

$$k_V = 3661 \times 10^{-6} = 1/273.15 \ (°C)^{-1}.$$

Apparently in the limit of low pressure when we reduce the the density of the gas while holding its temperature constant, k_V tends to a limiting, universal value. Therefore, it makes sense to assign to an ideal gas the property that its thermal expansivity is $k_V = 3661 \times 10^{-6} \ (°C)^{-1}$.

Gay-Lussac's Law and the Kelvin Temperature Scale

Setting $k_V = 3661 \times 10^{-6} = \frac{1}{273.15} \ (°C)^{-1}$ in Eq. 5 leads to one version of what is known as Gay-Lussac's Law (also known as Charles' Law) for the temperature dependence of the volume of a gas at constant pressure:

$$V = V_0(1 + \frac{1}{273.15} \ t) = V_0 \frac{273.15 + t}{273.15}. \tag{6}$$

Equation 6 is not the usual form of Gay-Lussac's Law because the temperature t is in degrees Celsius. Notice that Eq. 6 becomes much simpler if you define a new temperature scale T by shifting the zero point of the Celsius scale by 273.15 °C. To make the shift you define

$$T \text{ in kelvins} = 273.15 + t \text{ in } °C \tag{7}$$

This new scale is called the Kelvin temperature scale, and temperatures on the Kelvin scale are measured in "kelvins" abbreviated K. On the Kelvin scale water freezes at 273.15 K and boils at 373.15 K, i.e., 0 °C = 273.15 K and 100 °C = 373.15 K. A kelvin is the same size as a degree Celsius. The word "degree" and the symbol ° are not used with units of kelvins.

Rewritten in terms of the Kelvin scale, Eq. 6 becomes

$$V = \frac{V_0}{273.15} T$$

which is usually written as

$$\frac{V}{T} = \frac{V_0}{273.15} = \frac{V_0}{T_0} = \text{constant.} \qquad \text{Gay-Lussac's Law} \qquad (8)$$

The value of the constant depends on how much gas you have. Suppose you have 22.4 L of gas at 273.15 K. Then the constant is $\frac{22.4}{273.15} = 0.082$ L K^{-1}. If you have 27.3 L of gas at 27.3 K, the constant is 1 L K^{-1}. Given any volume of gas V_1 at a temperature T_1 and some pressure P_1, you can find the volume V at the same pressure and any other temperature T from

$$\frac{V}{T} = \frac{V_1}{T_1}.$$

Volume and temperature of an ideal gas exactly obey Gay-Lussac's law. The law is also a good approximation to the behavior of simple gases at low densities and at temperatures appreciably above the value at which the gas liquefies.

EXERCISES

12. Suppose you have 22.4 L of ideal gas at 273.15 K. What will be the volume of the gas at the same pressure but heated to 373.15 K?

13. Suppose the gas is kept at the same pressure and cooled to 27.3 K. What is its volume then?

14. What if it is cooled to 2.73 K?

15. What is the pressure of the gas in each case? Answer: You can't tell. You need to know how much gas is present, i.e. the number of moles.

16. Suppose you have 20 L of ideal gas at 0 °C. At what temperature in degrees Celsius will the volume double if the pressure at that temperature is the same as the pressure at 0 °C?

Gay-Lussac's law says that at constant pressure the volume of a gas is proportional to its temperature in kelvins (remember: use *kelvins* not °C). This proportionality says that as a gas cools to $-273.15\,°C = 0\,K$, its volume goes to zero. Although real gases don't behave that way, an ideal gas does. According to this hypothetical behavior there can be no temperature lower than $T = 0\,K$. For this and better reasons, the Kelvin scale is often called the "absolute value" scale of temperature and $T = 0\,K = -273.15\,°C$ is called absolute zero. Chap. 5 will show you another reason to believe in absolute zero.

The ideal gas offers an escape from circular reasoning about temperature and thermometers by providing a guide for choosing a standard thermometer. By definition an ideal gas has a thermal expansion coefficient of $3661 \times 10^{-6}\,K^{-1}$ that is constant for all temperatures. From Fig. 4.5 you can see that hydrogen and helium behave as ideal gases. Therefore, we can make from either of these real gases an excellent thermometer that is linear over a significant range of temperatures. Historically the hydrogen gas thermometer was used as a standard thermometer, and other thermometers, including the mercury thermometer, were calibrated against it.[8]

4.5   THE IDEAL GAS LAW

By combining Boyle's Law and Gay-Lussac's Law you get the ideal gas law. It provides an exact description of an ideal gas, and it is a good and useful approximate description of many ordinary gases at temperatures from somewhat below $0\,°C$ to some hundreds of $°C$, and at pressures from a few atmospheres downwards. Under these conditions Boyle's law and Gay-Lussac's law provide good representations of ordinary gases. In general, the higher the temperature and the lower the pressure, the better the two laws and their combined form, the ideal gas law, apply to real gases.

To put the two laws together, use the limiting behavior of Boyle's law: At a fixed temperature and in the limit of low pressure, the product of P and V of a given amount of any gas approaches the same constant value. For one mole of any gas at $t = 0\,°C$ experiment shows that as P gets small, PV approaches the value $2271\,kg\,m^2\,s^{-2} = 2271\,J$. This property is independent of the type of gas, and you can define an ideal

---

[8]The nonlinearity of the mercury thermometer turns out to be quite small for most practical purposes. Nevertheless, logically you can't verify Gay-Lussac's (Charles') Law with an instrument that has been calibrated by assuming the validity of the law.

gas as one for which $PV = 2271$ J for one mole of gas at a temperature of $0\,°C = 273.15$ K. This definition is equivalent to the one given above in terms of k_V.

As you will see later, it is not a coincidence that the units of PV are joules.

◼ EXERCISES

---

17. Show that the units of PV are joules.

18. Why does the coefficient of thermal expansion have the same value in units of $(°C)^{-1}$ as in units of K^{-1}.

19. Suppose you had 2 moles of an ideal gas at 273.15 K. What would be the value of PV?

20. Assume you have a mole of an ideal gas and that the mole weighs 32 g. At $T = 273.15$ K what would be the value of PV? What if you had only 16 g of this ideal gas?

21. Suppose you have two different ideal gases, one with a mass of 29 g per mole and the other with a mass of 17 g per mole. What would be the value of PV of a mixture of one mole of each gas?

Now you can combine Boyle's Law and Gay-Lussac's Law into a single law that connects all possible values of P, V, and T. To start, assume you have just one mole of gas. At $0\,°C$, i.e., at $T_0 = 273.15$ K, it will have some pressure P_0 and some volume V_0. Keeping the temperature fixed at T_0, you can get any pressure P you want by changing the volume V_0 to some volume V' so that Boyle's Law gives

$$PV' = P_0V_0. \tag{9}$$

Then holding the pressure constant at P you can choose a temperature T to get any volume V that you want according to Gay-Lussac's Law:

$$V' = V\frac{T_0}{T}. \tag{10}$$

Now replace V' in Eq. 9 with the expression for V from Eq. 10 and divide both sides by T_0. You get

$$\frac{PV}{T} = \frac{P_0V_0}{T_0} = \frac{2271}{273.15} = 8.314 \ \text{J K}^{-1}. \tag{11}$$

You can see from Exercise 19 that when you have n_m moles, the product PV increases by a factor of n_m. Using this fact you can generalize Eq. 11

to apply to any amount of gas, i.e., any number of moles, by defining the universal gas constant R to be

$$R \equiv 8.314 \ \mathrm{J\,K^{-1}\,mol^{-1}},$$

and writing

$$PV = n_{\mathrm{M}}RT, \qquad\qquad \text{ideal gas law.} \qquad\qquad (12)$$

This equation is the ideal gas law. Although dimensionless, the quantity n_{M} has units of "mol," the SI abbreviation of mole, and it can be any positive real number. The units of R are $\mathrm{J\,K^{-1}\,mol^{-1}}$, as they must be to make the units of Eq. 12 come out right.

■ EXERCISES

22. Give a convincing argument why multiplying the right hand side of Eq. 11 by n_{M} makes the ideal gas law correct for any amount of gas.

23. Suppose you have 1.5 mol of ideal gas at 20 °C sealed in a 10 L can. What is the pressure of the gas?

24. What will be the pressure if you raise the temperature to 100 °C? Give your answer in atmospheres.

What Underlies Such a Simple Law?

You have seen that to useful accuracy over a range of temperatures and pressures many gases obey the ideal gas law:

$$PV = n_{\mathrm{M}}RT.$$

But why? What is there about gases that makes them behave so similarly and so simply? The temptation is strong to model the gas from simple components. That is our cue to try to understand this law in terms of atoms. The next chapter shows how a simple atomic model of gases explains a wide variety of their physical properties.

PROBLEMS

1. Find the location of a nearby barometer. Go there and measure the pressure of the atmosphere. Make a careful drawing of the pointer and scale on the barometer showing the pressure as you read it.

If no barometer is available, you can go to almost any weather website and find the current atmospheric pressure for the nearest weather station.

Now, to run yourself through all these terms and units, make a table with two rows and eight columns. In the first row label each column with the name of a unit of pressure. Using the appropriate units, put the value of the pressure that you measured into each column of the second row.

Why is the number you measured different from the value of the "standard atmosphere"?

2. Suppose you have a piston with an area of $100\,\text{cm}^2$ sitting on a volume of $10\,\text{L}$ of nitrogen gas at $T = 20\,^\circ\text{C}$. If the gas is heated and warmed to $T = 30\,^\circ\text{C}$, by how much does the volume change? Give your answer accurate to $\pm 1\%$.

3. Suppose you have a pyrex flask that has a volume of $350\,\text{mL}$ at $0\,^\circ\text{C}$. What will be its volume after being heated to $100\,^\circ\text{C}$?

4. State the argument by which the ideal gas law is determined. Derive the ideal gas law from Boyle's Law and Gay-Lussac's Law.

5. State the ideal gas law, identify the variables in the law, and use it to find the pressure of 1 mole of gas contained in a volume of $11.2\,\text{L}$ at $0\,^\circ\text{C}$.

6. A 45-liter container at room temperature is known to have in it a total of $N = 15 \times 10^{23}$ diatomic molecules of oxygen gas.
   a. What is the number density n of the molecules of this gas? Give your answer in units of cm^{-3}.
   b. What is the number of moles, n_M, of this gas?
   c. What is the pressure of this gas?

7. An experimenter measures pressure of gas as its volume is changed and gets the following data:

Pressure (Pa)	Volume (m^3)
1000	1.5
1500	1.0
2000	0.75
2500	–

a. Do these data obey Boyle's law? Explain how you know.

b. What value of volume would you expect to measure for the missing entry in the above table?

8. Suppose the above data were taken at a temperature of 27 °C and then later the gas was cooled to −73 °C. At what volume would the cooled gas have a pressure of 1000 Pa?

9. Consider the following table of measurements of pressure vs. volume for O_2 ($M_{O_2} = 32\,u$) in a closed container at 300 K.

Pressure (kPa)	Volume (cm^3)
90.9	24.89
73.6	30.73
59.5	37.93

a. Is it reasonable to conclude that the gas obeys Boyle's law? How did you come to this conclusion? Discuss clearly.

b. How many molecules are contained in this volume? How do you know?

c. If the oxygen molecules are replaced by an equal number of helium gas molecules ($M_{He} = 4\,u$) at the same temperature, what would be the gas pressure when the volume $V = 30.73\,cm^3$? (Hint: See the above table.)

10. Boyle's measurement of the air pressure outside his apparatus (Fig. 4.6) was $29\frac{2}{16}$ inches of mercury (Hg). [The density of Hg is $\rho_{Hg} = 13.6\,g\,cm^{-3}$.]

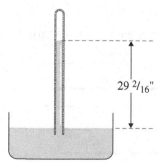

29 ²/₁₆"

FIGURE 4.6 Barometer for Problem 10.

a. What was the air pressure in mmHg?

b. What was the air pressure in Torr?

c. What was the air pressure in Pa?

11. An object floats (moving neither up nor down) in air when its "buoyancy force" equals its weight. Archimedes discovered that the buoyancy force on an object equals the weight of the volume of air displaced by the object. For example, a 10 m^3 block will experience a buoyancy force upwards equal to the weight of 10 m^3 of air.

    a. Imagine that you are holding a balloon with a volume of 10 liters (L) filled with He gas ($M_{He} = 4\,u$). The temperature is a pleasant 27 °C. Assuming that the He gas is at atmospheric pressure (P_0), what is its mass?

    b. Assuming that air is 20% O_2 ($M_{O_2} = 32\,u$) and 80% N_2 ($M_{N_2} = 28\,u$), what is the buoyancy force on the balloon?

    c. What volume of helium should the balloon contain in order to float carrying its own weight of 20 kg plus the weight of a 70 kg person?

12. Certain volumes of nitrogen gas and hydrogen gas react to form ammonia gas, according to the reaction shown below:

$$N_2 + 3H_2 \rightarrow 2NH_3.$$

The atomic masses of nitrogen *atoms* and hydrogen *atoms* are 14.0 u and 1.0 u, respectively. The mass of NH_3 formed in this reaction is 3.4 g.

    a. If the pressure and temperature of each gas (N_2 and H_2 before reaction, and NH_3 after reaction) are equal to 83.1 kPa and 300 K, find the final volume of NH_3.

    b. What would be the final volume if you arranged to double the final pressure?

13. The weight of air produces Earth's atmospheric pressure.

    a. What is the mass of a column of air 1 cm^2 in cross section rising from Earth's surface?

    b. How many molecules are there in that column?

    c. How many moles of molecules?

14. A glass bottle is almost filled with 0.75 L of wine. It is tightly sealed, leaving a bubble of air within at atmospheric pressure and with a volume of 0.09 L. The bottle is placed in a freezer and brought to a temperature of 250 K. The wine freezes, and in so doing, expands to occupy a volume of 0.79 L.

Assume that the volume of the bottle is unchanged. What is the pressure of the bubble now that it is squeezed into a volume of 0.05 L? What is the approximate force on the cork?

15. Imagine a glass of water is on the ground and that you are standing on the roof of a building drinking from the glass through a straw. What is the longest vertical straw with which you can hope to drink this way?

16. To pump water from a well to a faucet in a second-floor apartment, it's possible to put the pump either at the well or by the faucet. To pump water to the fifth floor, the pump must be located at the well, not on the fifth floor. Why?

CHAPTER 5

Hard-Sphere
Atoms

5.1   INTRODUCTION

Why do gases obey the simple ideal gas law so well? Even gases of more complicated entities such as CH_4 behave the same way. This fact suggests that the internal structure of the entities has little, if anything, to do with the gas law. The observed similarities of the properties of all gases must arise from properties that are common to all atoms and molecules. What then is the least we need to assume about the particles of gas that will explain the gas law?

To answer this question we try to imagine what atoms and molecules are like, and we make up a *model* to represent the atom or molecule. For gases we will use a model so simple that it does not distinguish between atoms and molecules, and therefore in this chapter we use the word "atom" to mean "atoms and molecules." We call our description a "model" to remind ourselves that it is surely incomplete, that we are abstracting only a few important features of atoms to explain a limited set of properties.

Assuming that a gas is made of atoms, how can we connect the ideal gas law $PV = n_{\mathrm{M}}RT$ to their behavior? The volume V is a geometric property fixed by the choice of container; the temperature T is somewhat mysterious; the pressure P seems the best place to look for connections between the gas law and atoms. Indeed, we know that pressure is related to forces, i.e., to changes in momentum, so this might be a good place to look for connections between the gas law and the Newtonian mechanics of little particles.

Chapter 4 showed you that while the volume of solid materials responds weakly to changes in temperature or external pressure, a gas responds much more dramatically. The solid acts as though its parts were in close

C.H. Holbrow et al., *Modern Introductory Physics, Second Edition*,
DOI 10.1007/978-0-387-79080-0_5, © Springer Science+Business Media, LLC 1999, 2010

contact with each other; the gas acts as though its atoms had plenty of space between them. A gas can expand indefinitely, but at any volume it will continue to resist compression. Something must be happening to keep its atoms apart. We are led to guess that because atoms don't pile up like sand grains on the floor of the container, they are in rapid, incessant motion. Other similar considerations lead to the following assumptions about gas atoms (and gas molecules):

- An atom or molecule is a tiny, hard sphere, too small to be observed by eye or microscope.

- An atom has mass.

- Every atom is in constant random motion.

- Atoms exert forces on each other only when they collide.

- When atoms collide, both momentum and kinetic energy are conserved.

Historically it was difficult to get direct evidence for atoms. Attempts to subdivide matter in straightforward ways did not show discreteness or granularity. This could be understood if atoms were so small that even very small chunks of matter contained vast numbers of them.[1] The accurate description of the observed behavior of the pressure of many different gases with this simple atomic model was strong, if indirect, confirmation that atoms exist.

5.2 GAS PRESSURE FROM ATOMS

If a gas is a jostling myriad of tiny atoms, it is reasonable to imagine that pressure arises from the force on a wall that comes from numerous collisions of the gas atoms with the walls. The average of the forces of these collisions gives rise to the observable pressure.

To see how this can occur, consider a cubical box full of gas at some pressure P. Assume there are N atoms, each of mass m, in a box (see Fig. 5.1a) that has a length L in the x direction and walls with area A at each end. To simplify the discussion, assume that the walls are smooth, so that they produce no changes in momentum of the particles in directions at right angles to x. While Fig. 5.1b shows one atom colliding with the right-hand wall, it is all the atoms continually and repeatedly hitting this

---

[1]Modern imaging techniques now provide direct observation of atoms, but there was certainly nothing so convincing in the nineteenth century.

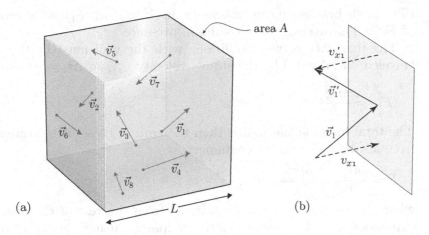

FIGURE 5.1 (a) Gas atoms move in random directions. (b) An atom's change in momentum upon colliding with a wall.

wall that exert the average force on the wall and produce the observed pressure. The average force exerted by the atoms on the wall is the sum of their changes in momentum divided by the time interval over which the collisions occur.

The force on the wall giving rise to the pressure comes from the collisions of many atoms moving with the many different values of velocity, but a simplified picture shows the essence of the effect. (A more thorough treatment based on ideas used later in this book is given in the appendix on page 135.) In the volume shown in Fig. 5.1a, consider an atom with a velocity component v_{x1} in the positive x direction. When it rebounds from the right wall, it must have its x component of velocity reversed, because by assumption there is no change in the other components and energy is conserved. This means that each time one of these atoms hits the wall, its momentum changes from mv_{x1} to $-mv_{x1}$. Therefore the atom's change in momentum, its final momentum p' minus its initial momentum p, is

$$p' - p = -2mv_{x1},$$

and the wall undergoes an exactly equal and opposite change of momentum $\Delta p = 2mv_{x1}$. These statements are illustrated in Fig. 5.1b.

You don't have to know how to describe the *instantaneous* force on the wall, because in the long run this atom will exert an *average* force given by the changes in momentum divided by the time Δt_1 between its successive collisions with the wall,

$$\langle F_1 \rangle = \frac{\Delta p}{\Delta t_1} = \frac{2mv_{x1}}{\Delta t_1}.$$

(The angle brackets mean "average value"; so read $\langle F_1 \rangle$ as "average value of F_1.") This average force results in pressure.

The time Δt_1 between collisions with the wall must be the distance traveled, $2L$, divided by the atom's velocity v_{x1}, so $\Delta t_1 = \frac{2L}{v_{x1}}$ and

$$\langle F_1 \rangle = \frac{m v_{x1}^2}{L}.$$

The total force on the wall is then the sum of these little average forces from all the atoms in the container:

$$F = \frac{m v_{x1}^2}{L} + \frac{m v_{x2}^2}{L} + \cdots,$$

where $\cdots$ represents the contributions from the rest of the atoms. This expression can be written more compactly using the notation for a summation

$$F = \frac{m}{L} \sum_{i=1}^{N} v_{xi}^2, \tag{1}$$

where v_{xi} is the x velocity of an arbitrary atom labeled i, and the symbol $\sum_{i=1}^{N}$ stands for the sum over all atoms with i ranging from 1 to N, the total number of atoms in the box.

Are you bothered by the assumption that the atoms travel freely back and forth between the walls? When the density gets large enough, there must be collisions between atoms. What then happens to the x momentum of each atom? The values must change. But conservation of momentum comes to the rescue: Whatever momentum is lost by one atom will be gained by another, so the total incoming momentum at the wall in any time interval is the same, regardless of the number of intervening collisions between atoms. (The treatment in the appendix avoids this problem.)

The summation in Eq. 1 is staggeringly large; for 1 mole of gas there will be 6×10^{23} terms in the summation. Physicists get around this difficulty by defining an average square speed:

$$\langle v_x^2 \rangle = \sum_{i=1}^{N} \frac{v_{xi}^2}{N}. \qquad \text{average square speed}$$

Then the sum in Eq. 1 is just N times the average square speed. The force can then be written

$$F = \frac{Nm}{L} \langle v_x^2 \rangle. \tag{2}$$

To find the pressure P, divide both sides of Eq. 2 by the area A of the end wall of the box, and note that the product LA is just the volume of the container. Then you obtain

$$P = \frac{m}{LA} N \langle v_x^2 \rangle = \frac{N}{V} m \langle v_x^2 \rangle. \tag{3}$$

The quantity N/V is the number density of the gas, which, as you saw in Chap. 3, tells you how many atoms or molecules there are in a unit volume. N/V is often given its own separate symbol n; n is very useful when we don't want to talk about some particular volume.[2]

▼ EXAMPLES

1. The International Union of Pure and Applied Chemistry (IU-PAC) defines standard temperature and pressure (STP) to be $0\,^\circ$C and 100 kPa. What is the number density of a gas at STP? Remember that at STP, one mole of ideal gas occupies 22.4 L, or 22 400 mL $= 22\,400\,\text{cm}^3$. Thus

$$n = (6 \times 10^{23})/(22\,400\,\text{cm}^3) = 2.7 \times 10^{19}\,\text{cm}^{-3}.$$

▪ EXERCISES

1. The U.S. National Institute of Standards and Technology (NIST) defines STP to be $20\,^\circ$C and 101.325 kPa. Calculate the number density of a gas at STP (NIST). By what percentage does it differ from the number density of a gas at STP (IUPAC)?

2. What is the number density for a gas in an ultra-high-vacuum chamber at a pressure of 10^{-12} Torr? Take the temperature to be $0\,^\circ$C.

3. It is useful to *remember* the number density of 1 Torr of ideal gas at $0\,^\circ$C. What is it?

It is more meaningful, as you shall see, to use the magnitude of the total velocity and not just the x-component. As with any vector magnitude in three dimensions,

$$v^2 = v_x^2 + v_y^2 + v_z^2,$$

---

[2]Have you got all these n's straight? n_M is the number of moles; N is the total number of atoms or molecules; $n = N/V$ is the number density.

which when averaged becomes

$$\langle v^2 \rangle = \langle v_x^2 \rangle + \langle v_y^2 \rangle + \langle v_z^2 \rangle.$$

Nature cannot tell the difference between the x direction and the y or z directions, so it must be true that

$$\langle v_x^2 \rangle = \langle v_y^2 \rangle = \langle v_z^2 \rangle,$$

from which it follows that

$$\langle v^2 \rangle = 3 \langle v_x^2 \rangle.$$

The quantity $\langle v^2 \rangle$ is called the "mean square velocity," where the word "mean" is a synonym for "average." You can rewrite Eq. 3 in terms of the mean square velocity and obtain

$$P = \frac{Nm}{V} \frac{\langle v^2 \rangle}{3},$$

which you can rearrange to get

$$PV = \frac{1}{3} N m \langle v^2 \rangle. \tag{4}$$

▪ EXERCISES

---

4. It is sometimes useful to express Eq. 4 in terms of the mass density ρ of the gas. Show that Eq. 4 implies

$$P = \frac{1}{3} \rho \langle v^2 \rangle. \tag{5}$$

5. Show that PV equals $\frac{2}{3}$ of the average kinetic energy of the gas.

Equation 5 is important because it tells you something about how fast atoms move. For example, nitrogen at STP has $\rho = 1.25 \, \text{kg m}^{-3}$ so

$$\langle v^2 \rangle = \frac{3 \times 101 \times 10^3 \, \text{Pa}}{1.25 \, \text{kg m}^{-3}} = 2.42 \times 10^5 \, \text{m}^2 \, \text{s}^{-2}. \tag{6}$$

From this result it follows that

$$\sqrt{\langle v^2 \rangle} = 493 \, \text{m s}^{-1}. \tag{7}$$

This quantity is called the "root mean square" (rms) velocity and is often written as v_{rms}. The rms velocity of N$_2$ molecules at 0 °C is 493 m s^{-1}. (Notice that to get this result you did not need to know the size or the number of the atoms.) Compare this to the $30\,\mathrm{m\,s^{-1}}$ of a fast traveling automobile, or to the $250\,\mathrm{m\,s^{-1}}$ of a jet plane; you see that atoms and molecules move quickly.

It is informative to rewrite Eq. 4 using the fact that $N\,m = n_{\mathrm{m}}M_{\mathrm{M}} = M$ the total mass of the volume V of gas, where M_{M} is the mass of one mole. Then you get

$$PV = \frac{1}{3}n_{\mathrm{M}}\,M_{\mathrm{M}}\langle v^2\rangle. \tag{8}$$

This form is especially convenient at STP. For example, you know from Chap. 4 that at $T = 273.15\,\mathrm{K}$, $PV = 2271\,\mathrm{J}$ for one mole of any gas that behaves like an ideal gas; then you can use the mass of a mole M_{m} of nitrogen in Eq. 6, and the calculation becomes

$$\sqrt{\langle v^2\rangle} = \sqrt{\frac{3 \times 2271\,\mathrm{kg\,m^2\,s^{-2}}}{1\,\mathrm{mol} \times 0.028\,\mathrm{kg\,mol^{-1}}}} = 493\,\mathrm{m\,s^{-1}}.$$

You can go a step further and use the fact that $PV = n_{\mathrm{M}}RT$. Then, in general,

$$v_{\mathrm{rms}} \equiv \sqrt{\langle v^2\rangle} = \sqrt{\frac{3PV}{n_{\mathrm{m}}M_{\mathrm{M}}}} = \sqrt{\frac{3RT}{M_{\mathrm{M}}}} \qquad \text{rms velocity} \tag{9}$$

where the symbol $\equiv$ means "equivalent to" or "defined to be."

5.3 TEMPERATURE AND THE ENERGIES OF ATOMS

The importance of Eqs. 8 and 9 goes far beyond their convenience. The equations point to a major new insight—that temperature is associated with atomic motion. The ideal gas law connects the right side of Eq. 8 to $n_{\mathrm{M}}RT$, so that

$$\frac{1}{3}n_{\mathrm{m}}M_{\mathrm{m}}\langle v^2\rangle = n_{\mathrm{M}}RT.$$

This equation conveys a big message: T, the gas temperature on the kelvin scale, is proportional to the mean square velocity of the atoms. This means that the temperature T is proportional to the average kinetic energy $\langle K\rangle$ of the gas:

$$n_{\mathrm{M}}RT = \frac{1}{3}M\langle v^2\rangle = \frac{2}{3}\frac{1}{2}M\langle v^2\rangle = \frac{2}{3}\langle K\rangle, \tag{10}$$

and so is PV.

The result shows you that a given volume of any gas has the same kinetic energy as an equal volume of any other gas at the same temperature and pressure. That's why at $T = 273.15$ K a mole of any (ideal) gas has $PV = 2271$ J. You also see why gases make good thermometers: gas temperature depends only on the average kinetic energy of the gas and not on any other properties, e. g., not on how many hooks the atoms have, not on how big the atoms are, not on how complicated the molecules are.

■ EXERCISES

> **6.** What is the value of RT at $0\,°C$?
>
> **7.** Why was it unnecessary to do a calculation to find the answer to the question?

The connection between the ideal gas and a simple model of hard, point-like atoms bouncing off walls suggests that you can understand temperature of gases to be a measure of the energy of random motion of the atoms. You probably already believe this because you have often been told that it is so. For the same reason, you probably already believe that atoms exist. Here you have seen evidence for both ideas: A simple model of atoms explains why the value of PV is the same for one mole of any gas at a fixed temperature; the model and Eq. 10 persuasively and quantitatively suggest that temperature is an effect of a basic physical property of atoms, their kinetic energy.

Energies of Atoms: Boltzmann's Constant

To explore the world of atoms, you need to be able to talk about the properties of individual atoms. For example, it's all very well to know that at STP a mole of atoms has a kinetic energy of 3405 J, but what is the average kinetic energy of a single atom? To answer this basic question, you have to know the value of Avogadro's number. To determine reliable values of N_A was a major challenge to physicists and chemists, and the first experimental measurement of N_A—described later in this chapter— occurred only in the second half of the nineteenth century. For now, let's just use the currently accepted value of $N_A = 6.02 \times 10^{23}$ and answer the question.

To find the answer take the kinetic energy in a mole and divide by the number of atoms in a mole. You get

$$\frac{3405 \text{ J/mol}}{6.02 \times 10^{23} \text{ atom/mol}} = 5.66 \times 10^{-21} \text{ J/atom.}$$

■ EXERCISES

8. What is the average kinetic energy of an O_2 molecule at room temperature (20 °C)?

9. What is the average kinetic energy of a CO_2 molecule at 20 °C?

10. For the preceding items, did you do complicated calculations, or did you scale your answers from 5.66×10^{-21} J/atom? Why?

In general, to rescale from the macroscopic world of moles to the microscopic[3] world of single atoms or molecules, you divide macroscopic quantities by the number of atoms N. It is usually convenient to use the relation (really a definition of n_M)

$$N = n_M N_A.$$

For example, to connect the total kinetic energy K_{total} of a collection of N atoms to $\langle K_{molecule} \rangle$ the average kinetic energy of a single molecule write

$$K_{total} = n_M N_A \langle K_{molecule} \rangle.$$

Then, since $K_{total} = \frac{3}{2} n_M RT$, the factor n_M cancels from both sides, and you get

$$\langle K_{molecule} \rangle = \frac{3}{2} \frac{R}{N_A} T.$$

The ratio R/N_A appears over and over again in atomic physics, and so it is given its own symbol, "k_B," and name. It is called "Boltzmann's constant" and has a value

$$k_B = \frac{8.314 \, \mathrm{J \, mol^{-1} K^{-1}}}{6.02 \times 10^{23} \, \mathrm{mol^{-1}}} = 1.38 \times 10^{-23} \, \mathrm{J \, K^{-1}}.$$

In terms of k_B you get the standard expression for the average kinetic energy of a molecule:

$$\langle K_{molecule} \rangle = \frac{3}{2} k_B T. \qquad \text{average kinetic energy of an atom} \qquad (11)$$

---

[3]Technically it is incorrect to call the world of atoms "microscopic," because atoms are much smaller than can be seen in a microscope. Nevertheless, it is common usage to mean "extremely small."

▽ EXAMPLES

2. What is the average kinetic energy of a hydrogen molecule at room temperature? Notice that to answer this question you do not have to know the mass or the velocity or any details; all you need to know is the temperature and Boltzmann's constant, which is known from measurements of the gas constant R and Avogadro's constant N_A. The average kinetic energy of a hydrogen molecule at room temperature is then

$$\frac{3}{2} \times (1.38 \times 10^{-23}\,\mathrm{J\,K^{-1}}) \times (293\,\mathrm{K}) = 6.1 \times 10^{-21}\,\mathrm{J},$$

a rather small number.

The Electron Volt (eV)

Because the energies of atoms are important in atomic physics and we refer to them frequently, it is convenient to have a unit of energy scaled to the size of atomic happenings. This unit is the "electron volt," abbreviated eV. It has what may seem to you a strange definition:

$$1\,\mathrm{eV} = 1.602 \times 10^{-19}\,\mathrm{J}.$$

However, a few chapters from now you will see that it is a very reasonable unit. *It is so reasonable that from now on you should always use the eV as the unit of energy when talking about phenomena at the atomic scale.* All the usual multiples are used: meV, keV, MeV, GeV, and even TeV for tera electron volts where T and tera stand for a U.S. trillion, i.e., 10^{12}.

▨ EXERCISES

11. Show that the average kinetic energy of the hydrogen molecule in the previous example is 0.038 eV.

12. What is the average kinetic energy of an O_2 molecule at room temperature? What about N_2? He?

13. How many different calculations did you do?

The number obtained in Exercise 5.11 is important. It tells you the magnitude of thermal energy associated with room temperature (usually taken to be 20 °C (293 K), but sometimes 300 K). Most physicists *remember* that at room temperature

$$k_BT = 0.025\,\text{eV} = \frac{1}{40}\,\text{eV}, \tag{12}$$

or

$$\frac{3}{2}k_BT = 0.038\,\text{eV} \approx \frac{1}{25}\,\text{eV}.$$

▮ EXERCISES

14. Hydrogen and helium atoms near the surface of the Sun are at a temperature of about 6000 K. What is the average kinetic energy of one of these H atoms? Of one of the He atoms?

You might think that you can determine the rms speed v_{rms} (pg. 115) of a single atom from its mass m and its kinetic energy $\frac{1}{2}m\langle v^2\rangle \equiv \frac{1}{2}mv_{\text{rms}}^2$. Well, you're right, but your calculation will be equivalent to using Eq. 9. To find the average kinetic energy of a single atom in a mole of atoms at some temperature T, you find the average kinetic energy $\langle K\rangle$ of the mole of atoms and then divide by N_A the number of atoms in a mole. And to find the mass m of a single atom in a mole take the mass M_M of a mole of the atoms (i.e. their molecular weight) and divide by N_A because $M_M = N_Am$. In effect, you rescale Eq. 9 as follows:

$$v_{\text{rms}} = \sqrt{\frac{3RT}{M}} = \sqrt{\frac{3N_Ak_BT}{N_Am}} = \sqrt{\frac{3k_BT}{m}}. \tag{13}$$

▼ EXAMPLES

3. The mass of a hydrogen atom is 1.67×10^{-27} kg. What is the rms velocity of hydrogen at room temperature? To answer this question you need to know that at room temperature hydrogen atoms are in H_2 molecules. Each molecule has a mass of 3.34×10^{-27} kg. Therefore, using the value for k_BT at room temperature,

$$v_{\text{rms}} = \sqrt{\frac{3 \times 0.025\,\text{eV} \times 1.60 \times 10^{-19}\,\text{J/eV}}{3.34 \times 10^{-27}\,\text{kg}}},$$

$$v_{\text{rms}} = 1906\,\text{m s}^{-1}.$$

▦ EXERCISES

> **15.** Do you need to know the mass of an individual molecule in order to find its rms velocity? Explain using Eq. 10.
>
> **16.** Calculate the rms velocity of oxygen at room temperature. What is it for nitrogen? Helium?
>
> **17.** Compare the speed of sound in air at room temperature to the rms velocity of nitrogen.
>
> **18.** An object moving with a velocity of 11 km s^{-1} will, if not deflected, leave the Earth and never return. At what temperature would hydrogen atoms have v_{rms} equal to this escape velocity?

5.4   SUMMARY THUS FAR

The results of the atomic model of an ideal gas are very satisfactory. The fact that many gases closely obey the ideal gas law suggests that in important respects real gases resemble an ideal gas.

The model also leads us to an important insight into the nature of temperature. The model predicts that the pressure of a gas depends only on the atoms' kinetic energy per unit volume. The connection of the atomic model to the ideal gas law strongly suggests that temperature is a property of the motion of atoms, that temperature is a measure of the atoms' average kinetic energy. Indeed, when measured on the kelvin scale, temperature is directly proportional to the average kinetic energy. In other words, the atomic model of an ideal gas predicts that equal volumes of gases at the same temperature all have the same kinetic energy and, therefore, the same pressure. Because this is close to what is observed, it is convincing evidence that atoms have the properties assumed for the model: A gas is composed of atoms; they are very small (in effect point masses); they move all the time; their collisions with the walls produce pressure.

Take a moment and think about the logic of these arguments. We have measurements of pressures and temperatures made on large collections of matter—liters of gas. We observe that the connections among these directly measurable quantities are fairly well described by the ideal gas law. Then we imagine that the large quantities of matter are made up of numerous particles too small to see; we ascribe certain properties to these tiny particles and show that particles with these properties will behave according to the ideal gas law. From this we conclude that a gas is made up of such particles. We use macroscopic measurements to infer the existence and properties of submicroscopic bits of matter. This is speculation; there is no logical guarantee that our conclusions are right.

Why then should you believe such conclusions? Because there are other macroscopic observations that can be nicely explained by the ideal gas atomic model of matter. For example, the ideal gas atomic model yields values of the root-mean-square velocities of atoms, a useful quantitative measure of how fast they travel. From this number you can predict how quickly gas will stream through a small hole into an evacuated vessel. Observations confirm the predictions. The more such confirmations we obtain, the more confident we are that atoms are as we imagine them.

But what do you do when observations disagree with predictions? This is an important question because for all our talk about how well the ideal gas law describes the behavior of real gases, it is apparent from the data that the description is only approximately right. Look at Fig. 4.5; it shows that real gases deviate from the predictions of the ideal gas law. The simplicity of the model of an ideal gas is appealing, but it is obviously too simple. For another example, if the hole mentioned in the previous paragraph is too small, no gas will stream into the vacuum. This is not surprising because the assumption that atoms are point particles can hardly be correct; atoms surely have size, and a hole can be too small for a real atom to fit through. In general, when observations disagree with predictions, we try adding some new feature of the atom to the model. Much of the time this works.

5.5   SIZE OF ATOMS

How big are atoms? This is a fundamental question. Its answer sets the scale of atomic phenomena; it tells us how many atoms are in a mole—Avogadro's number. But the question was not easy to answer. It took some fifty years after Dalton's proofs that atoms exist before physicists and chemists could deduce from experiments that atoms are $\sim 2 \times 10^{-10}$ m in diameter.

To deduce atomic sizes from experiments chemists and physicists used the pattern of inference described above (Sect. 5.4). They modeled atoms as rapidly moving, frequently colliding hard spheres and showed how the size of such spheres affects the rate at which they collide with one another. By showing that the collision rate would affect a gas's viscosity, they connected the model of hypothetical microscopic behavior of atoms to an observable macroscopic property of bulk matter. The improved model predicted the observed temperature and density dependence of viscosity.

It is convenient and customary to describe collisions among atoms in terms of their "collision mean free path." This is a measure of the average distance an atom travels before it collides with another; it is an idea that has many uses in physics. The next section explains the idea; the subsequent sections describe viscosity and show how the atomic model explains viscosity of gases and how physicists used laboratory measurements of viscosity to infer that the mean free path of molecules of air is about 60 nm at room temperature and pressure and then inferred values of atomic sizes and Avogadro's number from this value of the mean free path.

Colliding Atoms, Mean Free Path

For the atomic model to describe the effects of collisions, its atoms must have finite size. That collisions occur at all implies that atoms are not point particles. Point particles with no physical extent occupy no space, and collisions between them would be impossible. By adding to the model the radius r_m for each atom or molecule, you can extend the model to describe the likelihood of collisions.

A common way to describe the likelihood of collisions is to specify the average distance an atom travels before it hits another. A short distance means frequent collisions; a long distance means infrequent collisions. Of course the actual distance a particular atom travels between collisions varies greatly from collision to collision, but there will be some average distance that atoms travel before colliding. This average distance between collisions is called the "mean free path," and is often denoted as ℓ. The frequency of collisions also depends on how fast the atoms are moving. Fast moving atoms collide more often than slow moving atoms with the same mean free path.

▤ EXERCISES

> **19.** Atoms in a jar have a mean free path of ℓ and an average speed of $\langle v \rangle$. In terms of these parameters what, at least roughly, will be f_c, the frequency of collisions between atoms?

The mean free path of an atom also depends on its size. As you might expect, big atoms are more likely to run into each other than are small ones. To see how the mean free path relates to the size of an atom and its speed, consider Fig. 5.2. Figure 5.2a shows a molecule of radius r_m moving past other similar molecules. Obviously, two molecules cannot get closer than a distance $2r_m$ to each other without colliding.

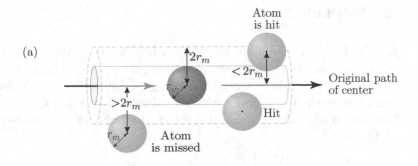

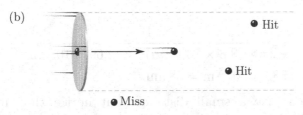

FIGURE 5.2 (a) Atoms will collide if their center-to-center distance is $<2r_m$. (b) Equivalent picture of an atom's cross section of radius $2r_m$ colliding with point-sized atoms.

It is convenient to describe this configuration by the equivalent arrangement shown in Fig. 5.2b, where one atom of radius $2r_m$ moves among point atoms. In effect we put all the physical size in one atom by treating it as a sphere of cross-sectional area $\pi(2r_m)^2$. Now imagine a disk of this area moving through a volume containing point atoms with a density of n points per unit volume. While moving a distance d, the disk will sweep out a volume of $\pi(2r_m)^2 d$. This volume contains $n\pi(2r_m)^2 d$ molecules, because n is the number of molecules per unit volume. If the molecules were all at rest and equally spaced, then the moving disk would suffer $n\pi(2r_m)^2 d$ collisions every time it traveled a distance d. Let ℓ be the distance for just 1 collision. Then

$$\ell = \frac{1}{\pi(2r_m)^2 n}.$$

This cannot be quite right because the point molecules are not all at rest. They are rushing toward and away from the moving molecule of radius $2r_m$. Instead of learning how to correct for this relative motion, just take the correction as given; it changes the above expression for ℓ by a factor of $1/\sqrt{2}$:

$$\ell = \frac{1}{\sqrt{2}\pi(2r_m)^2 n}. \qquad \text{mean free path} \qquad (14)$$

EXAMPLES

4. Use Eq. 14 to calculate some mean free paths. (This is the reverse of *measuring* the mean free path and using it to determine atomic sizes.) A nitrogen molecule has a radius of $r_m \approx 1.9 \times 10^{-10}$ m. What is the mean free path of a N_2 molecule in a volume of nitrogen gas at STP? To find the number density n use the fact that at STP one mole occupies 22.4 L, so

$$n = \frac{N_A}{V_m} = \frac{6.02 \times 10^{23}}{22.4 \times 10^{-3}\,\mathrm{m^3}} = 2.69 \times 10^{25}\,\mathrm{m^{-3}}.$$

Then

$$\ell = \frac{1}{\sqrt{2}\pi \times (3.8 \times 10^{-10}\,\mathrm{m})^2 \times (2.69 \times 10^{25}\,\mathrm{m^{-3}})}$$
$$= 5.80 \times 10^{-8}\,\mathrm{m} = 58\,\mathrm{nm}.$$

This may seem like a small distance, but notice that it is ≈ 300 molecular radii.

5. What happens to ℓ if you increase the temperature of the gas while keeping the pressure constant? You could go back to the basic equations and calculate new values for n and v_{rms}, but it is more insightful and more efficient to scale your result using the following observations. If P is constant, then V grows $\propto T$ and, therefore, $n \propto 1/T$. At the same time $v_{\mathrm{rms}} \propto \sqrt{T}$, so

$$\ell \propto \frac{1}{\left(\frac{1}{T}\right)\sqrt{T}} \propto \sqrt{T}$$

from which it follows that ℓ at room temperature is

$$58\sqrt{\frac{293}{273}} = 60\,\mathrm{nm}.$$

6. On average, how many collisions will a nitrogen molecule have in 1 s at STP? The average time between two collisions $\tau_{\mathrm{ave}} = \ell/v_{\mathrm{rms}} = (58 \times 10^{-9}\,\mathrm{m})/(493\,\mathrm{m\,s^{-1}}) = 1.2 \times 10^{-10}$ s. The number of collisions in 1 s is $1/\tau_{\mathrm{ave}} = 8.5 \times 10^9\,\mathrm{s^{-1}}$.

Viscosity

You have experienced viscosity. When you run in air or swim in water, you feel a resisting force. If you pole a raft along the surface of shallow still water, there is a resistance that is larger the faster you try to make the raft go. Close examination shows that right where the raft touches the water, water is dragged along at the same speed as the raft. A little below the surface, the water moves but not as fast as the raft. The deeper below the moving raft, the more slowly the water moves until at bottom it is not moving at all. If you imagine the water to be in layers, the layers nearer the raft move faster than the layers further below. The motion of each layer is resisted by the layer beneath it. This resistance of fluid layers to relative motion is called viscosity. You can also think of viscosity as resistance to pouring. Molasses in January is very viscous; so is motor oil; water is not as viscous as motor oil or honey; air and other gases are viscous but much less so than most liquids.

These ideas can be made quantitative. Move a flat plate at a velocity u on the surface of a liquid that is z deep above a stationary flat bottom. Just as for the raft, right at the upper plate the liquid will adhere to its surface and flow with it at a velocity u. The liquid at the bottom will have no net flow velocity. There has to be a smooth transition in flow velocity from u to 0, so you can expect that just below the top layer, the liquid will have a flow velocity a little slower than u. Further down it will be flowing more slowly yet, and so on, until on the bottom surface it will not be flowing at all. Figure 5.3 illustrates the situation.

The upper plate has a retarding force on it, a drag. Experiments show that this force depends on the kind of fluid and its temperature. We say that this drag is produced by the "viscosity" of the fluid.

Viscosity is a measurable quantity. In principle you might pull one plate past another and measure the force of the drag. This is difficult in practice. It is much easier to rotate one cylinder inside a larger one

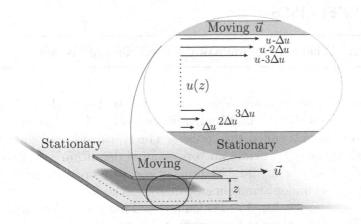

FIGURE 5.3 Change in velocity of fluid flow near a moving surface.

and measure the force resisting the rotation. If the gap z between the cylinders is small compared to their radii, the geometry is equivalent to the flat plate diagrammed in Fig. 5.3.

EXERCISES

20. Sketch for yourself a likely experimental setup of the two-cylinder arrangement for measuring viscosity.

From such measurements the force is found to be proportional to the area A of the cylinder and also proportional to the difference in velocity, u, divided by the width of the gap, z, between the two cylinders. We write

$$F = \eta A \frac{u}{z}, \tag{15}$$

where the coefficient of proportionality—conventionally represented by the lowercase Greek letter eta, η—is a measure of the strength of the viscous resistance and is called the "coefficient of viscosity" or often just "the viscosity" of the fluid in question.

When at $20\,°\text{C}$ a cylinder of area $50\,\text{cm}^2$ inside a similar cylinder and separated from it by $2\,\text{mm}$ of air is rotated to have a circumferential velocity of $10\,\text{m s}^{-1}$, it feels a drag of $450\,\mu\text{N}$. If water at $20\,°\text{C}$ is poured between the two cylinders, the drag force becomes $25\,\text{mN}$. If you use SAE 10 motor oil at $30\,°\text{C}$, the drag becomes $5\,\text{N}$. All of which seems to show that, as you might expect, gooey things have more viscosity than runny things.

EXERCISES

21. What are the units of the coefficient of viscosity?

In SI units viscosity is measured in pascal-seconds, i.e., Pa s. Although the Pa s is the official SI unit, many tables give viscosities in terms of a historical unit called the "poise." The conversion is $1\,\text{Pa s} = 10\,\text{poise}$.

Equation 15 provides an "operational definition" of viscosity. It defines the viscosity of any fluid in terms of specified operations: Put the fluid into the two-cylinder arrangement; measure the force; measure A; measure z; measure u; use Eq. 15 to calculate η. That's the viscosity.

Therefore, from the experimental data given above you can determine the viscosities of oil, air, and water.

▼ EXAMPLES

7. For SAE 10 motor oil we find that the viscosity is

$$\eta = \frac{F}{A}\frac{z}{u} = \frac{5}{0.005} \times \frac{0.002}{10} = 200 \, \text{mPa s}.$$

■ EXERCISES

22. Use Eq. 15 to find the viscosity of air at $20\,°\text{C}$. Do the same for water.

Values of the viscosity of various gases are given in Table 5.1.

How does viscosity arise? Why is it different for different fluids? How does it depend on density or temperature? A small extension of our model of the atom leads to answers to some of these questions—but only for gases. Keep that in mind: the model and arguments given below apply only to gases.

Because the viscosity of a gas can be understood to result from collisions of hard-sphere atoms or molecules with each other, we need first to develop a model of collisions; next we show how colliding hard spheres might carry away momentum and thus produce viscous drag; finally we connect the experimentally measurable quantity η to the collision behavior and estimate sizes of atoms using experimental measurements of η.

An Atomic Model of Viscosity

Now you can see what collisions of atoms and the mean free path ℓ have to do with the viscosity of a gas. The basic idea is that as a plate, for example, moves with speed u through a gas, the atoms of the gas, colliding with the plate, pick up momentum from it. Then the atoms,

TABLE 5.1   Viscosity in $\mu\text{Pa s}$ of some common gases at $20\,°\text{C}$

Gas	Viscosity	Gas	Viscosity
N_2	17.57	H_2	8.87
O_2	20.18	He	19.61
CO_2	14.66	Ne	31.38
Cl_2	13.27	Ar	22.29

moving with average speeds $\langle v \rangle$ and colliding among themselves, carry this momentum away. The resulting drain of momentum is a viscous force on the plate.

Notice that this explanation involves two different speeds. There is the velocity u with which the gas flows when it is dragged along by the moving plate, and there are the much larger and randomly directed speeds of the gas molecules. These speeds are quite varied, but we represent them by an average value $\langle v \rangle$. You know what magnitudes of $\langle v \rangle$ to expect, because you have seen that for N_2, $v_{rms} = 500\,\mathrm{m\,s^{-1}}$. Think of the molecules right next to the plate as moving randomly with the great speed $\langle v \rangle$ to which a small amount of velocity u has been added to give them a net overall flow with velocity u.

Here is the argument one more time. The viscous force arises from the transfer of momentum from faster-flowing gas to slower-flowing gas. Such a transfer must occur, because, as Fig. 5.3 suggests, when a plate moves with velocity u through a gas, the gas moves with the plate, but the gas at a stationary surface a distance z away is not flowing. Across the gap z, for every change Δz there is a change Δu in the flow velocity.

A molecule moving with a velocity $\langle v \rangle$ away from the surface of the upper plate in Fig. 5.3 goes from a region of flow velocity u to one of $(u - \Delta u)$; similarly, molecules moving toward the upper surface with velocity v move from a region where the flow velocity is $(u - \Delta u)$ to one where it is u. These slower-moving molecules get speeded up by collisions with faster-moving gas molecules, and momentum is transferred from the higher momentum gas to the lower-momentum gas. The rate at which the transfer occurs is governed by the average speed $\langle v \rangle$ of the atoms. This transferred momentum must come from the moving surface, which means that a force is being exerted on the surface, i.e., there is viscous drag.

To find the viscous force on the surface you need to find how much momentum Δp moves away from an area A of the plate in an interval of time Δt. Momentum leaves the surface because molecules with flow velocity u leave the surface and those with flow velocity $(u - \Delta u)$ come to it. On average a number of molecules proportional to $n\langle v \rangle$ will cross a unit area of surface each second.

To connect the viscosity η to n, $\langle v \rangle$, the mass m of a molecule, and its mean free path ℓ, imagine that at each layer half the molecules are moving directly toward the surface and that half are moving directly away. Also picture the flow velocity of the gas as divided into layers each of thickness ℓ. In each layer the flow velocity is Δu less than in the layer above. The idea is that a molecule with molecular speed $\langle v \rangle$ has to move approximately a distance ℓ before it collides and exchanges momentum with

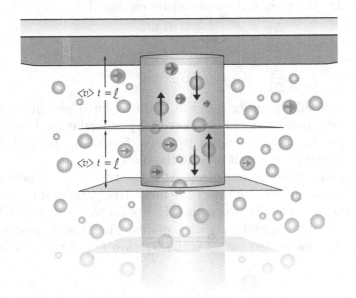

FIGURE 5.4 As a plate moves with speed u through the gas, it drags gas molecules with it at speed u. These molecules move with an atomic speed $\langle v \rangle$ a mean free path distance ℓ both into and out of the next layer a mean free path distance ℓ below. As a result there is a net flow of momentum $m\Delta u$ from the faster moving upper layer to the slower moving lower layer. This momentum transfer shows up as viscous drag.

another molecule. You build this condition into the model by imagining that a molecule moves a distance ℓ from one layer to the next, collides with a molecule in the next layer, and contributes its momentum to that layer. The model is illustrated schematically in Fig. 5.4.

Imagine a volume of gas of area A extending two layers of thickness ℓ down from the moving plate. For simplicity take all molecules in each layer to have the same speed $\langle v \rangle$ and half are moving up and half are moving down. Then in a time $\Delta t = \frac{\ell}{\langle v \rangle}$, $\frac{n}{2} A \ell$ molecules with flow speed u move down and are replaced by $\frac{n}{2} A \ell$ moving up and carrying flow speed $(u - \Delta u)$. The result is that half the molecules in the layer next to the plate have each lost an amount of momentum $m\,\Delta u$, where m is the mass of a molecule. Collisions with the plate will increase the flow speed of these molecules to u, but this means that in the time interval Δt an amount of momentum $\Delta p = \frac{n}{2} A \ell m \Delta u$ will be extracted from the motion of the plate. The loss of this momentum means that the plate experiences a force, i.e. a viscous drag. To keep the plate moving at u an external force must replace the lost momentum. The magnitude of this force, i.e. the rate of the momentum change, is

$$\frac{\Delta p}{\Delta t} = \frac{n}{2} A \ell m \Delta u \frac{\langle v \rangle}{\ell}$$

$$\frac{1}{A} \frac{\Delta p}{\Delta t} = \frac{F}{A} = \frac{n}{2} m \langle v \rangle \Delta u. \tag{16}$$

Although this quantity has the units of force per unit area, it is not pressure. Pressure would be a force perpendicular to the surface, but this force per unit area is tangent to the surface of the plate. It is the viscous force dragging on a unit area of the plate.

In addition to assuming that Δt is the time interval between collisions, i.e., the time it takes to travel one mean free path, assume for convenience that the flow velocity drops linearly from u to 0 across the distance z between the moving and stationary plates. Then the change in flow velocity across one mean free path ℓ is the fraction ℓ/z of the total u,

$$\Delta u = \ell \frac{u}{z}.$$

To connect this result to the experimentally measurable coefficient of viscosity η, rewrite Eq. 16 as

$$\frac{F}{A} = \frac{n}{2} m \langle v \rangle \ell \frac{u}{z}.$$

Now compare this result with Eq. 15, and you see that

$$\eta = \frac{n}{2} m \langle v \rangle \ell = \frac{1}{2} \rho \langle v \rangle \ell, \tag{17}$$

because $nm = \rho$, the mass density of the gas.

Several aspects of our model are unrealistic. The molecules move in all directions not just directly toward or away from the plate; the molecules have a wide range of speeds; and their collisions are more complicated than those of hard spheres. But our mixture of simplifications accidentally leads to a good result. A more complete model and calculation gives a factor of 0.499 (which you can round off to 0.50 in calculations if you wish):

$$\eta = 0.499 \, nm \langle v \rangle \ell,$$

or

$$\eta = 0.499 \, \rho \langle v \rangle \ell. \tag{18}$$

Using Eq. 18 you can find the mean free path ℓ of a gas molecule from measurements of the gas's viscosity. Historically this was an important advance in understanding gases. For air at room temperature you know $\langle v \rangle$

is around $500 \, \mathrm{m\,s^{-1}}$ (see Eq. 7), and its density is $1.29 \, \mathrm{kg\,m^{-3}}$. Therefore, if the viscosity is approximately that of N_2, the mean free path must be about

$$\ell = \frac{18 \times 10^{-6}}{0.499 \times 1.29 \times 500} = 56 \, \mathrm{nm}.$$

■ EXERCISES

23. What is the mean free path of an atom in a chamber evacuated to 1 Torr at room temperature?

24. Ultrahigh vacuum is considered to be around 10^{-11} Torr. What is the mean free path of a molecule in such a vacuum?

A remarkable feature of viscosity appears when you substitute the value of ℓ from Eq. 14 into Eq. 18. You get the dependence of viscosity upon atomic parameters:

$$\eta = \frac{0.499 m \langle v \rangle}{\sqrt{2}\pi (2 r_m)^2}, \tag{19}$$

and you find that the viscosity of our model gas does not depend upon its density. "Do you mean to say if you double or triple the density of air, its viscosity remains the same?" Yes, that's what the equation says. James Clerk Maxwell, one of the greatest physicists ever, was the first person to derive the result. He did not believe it, so he built an apparatus and tested the prediction. Viscosity of a gas is essentially independent of its density as long as the mean free path does not become too large, i.e., comparable to the dimensions of the container, or too small, i.e., of the size of the molecules. Experimental confirmation of this surprising prediction built confidence in the model.

■ EXERCISES

25. What will happen to the viscosity of a gas as you increase the pressure? Justify your answer.

26. Suppose you cool the gas. Will its viscosity increase or decrease or stay the same? [Hint: Equation 19 shows that η is proportional to $\langle v \rangle$. Does $\langle v \rangle$ increase or decrease with temperature?] Do you expect motor oil to become more viscous or less as you cool it? Compare this with the behavior of the viscosity of gases.

For a gas $\langle v \rangle \propto \sqrt{T}$, so you might expect $\eta \propto \sqrt{T}$. This is not the case because the interaction between colliding real molecules is more complicated than for hard spheres. From experiment $\eta \propto T^s$ where for air $s \approx 0.7$.

27. On a hot summer day air temperatures may rise to $40\,°\text{C}$. By what percentage will the viscosity of this air change relative to room temperature?

5.7   THE SIZE OF ATOMS

Now that you have a value $\ell \approx 60\,\text{nm}$ for the mean free path of molecules of air, you can estimate the size of an "air molecule" the way J. J. Loschmidt of the University of Vienna did in 1865. He used Maxwell's value of the mean free path of molecules of air deduced from measurements of viscosity as described above. By "air molecules" Loschmidt and Maxwell meant either N_2 or O_2, because the precision of their numbers did not distinguish between them.

Radius of a Molecule

In the mean free path equation

$$\ell = \frac{1}{\sqrt{2}\pi n (2r_m)^2} \tag{14}$$

the number density n and the atom radius r_m are both unknown. Therefore, knowing the value of ℓ is not enough to find r_m; you need another independent relation between n and r_m. You can find it using a version of an argument by Loschmidt. In a unit volume containing n molecules with radius r_m, the molecules take up a volume of $n\frac{4}{3}\pi r_\text{m}^3$. Loschmidt argued that this is the volume the molecules would occupy if they were condensed into the solid state. However, this assumption leaves no space at all between the molecules, and it is more realistic to assume instead that when the gas is condensed, each sphere occupies a cube $2r_\text{m}$ on a side. This model seems plausible especially for a liquid. The liquid's volume would then be built up out of these cubes as shown in Fig. 5.5. Because each block contains one atom of mass m, the mass density of each block is

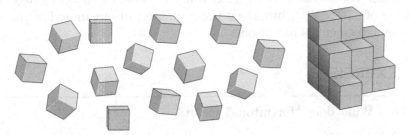

FIGURE 5.5 N blocks each of volume $(2r_m)^3$ add up to a total volume of $N(2r_m)^3$.

$m/(2r_m)^3$, but this is also the mass density of the liquid: $\rho_\ell = \frac{m}{(2r_m)^3}$. On the other hand, the density of the gas is $\rho_g = nm$. The ratio of the two densities ρ_g/ρ_ℓ is a measurable quantity that Loschmidt called ϵ:

$$\epsilon \equiv \frac{\rho_g}{\rho_\ell} = \frac{nm}{\rho_\ell} = n(2r_m)^3. \tag{20}$$

Multiplying Eq. 14 by Eq. 20 eliminates n and gives you

$$r_m = \frac{\pi}{\sqrt{2}} \epsilon \ell. \tag{21}$$

What is ϵ for air? The average density of 78% liquid nitrogen and 21% liquid oxygen is $\rho_\ell = .87 \times 10^3$ kg m^{-3}, and the average density of gaseous air is 1.29 kg m^{-3}. Therefore $\epsilon = 1.48 \times 10^{-3}$ and, from Eq. 21,

$$r_m = \frac{\pi}{\sqrt{2}} \times \left(1.48 \times 10^{-3}\right) \times \left(56 \times 10^{-9}\,\text{m}\right) = 1.8 \times 10^{-10}\,\text{m}.$$

The radius of a molecule of air is thus 0.18 nm. Modern experiments confirm this result and show that the radii of single atoms generally range between 0.1 and 0.2 nm.[4]

Avogadro's Number

Having estimated the atomic radius, you can now go a step further than Loschmidt did and find Avogadro's number N_A. Substitute the value of r_m back into Eq. 14 or into Eq. 20; you will get $n = 2.9 \times 10^{19}$ cm^{-3}. Multiply n by 22 400, the number of cm^3 in a mole of gas at STP, and

---

[4]Or between $\sim$1 and $\sim$2 Å where 1 Å = 10^{-10} m = 0.1 nm. Å stands for Ångstrom, a unit used in the past for lengths at the atomic scale. The Ångstrom was named for a Swedish physicist who made the first precise measurements of atomic and molecular sizes. It is not an officially sanctioned SI unit, but it is still used by physicists too old to change their habits.

you get $N_A = 6.6 \times 10^{23}$. This value is close to the currently accepted value of 6.02×10^{23}, but given the crudeness of our model of the solid the close agreement is fortuitous.

■ EXERCISES

28. What does "fortuitous" mean?

29. At $-260\,°C$ the density of solid hydrogen is $0.0763\,\mathrm{g\,cm^{-3}}$.
(a) What is the radius of a hydrogen molecule estimated by Loschmidt's method?
(b) What is the value of N_A that you obtain for this case?

30. Compare the mean free path of nitrogen at STP with its molecular radius. What is the mean free path measured in molecular radii?

5.8   CONCLUSIONS

This chapter is intended to give you the flavor of physical reasoning. It shows a simple model and some of the arguments and experimental data that established the atom as a physical entity. There are now more complete models, more sophisticated arguments, and more accurate and precise data, but the conclusions remain unchanged:

- Atoms exist.

- Temperature is a measure of atoms' kinetic energy. At room temperature, $T = 293\,\mathrm{K}$, any gas atom has an average kinetic energy of $0.038\,\mathrm{eV}$,

- and it moves with an rms speed of hundreds of meters per second.

- At atmospheric pressure atoms travel tens of nanometers between collisions.

- Atoms have diameters of around 0.2–$0.4\,\mathrm{nm}$.

- At STP a cubic centimeter of gas contains 2.6×10^{19} atoms.

- Over a large range of pressures and temperatures the ideal gas law describes real gases quite well.

- The model of atoms as tiny hard spheres explains a variety of physical properties of real gases. It predicts that viscosity (and diffusion, thermal conductivity, and the speed of sound)
  - Will be independent of gas density,
  - Will increase roughly proportional to the square root of temperature measured on the Kelvin scale, and
  - Will vary as the inverse of the square root of the molecular weight.
  - These predictions are accurate over a range of pressures and temperatures where the mean free path is large compared to the molecule's size and small compared to the dimensions of the container.

The model of hard-sphere atoms is too simple. The physical size of real atoms affects their behavior in ways other than the ones described above, and the assumption that atoms interact only when in contact is approximate, as becomes especially apparent at high pressures. Notice also that we have said very little about the distribution of velocities of atoms. At any instant gas atoms have a range of velocities, and it is possible to predict how many atoms will have which velocities. The deduction and prediction are well supported by experimental evidence.

Toward the end of the nineteenth century it became clear that matter is electrical in nature. This discovery showed that atoms have internal structure; it also led to a precise value for N_A. So that you can understand these developments, the next chapter presents some basic phenomena and ideas of electricity and magnetism.

APPENDIX A: AVERAGES OF ATOMIC SPEEDS

Introduction

The main part of this chapter argued that gas pressure P is the result of the average effect of many collisions of many different atoms with the walls containing the gas. The argument assumed that all the atoms had the same speed; this is extremely implausible. This appendix presents an argument that takes into account the fact that the atoms have a wide range of velocities. This more complete argument also is an opportunity for you to see how to compute averages in complicated situations. The computation of averages will be important later in the book.

TABLE 5.2   Names, ID numbers, and ages of 16 hypothetical students

Name	ID number	Age	Name	ID number	Age
Aimée	1	18	Brian	2	17
Ty	3	19	Kevin	4	19
Mannie	5	18	Doug	6	18
Yoko	7	18	Max	8	18
Sean	9	20	Laura	10	19
James	11	17	Ann	12	21
David	13	20	Jan	14	19
Kate	15	18	Luis	16	18

Sums and the $\sum$ Notation

You often need to add up long strings of numbers. For example, suppose you wanted to know the average age of the students in a laboratory section that has 16 students with the names and ages shown in Table 5.2.

To find the average age $\langle A \rangle$ you just add all the ages and divide by the number of students:

$$\langle A \rangle = \frac{18 + 17 + 19 + 19 + 18 + 18 + 18 + 18 + 20 + 19 + 17 + 21 + \cdots}{16}$$
$$= 18.56 \, \text{y}.$$

This is too tedious to write out. We need a notation that is both compact and general.

We start by giving a symbolic name to the quantity we are averaging. It is "age," so let's call it a. To talk about the age of a particular student we can use the ID number as a subscript, often called an "index." Then Aimee's age is a_1, while Jan's is a_{14} and Sean's is a_9. We can write the average in terms of these symbols as

$$\text{average age} = \frac{a_1 + a_2 + a_3 + a_4 + \cdots + a_{15} + a_{16}}{16}.$$

This form has the advantage of generality; it represents the average of any set of 16 ages. It is, however, just as tedious to write out as the previous form.

The $\sum$ notation provides a compact representation of a sum like the one above. We write

$$\sum_{j=1}^{16} a_j \equiv a_1 + a_2 + a_3 + a_4 + \cdots + a_{15} + a_{16}.$$

The idea is that whenever you see

$$\sum_{j=1}^{N} a_j$$

you know that it represents a sum of N quantities with the names a_1, a_2, $a_3, \ldots, a_{N-2}$, a_{N-1}, a_N.

In this notation the average age of the students of Table 5.2 can be written

$$\text{average age} = \frac{1}{16} \sum_{j=1}^{16} a_j.$$

The notation enables us to say that the average of any N quantities a_1, $a_2, \ldots, a_N$ is

$$\text{average of } a = \frac{1}{N} \sum_{j=1}^{N} a_j. \tag{22}$$

Equation 22 is the general definition of an average.

There are many other useful applications of the $\sum$ notation. Sometimes the quantities being summed may be expressed algebraically in terms of the indices. For example, if you want to write down the sum of all the odd integers from 1 to 17 you can write

$$\text{sum of odd integers} = \sum_{j=1}^{9} (2j - 1) = 81.$$

Distributions and Averages

Because it conveys only limited information, the average does not always provide a useful description of a set of quantities. For example, a class consisting of 8 one-year-olds and 8 thirty-five-year-olds has the same average age as a class of 16 eighteen-year-olds. Knowing the average age would not tell you which of these two classes you would prefer to be in. Knowledge of the "distribution" of a set of quantities gives much more complete information about the set of data.

Let's reorganize the data of Table 5.2 to exhibit its distribution. We group the data by age. Then we tabulate the number of students in each group. Table 5.3 shows the result of this grouping. Because there are five different ages in this particular case, we created five groups or "bins." We have labeled each bin by assigning sequential identification numbers from 1 to 5.

TABLE 5.3   Distribution of ages of 16 hypothetical students

ID index i	Age group	Number of students in age group
1	17	2
2	18	7
3	19	4
4	20	2
5	21	1

Now we can describe the set of numbers by saying that in bin 1 there are 2 occurrences of the number 17; in bin 2 there are 7 occurrences of the number 18; in bin 3 there are 4 occurrences of the number 19; in bin 4 there are two occurrences of the number 20; and in bin 5 one occurrence of the number 21. Such sets of data are called "distributions" because they show how a property is distributed over a set of entities. Our example shows how the property of age is distributed over the set of students in a class.

Implicit in our example is the important idea that a bin has width. None of the students is exactly 17 or 18 or 19, etc. Indeed, we need to say what we mean when we say that someone is 18. Most likely, we mean that person has passed her eighteenth birthday and not yet reached her nineteenth, but life insurance companies usually mean that you are somewhere between 17.5 and 18.5 years old. In either case, we are grouping our data into bins that are 1 year in width. We could use a finer grouping, say in terms of months, or we could use a coarser grouping, e. g., two-year wide bins. Choosing the size of the bin for a distribution is an important decision based largely on common sense about how much information there is in the data set and on your judgment about how much of that information will be useful.

■ EXERCISES

31. In the example above why wouldn't you choose bins one month wide? Why not choose bins two years wide?

It is very helpful to represent a distribution graphically. This is usually done as a vertical bar graph. Each bar represents a bin. The width of

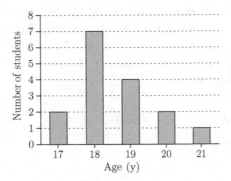

FIGURE 5.6 Distribution of student ages in our hypothetical class.

the bar corresponds to the width of the bin, and the height of the bar represents the number of entities with values in the bin. Such a graphical representation is called a "histogram." You will use them frequently in this book. Figure 5.6 presents a histogram of the distribution of the data in Table 5.3.

■ EXERCISES

32. Find the ages of the students in your class and plot their distribution histogram.

We often want to calculate an average from a distribution. It is much easier to do this from the distribution data of Table 5.3 than from the complete data set in Table 5.2. In long form we can compute

$$\text{average age} = \frac{2 \times 17 + 7 \times 18 + 4 \times 19 + 2 \times 20 + 1 \times 21}{2 + 7 + 4 + 2 + 1} = 18.56\,\text{y}.$$

In the numerator and in the denominator there are as many terms as there are bins. In the numerator each term is the product of the quantity being averaged and the frequency with which it occurs. In the denominator each term is just the frequency of occurrence. We can put this in a compact and general form using the $\sum$ notation if we properly organize the names and labeling of the quantities involved.

You can see by referring to Table 5.3 that you must keep track of three different things when describing a distribution using the summation notation. First, you must know which bin you are talking about, so you need a bin number i. Second, you need the numerical value X_i that corresponds to bin i; this is the age in column 2 of Table 5.3. Third, you need the number of times that each value of X_i occurs; this is designated n_i here and is often called the frequency. In terms of these symbols we can rewrite the average as

$$\langle X \rangle = \frac{\sum\limits_{i=1}^{i_0} n_i X_i}{\sum\limits_{i=1}^{i_0} n_i}, \tag{23}$$

where i is the bin number; i_0 is the total number of bins; and n_i is the number of occurrences of the quantity X_i. Notice that $\sum\limits_{i=1}^{i_0} n_i = N$ will always be the total number of quantities, 16 in the above example.

By convention the angle brackets denote "average value." Other notations used for average value are X_{ave} or $\bar{X}$ (read as "ex-bar"). In this book we usually use the angle brackets.

Equation 23 is the general form for the average of any distribution of quantities. When you see summations that look like this, you know they are averages.

A Distribution of Velocities

Let's apply these ideas to the data in Table 5.4. The table presents a distribution of velocities of some nitrogen molecules. There are really 29 bins in the distribution, numbered from -14 to $+14$. Each bin is $50\,\text{m/s}$ wide. The bins from -14 to -1 are not shown because they are identical to the bins from 1 to 14. The velocity given in the table is the velocity at the midpoint of a bin.

■ EXERCISES

33. Fill in the blanks in Table 5.4.

Momentum Transfers by Collision

Figure 5.1 represents an atom rebounding from a wall. Assume that it has an x-component of velocity v_{xi} and that in a unit volume of the gas there are $n_i/2$ atoms that have this same velocity. For these atoms the

TABLE 5.4   Distribution of x-Component velocities of 1000 N_2 atoms

x-velocity		Number per cm³ in the velocity interval $\delta v = 50\,\text{m/s}$	Bin number i
v_{xi}	(m/s)	n_i	
v_{x0}	0	68	0
v_{x1}	50	67	1
v_{x2}	100	64	2
v_{x3}	150	60	3
v_{x4}	200		
	250	51	5
v_{x6}		41	
	350	34	7
v_{x8}	400		8
v_{x9}	450	21	9
v_{x10}	500	16	10
v_{x11}	550	12	11
v_{x12}	600	9	12
v_{x13}	650	6	13
v_{x14}	700	4	14

amount of momentum transferred to the wall in a time Δt is just

$$\delta p_i = (2m\,v_{xi})(n_i/2)(A\,v_{xi}\,\Delta t) = n_i\,A\,m\,\Delta t\,v_{xi}^2.$$

This is the same result worked out in the text, because the atoms in a single bin are all taken to have the same magnitude of velocity in the x-direction. The difference is that we now have to add together the contributions of atoms that have different velocities, i.e., atoms in other bins. In other words, we propose to find the total change of x-momentum Δp by summing δp_i over all the different velocity bins indicated by i. Formally, this is

$$\Delta p = \sum_{i=1}^{i_0} \delta p_i = mA\Delta t \sum_{i=1}^{i_0} n_i v_{xi}^2,$$

where factors common to every term have been taken outside the sum. $\Delta p/\Delta t$ is the total force F exerted by the wall on the gas. By Newton's third law of motion, the gas must exert an equal but opposite force on the wall. Hence, F is the force on the wall, and F/A is the pressure P of the gas on the wall. We get

$$P = m \sum_{i=1}^{i_0} n_i v_{xi}^2.$$

This looks almost like an average. If it were divided by $\sum_{i=1}^{i_0} n_i$, the sum of the number of occurrences of each velocity, it would be an average. We multiply and divide by that sum, recognizing that the sum of the number of atoms per unit volume with a particular velocity v_{xi} must be the total number of atoms per unit volume, which we have been writing as n, i.e.,

$$P = mn \frac{\sum_{i=1}^{i_0} n_i v_{xi}^2}{\sum_{i=1}^{i_0} n_i}.$$

Referring back to Eq. 23, we see that

$$P = nm \langle v_x^2 \rangle,$$

where $\langle v_x^2 \rangle$ is the average of the squares of the x-components of the atomic speeds.

The argument on page 114 shows that $\langle v^2 \rangle = 3 \langle v_x^2 \rangle$. This result enables us to obtain the final result

$$P = \frac{nm \langle v^2 \rangle}{3}.$$

Velocity Bins

The above calculation assumes that the atoms have been sorted into i_0 bins. We never said how big i_0 is, and in fact, since no two atoms have exactly the same velocity, i_0 could be infinite. Calculus enables us to deal with that problem, but a little fudging does just as well. We define a bin width, a range of speeds, and we count as being in a particular bin all those atoms that have speeds within the specified range. For example, in Table 5.4 the bin width is $\delta v_{xi} = 50\,\text{m/s}$. The table says that in a collection of 1000 atoms in a cubic centimeter, about 64 of them have velocities between 75 and 125 m/s.

■ EXERCISES

34. Calculate the average velocity of these atoms.

35. If there are 1000 atoms in a cubic centimeter, what should be the sum of n_i in Table 5.4? What is it? Explain your answer.

36. What is the average of the square of these velocities? Calculate it directly from Table 5.4 (you might use a spreadsheet), and calculate it assuming that the gas is at room temperature. How do the two answers compare?

37. What will be the total change in the average momentum of all of these atoms that strike a wall in a time interval of 1 s?

38. For Table 5.4 evaluate $\displaystyle\sum_{i=-14}^{14} n_i$.

39. For Table 5.4 evaluate $\displaystyle\sum_{i=-14}^{14} n_i\, v_{xi}$

40. For Table 5.4 evaluate $\displaystyle\sum_{i=-14}^{14} n_i\, m\, v_{xi}^2$.

41. Make a graph of n_i vs. v_{xi}. How would the values of n_i change if you used $\delta v_{xi} = 25\,\text{m/s}$ instead of $50\,\text{m/s}$? How would the graph change under these circumstances? How many velocity groupings would you need now? Can you see that if instead of plotting n_i you plotted $n_i/\delta v_{xi}$, you would get graphs that were nearly the same in the two cases? And can you see that the smaller you made δv_{xi}, the smoother the curve would become?

The limiting ratio of the number in a velocity bin to the width of the bin is called the distribution function. In the limit of vanishingly small δv_{xi} the ratio is a continuous function of v and is often written $n(v)$, although it is really just the derivative of the occurrence frequency with respect to the velocity. Every distribution function is a derivative. The notation $n(v)$ is also confusingly close to the n_i in Table 5.4, but it is different. For one thing, the units of $n(v)$ are number per m^3 per m/s (which is $\text{m}^{-4}\,\text{s}$); the units of n_i are just m^{-3}.

PROBLEMS

1. What is meant by the "mean free path" of a molecule?

2. How is viscosity η defined? What is a typical value of the viscosity of a gas?

3. a. Derive a relation between number density n, average speed $\langle v \rangle$, and mean free path ℓ. Take $\langle v \rangle \approx v_{rms}$.

    b. What is the mean free path of a nitrogen molecule in air at room temperature?

4. In a lab experiment, N_2 molecules escape into a vacuum through a hole 0.051 cm in diameter. At what pressure will the mean free path of nitrogen molecules become comparable to the size of the hole?

5. Describe Loschmidt's method for estimating the radius of an "air molecule." Show how to use his results to estimate Avogadro's number, N_A.

6. Given that solid CO_2 at $-79\,°C$ has a density of $1.53\,g\,cm^{-3}$ and that CO_2 vapor at STP has a viscosity $\eta = 13.8\,\mu Pa\,s$, estimate Avogadro's number. How might you modify Eq. 20 to better represent the condensation coefficient of solid CO_2?

7. Given that $N_A = 6.02 \times 10^{23}$, estimate the size of an atom.

8. Suppose N_2 molecules are known to have an rms velocity of $500\,ms^{-1}$. What would be the rms velocity of hydrogen molecules at the same temperature and pressure? Explain your reasoning.

9. We found that the viscosity of an ideal gas is $\eta = 0.499\rho\langle v\rangle\ell$, where ρ is the density of the gas and the mean free path

$$\ell = \frac{1}{\sqrt{2}\pi(2r_m)^2 n},$$

where n is the number per unit volume of molecules or atoms of radius r_m. At STP, nitrogen gas in a $10\,m^3$ container has a density of $1.25\,kg\,m^{-3}$, a viscosity of $17\,\mu Pa\,s$, and a mean free path of 60 nm. Suppose that without adding gas the volume of the container is doubled while keeping the temperature T constant. What then is

   a. the density?

   b. the rms velocity?

   c. the mean free path?

   d. the viscosity?

10. Assume you can prove that for our simple atomic model of an ideal gas the pressure $P = \frac{1}{3}\rho\langle v^2\rangle$, where ρ is the density of the gas and $\langle v^2\rangle$ is the average of the square of the velocity of the gas molecules.

   a. Use the ideal gas law to derive a relation between the temperature of this gas (measured in kelvins) and the average kinetic energy of a gas molecule.

   b. If the gas is helium ($M = 4\,\mathrm{u}$), what is the value of the average kinetic energy of a helium atom at room temperature?

   c. If the gas consisted of oxygen molecules ($M = 32$), what would be the average kinetic energy of a molecule at room temperature? Explain your answer.

11. Consider the following table of measurements of pressure vs. volume for air in a closed volume at room temperature,

Pressure	Volume
(kPa)	(cm^3)
90.9	24.89
73.6	30.73
59.5	37.93

   a. Show that these data obey Boyle's law.

   b. How many molecules are there in this volume? Show how you got your answer.

   c. What is the rms velocity of the oxygen molecules in this sample of air? Explain how you got your answer.

   d. If the temperature is increased by 15%, by how much does the rms velocity of the oxygen molecules change? Explain.

12. A quantity of oxygen gas is contained in a vessel of volume $V = 1\,\mathrm{m}^3$ at a temperature of $T = 300\,\mathrm{K}$ and a pressure of P. The vessel is connected

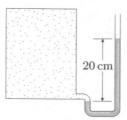

FIGURE 5.7 Apparatus for Problem 12.

to a mercury-filled tube as shown in Fig. 5.7. Note that the upper end of the tube is *open* to the atmosphere.

  a. Is P greater or lesser than 1 atm? Calculate P in units of *Torr* and also *pascals*.

  b. If the temperature of the gas is doubled, keeping V constant, by what factor does each of the following change?

  i. density (g/cm^3)

  ii. average kinetic energy of a molecule

  iii. rate of molecular collision with the walls

  iv. v_{rms}

  v. mass of 1 mole of gas

  c. By means of a small pump the gas pressure is reduced to 100 Torr, while the temperature and volume remain fixed at $300\,K$ and $1\,m^3$. What then is the average kinetic energy of 1 molecule of the gas? Express your final answer in eV.

13. A box of H_2 gas is at STP ($0\,°C$ and 101.3 kPa).
  a. What is the pressure of the gas in

  i. Torr

  ii. pascals

  iii. atmospheres

  b. What is the root-mean-square velocity of the H_2 molecules?

  c. The temperature of the gas of H_2 is tripled and the number of molecules is halved. All else remains the same.

  i. What is the new pressure in the box?

  ii. How does each molecule's average kinetic energy change from its value at STP?

14.a. If 1 liter of oxygen gas at $0\,°C$ and 101 kPa pressure contains 2.7×10^{22} molecules, how many molecules are in 1 liter of hydrogen gas at the same temperature and pressure? How do you know?

b. If at room temperature the average kinetic energy of an oxygen molecule ($M_{O_2} = 32$) is 0.04 eV, what is the average kinetic energy of a hydrogen molecule at the same temperature? Why?

c. When hydrogen and oxygen gas combine to form water, about 2.5 eV of energy is released as each water molecule forms. Assume that the hydrogen and oxygen combine inside a closed tank and that 10% of the released energy goes into kinetic energy of the molecules. Estimate the rise in temperature of the water vapor.

15. Suppose you have a small cube with "3" painted on 3 sides, "4" painted on 2 sides, and "1" painted on the sixth side.

a. What is the frequency distribution of the numbers?

b. What is the average value of the **squares** of the numbers?

16. What is the average total kinetic energy of one mole of oxygen molecules at 27 °C?

17. A beach ball with a radius of 0.5 m is rolling along the sidewalk with a speed of $2 \, \mathrm{m \, s^{-1}}$. There is a light mist consisting of 3.1 small droplets of water in every cubic meter of air. Estimate to within a factor of 2 the mean distance the ball travels between collisions with droplets of the mist.

18. If the mean free path of a nitrogen molecule in a bell jar is 100 nm at atmospheric pressure, what is its mean free path at 0.76 Torr? Show how you get your answer.

19. If in the previous question you raise the temperature of the gas from 300 K to 600 K,

a. By how much does the collision time change?

b. Explain why the mean free path does not change when the temperature goes up.

20. A 3 L vessel ($1 \, \mathrm{L} = 10^{-3} \, \mathrm{m^3}$) contains 16 g of O_2 gas at a pressure of 4 atm.

a. How many moles of O_2 are contained in the vessel?

b. What is the temperature of the gas?

c. How much energy is needed to raise the temperature of the gas by 100 K?

21.   Nitrogen gas (N_2, M $=$ 28 u) at room temperature (300 K) and atmospheric pressure has a density of $1.25\,\mathrm{kg\,m^{-3}}$. The measured root mean square velocity of a N_2 molecule is $v_{rms} \approx 500\,\mathrm{m/s}$.

    a. What is the average energy of a N_2 molecule under these conditions? Express your answer in J as well as eV.

    b. At the same temperature and pressure, what is the average energy of a helium atom ($M = 4.0\,\mathrm{u}$)?

    c. Under these conditions, what is v_{rms} for a helium atom?

    d. Figure 5.8 shows a plot of the logarithm of pressure ($\ln P$) in a jar vs. time for an experiment in which gas leaks out of the jar through a tiny hole into a vacuum. The heavy solid line shown is for N_2. Which of the two dotted lines (A or B) best approximates the behavior of helium under the same experimental conditions? Explain briefly.

22.   We showed that the pressure of a gas is related to its density ρ (mass/volume) of the gas and the average of the square of the velocity of the gas molecules:

$$P = \frac{1}{3}\rho\langle v^2\rangle.$$

    a. Starting from the above equation, prove that the total kinetic energy of a gas is

$$KE_{tot} = \frac{3}{2}PV.$$

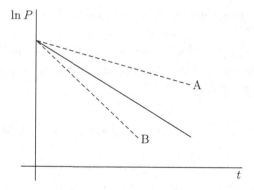

FIGURE 5.8 Logarithm of pressure vs. time for gases leaking into a vacuum (Problem 21).

b. Starting from the last equation, prove that the average kinetic energy of a single molecule is

$$KE(\text{one molecule}) = \frac{3}{2}k_{\text{B}}T,$$

where k_{B} is Boltzmann's constant.

c. If the gas is H_2, what is the value of the average kinetic energy of a molecule at room temperature?

d. Find the ratio between the rms velocities (v_{rms}) for O_2 ($M = 32$) and H_2 molecules at the same temperature.

23. Present-day x-ray analysis of silicon (the element from which modern electronic devices are constructed) shows that atoms in the solid state of Si are arranged as shown in Fig. 5.9. The cubic box shown contains 8 Si atoms, and its side length is 543.1 pm. (1 pm = 10^{-12} m)

a. What is the volume of the box in m^3?

b. The measured density of Si is 2330 kg/m^3. Calculate the mass of a single Si atom.

c. The atomic weight of Si is 28.086 g (=? in kg). From this and your answers to (a) and (b), find Avogadro's number N_A. (This is presently the most accurate way known to find N_A.)

24. A spherical balloon with negligible elastic force is filled at room temperature (20 °C with enough helium gas that it just floats in the surrounding air at standard atmospheric pressure. If the balloon is 2 m in diameter,

a. What is the pressure of the gas in the balloon?

b. What is the mass of the gas it contains?

c. What is the density of the gas in the balloon?

d. What is the density of the air outside, taking it to be 20% O_2 and the rest N_2?

e. What is the mass of the balloon without the gas?

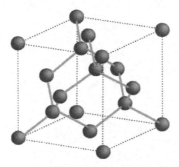

FIGURE 5.9   Arrangement of Si atoms in a crystal of silicon referred to in Problem 23.

C H A P T E R

6.

Electric Charges and Electric Forces

6.1 INTRODUCTION

By imagining a gas to be a collection of tiny spheres, physicists and chemists were able to explain many features of the behavior of gases and estimate the number and size of molecules. Their results gave further credibility to the idea that atoms exist, but their numbers were imprecise, yielding estimates of Avogadro's number and atomic sizes accurate only to within an order of magnitude, i.e., only to within a factor of ten.

Greater precision came as physicists better understood electricity and magnetism. Along with increasing precision came a growing understanding of the inner structure of the atom. During the two decades bracketing the start of the twentieth century physicists discovered the electrical nature of the atom, measured the electron's minute electric charge to better than 1%, and began to understand its behavior inside the atom.

Before you can understand these advances, you need to review some electric and magnetic phenomena and the rudiments of how they are described and understood. In particular you need working ideas of

- electric charge,

- electric field and electric force,

- electric potential,

- electric current,

- magnetic field and magnetic force.

These are the topics of the next three chapters.

C.H. Holbrow et al., *Modern Introductory Physics, Second Edition*,
DOI 10.1007/978-0-387-79080-0_6, © Springer Science+Business Media, LLC 1999, 2010

6.2   ELECTRIC CHARGE

Unlike the ideas of mass, length, and time, which are ancient, the concept of electric charge is modern. It was developed during the eighteenth and early nineteenth centuries. The idea of electric charge is needed to explain the curious but simple experiments described below.[1]

Experiments with Electroscopes

First take a thin, flexible strip of gold or aluminum foil and hang it over a metal hook as shown in Fig. 6.1. The foil must be metallic, light in weight, and so flexible and free to move that it must be kept inside a glass case to protect it from air currents. (Foils of gold leaf can be made so thin that they transmit light.) Then the hook must be connected by a metal rod to a knob on top of the glass case. This device is called an "electroscope."

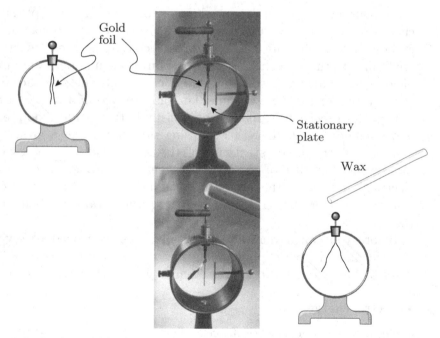

FIGURE 6.1   (a) A discharged electroscope; (b) a charged electroscope.

---

[1]Albert Einstein and Leopold Infeld in their fine book *The Evolution of Physics*, Simon & Schuster, NY 1938, point out that these experiments exhibit especially well the interconnection of theory and experiment. No one would dream of doing these experiments deliberately unless one already had a reasonably clear idea of the theory that is to be used to interpret them. Only then does the experimental plan make any sense.

Next take a rod made of sealing wax[2] or hard rubber and wipe it firmly with cat fur. Now hold the rod quite near the electroscope but without touching it. The two hanging leaves of foil will spread apart. They repel each other, as shown in Fig. 6.1, and we say that the leaves of the electroscope have become "charged."[3]

If you move the rod away from the electroscope, the two leaves fall back together. The leaves become "discharged."

If you touch the sealing wax to the metal knob connected to the electroscope's leaves, something new happens. Now when you move the sealing wax away, the leaves do not come entirely back together. Some mutual repulsion remains.

If you repeat the experiment using a glass rod that has been rubbed with silk, exactly the same sequence of events is seen. The leaves spread when the rod comes near; they fall back together when the rod is moved away. If the electroscope knob is touched with the rod, the leaves remain apart when the rod is moved away. This sequence of events is shown schematically in Fig. 6.2.

Now here is something new and curious: Prepare the glass rod (by rubbing it with silk) and the sealing wax (by rubbing it with fur). First charge the electroscope by touching it with the glass rod and then bring the sealing wax near. As the sealing wax approaches, you'll see the

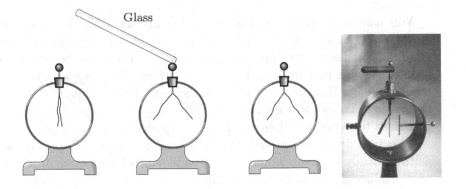

FIGURE 6.2 An electroscope charged by a glass rod rubbed with silk.

---

[2]If you don't seal your letters as eighteenth century correspondents did, you may not know that sealing wax is a mixture of gum-lac, melted and incorporated with resins, and then colored with some pigment—red for business correspondence, black for mourning.

[3]Originally, the word "charged" was used simply in the sense that the electroscope was being loaded with something, as we say a gun is charged and refer to its load as a "charge."

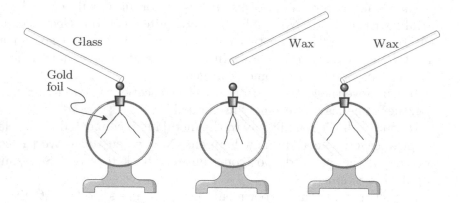

FIGURE 6.3 Sequence of events as rubbed sealing wax is brought nearer and nearer to an electroscope charged by a rubbed glass rod.

leaves of the electroscope fall back together. The electroscope appears to discharge. If the sealing wax has been rubbed vigorously, you may see the electroscope leaves discharge as the wax is brought near and recharge as the wax comes closer yet (Fig. 6.3).

Apparently the glass rod and the sealing wax charge the electroscope differently. Even though individually each causes the leaves to separate when their chargings are combined, they somehow cancel each other out.

You can make sense of these experiments with the following model. Assume there exist two "fluids"—"electric fluids" you might call them. In combination they exert no effect, and we say they are "neutral." When separated, the two fluids attract each other. They appear to strive to combine, each fluid tending to draw enough of the other to achieve neutralization. To cause electrical effects you need to separate the two fluids; you need to overcome their tendency to neutralize.

The two fluids are called "positive" and "negative" partly because arithmetically a given amount of + will cancel an equal amount of − and produce a null value corresponding to neutrality. The + and − kinds of electric fluid attract each other, but + repels + and − repels −. Unlike kinds attract; like kinds repel.

With this model you can interpret the above experiments as follows. Rubbing the sealing wax with cat fur removes some of one of the "fluids" and leaves an unneutralized remnant of the other. Similarly, the glass rod rubbed with silk gets left with an unneutralized amount of one of the fluids. These experiments with the electroscope show that the sealing wax and the glass rod have been given excesses of unlike fluids.

Let's stop calling them "electric fluids." That terminology was acceptable in the eighteenth century,[4] but nowadays we call them "electric charge."

Using this terminology we say that positive and negative charges attract each other, but positive charges repel positive charges, and negative charges repel negative charges. In other words, *unlike charges attract; like charges repel.*

Which charge is on the glass rod? Which is on the sealing wax? The answer depends on how we choose to name them, and that choice is a matter of convention.

The universally accepted definitions of the names are as follows.

positive charge: any charge repelled by a glass rod that has been rubbed with silk (so there must be some unneutralized amount of + charge on the glass rod);

negative charge: any charge repelled by sealing wax that has been rubbed with cat fur (so there must be some unneutralized amount of − charge on the sealing wax).

Thus, the proper designation of + and − on batteries or electrical power supplies throughout all the world's great electrical industries depends ultimately on cat fur.

Now you can explain the behavior of the electroscope when the sealing wax comes near as in Fig. 6.4. This is the same drawing as Fig. 6.1, except that the distribution of charges is shown. The sealing wax with its negative charge repels negative charge on the electroscope. The charge moves down onto the leaves of the electroscope; these now repel each other, and they spread apart. When you move the sealing wax away, the negative charge moves back to the portions of the electroscope from which it had been displaced; the leaves become neutral again and collapse together.

If the sealing wax touches the electroscope, some of the negative charge on the sealing wax transfers to the electroscope, leaving it with a net negative charge. Now when you move the sealing wax away, the electroscope cannot become neutral, and the leaves remain separated.

A similar description applies to the glass rod and the electroscope, except that now the rod's charge is positive.

---

[4]When the American physicist Benjamin Franklin worked out these ideas.

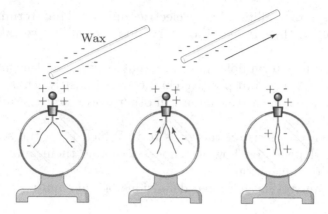

FIGURE 6.4 Negative charge on wax rod induces negative charge on electroscope leaves. As the rod is withdrawn, the charges redistribute themselves, and the leaves of foil collapse.

Notice that to explain the behavior of the electroscope you do not need to assume that the negative charges move onto the electroscope. The explanation works just as well when you assume that positive charges move off from the electroscope. Electroscope experiments can not distinguish between negative charge flowing onto the electroscope and positive charge flowing off. From other kinds of experiments we now know that in metals negative charges move freely while positive charges remain fixed. In liquids both positive and negative charges move.

■ EXERCISES

1. Assuming that the negative charges can move freely and that positive charges cannot, explain with both a sketch and words what happens when a glass rod that has been rubbed with silk is brought near an electroscope and then touches it.

2. Imagine that it is the positive charges, not the negative charges, that are free to move. In that case, re-write the explanation of the behavior of an electroscope when sealing wax is brought near. Include a sketch similar to Fig. 6.4.

3. If you understand this model of charge and its properties, you should be able to explain why an electroscope charged by a glass rod is first discharged and then charged as a piece of sealing wax is brought closer and closer. (See Fig. 6.3.) Explain it.

Conductors and Insulators

The electroscope's metal rod and leaves are particular examples of substances called electrical "conductors." An electrical conductor (or, more often, simply a conductor) is any substance on or in which electric charge moves easily. The outstanding example of conductors are the metals, but salt water and other solutions are also conductors of electricity.

The wax and the glass illustrate the opposite property. They are called "insulators," and charge does not move easily on or in them. You can see why the foils of the electroscope must be conductors if the device is to work well. Why must the wax and the glass rod be insulators? One reason is that if they were not insulators, the charge that you managed to rub onto them would immediately flow along them into your body and on out to the ground. The human body, as a bag of salt water, is a pretty good conductor.

The existence of materials with these two quite different properties is basic to all of electrical technology. Wires are possible only because they can be made of conducting material such as copper or aluminum. But control of the flow of electricity along a wire is possible only because you can wrap the wires in a material, such as rubber, fiberglass, or plastic, that does not conduct electricity.

Quantitative Measures of Charge

As always in physics, we want to make this new concept quantitative. To do this we arrange pieces of charged matter so that their attractive or repulsive forces act to produce some observable mechanical effect such as motion, compression of a spring, deflection of a lever, or twisting of a fiber. The amount of the mechanical effect can then be used as a measure of the amount of charge present. One early method used the angle of separation between the leaves of the electroscope as a measure of the amount of charge on them.

The French engineer and physicist Charles Augustin Coulomb[5] developed a more precise measure. He built a device called a "torsion balance," shown in Fig. 6.5, that used the amount of torsion (twist) induced in a very fine wire or fiber to measure forces smaller than 10^{-8} N.

To make his torsion balance, Coulomb clamped a very fine, delicate torsion wire to a rotatable cap that he placed at the top of a narrow glass cylinder. He let the wire hang down through the narrow cylinder and into a short, large-diameter cylinder about 30 cm across and 30 cm

---

[5]Born in Angoulême, France on 14 June 1736; died in Paris on 23 August 1806. Had two sons, one born in 1790, the other in 1797. He married in 1802.

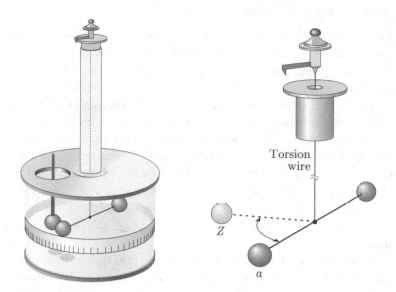

FIGURE 6.5 Coulomb's torsion balance.

high. He attached the midpoint of a lightweight, thin insulating rod, about 20 cm long, to the lower end of the fine wire. The cylindrical containers shielded the sensitive apparatus from air currents.

To each end of the rod he attached a small conducting ball, forming the dumbbell shown in Fig. 6.5. Another ball (ball Z in Fig. 6.5) could be inserted through a hole in the cover of the larger cylinder. With both balls uncharged, he rotated the cap to bring ball a almost into contact with ball Z; then he removed Z, charged it, and put it back next to a. When ball a lightly touched the charged ball Z, a picked up some of Z's charge and was then repulsed. The force exerted by ball Z on ball a caused the dumbbell to rotate and twist the fine wire. If you and a friend have ever wrung out a wet towel by twisting each end in opposite directions, you know that the more you twist the towel, the more it resists. In the same way, Coulomb's wire twisted until it just balanced the effect of the electric repulsion between the two charged balls. The amount of twist in the wire was a measure of the amount of force between balls a and Z.

By rotating the cap to which the wire was fastened at the top of the column, Coulomb could twist the wire more and increase the amount of force it exerted on the dumbbell and move ball a closer to Z. By twisting the cap enough, he could choose how close a was to Z.

Just as for wringing out a towel, the total twist in the torsion wire is the amount of twist from turning the cap at one end plus the amount of twist in the opposite direction produced at the other end when Z pushed on a and rotated the dumbbell through some angle. In one experiment, the angle was 36°, so the motion of the dumbbell twisted the torsion wire through 36°.

To measure how the electric force depended on the distance between the charged balls, he rotated the cap to bring the balls closer together. To move ball a to a position 18° from ball Z, he had to turn the cap and twist the wire by an additional 126°, making a total twist of 144°. He then observed that by turning the cap through 567° the balls moved to within 8.5° of each other, making a total twist in the fiber of 575.5°. (See Problem 5.)

■ EXERCISES

> **4.** Show that these results imply that the force between the two balls depends inversely on the square of the distance between them.

With this apparatus he showed that to within a few percent the magnitude of the force between charges was inversely proportional to the square of the distance between them:

$$F \propto \frac{1}{r^2},$$

where r is the distance between the centers of the two charged balls. His result is strictly true if the distribution of charge on or in the ball is spherically symmetric, and it is an excellent approximation for charged objects that are small compared to their separation—so-called "point charges."

Coulomb also found that the force is proportional to the product of the amount of the two interacting charges.[6] If we call those charges q_1 and q_2, then we have

$$F \propto q_1 \, q_2.$$

Combining these two equations and letting k_c be some constant of proportionality, we have what is known as Coulomb's law for the force between two spherical charges:

$$F = k_c \, \frac{q_1 \, q_2}{r^2}. \tag{1}$$

You can use this equation to define a unit of charge. By convention we agree that two identical charges separated by 1 m and exerting a force

---

[6]These are rather casual sounding statements. You can get a deeper insight into Coulomb's accomplishments if you ponder how he might have been able to arrive at such generalizations without having ways of measuring amounts of charge.

of 8.98755×10^9 N on each other each have a charge of 1 coulomb. This statement is a definition of the "coulomb" as a unit of charge. You might object that this is a strange way to do things. Why not just let $k_c = 1$? In fact that was once the way it was done. However, some awkwardness in combining electric and magnetic units led to the creation of the currently used definition.[7] The abbreviation of the coulomb is C. All the SI multiples can be used, but the small ones like pC, nC, and μC are especially common.

Adopting this definition for our unit of charge means that we have agreed to assign to k_c, the constant in Eq. 1, the value 8.987552×10^9 N m^2 C^{-2}. In this book, as in much practical work, you can round off k_c and use the value $k_c = 9 \times 10^9$ N m^2 C^{-2}.

For various reasons the constant k_c is often written as $k_c = 1/(4\pi\epsilon_0)$, where ϵ_0 has the mystifying name of "permittivity of free space" and has the precise value of $8.854187818 \times 10^{-12}$ C^2 N^{-1} m^{-2}. We will be using k_c in this book, but in many places you may see the other form, $1/(4\pi\epsilon_0)$.

■ EXERCISES

5. What is the force between a charge of $1.1\,\mu$C and one of $1.3\,\mu$C when they are separated by 1 cm? By how much does the force change if they are moved to a separation of $1\,\mu$m?

6. Two point charges of equal magnitude are placed 2 m apart. Each attracts the other with a force of $9\,\mu$N. What is the amount of each charge? What is the sign of each charge?

7. Suppose the two charges of the previous problem are moved until they are 1 m apart. What now is the force between them?

6.3   ELECTRIC CURRENT

A stream of moving charges is called "electric current." Imagine a surface through which electric charge is flowing, e. g., the cross section of a wire. Electric current is defined as the amount of charge flowing through

---

[7]The official SI definition of the coulomb is actually cast in terms of an experimental method that is capable of much more precision than is possible by an experiment derived from Coulomb's law, but the results are the same. The official coulomb is defined as the amount of charge corresponding to the passage of 1 ampere of current for 1 s. The ampere of current is defined in terms of the force between two wires carrying that amount of electric current. We don't need to go into those details now.

that surface in 1 s. Electric current is, therefore, $\Delta q/\Delta t$, and its units are coulombs per second (C s^{-1}).

There is a special SI name for this group of units; it is called the "ampere" and is abbreviated "A." All the standard SI multiples are used: e. g., pA, nA, μA, mA, A, kA, MA.

▼ EXAMPLES

1. If $1\,\mu$C passes through the cross section of a wire every second, the current I is $1\,\mu\text{C s}^{-1}$ or, equivalently, $1\,\mu$A.

If $1\,\mu$C passes through the cross section of a wire every 0.1 s, then $I = 10\,\mu$A.

For historical reasons the direction of electric current is taken to be the direction positive charges would move to produce the observed transport of electricity, even though quite often the transport is caused by the motion of negative charges in the opposite direction.

▮ EXERCISES

8. Two carbon rods are placed in a solution of H_2SO_4. A current of 12 A is passed between the two rods for 2 h 14 min. This results in the passage of an amount of charge that releases 1.008 g of H_2. How much charge is transported in this time?

Speed of Charges in a Current

If electric current involves the motion of electric charges, how fast do they move? Let's consider electric current flowing in copper, a metal easily drawn out into wire. It has an atomic mass of 63.55 u and a density of $8.9\,\text{g cm}^{-3}$. Copper wire is sold in standard sizes specified by a number called its gauge. In the U.S. we use American Standard Gauge. As the gauge number gets larger, the wire gets thinner. For example, #18 copper wire has a diameter of 1.024 mm; #20 has a diameter of 0.812 mm; and #22 has a diameter of 0.644 mm.

▽ EXAMPLES

2. Show that a 1 m length of #18 wire has a mass of about 7 g.

This wire is a cylinder. To find the mass of a 1 m long cylinder of #18 wire, first find its volume and then multiply by the density of copper. The volume of a cylinder is its length L times its cross-sectional area A.

The cross-sectional area A can be found from the diameter $D = 0.1024$ cm using the fact that $A = \frac{\pi}{4}D^2$. This gives

$$A = \frac{\pi}{4} \times (0.1024\,\mathrm{cm})^2 = 8.23 \times 10^{-3}\,\mathrm{cm}^2,$$

and the volume V of a 1 m length is then

$$V = (8.23 \times 10^{-3}\,\mathrm{cm}^2) \times (100\,\mathrm{cm}) = 0.823\,\mathrm{cm}^3.$$

The mass m_1 of a 1-meter length of wire is then its volume V times the density of copper. This gives $m_1 = 0.823\,\mathrm{cm}^3 \times 8.9\,\mathrm{g\,cm}^{-3} = 7.33\,\mathrm{g}$.

Metallic copper is a good conductor of electricity because some of its electrons, approximately one from each atom, move freely and swiftly ($\sim 10^6\,\mathrm{m\,s}^{-1}$), like particles of a gas, among the metal's copper atoms. This goes on whether the metal is a chunk of copper or whether it is drawn out into a wire.

If one end the wire is connected to the $+$ end of a battery and the other end to the $-$ end, then in addition to their rapid, random motion, the electrons acquire a small average velocity $\langle v \rangle$ that carries them along the wire toward the battery's positive terminal. The motion of the negatively charged electrons means that an electric current flows. It is a convention that electric current flows in the direction that positive charge would move, i.e., from the $+$ end of a battery to its $-$ end; therefore, the electric current flows in the direction opposite to the motion of the electrons.

▽ EXAMPLES

3. If, as is the case, each atom of copper contributes one electron to the flow of charge along a wire, how fast, on the average, do electrons move along a piece of #22 wire (a size often used in electric circuits) when the electric current is 1 A?

To answer this question you need to know the cross-sectional area of the wire and how many electrons pass across that area each second. This is essentially the same problem as finding how many molecules are in a volume swept out by a disk of area A in 1 s (p. 123) or how many molecules hit an area A of a container's wall in 1 s.

If a container holds n particles per unit volume and each of them is moving toward A with a speed $\langle v \rangle$, then in a time interval Δt the number ΔN that strike the area A is the number of particles in a cylinder of base area A and length $\langle v \rangle \Delta t$. You can write this as

$$\Delta N = A\, n\, \langle v \rangle \Delta t,$$

so the rate at which the molecules strike A is

$$\frac{\Delta N}{\Delta t} = n \langle v \rangle A. \tag{2}$$

As already noted, electric current is the rate at which electric charge passes through an imaginary surface that cuts across the wire. If ΔN particles pass through the surface in a time interval Δt, and if each of these particles has a charge e, then the amount of charge passing in time Δt is $\Delta Q = e\, \Delta N$, and the electric current is

$$I = \frac{\Delta Q}{\Delta t} = e \frac{\Delta N}{\Delta t} = e\, n \langle v \rangle A.$$

Solving for $\langle v \rangle$ you get

$$\langle v \rangle = \frac{I}{e\, n\, A}. \tag{3}$$

The current I is just 1 A. What is the value of the number density n; what is the value of the area A? The diameter of #22 wire is 0.0644 cm, so the area A is $\pi/4 \times 0.0644^2 = 3.26 \times 10^{-3} \text{ cm}^2$.

To find n use the facts that a mole of copper has a mass of 63.55 g and the density of copper is $8.9\,\mathrm{g\,cm^{-3}}$. These facts tell you that $1\,\mathrm{cm}^3$ of copper contains $8.9/63.55$ mol of Cu atoms, from which it follows that in copper the number density of the electrons that are free to move around and cause an electric current is

$$n = \frac{8.9}{63.55} \ \mathrm{mol\,cm^{-3}} \times 6.02 \times 10^{23}\ \mathrm{mol^{-1}} = 8.43 \times 10^{22}\ \mathrm{cm^{-3}}.$$

Now you can compute the value of the average velocity of the electrons carrying 1 ampere of current along a #22 wire.

$$\langle v \rangle = \frac{I}{e\, n\, A}$$

$$= \frac{1\,\mathrm{A}}{(1.60 \times 10^{-19}\ \mathrm{C})(8.43 \times 10^{22}\ \mathrm{cm^{-3}})(3.26 \times 10^{-3}\ \mathrm{cm^2})}$$

$$= 0.0227\ \mathrm{cm\,s^{-1}}.$$

This is a remarkable result. The electrons are moving $\sim 10^6 \, \mathrm{m\,s^{-1}}$ in random directions, but the net flow of electric current arises from the average drift of the electrons along the wire at a speed of only a fraction of a millimeter per second!

How long will it take them to travel a distance of 1 cm along a piece of #22 wire? How does your answer compare to the amount of time that it takes a ceiling light to come on after you flip the wall switch?

6.4   SUMMARY: ELECTRIC CHARGES

There is a physical entity called electric charge. There are two kinds of charge that by convention are labeled $+$ and $-$. Like kinds of charge repel; unlike kinds of charge attract. Small point-like charges q_1 and q_2 exert a force on each other proportional to the magnitude of each charge and inversely proportional to the square of the distance r between them:

$$F = k_c \, \frac{q_1 q_2}{r^2}. \hspace{2cm} \text{Coulomb's Law (Eq. 1)}$$

Choosing the constant k_c in the above equation to be $9 \times 10^9 \, \mathrm{N\,C^{-2}\,m^2}$ defines the unit of charge called the "coulomb." In other words, if two identical point charges separated by 1 meter exert a force of $9 \times 10^9 \, \mathrm{N}$ on each other, they each carry a charge of 1 coulomb, i.e., $1 \, \mathrm{C}$.

Moving streams of charge are called electric currents. The direction of flow of electric current is always in the direction that positive charge would move even if the actual current is negative charge flowing in the opposite direction. The magnitude of electric current is the rate $\Delta Q / \Delta t$ at which charges cross a surface that you imagine to be in their path. The SI unit of electric current is the "ampere." It is the current of $1 \, \mathrm{C\,s^{-1}}$ crossing a surface. The abbreviation for ampere is "A."

PROBLEMS

1. The force of gravity due to a spherical mass falls off proportionally to $1/r^2$ just like the electrical force due to a charge of $5 \, \mathrm{C}$. What would be the value of g (acceleration due to gravity) one Earth radius above Earth's surface?

2. The force between two charges 2 m apart is measured to be 3 N. What will be the force between the charges if they are moved to be

    a. $\frac{2}{3}$ m apart?

    b. 6 m apart?

3. Coulomb calibrated his torsion fiber and found that it required a force of 1.53 μN to twist it through 360°. Using the data given in problem 5 and assuming that the pith balls were equally charged, calculate how much charge was on each.

4. Coulomb measured the force between two charges by the amount of twist in a fiber from which a small charged pith ball hung at the end of a 10 cm long balanced arm, as shown in Fig. 6.6.

    a. If $q_1 = -10$ pC and $q_2 = -20$ pC, what is the electric force between the two "point" charges when the fiber has twisted through 0.2 radians, as shown in Figs. 6.6 and 6.7?

    b. How much twist in the fiber will be needed to make the angle between the two lines from the fiber to the two charges become 0.1 radians, as shown in Fig. 6.7?

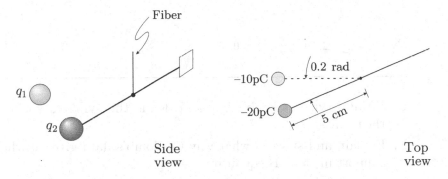

FIGURE 6.6   Positions and charges of pith balls (Problem 4).

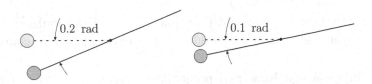

FIGURE 6.7   Before and after twisting the fiber (Problem 4).

c. In part (a) what is the magnitude of the electric field produced by q_2 at the location of q_1?

d. By what factor will the electric field at q_1 due to q_2 change when the angle goes from 0.2 to 0.1 radians?

5. Looking down onto Coulomb's torsion balance, we see a rod 2 cm long suspended from a thin filament as shown in the diagrams of the following table:

Some data from Coulomb's torsion balance

View from above	Separation between charges (cm)	Rotation of needle (deg)	Twist of cap (deg)	Total twist of filament (deg)
(i)	0.62	36	0	36
(ii)	0.31	18	126	144
(iii)	0.15	8.5	567	575.5

a. Write down Coulomb's law and define the symbols you use and their units.

b. Explain and show in what way Coulomb's data given in the table support his law. Be quantitative.

c. If the charge on each tiny sphere was 1 nC, what would be the force between the two charges in diagram (i) of the table?

d. What would be the electric field acting on the charge at the end of the needle in (i)?

e. By how much would that field change if we doubled the charge at the end of the needle but left the other charge unchanged?

6. Suppose you have two point-like charges $q_1 = 0.5\,\mathrm{C}$ and $q_2 = 1\,\mathrm{C}$ separated by 0.5 m. What is the magnitude of the force on the smaller of the two charges? Compare this force to the force on the larger charge.

7. Each of two small balls has a mass of 16 g, and each is charged with +8.5 nC. At what distance apart will the electric force on a ball have the same magnitude as the ball's weight?

8. Suppose you decide to measure charge in some new unit. Call it the esu, or "electrostatic unit," and choose it so that Coulomb's Law reads

$$F = \frac{q_1 \, q_2}{r^2},$$

where q_1 and q_2 are two charges, and r is the distance between them. (See how cleverly you are making the Coulomb force constant be unity.) Now define 1 esu by specifying that $F = 10^{-5}$ N when $q_1 = q_2 = 1$ esu and $r = 1$ cm. How many esu are in 1 coulomb? (Note that the esu is actually used in some old systems of electrical units.)

9. Charges Q_1 and Q_2 are placed 15 cm apart. When charge Q_3 is placed between Q_1 and Q_2, 5 cm from Q_1 (see Fig. 6.8), the total force on Q_3 is zero.

   a. If $Q_1 = +20$ nC, what is Q_2?

   b. When Q_3 is moved to the right (closer to Q_2), it feels a net force pushing it to the left, toward Q_1. Is Q_3 a positive or a negative charge? Explain carefully.

   c. When Q_3 is displaced to the right by 1 cm, as in part 0b, the force on it is 10^{-3} N (to the left). Determine the charge Q_3.

10. You may have already learned the equation for the power delivered by an electrical circuit: $P = IV$, where P is the power delivered (in watts = joules/second); I is the current (in amperes); and V is the voltage drop across the load, which is also equal to the voltage provided by the source. Derive this equation in terms of the rate of energy given to the electrons by the source (which must be equal to the rate at which electrons deliver energy to the load).

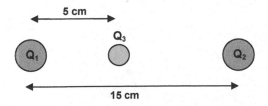

FIGURE 6.8 Arrangement of charges in Problem 9.

11. Solid metallic aluminum has an atomic mass of 26.98 u and a density of 2.73 g cm^{-3}. If each Al atom contributes 3 electrons to current flow, what is the number density of these "conduction" electrons? In an Al wire of diameter 1 mm, what is the average speed of the electrons when the current is 1 A?

12. Consider two identical conducting balls B_1 and B_2. B_1 initially carries a charge Q, and B_2 is initially uncharged. When the balls are placed in contact with each other, charge q is transferred from B_1 to B_2, and the balls repel each other. What fraction of the initial charge, $\frac{q}{Q}$, must be transferred to maximize the force between the balls? Justify your answer.

7.

Electric Fields and Electric Forces

This chapter introduces you to the electric field—an important and useful way to describe electric forces.

Coulomb studied the force exerted by one charge on another. According to his description, one charged object pushes or pulls another charged object located some distance away. This is called "action at a distance." Many people don't like this idea. They feel intuitively that the force on an object should come from something located right where the object is. To accommodate this view, physicists invented the idea of a "field of force." We say that electric charges produce something called an "electric field" that extends throughout all space. When another charge is placed anywhere in this electric field, the field exerts a force on the charge.

Although at first it may seem an abstract complication introduced only to comfort people who don't like action at a distance, as you learn more about the electric field, it will become more real. A field has energy; it has mass; by disturbing an electric field you can produce light; you can cause particles of matter (and anti-matter) to materialize from intense electric fields. The field concept also applies to other forces than electric; it is one of the most fruitful ideas in physics.

7.1 ELECTRIC FIELD: A LOCAL SOURCE OF ELECTRIC FORCE

In the simple case Coulomb studied, the electric field was produced by a single charge q_1. You know from Coulomb's Law that at any distance r from a localized point-like charge q_1, a point charge q_2 has a force acting on it, $F = k_c \frac{q_1 q_2}{r^2}$. Now change how you think about where this force comes from. Rather than saying that q_1 acts over the distance r to produce a

C.H. Holbrow et al., *Modern Introductory Physics, Second Edition*,
DOI 10.1007/978-0-387-79080-0_7, © Springer Science+Business Media, LLC 1999, 2010

force on q_2, imagine instead that q_1 fills all of the space around it with something we call "electric field." There will then be electric field E at the location of q_2, and it is this electric field that exerts a force on q_2. The charge q_2 produces an electric field that also fills the space around the charge; the part of this field at the location of q_1 exerts a force on q_1. In other words, in the space around q_1 and q_2 the electric field is the combination of the fields from q_1 and q_2. Note that because force has both magnitude and direction, so does electric field.

If many charges are present, each makes a separate contribution to the total electric field, but you can measure the field's magnitude and direction at any point in space without knowing anything about the distribution of charges producing it. For example, you might use the following two-step procedure to test if there is any E present at a point in space and to measure its value there. First, place a small neutral particle where you wish to test for the presence of electric field. Although electrically neutral, the particle will accelerate because of gravity. Now put a tiny charge q_t (subscript 't' for 'test') on the particle. If, after becoming electrically charged, the particle's acceleration changes, an electric field must be acting on it. The force due to the electric field acting on q_t is used to define the strength (magnitude) of the field:

$$E = \frac{F}{q_t}. \tag{1}$$

An important feature of E is that it is a property of the locality and independent of q_t.[1] Conceptually we have separated the thing that acts— the electric field E—from the thing on which it acts—the electric charge q_t. Knowing E at some location in space, you can immediately calculate the force exerted on *any* charge Q placed at that location. It will be

$$F = QE. \tag{2}$$

It follows from Eq. 1 that the units of electric field are N C^{-1}, i.e., newtons per coulomb.

Because electric field is a vector quantity, it has direction as well as magnitude. By definition the direction of E is the direction in which E would push a tiny *positive* charge q_t. Therefore, if you have a large ball of negative charge Q and you place a tiny positive q_t near it, you know q_t will be attracted. This tells you that E, the electric field produced by a negative charge, points toward it. To find the direction of an electric field

---

[1]It is strictly true that E is independent of q_t, but q_t produces its own electric field that will exert force on the distant charges that are producing the electric field felt by q_t. Then q_t's field will rearrange distant charges and so result in a change in the E that they produce at the location of q_t. That's why to measure an E you use a tiny q_t. "Tiny" means small enough to produce negligible change in the E that you are measuring.

just imagine how a *positive* charge will move if placed in the field. Notice, also, that if q_t were negative, it would move in the direction opposite to that of E.

Two Useful Electric Fields

Electric fields are produced by electric charges. Different arrangements of charge produce in the surrounding physical space different patterns of electric field. These can be quite complicated, but in this book we use mainly two simple electric fields. One is the electric field in the space surrounding a point or any spherically symmetric distribution of charge, e.g., a ball of charge; the other is the uniform electric field produced in the space between two parallel, oppositely charged conducting plates.

Electric Field Outside a Point or Sphere of Charge

EXAMPLES

1. A point charge produces an electric field everywhere in space. What is the magnitude of the electric field that a point charge $q_2 = 0.5\,\mathrm{C}$ produces at a location 2 m away from itself?

To answer this question place a small charge, let's say $q_t = 1\,\mu\mathrm{C}$, at a point 2 m from the 0.5 C charge. Now calculate the force on it. From Eq. 1 the force on q_t will be

$$F = \frac{9 \times 10^9 \times 0.5 \times 10^{-6}}{2^2} = 1.125 \times 10^3\,\mathrm{N}.$$

What then is the value of the electric field at the position of the charge? Using the definition of Eq. 1, you get

$$E = \frac{F}{q_t} = \frac{1.125 \times 10^3}{1 \times 10^{-6}} = 1.125 \times 10^9\,\mathrm{N\,C^{-1}}.$$

Notice that you first multiplied by $q_t = 1 \times 10^{-6}\,\mathrm{C}$ to find the force, and then divided by the same quantity q_t to find the electric field. The essential idea is that the electric field is independent of the charge upon which it acts; it depends only on other charges that are the sources of the field. For the particular case of a source that is itself a point charge q_s, the force on a particle with charge q_t a distance r from q_s is, according to Coulomb's Law, $F = k_c\,q_t\,q_s/r^2$. From the definition of electric field $E = F/q_t$, you see that the electric field in the space surrounding q_s is

$$E = \frac{k_c q_s}{r^2}. \qquad \text{Electric field of a point charge} \qquad (3)$$

What is the direction of this electric field? Because both q_t and q_s are positive, the force on q_t pushes it away from the source q_s; therefore the electric field points radially outward from q_s. If $q_s < 0$, then a positive charge $q_t > 0$ will be attracted to it, meaning that E points toward q_s. In general, electric fields point away from positive source charges and towards negative source charges.

Don't think that a "point charge" is an artificial idealization. It turns out that the electric field *outside* a spherically symmetric distribution of charge (one or more nested spherical shells of charge) is exactly the same as if all that charge were concentrated in a point at the center of the sphere.[2]

Constant Electric Field

Another particularly useful electric field is the uniform constant field. When two parallel, conducting plates are oppositely charged, the charges spread over the facing surfaces of the plates and produce an electric field that is nearly uniform in the space between the plates if the separation between the plates is small (say less than 10%) compared to the lengths of their edges. Such an arrangement is shown schematically in Fig. 7.1. In the space between the plates the electric field E has the same value at every point. As a result, a charged particle will experience the same force and acceleration, no matter where you put it between the plates.

As before, you find the direction of a uniform electric field by imagining a small positive charge between the plates. It will be attracted to the

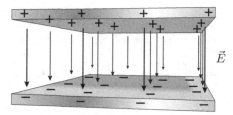

FIGURE 7.1 A common arrangement for producing a uniform electric field. You are looking edge on at a pair of parallel plates with inner surfaces uniformly covered with charge as indicated. The arrows represent the value and direction of E at various points in the space between the plates.

---

[2]Even more surprising, at any radial point r_i *inside* a spherically symmetric charge distribution, there is no electric field from any shell of charge with radius greater than r_i. The fields from different parts of a shell of charge cancel out inside the shell, so a set of nested shells of charge produce no electric field in the space inside them, i.e., at $r \leq r_i$. Only the charged spherical shells with radii less than r_i produce electric field at $r = r_i$. We will use these assertions (without proving them) when considering the internal structure of atoms.

negatively charged plate and repelled by the positively charged plate. Therefore, because the electric field points in the same direction as the force on a positive charge, the electric field points from the positively charged plate to the negatively charged one as shown in Fig. 7.1.

▼ EXAMPLES

2. Suppose you have a charge of $1\,\mu$C on a small sphere of mass $1\,\mu$g. When this charged mass is placed in a uniform (constant) electric field of $3\,$N$\,$C^{-1}, how much acceleration does the electric field produce? From the fact that

$$F = qE = ma,$$

it follows that *anywhere* in the volume occupied by the electric field E

$$a = \frac{qE}{m} = \frac{1 \times 10^{-6} \times 3}{1 \times 10^{-9}} = 3 \times 10^3\,\mathrm{m\,s}^{-2}.$$

■ EXERCISES

1. What size electric field would you have to apply to the above particle in order to exactly counteract the force of gravity? What would be the direction of this field?

2. A particle called an electron has a charge of $-1.60 \times 10^{-19}\,$C and a mass of $9.11 \times 10^{-31}\,$kg. Suppose it is placed in an electric field that points downward and has a magnitude of $1000\,$N$\,$C^{-1}. What will be the magnitude and direction of its acceleration?

7.2   ELECTRIC POTENTIAL ENERGY AND ELECTRIC POTENTIAL

A charged particle in an electric field has electric potential energy much as a mass near Earth has gravitational potential energy. The concept of electric potential energy is useful; it simplifies the calculation of important effects of an electric field on a charged particle; it also prepares you to understand the important related concept of "electric potential"—often called "voltage."

When a charged particle is in an electric field, the field exerts a force on the charge. If no other forces are acting, the particle accelerates and acquires kinetic energy. Where does this kinetic energy come from? Just as when a mass falls toward Earth and its gravitational potential energy becomes kinetic energy, so when a charged particle is accelerated by an electric field, its electric potential energy becomes kinetic energy. The charged particle has potential energy associated with its location in the electric field; this potential energy is converted into kinetic energy. In general, as the particle moves through an electric field, the sum of its kinetic and electric potential energy stays constant, i.e., its total mechanical energy is conserved.

Electric Potential Energy in a Constant Electric Field

From the following example you can see how the electric potential energy of a charged particle in a constant electric field is analogous to the gravitational potential energy of a mass near the surface of Earth.

▼ **EXAMPLES**

3. Imagine a charge Q of $2\,\mathrm{C}$ in a uniform constant electric field E of $10\,\mathrm{N}\,\mathrm{C}^{-1}$ pointing in the positive x-direction. You see that the field exerts a force of $20\,\mathrm{N}$ on the charge. Now push the charge so that it travels a distance $d = 0.1\,\mathrm{m}$ directly against the field and is at rest when you are done pushing. As you push, that is, as you exert a force over a distance, you are doing work. Because in this example E is constant, the force is constant, and the work you do is $W = F\,d = Q\,E\,d = 2\,\mathrm{J}$. The work you do increases the energy of the system.

Where is this additional energy? After you push the charged particle $0.1\,\mathrm{m}$, it is at rest, so it has no kinetic energy. The only thing different about the particle is its position. Therefore there must more be energy associated with being at its new location than at its starting location. The energy associated with the position of the charged particle in an electric field is called the particle's "electric potential energy." By moving the particle you increased its electric potential energy by $2\,\mathrm{J}$.

■ **EXERCISES**

3. Does the change in electric potential energy depend on the size of Q?

To see that the energy described in Example 3 really is present, let go of the particle. The electric field will then accelerate it, and after it moves the distance $d = 0.1\,\text{m}$ in the direction of the field, the particle's kinetic energy will increase by 2 J. This increase in kinetic energy comes from the decrease of the particle's electric potential energy. The work that you put into the system increased its total energy by 2 J. After that, as long as no external agents act, the total energy, the sum of the potential and kinetic energy of the charge, remains constant.

What happens if the particle continues to accelerate over a distance beyond d? Suppose it goes an additional distance of 0.2 m. For the case of a constant field E the potential energy of Q continues to decrease. The total decrease would be $E\,Q\,(0.1 + 0.2) = 6\,\text{J}$. For a particle carrying charge Q in a constant electric field, the change in potential energy ΔU is

$$\Delta U = -QE\Delta x \qquad \text{for constant } E, \tag{4}$$

where Δx is the total distance the particle travels in the direction of E; the minus sign means that when E accelerates positive charge, electric potential energy decreases.

Although the constant electric field is a simple special case, there is one feature of Eq. 4 that is true for all cases no matter how complicated: The *difference* between the potential energies at two different points in space is what has physical meaning. Nothing physical changes if you increase or decrease the potential energy by the same amount everywhere. Consequently, you can set the potential energy to be zero wherever you want. The choice of zero point is arbitrary although some choices may be more convenient than others.

As an example, consider two parallel plates a distance of 0.20 m apart with a constant electric field of $20\,\text{N C}^{-1}$ pointing from the top plate toward the bottom. You could choose to have $U = 0\,\text{J}$ at the top plate. Then a 1 C charge would have an electric potential energy of $-4\,\text{J}$ at the bottom plate. Or you could choose $U = 0$ at the points half way between the two plates. Then the charge would have $U = -2\,\text{J}$ at the bottom plate and $U = +2\,\text{J}$ at the top plate.

You can always put the zero anywhere you like, but for our example here it seems nice to have $U = 0\,\text{J}$ at the bottom plate. For this choice the potential energy of the charge is positive everywhere between the plates, and, if you set up your coordinate system with $y = 0$ at the bottom plate, the equation for $U(y)$ of the charge Q resembles the equation for $U(y)$ of a mass m near Earth, because you can then write

$$U(y) = -Q\,Ey \tag{5}$$

where y is the vertical distance of the charge Q above the bottom plate. But notice, no matter where you set $U = 0$, $\Delta U = U(y_2) - U(y_1)$ between two given points y_2 and y_1 has the same value.

Writing the potential energy as $U(y)$ in Eq. 5 reminds you that U is a function of position in space; in this case it depends only on vertical position y. The minus sign is in Eq. 5 to show that potential energy increases in the direction opposite to E's direction, and, in the example just given, the numerical value of E will be negative because E points opposite to the direction of increasing y.

■ EXERCISES

4. If $U(\frac{d}{2}) = 0$, i.e., at any point half way between the two plates, what will be the value of $U(d)$, i.e., at any point at the top plate?

5. For each of the above choices of where $U = 0$, what will be the change in its electric potential energy when a $1\,\mathrm{C}$ charge moves from the top plate to the bottom one?

6. How would you rewrite Eq. 5 to have an equation for $U(y)$ if the bottom plate was positively charged and the upper plate was negatively charged?

7. Why does Eq. 5 remain valid if Q is negative rather than positive?

Rewriting Eq. 4 as

$$Q\,E = -\frac{\Delta U}{\Delta x} \tag{6}$$

exhibits a fundamental connection between $U(x)$ and $E(x)$. It says that the component of the force qE in the direction of x is the negative derivative of U with respect to x. This is a general feature of a potential energy function. It means that if the potential energy decreases in some direction in space, there is a force in that direction; the steeper the decrease of the potential energy over a given small distance, the greater is the force.

Electric Potential Energy of a Positive Charge Q in the Electric Field Produced by a Point Charge q_s

Suppose you are moving a charge Q around in an electric field E produced by a point (or sphere of) positive charge q_s. Imagine you start from far away and push Q directly against E closer to q_s. Because you are pushing against the force exerted on Q by the electric field $E = \frac{k_c\,q_s}{r^2}$ (Eq. 3), you are increasing the potential energy of Q in the field E. This is like the explanation given on page 174 for a charge Q in a constant uniform electric field.

For the constant field you pushed Q against a constant force, and it was easy to calculate the work you did. However, the electric field E around a point charge q_s is not constant; as Eq. 3 shows, E varies as $1/r^2$, where r is the distance from the center of the source charge q_s. As a result, the work you do pushing Q a small distance Δr closer to q_s is different at different distances r. Using calculus you can calculate the work done pushing Q from one value of r, say r_1, to another, say r_2; your answer will be

$$\Delta U = U(r_2) - U(r_1) = Q \, k_c \, q_s \left(\frac{1}{r_2} - \frac{1}{r_1} \right). \tag{7}$$

For this case it is convenient and usual to call the potential energy of Q zero when Q is far, far away from the source (we say "at infinity"). Then the potential energy of Q at r is the work you would have done pushing Q in from $r_1 \approx \infty$ to r

$$U(r) = Q \, k_c \, q_s \frac{1}{r}. \qquad \text{P.E. of } Q \text{ in the electric field of } q_s \tag{8}$$

▮ EXERCISES

8. Show that Eq. 8 follows from Eq. 7 when you assume that the electric potential energy of the charge Q is zero at $r_1 \approx \infty$.

▽ EXAMPLES

4. Suppose you have a charge $Q = 0.5 \, \mu C$ in the electric field of a charge $q_s = 3 \, mC$.

 a. What is the electric potential energy of Q when it is 1.5 m away from q_s?

 b. When it is 0.5 m from q_s?

 c. How much work must be done to move Q from 1.5 m to 0.5 m away from q_s?

 d. Suppose you choose to set the potential energy of Q to be zero when Q is 1 m away from q_s.

  i. What then would be the electric potential energy of Q very far away from q_s?

  ii. How would your answer to (c) change?

From Eq. 8 it follows that Q's electric potential energy at $r = 1.5$ m is

$$U = 0.5 \times 10^{-6} \times 9 \times 10^9 \, \frac{3 \times 10^{-3}}{1.5} = 9.0 \text{ J}.$$

To answer (b) you could redo your calculation replacing 1.5 m with 0.5 m. Don't do that. Notice instead that the electric potential energy varies as $1/r$, and when you go from 1.5 m to 0.5 m you have changed r by a factor of $1/3$ so that the potential energy must have increased by a factor of 3. Therefore, the answer is 3 times your answer to (a), i.e., $3 \times 9 = 27$ J.

And to move Q from $r_1 = 1.5$ m to $r_2 = 0.5$ m will require $27 - 9 = 18$ J of work.

With Q's potential energy chosen to be zero at $r \approx \infty$, its potential energy at $r = 1$ m will be 13.5 J. (Show this is so using the kind of scaling argument illustrated above.) To make U be zero at $r = 1$ m, subtract 13.5 J from every value everywhere; then at $r = \infty$, the new value will be -13.5 J.

At $r_1 = 1.5$ J the new value of U would be $9 - 13.5 = -4.5$ J and at $r_2 = 0.5$ m the new value would be $27 - 13.5 = 13.5$ J. The work done to go from r_1 to r_2 will be $13.5 - (-4.5) = 18$ J. This difference in electric potential energy is the physically significant quantity; it is independent of where you set the zero. Your answer to Example 4(c) does not change.

You will often see the words "potential energy" with no modifier. This is a common way of talking about it, leaving you to infer from the context whether the potential energy being discussed is gravitational, electric, spring, or some other kind. Most of the time the correct inference will be clear.

7.3   ELECTRIC POTENTIAL

Equations 5 and 8 are true only for the corresponding particularly simple cases of a constant electric field and a $1/r^2$ electric field. However, the equations illustrate an important property that is always true: The electric potential energy of a charge Q is proportional to the value of Q. This property means that you can think of the electric potential energy of a charge as arising from two independent factors—one is the size and sign of the charge; the other is a property of the space where the charge is situated. This separation is possible because these two properties—the size (and sign) of the charge Q itself and the electric state of the (x, y, z) point in space where the charge is located—are independent of each other.

For discussions of energy this separation of the electric state of space from the charge placed in that space has advantages, so physicists have invented a new concept to describe the electric state of space at (x, y, z). They call it the "electric potential" and often represent it by the function $V(x, y, z)$.

Although they are related, electric potential and electric potential energy are distinctly different entities. Electric potential is a property of space; electric potential energy is a property of a charge placed in that space. The connection between the two is that V is numerically equal to the electric potential energy of a unit (1 C) of charge. Knowing the electric potential V at a point (x, y, z) in space, you can calculate the electric potential energy $U(x, y, z)$ of any charge q at that point (x, y, z) by multiplying the value of q by the value of V at that point.:

$$U(x, y, z) = qV(x, y, z). \tag{9}$$

Both U and V are functions of position (x, y, z), but it is quite usual to refer to them without explicitly showing that dependence. In that case Eq. 9 would be written as

$$U = qV. \tag{10}$$

You can see from Eq. 10 that the units of V have to be joules per coulomb, $J\,C^{-1}$. This group of units is called the "volt." "Volt" is a synonym for $J\,C^{-1}$ in the same way that "joule" is a synonym for $N\,m\,s^{-2}$. The abbreviation for volts is V. You see it written on the 1.5-V AA, AAA, C and D batteries you use in electric gadgets. The manufacturer is telling you that the end of the battery with the little knob on it, usually marked $+$, has an electric potential that is 1.5 volts higher than the flat $(-)$ end. Notice that we use the letter "V" in two different ways when describing electric potential. First, it is often the algebraic symbol or variable representing the potential at points in space; in this case the $V(x, y, z)$ is *italicized*. Second, the letter V is the abbreviation for volts, the units in which magnitude of the electric potential is normally given; in this case the letter V is not italicized.

It may help you distinguish between the two concepts if you notice that the word potential is used as a noun when you are talking about electric potential, but it is used as an adjective when you are talking about electric potential energy. Electric potential is a property of space; electric potential energy is a property of a charge situated in that space.

People often refer to electric potential as "voltage." Rather than ask: "What is the electric potential at (x, y, z)?," they will ask: "What is the voltage there?" It is also usual to refer to differences of electric potential simply as "potential differences" or "voltage differences." These mean the same thing.

Similar to the case of potential energy, only the *difference* of electric potential ΔV has physical significance; also, as for U, the location you choose to correspond to $V = 0\,\text{V}$ is arbitrary. (But some choices are more convenient than others.) Therefore, it is useful to write Eq. 9 as

$$\Delta U = Q\,\Delta V. \tag{11}$$

In this form Eq. 11 explicitly reminds you that the physically important information is the difference between the electric potential at two different points. If you know ΔV, you can easily find the change in potential energy of a charge Q that moves between those two points.

Do the following exercises to see how E and V are related.

■ EXERCISES

---

9. Write out Eq. 11 in a form that shows explicitly the position dependence of ΔU and ΔV.

10. Use Eq. 6 to show that if the electric potential is $V(x, y, z)$ at some point in space, the value of the electric field E in the x direction at that point is

$$E_x = -\frac{dV}{dx}. \tag{12}$$

The minus sign means that E points in the direction in which V is decreasing.

When the electric field is constant, Eq. 12 becomes especially simple:

$$E = -\frac{V(y_2) - V_1(y_1)}{y_2 - y_1}. \tag{13}$$

It is common practice to make a constant electric field by applying a given voltage difference across two parallel conducting plates. For this case, you can use Eq. 13 to find the value of the electric field at any point in the volume between the plates.

▽ EXAMPLES

5. Imagine two metal plates parallel to each other and 0.02 m apart. If you connect a 3 V battery between them, you get an electric field of

$$E = \frac{3}{0.02} = 150 \, \text{V m}^{-1}.$$

6. You can generalize the previous example. In the case of the parallel plates where E is everywhere constant pointing from the top plate toward the bottom, take $V = 0$ and $y = 0$ at the bottom plate. Then $V(y) = -E\,y$. (Remember, this is only true because E is constant.)

Notice that the units of electric field have just been given as "volts per meter." Those units follow from Eq. 12. But, you ask, according to Eq. 1 aren't the units of E "newtons per coulomb"? Yes, they are. The two different combinations of units are equivalent: $1 \, \text{N C}^{-1} \equiv 1 \, \text{V m}^{-1}$. Values of electric field are usually given in units of volts per meter.

▩ EXERCISES

11. In Example 5, what would be the value of E in N C^{-1}?

12. Roughly how large should the plates be for the calculation in Example 5 to be reasonably valid? Hint: See page 172.

13. What would be the electric field between the plates if they were only 1 mm apart?

14. Describe two different things you could do to double the value of the electric field in Example 5.

The following examples illustrate how to discuss the energy of a charge in a constant electric field or in a $1/r^2$ electric field in terms of voltage.

▽ EXAMPLES

7. What is the change in kinetic energy of a charge of 3 C that moves through a potential difference of 7 V? Because $1 \, \text{V} = 1 \, \text{J C}^{-1}$, it is $3 \times 7 = 21 \, \text{J}$.

8. Suppose you have two parallel conducting plates spaced 1 cm apart. Imagine you have applied a potential difference of 12 V between them. (You could do this by connecting the lower plate to the negative terminal and the upper plate to the positive terminal of your car battery.[3]) What would be the change in electric potential energy of a $2\,\mu C$ charge Q after it has been moved from the bottom plate to a point 5 mm closer to the upper plate?

The potential difference over the 1 cm distance from bottom to top is 12 V. Over half that distance, i.e., over 5 mm, the potential difference will be half of 12, $\Delta V = 6$ V. The change in potential energy of Q will be

$$\Delta U = \Delta V\, Q = 6 \times 2 \times 10^{-6}\ \text{J} = 12\ \mu\text{J}.$$

9. How would your answer change if the charge Q was increased to $7\,\mu C$? If the apparatus and the applied voltage are unchanged,

$$\Delta U = 6 \times 7 = 42\ \mu\text{J}.$$

This example is yet another illustration that the amount of charge and the amount of potential difference contribute separately to the change in the particle's potential energy.

10. What if the question were: How much does the potential energy of $Q = 3\,\mu C$ change if it moves from $y = 1$ mm to $y = 8$ mm? First find $\Delta V = 12\,\frac{8-1}{10} = 8.4$ V; then multiply by Q to get $\Delta U = 3 \times 8.4 = 25.2\ \mu\text{J}$.

11. Consider a charge Q in the electric field outside a small ball of charge q_s (where, as usual, the subscript 's' is to remind you that this charge is the source of the electric field and the electric potential). Suppose $Q = 2\,\mu C$ and $q_s = 5\,\mu C$. What is the change in potential energy of Q when it moves from $r_1 = 1.5$ m to $r_2 = 0.5$ m away from q_s?

From now on always answer such a question by first calculating the electric potential at the two locations. In this case that means finding the values of q_s's electric potential at 1.5 m and 0.5 m away from q_s.

$$V(1.5) = 9 \times 10^9 \frac{5 \times 10^{-6}}{1.5} = 30\ \text{kV}.$$

---

[3]Don't do this at home!

Do you see that $V(0.5) = 3 \times V(1.5) = 90$ kV? The potential difference between the two points is $90 - 30 = 60$ kV.

The change in Q's potential energy is then $Q \, \Delta V = 2 \times 10^{-6} \times 60 \times 10^3 = 0.12$ J.

EXERCISES

15. Redo Example 11 assuming $Q = 7 \; \mu$C.

16. Redo Example 11 assuming $q_s = 9 \; \mu$C.

17. What is the electric potential a distance of 3 mm from a point charge of 18 μC?

18. How much work would be required to bring a charge of 1 C from far away to a distance of 3 mm from a charge of 18 μC?

19. What is the strength of the electric field 1.5 m away from $q_s = 18 \, \mu$C? What is the strength of the electric field at half that distance? What is the value of the electric potential there?

20. Suppose you double the value of Q. By how much do your answers to the previous three exercises change?

Acceleration of Charged Particles Through a Difference of Potential

Many experimental techniques for studying electrons, ions, atomic nuclei, or other charged particles speed them up or slow them down by passing them through a region of space in which there is a change of electric potential ΔV. It is important for you to know how to find the change in velocity of a charged particle after it has passed through a potential difference. The following example shows how to do this.

EXAMPLES

13. For arbitrary combinations of charge and field the mathematical form of the potential energy function can be complicated. However, you can do without any formula at all if you can directly measure the *numerical values* of the electric potential at points in space. Suppose the potential at point A is 250 V, and at point B it is 150 V. What will be the velocity v_B at B of a dust particle of mass $m = 1 \ \mu$g carrying a charge of $Q = 2 \ \mu$C that has started from rest at A and moved to B?

First find the change in potential between those two points by subtracting the value of the potential at the starting point A from the value at the ending point B: $V_B - V_A = -100$ V.

Then the change in the particle's potential energy ΔU of Q is just $Q \, \Delta V = 2 \times 10^{-6} \times (-100) = -2 \times 10^{-4}$ J. By conservation of energy the decrease in potential energy has gone to increase the particle's kinetic energy from 0 J to 2×10^{-4} J, i.e., the change in kinetic energy is

$$\frac{1}{2} m v_B^2 - \frac{1}{2} m v_A^2 = -\Delta U = 2 \times 10^{-4} \text{ J}.$$

Since for this case $v_A = 0$, it is simple to solve for v_B:

$$v_B = \sqrt{\frac{2 \, \Delta U}{m}} = \sqrt{\frac{4 \times 10^{-4}}{10^{-9}}} = 632 \text{ m s}^{-1}.$$

(Remember 1 μg = 10^{-9} kg.)

Did you notice how similar this electric case is to the gravitational one discussed on page 44 in Chap. 2? The principal difference is that now your first step is to find the change in electric potential ΔV. Next, you use the value of the charge Q to determine its change in electric potential energy $\Delta U = Q \Delta V$. From conservation of energy you find $\Delta K = -\Delta U$, then from the mass m of the particle and its initial velocity, you can find its final velocity at anywhere you know the electric potential.

Unlike the gravitational potential energy of a mass m, which always decreases as the height of m decreases, the direction in which the electric potential energy of a charge Q decreases depends on the sign of Q. You can get correct answers if you are careful with signs, but it is easier to use physical intuition. In whichever direction E accelerates Q, Q's potential energy decreases. For a positive charge, the electric force is in the direction of the electric field, so the charge's electric potential energy decreases as it moves with the electric field, from high V to lower V. For a negative charge, the electric force accelerates the charge opposite to the electric field, and the charge's electric potential energy decreases as it moves opposite the electric field, from low V to higher V.

EXERCISES

21. An electron is a particle with a mass of 9.11×10^{-31} kg and a charge of $q = 1.60 \times 10^{-19}$ C. If it is initially at rest, what is its velocity after it goes from -10 V to $+10$ V?

22. Suppose the electron of the previous exercise initially has a kinetic energy of 16×10^{-19} J. After going from -10 V to $+10$ V, what is its new velocity?

Energy, Electric Potential, and Electric Current

You can use these ideas of electric potential and electric potential energy to understand that an electric battery is a reservoir of energy that is drained as current flows out of the battery from one electrode and along a wire and back into the battery through its other electrode. (Such a closed path is called an "electric circuit" or just a "circuit").

You have just seen that when charges move through a difference of potential, their kinetic energy increases. But that is only true if no other forces act. You have already seen (p. 163) that in a wire charges move with a constant speed; their kinetic energy is not increasing even though they are moving through a difference of electric potential and, therefore, their potential energy is decreasing. This happens because energy leaves the system at the same rate that potential energy is converted into kinetic energy. The modern atomic picture is that electrons moving in a wire lose energy as they collide with impurities and irregularities. This energy appears as heat. The rate at which this heat is generated is the product of the current I and the voltage difference V, i.e., power dissipated $= IV$.

EXAMPLES

14. How much energy does a 1.5 V "D" battery expend when it pushes $10 \, \text{mC} \, \text{s}^{-1}$ through a wire for $20 \, \text{min}$? To answer you need to know the total amount of charge ΔQ moved through the 1.5 V potential difference. The steady current of $10 \, \text{mA}$ corresponds to $\Delta Q / \Delta t = 10 \times 10^{-3} = 0.01 \, \text{C} \, \text{s}^{-1}$. In $20 \, \text{min}$, which is $1200 \, \text{s}$, the battery moves $1200 \times 0.01 = 12 \, \text{C}$ through 1.5 V. To do this the battery must expend $1.5 \times 12 = 18 \, \text{J}$ of energy.

15. The energy capacity of a battery is usually specified by two quantities, the battery voltage and the total amount of charge the battery will deliver. The total charge is measured by drawing a current from the battery until it runs down. The amount of charge taken from the battery is usually specified in units of ampere-hours, i.e., A h. An alkaline D cell will deliver 0.5 A for about 20 h, so its capacity would be given as 10 A h or 36 kC.[4] The energy extracted from this battery would then be $1.5 \times 36 = 54$ kJ. Such batteries cost about $1.50 apiece, so you pay $1.50/0.054 \approx$ $30 for a megajoule (MJ). Electric power off the grid costs you around $0.20 for a kW h or 3.6 MJ. The cost per MJ is therefore $0.2/3.6 = $0.06. The difference between $30 and $0.06 is what you pay for the convenience of portable energy. You can see that recharging batteries might be a good idea.[5]

Visualizing Electric Potential

It is often useful to envision electric potential $V(x, y, z)$ in terms of equipotential surfaces—surfaces on which the potential V is everywhere the same.

Between uniformly charged parallel plates the equipotential surfaces are planes because in a constant electric field the potential V varies linearly with y, the coordinate perpendicular to the plates. As a result, the potential is the same at all points that have the same value of y. The set of such points forms a plane parallel to the surfaces of the plates and perpendicular to the direction of the electric field. Figure 7.2a illustrates the equipotential surfaces for values of $V = 2$, 4, 6, and 8 volts.

Around a point charge q_s the electric potential $V(r)$ is the same at all points where r is the same. These points form the surface of a sphere, so the equipotential surfaces around a point or ball of charge are spherical shells centered on the source charge q_s. Figure 7.3a shows equipotential surfaces for $V = 2$, 4, 6, and 8 volts. Because the potential around a point charge varies as $1/r$, the distance between equipotential surfaces separated by the same ΔV gets smaller closer to the source charge and larger farther away.

---

[4] At a steady current of 3 A, a D cell delivers a total charge of only about 1.5 A h; high currents exhaust batteries disproportionately more quickly than low currents.

[5] You can recharge alkaline batteries, but to avoid dangerously overheating (and perhaps exploding them) don't charge alkaline batteries with a charger intended for NiCad batteries.

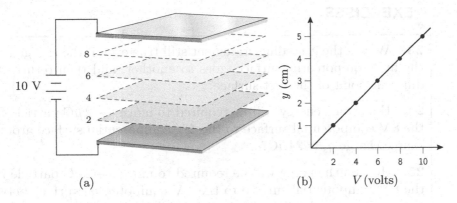

FIGURE 7.2 (a) Equipotential surfaces in the constant electric field between two oppositely charged parallel conducting plates; (b) a plot of the potential function along any line perpendicular to the plates.

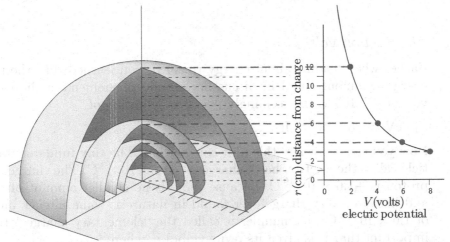

FIGURE 7.3 (a) Spherical equipotential surfaces in the electric field around a point charge; the portions of these surfaces below the plane are not shown. (b) A graph of the potential function along any line extending radially out from the source charge.

For our two special cases the electric potential function V is a function of only one variable, so you can plot a graph of V. Such plots are shown in Figs. 7.2b and 7.3b. They are marked to show the locations and values of the equipotential surfaces illustrated in the adjacent Figs. 7.2a and 7.3a.

Keep in mind that no work, i.e., no input or output of energy, is required to move a charged particle between any two points on the same equipotential surface.

■ EXERCISES

23. Why is the preceding statement still true even if the charge leaves the first equipotential surface, goes to another, and then returns to a different point of the first surface?

24. How much energy will be required to move a $-2\,\mu\mathrm{C}$ particle from the 8 V equipotential surface to the 4 V equipotential surface around a source charge $q_s = 24\,\mu\mathrm{C}$?

25. How much energy will be required to move a $-2\,\mu\mathrm{C}$ particle from the 8 V equipotential surface to the 4 V equipotential surface between a pair of oppositely charged parallel conducting plates?

26. How would your answer to Exercise 25 change if the potential difference was caused by a point charge rather than by oppositely charged parallel plates?

The Electron Volt

Earlier when reading about energy (p. 118), you learned that the unit of energy commonly used in discussing atomic phenomena is the electron volt or eV. It has the seemingly peculiar definition of

$$1\,\mathrm{eV} = 1.6 \times 10^{-19}\,\mathrm{J}.$$

Now you can see where this unit comes from. The tiny fundamental particle called the electron has a charge of -1.6×10^{-19} C. The charge on the proton is $+1.6 \times 10^{-19}$ C. Because there is no smaller amount of charge and because every charge ever found in nature is some integer multiple of 1.6×10^{-19} C, this number is called the "elementary charge." It is so important that it is given its own symbol e, where

$$e = 1.6 \times 10^{-19}\,\mathrm{C}.$$

The charge on an electron is $-e$. The charge on a proton is e. (People carelessly call e without the negative sign "the electron charge," and you have to remember to put in the minus sign when you do calculations.)

By how much does the potential energy of an electron change when it moves through a potential difference of $V = -1$ volt? From Eq. 11

$$\Delta U = e\,V = 1.6 \times 10^{-19}\,\mathrm{C} \times 1\,\mathrm{V} = 1.6 \times 10^{-19}\,\mathrm{J},$$

which is the same as 1 electron volt, or 1 eV. In other words, the conversion factor between J and eV is 1.6×10^{-19} J eV^{-1}. Now suppose a charge e is accelerated through a potential difference V. Its change of energy ΔU is

$$\Delta U = e\,[\mathrm{C}]\,V\,[\mathrm{V=J\,C^{-1}}] = eV\,\mathrm{J}.$$

(The symbols in square brackets are the units of the preceding quantity.) To express ΔU in eV, divide the above equation by the conversion factor from J to eV:

$$\Delta U = e\,V \frac{[\text{C J C}^{-1}]}{1.6 \times 10^{-19}\,[\text{J eV}^{-1}]}$$

$$= e\,V \frac{1}{1.6 \times 10^{-19}} \frac{[\text{C J eV}]}{[\text{C J}]}$$

$$= \frac{e\,V}{1.6 \times 10^{-19}}\,\text{eV} = V\,\text{eV}.$$

Do you see that in these units when a particle carrying a charge e moves through a potential difference of V volts, the change in the particle's potential energy is numerically equal to V eV? Because of the way the eV is defined, the task of calculating the change in kinetic energy of an electron or other fundamental particle accelerated through any potential difference is trivial when you use units of electron volts. The eV is also convenient for other calculations as we will show you in Chap. 12.

▮ EXERCISES

---

27. What is the change in kinetic energy of an electron accelerated through a potential difference of 1 kV? Give your answer in eV.

Of course, you still have to do some figuring to find the velocity.

▽ EXAMPLES

---

16. What is the velocity of an electron accelerated from rest (initial kinetic energy = 0) through a potential of 1 kV?

Clearly, its energy changes by 1 keV, or 1.6×10^{-16} J. Since it was at rest initially, its kinetic energy is now

$$\frac{1}{2}mv^2 = 1.6 \times 10^{-16}\,\text{J},$$

so we find that

$$v = \sqrt{\frac{2 \times 1.6 \times 10^{-16}\,\text{J}}{9.11 \times 10^{-31}\,\text{kg}}}$$

$$= 1.85 \times 10^7\,\text{m s}^{-1}.$$

This is 6.2% of the speed of light; this electron is moving right along.

7.4   SUMMARY: ELECTRIC FIELD
AND ELECTRIC POTENTIAL

It is convenient to attribute the electric force to two separate factors: to a property of space and to a property of the body to be placed in that space. The property of space we call "electric field." The property of the body is its "electric charge." The magnitude of the electric field E at a point in space equals the ratio of the force exerted on a small test charge q_t placed at that point to the magnitude of the charge:

$$E = \frac{F}{q_t}.$$ Electric field defined—p. 170

Electric field has direction; it points in the direction that a small, positive charge will move when placed in the field. The units of electric field are $N\ C^{-1}$ or, equivalently, $V\ m^{-1}$.

The force F on a charge Q placed in an electric field $E(x, y, z)$ is

$$F = QE.$$ Force a field exerts on a charge—p. 170

If the field E is everywhere constant, the charge Q will experience the same constant force when placed anywhere in such a field. The constant electric field is of practical importance because it is a good approximation to the electric field between two flat metal sheets close together and charged with opposite signs of charge. This arrangement is easy to make in the laboratory.

It follows from the definition of electric field and Coulomb's law that a point charge q_s will create at a distance r from itself an electric field E given by the following expression:

$$E = \frac{k_c q_s}{r^2},$$ Electric field of a point charge—p. 171

where k_c is the Coulomb force constant closely equal to $9 \times 10^9\ N\ m^2\ C^{-2}$.

Electric potential V is often easier to work with than electric field. An amount of charge q moved from one point in space to another undergoes a change in potential energy ΔU that is related to the difference between the electric potentials ΔV at the two points:

$$\Delta U = q\,\Delta V.$$ Change of q's potential energy when q is

moved through a potential difference—p. 180

You will often use conservation of energy to calculate the change in kinetic energy of a charged particle from its change in potential energy:

$$\Delta K = -\Delta U,$$

$$\frac{1}{2}\,mv_2^2 - \frac{1}{2}\,mv_1^2 = -q(V_2 - V_1).$$

The units of electric potential are "volts"; 1 volt = 1 J C^{-1}.

When the electric field is constant, there is a particularly simple relationship between electric field and electric potential. Two points separated by a distance Δy along the direction of a constant electric field have a potential difference ΔV given by the expression

$$\Delta V = -E\Delta y.$$

For this special case the potential difference is the product of the electric field E and the distance between the points.

Often you want to reason in the opposite direction and find E from V and d. For parallel charged plates separated by a distance d small compared to the dimensions of the plates and with a potential difference V between them, E is nearly constant, and you can find the strength, i.e., the magnitude, of the field, from the above equation:

$$E = \frac{V}{d}.$$

You can always find the direction of E by seeing in which direction V decreases most steeply, but for uniformly charged parallel plates the direction is from the plate with the higher potential toward the one with the lower potential.

The electric potential that a point charge q_s produces at a distance r away from itself is

$$V(r) = \frac{k_c q_s}{r}.$$

An electric field always points in the direction from high electric potential to low electric potential. (Why? You should be able to explain this from your knowledge about the electric field.)

PROBLEMS

1. What property of matter gives rise to electric fields?

2. A charge $q = 0.2\,\text{nC}$ is placed 10 cm from a charge $Q_a = 20\,\text{nC}$ as shown in Fig. 7.4a.

    a. What is the magnitude and direction of the electric field at the position of q?

FIGURE 7.4   (a) A charge $q = 0.2\,nC$ is placed in the electric field of $Q_a = 20\,nC$; (b) an additional charge of $Q_b = -80\,nC$ contributes additional electric field. (Problem 2).

    b. A second charge $Q_b = -80\,nC$ is added at a point 20 cm to the right of q as shown in Fig. 7.4b. What is the total electric field experienced by q?

    c. If Q_b is replaced by a positive charge of $80\,nC$, what is the electric field at the location of q?

    d. If q is changed to $q = 0.4\,nC$, what will then be the correct answers to a., b., and c.?

3. A charge of $1\,pC$ and mass $1\,pg$ is put at a certain point in space. As soon as it is released it begins to accelerate at $10^8\,m\,s^{-2}$.

    a. What is the magnitude of the electric field?

    b. If you are observing this particle near Earth, how do you know the acceleration is due to an electric field and not gravity?

4. An electron accelerates through a potential difference of 250 V.

    a. How much kinetic energy does the electron gain? Give your answer in electron volts and also in joules.

    b. If the electron was accelerated from rest, what would its velocity be?

5. The diagram in Fig. 7.5 shows two charged parallel plates separated by 5 mm. A device, which might be a battery or a power supply, has been attached to the two plates and has established a difference of electric potential of 10 V between the plates.

The symbol $\stackrel{\perp}{=}$ for "ground" or "common" means that the potential of the bottom plate is defined to be 0 V.

    a. What is the electric potential anywhere in the plane parallel to and halfway between the two metal plates?

    b. What is the electric potential energy of a $+2\,\mu C$ charge when it is halfway between the two plates?

    c. What is the electric potential energy of a $-2\,\mu C$ charge when it is halfway between the two plates?

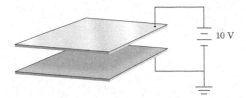

FIGURE 7.5 A pair of metal plates 5 mm apart, parallel to each other, and connected to a 10 V battery (Problem 5).

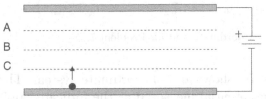

FIGURE 7.6 Parallel plates 8 cm apart and a 5 nC charged particle in an electric field of $E = 2.5 \times 10^4$ N C^{-1} (see Problem 6).

   d. For the previous two cases:

      i. Which charge has the larger electric potential energy?

      ii. Which charge is at the larger electric potential?

   e. If the two charges are moved to the bottom plate, which of them undergoes the largest increase in potential energy? How much?

   f. For the negative charge, what is the direction in space in which its potential energy will decrease most steeply? Why or why not is this surprising?

   g. In what direction does the potential energy not change?

   h. What is the value of E between the plates shown in Fig. 7.5?

6. A positive charge $q = 5$ nC is between two oppositely charged parallel conducting plates that are 8 cm apart (see Fig. 7.6). The plates are held a potential difference of V by a battery as shown. The electric field between the plates is $E = 2.5 \times 10^4$ N C^{-1}.

   a. What is the battery voltage?

   b. Assume the potential of the bottom plate to be $V = 0$. What is the electric potential on each of the planes A, B, and C?

   c. Again assume $V = 0$ at the bottom plate. What is the potential energy of the charge when it is at plane A? Plane B? Plane C?

   d. If the charge leaves the bottom plate with kinetic energy of $8\,\mu$J, what is its kinetic energy when it reaches plane C? Will it reach the top plate? If so, what will its kinetic energy be just before it hits the top plate? If not, how high will it go?

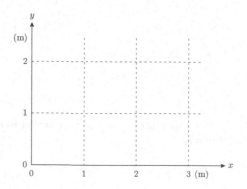

FIGURE 7.7   Coordinate grid for Problem 7.

7.   Figure 7.7 shows an x-y coordinate system. The table gives values of the electric potential measured at the indicated points.

(x, y)	Potential
(m)	(V)
$(0, 0)$	0
$(0, 2)$	4
$(1, 0)$	2
$(1, 1)$	8
$(1, 2)$	8
$(3, 1)$	10
$(3, 2)$	20
$(2, 2)$	10

a. Mark the points $(1, 0)$, $(2, 2)$, $(3, 2)$, and $(0, 2)$ on the graph

b. Label each point with the value of the electric potential at that point

c. A droplet containing a net charge of $-7\,e$ is released at $(0, 0)$ and moves freely to $(3, 1)$. By how much does its kinetic energy change? Give your answer in units of electron volts.

d. The particle then continues moving and goes from $(3, 1)$ to $(1, 2)$. Does its kinetic energy increase or decrease? How do you know?

8.   The apparatus diagrammed in Fig. 7.8, called a "tandem accelerator," is used to accelerate ions to high kinetic energies.

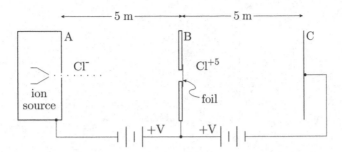

FIGURE 7.8 Problem 8 tandem accelerator.

At the left, labeled A, is a source of negatively charged chlorine ions Cl^- (i.e., each ion carries a charge $-e$). A 1-MV potential difference is applied between plates A and B, as shown.

    a. Find the electric field in the space between A and B. Specify both direction and magnitude.

    b. Find the kinetic energy of the ions when they reach plate B. Use appropriate units.

In the center of plate B is a hole covered by a very thin foil. Each Cl^- ion passing through this foil is stripped of six electrons while its velocity hardly changes (really true!). As a result, chlorine ions of charge $+5\,e$ emerge from plate B and are accelerated further in the region between B and C. The potential difference between B and C is $1\,MV$.

    c. On a coordinate system like that in Fig. 7.9 graph the kinetic energy vs. position from A to C. Scale the vertical axis in appropriate energy units. Your graph should be as quantitatively correct as possible.

    d. The atomic mass of Cl is 35. Show that the mass of a chlorine ion is 5.8×10^{-26} kg. What is the velocity of the ions reaching plate C?

9.   A voltage of 400 V is applied to the plates shown in Fig. 7.10.

    a. If the plates have a separation of 20 cm, calculate the magnitude and direction of the electric field.

    b. Imagine you now send a particle of charge $q = +2\,e$ and mass $m = 6.4 \times 10^{-27}$ kg through the hole in the left plate, as shown in the figure. What would be the particle's initial velocity such that it slows down and stops right at the other plate (without crashing)?

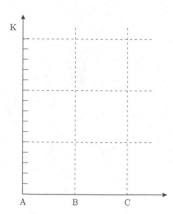

FIGURE 7.9 Coordinate grid for Problem 8.

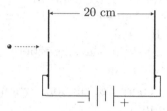

FIGURE 7.10 Region of electric field for Problem 9.

    c. Describe qualitatively the motion of the particle if you change the voltage to 200 V (same polarity).

    d. Now restore the voltage to 400 V. Describe qualitatively the motion of the particle if you decrease its initial velocity from the value found in (b).

10. Two parallel plates separated by 50 cm have a potential difference of 10 V between them, as shown in Fig. 7.11. A positively-charged ion with $q = 1.6 \times 10^{-19}$ C, initially at rest at the left plate, is accelerated between the plates and passes through a small hole in the right plate.

    a. What is the electric field between the plates, and what is its direction? (Don't forget units!)

    b. If the ion's speed when it passes through the hole in the right plate is $v = 2.89 \times 10^4$ m s^{-1}, what is its mass?

11. Suppose that to create an electron filter you modify the setup shown in Fig. 7.11 by drilling another small hole in the left-hand plate opposite the hole in the right-hand plate. (The holes are too small to have much effect on the electric field between the plates.) Suppose that you want to set up this apparatus so that when electrons are fired towards the holes,

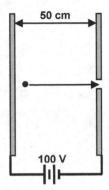

FIGURE 7.11 An arrangement of charged plates for accelerating positively charged ions.

any electrons with speeds less than $10^7 \, \mathrm{m \, s^{-1}}$ will be slowed down by the electric field between the plates and will not emerge from the other side, while electrons with higher speeds will emerge through the hole. Assuming your voltage source can be turned up as high as $500 \, \mathrm{V}$, how would you adjust your apparatus to achieve your goal? Be sure to specify the distance between the plates and the polarity of the voltage.

CHAPTER 8

Magnetic Field and Magnetic Force

8.1 MAGNETIC FIELD

In this chapter you meet another field of force, the magnetic field. It is quite different from the electric field. Electric fields produce forces on electrical charges whether they are moving or sitting still. The magnetic field exerts a force on an electric charge *only* if the charge is moving. Equally strange, the strength of the exerted force depends upon the direction of the charge's motion. Whenever you see such peculiar behavior, you know there is a *magnetic* field present.

Magnetic Force on a Moving Charge

Figure 8.1 shows different states of motion of an electric charge in a magnetic field B. If the charge is at rest and free to move, but does not accelerate, then no electric field E is present. If you now put the charge in motion, then in the presence of B you will see a force act on the charge. If you experiment by moving the charge in different directions, you will observe some curious things.

- When the charge moves in one particular direction, no force occurs.

- When the charge moves in other directions, a force does act on it but in a direction perpendicular to the charge's line of motion *and* perpendicular to that line of motion along which no force occurred.

- The strength of the force is different for different directions of the charge's motion; the force is a maximum when the charge is moving perpendicular to the line of motion for which the force is zero.

C.H. Holbrow et al., *Modern Introductory Physics, Second Edition,*
DOI 10.1007/978-0-387-79080-0_8, © Springer Science+Business Media, LLC 1999, 2010

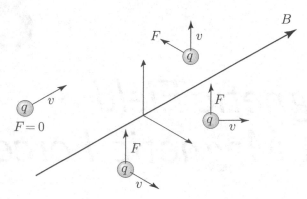

FIGURE 8.1   A set of pictures showing how the magnetic force on a positive charge q varies in magnitude and direction for various directions of motion with respect to the line of motion along which there is no force. The direction of the magnetic field is taken to be parallel to that line.

Clearly, some new kind of force is present. Physicists attribute this new force to a new kind of field that they call the "magnetic field." This is the kind of field that surrounds a bar magnet. This same kind of field surrounds the Earth and makes the little magnet that is a compass needle point northwards. A magnetic field also occurs around any electric current or moving charge.

If you do a series of experiments that measure the force on a succession of different charges all moving in the same direction at the same location in space, you find that the force due to a magnetic field is proportional to the size of the charge:

$$F \propto q.$$

If you move a charge q in the direction such that q experiences the maximum force F, and then vary the speed v along this direction, you find that the force is proportional to v, so

$$F \propto q\,v.$$

As with the electric field, the force arises from properties of the particle—its charge and its velocity—and from this new entity filling the space through which the charge is moving—the magnetic field. It is customary to represent the magnetic field by the symbol B. In a manner analogous to the case of the electric field, the magnitude of the magnetic force is proportional to the product of the particle properties and the field strength

$$F \propto qvB.$$

The units of magnetic field B are defined to make the constant of proportionality in this equation be 1 when the direction of v makes the force a maximum. Then the numerical value of B, i.e., the strength of the magnetic field, is defined by the equation

$$F_{\max} = q\, v\, B.$$

Thus, if a charge of 1 C moving with a speed of 1 m/s experiences a maximum force of 1 N, the magnetic field must have a value of 1 unit of field strength.

The SI units of B are $N\, C^{-1}\, m^{-1}\, s$, and they are given the name "tesla," abbreviated T. A magnetic field of 1 tesla (1 T) can exert a maximum force of 1 N on a charge of 1 C moving with a speed of 1 m/s. Earth's magnetic field is around 0.5×10^{-4} T. In the United States the field strength varies from nearly 60 μT at the northern border to close to 48 μT at the southern border. In the past the unit of magnetic strength was the "gauss" (abbreviated G). This unit is still often used: 1 gauss = 10^{-4} tesla. It's useful to remember that the intensity of Earth's magnetic field is about 0.5 G, a half of a gauss.

Magnetic field has direction as well as magnitude, but the direction of B is defined differently than that of electric field E. As you learned in Chap. 7, electric field has the direction of the force it exerts on a positively charged particle. Such a straightforward approach will not work for magnetic field, because the force it exerts is different in both direction and magnitude for different directions of the charged particle's motion.

As already mentioned, there is a special direction of motion along which any charged particle feels zero force regardless of the size of its charge or how fast it is moving. The direction of magnetic field B is taken to be along this zero-force direction of motion, i.e., the magnetic field B points along this line. As a result of this definition, a charge q moving with a speed v *perpendicular* to the direction of a magnetic field B feels the maximum force given by $F_{\max} = qvB$; a charge moving parallel or antiparallel to the direction of a magnetic field feels no force. These behaviors are illustrated by Fig. 8.1 which also shows that the line of B is perpendicular to the direction of motion that gives rise to a maximum force, and that the force is perpendicular to both of these directions, i.e., the force is perpendicular to the plane formed by the vectors $\vec{v}$ and $\vec{B}$.

One consequence of these properties of B is that the needle of a magnetic compass naturally lines up with this zero-force direction. (Why this is so may become clear later.) This behavior of the compass needle resolves the remaining ambiguity in the definition of the direction of B: Which of the two directions a charge can move along the line of zero force is the direction of B? As a matter of convention, we agree that B points in the same direction that the northward-pointing pole of a compass needle points when placed in the magnetic field B.

This convention establishes a fixed relationship between the directions of the charge's motion v, the magnetic field B, and the force F on the charge. If you know any two of these three quantities, you can figure out the direction of the third. This means that by observing the direction of motion of a charge and the direction of the force that a magnetic field exerts on the charge, you can determine the direction of B. This is the usual way to find the direction of B, and you will need to do this often. You will also need to be able to predict the direction of the force exerted by a known B acting on a charge moving in a known direction with a speed v. It helps if you can imagine in your head a picture like that of Fig. 8.1.

There are several recipes for figuring out the direction of a force exerted by a magnetic field on a moving charge. Figure 8.2 shows how to determine the direction of the force when a positively charged particle is moving with velocity v at some angle θ relative to B. Remember the force is always perpendicular to the plane determined by the directions of v and B. You can decide which way F acts, if you imagine that you stand on that plane at the vertex of the angle formed by v and B. Then if the line of v is to the right of B, the force is in the direction of your head. Otherwise, the force is toward your feet. If the charge is negative, the forces are reversed. There are other ways to determine the direction of the force that involve only moving your hand, rather than standing on a plane. All of these recipes are versions of what is known as "the right-hand rule."[1]

The magnitude of the force on a moving charge q is less than its maximum value when it moves in a direction that is not perpendicular to B. The force is found to be proportional to the sine of the angle between v and B. Therefore, the magnitude of the magnetic force on a charge q moving with velocity v at an angle θ relative to B is

$$F = qvB \sin \theta. \tag{1}$$

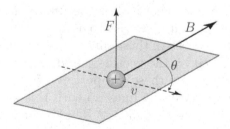

FIGURE 8.2 The force on a positively charged particle moving with velocity v at an angle θ relative to B.

---

[1] If you know about the vector cross product, then you can just remember that all these rules are equivalent to the vector equation $\vec{F} = q\vec{v} \times \vec{B}$.

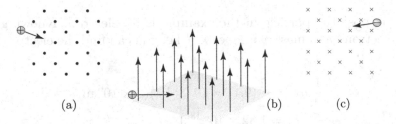

FIGURE 8.3 Three examples of charged particles moving in a uniform magnetic field. The dots mean that the magnetic field B points out of the page; the ×'s mean that the magnetic field B points into the page.

The direction is given by the right hand rule. This force of a magnetic field on a moving particle is often called the "Lorentz force."

▉ EXERCISES

1. The three diagrams in Fig. 8.3 show a positive charge q moving with velocity v in different directions relative to the magnetic field B. The diagram uses the customary convention of dots representing B pointing out of the plane of the page (tips of arrowheads), and ×'s representing B pointing into the plane of the page (tail feathers of arrows). For each case, what is the direction of the force?

Another way to describe how the magnitude of the force depends on the angle between the directions of v and B is to observe that $v \sin \theta$ is the component of the velocity that is perpendicular to B. Therefore, F is proportional to $v_{\text{perp}} = v_\perp = v \sin \theta$. For most of the cases of interest to us, v will be perpendicular to B and $\sin \theta$ will equal 1. However, when they are not perpendicular, you must include the factor of $\sin \theta$.

▽ EXAMPLES

1. Suppose a particle with a charge of -1.6×10^{-19} C moves with a speed of 2×10^6 m/s perpendicular to Earth's magnetic field of $55 \, \mu$T. What is the force on this particle? This is a straightforward calculation:

$$F = (1.6 \times 10^{-19} \, \text{C})(2 \times 10^6 \, \text{m/s})(55 \times 10^{-6} \, \text{T})$$
$$= 1.76 \times 10^{-17} \, \text{N}.$$

Suppose the particle in the example is an electron. Knowing that an electron has a mass of 9.11×10^{-31} kg, you can find the electron's kinetic energy:

$$\frac{1}{2}mv^2 = \frac{1}{2}(9.11 \times 10^{-31}\,\text{kg})(2 \times 10^6\,\text{m/s})^2$$
$$= 1.82 \times 10^{-18}\,\text{J}$$
$$= 11.4\,\text{eV}.$$

From this you see that because of its very small mass, it takes only about 11 V to get an electron moving as fast as $2 \times 10^6\,\text{m s}^{-1}$.

EXERCISES

2. What is the acceleration of the electron by the force in the above example? How does your answer compare to g, the acceleration due to gravity?

A Moving Charge in a Uniform Magnetic Field

The motion of a charge in a uniform magnetic field will be quite important to us. Such magnetic fields are frequently used to measure a charged particle's momentum.

Figure 8.4 shows the paths of three positively charged particles moving in a plane perpendicular to a uniform magnetic field B. As before, the little dots in the figure indicate that the magnetic field is pointing up out of the plane of the page. One of the particles, labeled by its charge q, enters the region with an initial velocity directed to the right. From the rules for finding the direction of the force on q, you should be able to show that at this initial instant, the force on q is toward the bottom edge of the page. (Try it!)

But the force changes the direction of the velocity of the charge, and as soon as the direction of v shifts, so does the direction of the force. These directions remain at right angles to each other (because F is always perpendicular to v), with the result that the charge moves in a circle of some radius R at a steady speed v. In other words, in a constant field B the charge moves with uniform circular motion.

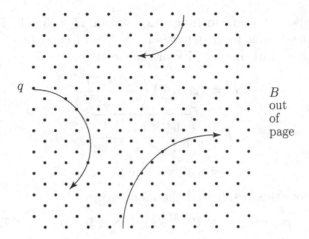

FIGURE 8.4 Circular trajectories of positively charged particles entering a uniform magnetic field pointing out of the page.

You saw in Chap. 2 that a force that makes a particle of mass m move at a uniform speed v in a circle of radius R always has a magnitude of

$$F = m\,a = m\frac{v^2}{R}.$$

For the situation here this force is the Lorentz force qvB, so you can equate the two expressions,

$$qvB = \frac{mv^2}{R},$$

and rearrange them to get

$$mv = p = qBR, \tag{2}$$

where, as in previous chapters, the symbol p represents momentum mv.

Equation 2 is important because it allows you to find the momentum of a fast moving charged particle by measuring the radius R of the particle's circular path as it bends in a uniform magnetic field of known magnitude B. If, as is often the case, you know v from some other measurement, you can then use Eq. 2 to find the particle's mass. Such measurements permit very precise determination of very small masses.

▼ **EXAMPLES**

2. If electrons accelerate from rest through $200\,\mathrm{V}$, they acquire a kinetic energy of $200\,\mathrm{eV}$. Assume that a beam of these electrons enters a region of uniform magnetic field of $B = 5\,\text{mT}$. If the beam enters perpendicular to B, what is its radius of curvature?

First, find the velocity of the electron.

$$\frac{1}{2}mv^2 = (200\,\text{eV})\left(\frac{1.6 \times 10^{-19}\,\text{J}}{1\,\text{eV}}\right)$$

$$v = \sqrt{\frac{64 \times 10^{-18}\,\text{J}}{9.11 \times 10^{-31}\,\text{kg}}}$$

$$= 8.38 \times 10^6\,\text{m\,s}^{-1}.$$

From this result find the momentum

$$p = mv = (9.11 \times 10^{-31}\,\text{kg})(8.38 \times 10^6\,\text{m\,s}^{-1}) = 7.64 \times 10^{-24}\,\text{N\,s},$$

from which you can get

$$R = \frac{p}{qB} = \frac{7.64 \times 10^{-24}\,\text{N\,s}}{(1.6 \times 10^{-19}\,\text{C})(5 \times 10^{-3}\,\text{T})}$$

$$= 9.54 \times 10^{-3}\,\text{m} = 9.54\,\text{mm}.$$

These units work out because, either from Eq. 1 or from the definition on p. 201, $1\,\text{T} = 1\,\dfrac{\text{N·s}}{\text{C·m}}$.

3. Consider an electron (mass 9.11×10^{-31} kg) in a uniform magnetic field of $0.1\,\text{mT}$. If the electron's path is observed to bend with a radius of $10\,\text{cm}$, what is the electron's momentum? What is its kinetic energy?

Its momentum is

$$p = qRB = (1.6 \times 10^{-19}\,\text{C})(0.1\,\text{m})(1 \times 10^{-4}\,\text{T}) = 1.6 \times 10^{-24}\,\text{N\,s},$$

and its kinetic energy is

$$\frac{1}{2}mv^2 = \frac{p^2}{2m} = \frac{(1.6 \times 10^{-24}\,\text{N\,s})^2}{2 \times (9.11 \times 10^{-31}\,\text{kg})}$$

$$= 1.41 \times 10^{-18}\,\text{J}$$

$$= 8.78\,\text{eV}.$$

■ EXERCISES

3. Show that $R = \sqrt{2mK}/qB$ and then redo Example 2. Hint: From Chap. 2 you know that kinetic energy $K = mv^2/2$ and $p = mv$.

4. What are the momentum and kinetic energy of an electron that bends through a radius of 5 cm in a magnetic field of 2 mT?

Sources of Magnetic Fields

Electric currents are the sources of magnetic fields. Whenever a charge moves, it produces a magnetic field. Notice the remarkable reciprocity: Magnetic fields exert forces on moving charges; moving charges, i.e., electric currents, produce magnetic fields. Just as electric charges produce electric fields that act on other electric charges, *electric currents produce magnetic fields* that exert forces on other currents.

An electric current I flowing in a long straight wire produces a magnetic field around the wire. The field's direction is always tangent to circles centered on the wire. To find the direction tangent to these circles, imagine that you grasp the current-carrying wire with your *right hand* so that the direction of your thumb points in the direction of the flow of the current (the direction that positive charge would have to be flowing to create the observed current). Then your fingers curl around the wire in the direction of the magnetic field. This rule, also called the "right-hand rule," is illustrated in Fig. 8.5; it is *not* the right-hand force rule.

The strength of the magnetic field B a distance r from a long, straight wire is

$$B(r) = \frac{\mu_0 I}{2\pi r}, \tag{3}$$

where μ_0 is a constant equal to $4\pi \times 10^{-7}\,\mathrm{T\,m\,A^{-1}}$ and I is the current flowing in the wire.

The magnetic field produced by the current in one wire will add (vectorially) with the magnetic field produced by a current in another

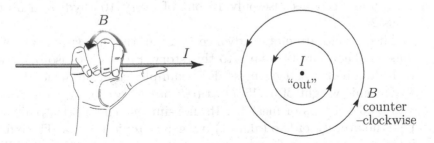

FIGURE 8.5 The right-hand rule for finding the direction of the magnetic field surrounding a long, straight, current-carrying wire.

wire. By using many wires or one wire bent into many loops it is possible to design different arrangements and strengths of magnetic field. In particular, it is possible to produce a magnetic field that is essentially uniform in a limited volume of space. Such uniform B fields are used in actual experiments and practical devices as well as for generating a wide variety of physics homework problems, exercises, and exam questions.

8.2   MAGNETIC FIELDS AND ATOMIC MASSES

Physicists have used magnetic deflection of charged particles to make two major advances in our understanding of atoms.

First, magnetic deflection provided conclusive evidence that the atoms of a chemical element may not all be identical. As early as 1912, J.J. Thomson used magnetic deflection to show that there are at least two kinds of neon atoms occurring in nature, one of atomic mass 20 and another of mass 22. (There is also a third naturally occurring neon atom of mass 21.) Atoms that have the same chemical behavior but different atomic weights are called "isotopes." Thus, chemically pure neon gas is a mixture of three different isotopes. Throughout the 1920s and 1930s Francis W. Aston used magnetic deflection to discover that most elements have more than one isotope. Physicists and chemists have now identified around 335 naturally occurring isotopes of the approximately 90 elements found on Earth.

For example, the two familiar chemical elements carbon and oxygen each have different isotopes. Most carbon atoms have mass 12, but about 11 out of every 1000 carbon atoms have mass 13. Most oxygen atoms have mass 16, but out of every 10^5 oxygen atoms 39 have mass 17, and 205 have mass 18. In 1931 Harold Urey discovered that familiar, much studied hydrogen has an isotope of mass 2 (called "deuterium"). It had gone unnoticed because only 15 out of every 10^5 hydrogen atoms have mass 2.

The second dramatic advance was in the measurement of atomic masses. There are two parts to this story. First, every isotope was found to have a nearly integer mass. For example, chlorine, which has a *chemical* atomic weight of 35.4527 u, turns out to be 75.4% atoms of mass 35 and 24.6% atoms of mass 37. Its non-integer chemical atomic weight is just the average of two (almost) integer isotopic masses. The existence of integer masses strongly suggests that there is an atom-like building block within the atom, another level of atomicity.

■ EXERCISES

5. Show that the above relative abundances of the two chlorine isotopes explain the observed non-integer chemical atomic weight.

Second, when measured precisely enough, the mass of every isotope turns out to be slightly different from an exact integer. And these small differences are very important. Using magnetic deflection it became possible to measure these differences with accuracies of parts per ten thousand and in this way find atomic masses of isotopes to a precision of parts per million. This was an enormous advance over the traditional chemical methods of determining atomic weights (Chap. 3), which even when pushed to their limits yielded values accurate to no better than a few tenths of a percent. It had dramatic consequences well beyond advancing our understanding of atoms. The new knowledge helped to explain how stars generate their energy, and it contributed significantly to the development of nuclear weapons.

Magnetic Mass Spectrometry

The discovery of isotopes and the precise measurement of their atomic masses were made possible by devices called "magnetic mass spectrometers." These devices bend ions in a magnetic field, and the diameters of their orbits provide a measure of the ions' masses. Let's look at one such device used by K.T. Bainbridge in the 1930s that combines electric and magnetic forces in a clever way. Its design is shown schematically in Fig. 8.6. There are two parts: a device that produces ions of a given charge and sends them down a channel in which there are both electric and magnetic fields exerting forces in opposite directions on the ions; and a region that contains only magnetic field that causes the ions to bend in circular paths.

The channel is an example of a "Wien velocity filter," named after the physicist Wilhelm Wien who invented it. Because the magnetic field pushes the ions to one side of the channel while the electric field pushes them to the other, only those ions for which the two opposing forces are equal can reach the exit slit S_3 without hitting the walls of the channel. The electric force is qE; the magnetic force is qvB. The two forces will be equal only for ions moving with the particular velocity v such that

$$qE = qvB,$$

which reduces to

$$v = \frac{E}{B}.$$

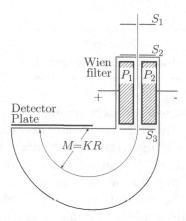

FIGURE 8.6 Bainbridge's apparatus for measuring the ratio of the charge to mass of ions. A voltage between the plates containing the slits S_1 and S_2 accelerates ions into the Wien filter. An electric field E, produced by a voltage between the plates P_1 and P_2, deflects the ions to their right; a magnetic field B at right angles to E deflects ions to their left. Beyond slit S_3 there is only a uniform magnetic field, and it bends ions in circular arcs.

In other words, only those ions with velocity $v = E/B$ can pass through the channel into the bending region. The other ions are filtered out, and the ions entering the bending region all have the same velocity.

■ EXERCISES

6. Show that E/B has units of $\mathrm{m\,s^{-1}}$ when E and B are expressed in SI units.

7. In Fig. 8.6, the uniform magnetic field points out of the page. What is the direction of the magnetic force on a positive charge traveling through the filter towards the bending region? What is the direction of the electric force on a positive charge in this region? Suppose the charges are negative. Would they still pass through the filter undeflected? What if you reversed their direction through the filter? Would they move from S_3 to S_1 without deflection?

8. Suppose that instead of the orientation shown in Fig. 8.6, the magnetic field points to the right and the particles are moving into the page. Which way should the electric field point in order to make a Wien filter? Sketch the orientation of the plates and show which one will be at the higher voltage.

If the ions entering the constant B field that fills the bending region all have the same charge and the same velocity, those ions that have the same mass m_{ion} will bend with the same radius of curvature R. If a photographic plate is placed as shown in Fig. 8.6, the ions will curve around and strike it at a distance from the entrance slit equal to the diameter of the ions' circular path. When the plate is developed, there will be a dark line where the ions struck. If ions with two different masses come through the apparatus, there will be circular paths with two different radii, and there will be two different lines on the developed photographic plate. Because $m_{ion}v = qRB$ and $v = E/B$, it follows that

$$m_{ion} = \frac{qB^2}{E}R. \tag{4}$$

For fixed values of E and B the mass of the ion m_{ion} is directly proportional to R, which can be determined by measuring the position of the dark line that the ions make on the photographic plate.

The above calculation assumes that B is the same in the bending region as in the Wien filter. This is correct for the Bainbridge apparatus, and so we used the same symbol for both regions. But clearly the two magnetic fields don't need to be the same. If they are not, you must create separate symbols for them (say B_1 and B_2), and your final result will be $m_{ion} = qB_1B_2R/E$. This modification doesn't make much difference to the result—the mass is still proportional to the radius, but it does make a point: Be alert about what your symbols mean, and don't use the same symbol for two different quantities.

To operate this mass spectrometer, choose B and E to select some particular velocity $v = E/B$. Then vary the accelerating voltage from, say, 5 kV to 20 kV. If in that range of accelerating voltages any ions acquire the correct velocity to pass through the filter, they will enter the magnetic field and be bent proportionally to their masses. Let's see how this might work for isotopes of neon.

▼ EXAMPLES

4. Suppose that when the accelerating voltage is set to 10 kV, singly charged neon ions are able to pass through crossed electric and magnetic fields of strength 4.66×10^5 N C^{-1} and 1.5 T, respectively. If, after traveling through the spectrometer, the neon ions hit the detector 8.58 cm away from where they exited the Wien filter, what is the atomic mass of any one of these ions?

The values of E and B tell you the ion velocity: $v = E/B$; the distance $d = 8.58$ cm specifies the radius of their circular path: $R = 0.0429$ m; and "singly charged" means each ion has a charge of $q = 1.602 \times 10^{-19}$ C. From this knowledge and Eq. 2 it follows that

$$m_{\text{ion}} = qR\frac{B^2}{E} = \frac{1.602 \times 10^{-19} \times 0.0429 \times 1.5^2}{4.66 \times 10^5} = 3.319 \times 10^{-26} \text{ kg}.$$

Dividing this result by 1.66×10^{-27} kg/u to convert from kilograms to atomic mass units, you get $m_{\text{ion}} = 19.992$ u. These are mass-20 ions of neon; the accepted value of their atomic mass is 19.992435 u.

■ EXERCISES

9. Show that if the neon isotope of mass 22 is present, the accelerating voltage will have to be 11 kV for these ions to pass through the same combination of E and B-fields that mass-20 ions passed through when V was 10 kV. Note that because the two isotopes are chemically identical, the charge of their ions will be the same.

10. Show that once these mass-22 ions enter the bending region, they will strike the photographic plate 9.44 cm from the exit of the Wien filter.

11. Where would you look on the photographic plate for evidence of a mass-21 isotope of neon? And what accelerating voltage would allow these mass-21 ions to pass through the Wien filter?

It is always important to understand the precision of measurements. How precisely can mass be measured with Bainbridge's mass spectrometer? Clearly, the answer will depend on how precisely you can measure E, B, and R. It is not difficult to measure R to a fraction of a millimeter, say ± 0.1 mm. Out of 8.2 cm this would be a fractional precision of

$$\pm\frac{\Delta R}{R} = \pm 1.2 \times 10^{-3} \approx 0.1\%,$$

or 12 parts per 10 000. This is not good enough to show that mass-20 neon's mass deviates slightly from being an integer, because that deviation turns out to be only $\approx \Delta M/M = 4 \times 10^{-4}$, i.e. 4 parts per 10 000.

There is a tactic often used by experimental physicists to improve the precision of their measurements. Rather than measure the value of a quantity directly, measure its difference from something that is already

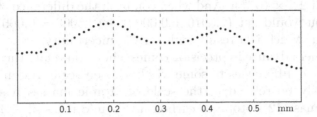

FIGURE 8.7 Two lines on a photographic plate of a mass spectrometer. The separation between the lines corresponds to the mass difference between an ionized hydrogen molecule H_2^+ and an ionized atom of mass-2 hydrogen.

known accurately. And if you don't have an accurately known reference, set one up. You have already seen one example of this technique: When chemists and physicists did not know how to determine the actual masses of atoms of the chemical elements, they made up a scale by assigning a mass of 16 to oxygen (today we use carbon 12) and measured all other atomic masses relative to this standard. The magnetic mass spectrometer led to another version of this technique based on measurements of "mass doublets."

The idea is to send through the mass spectrometer two different ions composed of different types of atoms so that they have nearly the same molecular weight. For example, Aston was able to measure the difference in mass between the mass-2 isotope of hydrogen and the two-atom H_2^+ molecule of ordinary mass-1 hydrogen atoms by sending both through his mass spectrometer at the same time. They are a doublet because they have nearly the same mass ($2\,u$ in this case), but their masses are slightly different. Figure 8.7 shows a tracing of the darkening of the photographic plate that occurred in Aston's spectrometer. The separation of the two peaks is quite clear and is measurable to within $\pm 0.01\,mm$. This separation corresponds to a mass *difference* of $0.00155\,u$.

▪ EXERCISES

12. Show that the fractional uncertainty in this mass difference is $\approx 4\%$ and that the uncertainty in the mass difference is therefore $\pm 6 \times 10^{-5}\,u$.

Referring back to Example 7.4 you can see that this result is better than would have been obtained from two independent measurements of the hydrogen masses with fractional uncertainties of $\Delta M/M = \Delta R/R \sim 0.1\%$, because then you would measure $2\,u \pm 2 \times 10^{-4}\,u$ and

2.0016 u $\pm 2 \times 10^{-4}$ u. And when you took the difference of these two numbers you would get $(2.0016 \pm 0.0002) - (2.0000 \pm 0.0002) = 0.0016$ u $\pm$ 0.0003 u, which is a result with a 19% uncertainty.

Further advance in precision comes when you relate these measurements of relative differences to some wisely chosen scale. Aston's measurements can easily be related to the scale of atomic masses based on assigning to the mass-12 isotope of carbon an atomic mass of *exactly* 12, because he also showed that one could measure the mass difference between C^{++} and a 3-atom molecule of mass-2 hydrogen. Notice from Eq. 4 that a doubly charged mass-12 C ion bends with the same radius of curvature as a singly charged mass-6 ion, and therefore it will strike the same part of the plate as a singly charged ion consisting of three atoms of mass-2 hydrogen. The observed separation between the two peaks corresponded to a mass difference between the two ions of 0.0423 u.

These results yield a very precise value of the mass of the hydrogen atom. To see how this works, write the two mass differences as linear equations:

$$2M_1 - M_2 = 0.00155\,\text{u},$$

$$3M_2 - \frac{M_{12}}{2} = 0.0423\,\text{u},$$

where M_1 is the mass of mass-1 hydrogen, M_2 is the mass of mass-2 hydrogen, and M_{12} is the mass of mass-12 carbon. By agreement, this last value is chosen as the standard of the atomic mass scale and is assigned the value of exactly 12.

To solve the two equations, multiply the top one by 3 and add it to the bottom one. Solve for M_1. You should get $M_1 = 1.00782$ u, and because the above mass differences are modern values, this is the currently accepted result. It is confirmed in many ways.

■ EXERCISES

---

13. Use these mass differences to find the mass of mass-2 hydrogen.

Because of mass spectrometry and other techniques, we now know the atomic masses of all naturally occurring isotopes accurately to six decimal places. Mass spectrometry also tells us the relative abundances of the different isotopes. For example, comparison of the areas of the peaks in Fig. 8.8 (taken by A.O. Nier in 1938) shows the relative abundance of the different isotopes of lead found in nature.

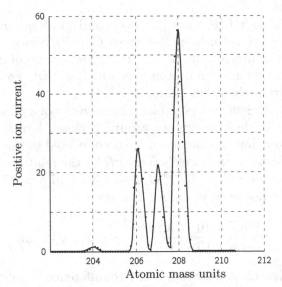

FIGURE 8.8 Mass spectrum of lead isotopes.

8.3 LARGE ACCELERATORS AND MAGNETIC FIELDS

We have described the forces exerted by magnetic fields on moving charged particles and shown that when a particle of charge q moves with momentum p perpendicular to a magnetic field B, the magnetic field will bend the particles in a circle of radius R such that

$$p = qBR.$$

When engineers and physicists design particle accelerators, they often use this property of magnetic fields to bend and guide fast-moving charged particles around a closed path.

The following two examples show how the strength of the magnetic field determines the size of the accelerator. As the momentum of the accelerated particles becomes greater and greater, it is necessary to make the accelerator very large.

▼ EXAMPLES

5. With modern superconducting materials it is feasible to build large magnets that operate reliably at magnetic fields as high as 8.3 T. Such magnets are at the edge of what is technically possible. The Large

Hadron Collider(LHC) has just been built underground at CERN, the European accelerator laboratory near Geneva, Switzerland. Two beams of protons of momentum $3.7 \times 10^{-15}\,\mathrm{kg\,m\,s^{-1}}$ circulate in opposite directions; occasionally a proton from one beam will collide head-on with a proton from the other beam.

What is the smallest possible circumference of a ring of 8.3 T magnets that could guide such protons around a closed loop?

You know that the momentum p is connected to the magnetic field B, particle charge q, and circle's radius R by the relationship $qBR = p$. For the LHC, $q = e = 1.6 \times 10^{-19}$ C, $B = 8.3$ T and $p = 3.7 \times 10^{-15}\,\mathrm{kg\,m\,s^{-1}}$. Solving the above equation for R gives

$$R = \frac{p}{eB} = \frac{3.7 \times 10^{-15}\,\mathrm{kg\,m\,s^{-1}}}{(1.6 \times 10^{-19}\,\mathrm{C})(8.3\,\mathrm{T})} = 2.8 \times 10^3\,\mathrm{m},$$

which is the same as 2.8 km. The circumference of a circle of this radius is about 18 km.

The actual distance around the LHC is almost 27 km because its shape is that of a race track rather than a circle. Between the arcs of bending magnets, there are straight sections where physicists install very large assemblies of equipment for studying collisions between the high-momentum particles.

6. On eastern Long Island at Brookhaven National Laboratory is an accelerator called the Relativistic Heavy Ion Collider (RHIC for short). Its 3.45 T magnets are designed to bend into a closed loop a gold ion with a charge of $79e$ (all its electrons removed) and a momentum of $1.05 \times 10^{-14}\,\mathrm{kg\,m\,s^{-1}}$.

What is the circumference of the smallest possible ring of magnets for RHIC?

This is the same problem as the previous one, and you can work it the same way. But it is good practice and more instructive to compare the parameters of the two accelerators and then scale the answer of 18 km from the LHC example.

How do you scale the answer? Notice that the maximum momentum of a gold ion in RHIC is 2.84 times that of a proton in the LHC. This means it will be 2.84 times harder for the B field to bend a Au ion than to bend a proton. On the other hand, because the gold ions are completely ionized and each ion has a charge of $79e$, it is 79 times easier for the B field to bend them. Finally, because the B field of RHIC is weaker than the B field of the LHC by a factor of $8.3/3.45 = 2.4$, the radius (and, therefore, the circumference) will be 2.4 times larger.

Put these together to get your answer: the smallest possible circumference of RHIC's ring of magnets will be $\frac{2.84 \times 2.4}{79} 18 = 1.6 \, \text{km}$.

If you found the argument hard to follow, look at the equation $p = qBR$. It tells you that for a given momentum p the radius R will decrease when either B or q is made larger, but it will increase if p gets larger. Therefore, the smallest possible circumference C of RHIC will be

$$C_{\text{RHIC}} = \frac{B_{\text{LHC}}}{B_{\text{RHIC}}} \frac{q_{\text{LHC}}}{q_{\text{RHIC}}} \frac{p_{\text{RHIC}}}{p_{\text{LHC}}} C_{\text{LHC}} = \frac{8.3}{3.45} \frac{1}{79} \frac{10.5}{3.7} 18 \, \text{km} = 1.6 \, \text{km}.$$

RHIC's actual circumference is 3.8 km. Figure 8.9 shows you that this is because between the bending regions of the ring there are six long straight sections where experiments are done.

The roughly hexagonal ring is very prominent from the air. Look for it if you are flying over eastern Long Island.

8.4   A SUMMARY OF USEFUL THINGS TO KNOW ABOUT MAGNETISM

In the chapters to come you will need to know the following things about magnetic fields.

Magnetic fields are produced by moving charges, e. g., currents in wires or beams of charged particles.

A magnetic field exerts a force only on a moving charge. The magnetic force is often called the Lorentz force; it is proportional to the charge, q, to that part of the charge's velocity, $v_{\perp}$, that is perpendicular to the magnetic field, and to the strength of the magnetic field, B. These relations are expressed in the equation

$$F = qv_{\perp} B = qvB \sin \theta. \qquad \text{Lorentz force, p. 202}$$

This equation defines the unit of magnetic field strength called the tesla, abbreviated T, which has the units of $\frac{\text{N·s}}{\text{C·m}}$.

The direction of a magnetic field is the direction that the northward-pointing end of a compass needle points when placed in the field. This direction is parallel to the line of motion of a charge on which B exerts no force.

The force exerted by a magnetic field on a moving charge is perpendicular both to the velocity of the charge and to the magnetic field. To find the direction of the force on a positive charge, imagine two arrows, one

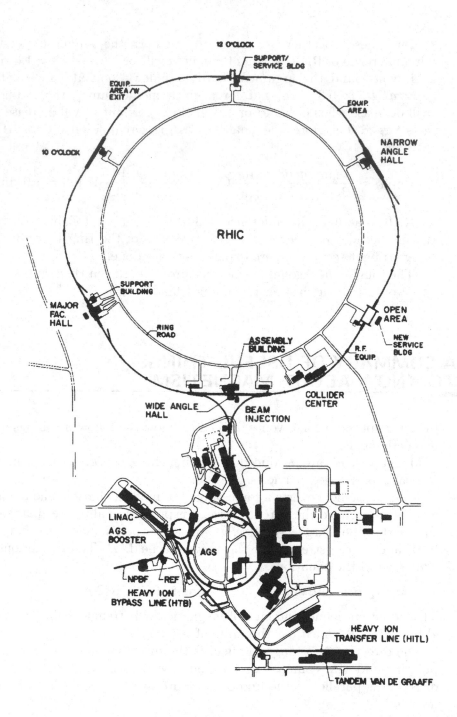

FIGURE 8.9   A map of RHIC. The large hexagonal ring is the main accelerator and storage ring containing the 3.45 T bending magnets. *Drawing provided courtesy of Brookhaven National Laboratory.*

pointing in the direction of the velocity and the other in the direction of the magnetic field. Now imagine the two arrows put tail to tail and that you are standing at the point where the tails join. Point your right arm in the direction of v and your left arm in the direction of B. If your arms are crossed, then the force is down, toward your feet; if your arms are not crossed, the force is up, toward your head.

This book often uses the special case of a charge q moving with a constant speed v perpendicular to a uniform magnetic field B. In this case the charge moves in a circle of radius R such that

$$mv = p = qRB.$$ Uniform magnetic fields bend moving charges in circles; p. 205

The deflection of charged particles by magnetic fields provides a way to measure masses of ions with great precision. These measurements show the existence of isotopes, atoms of the same chemical element with different masses. Magnetic fields are basic to the guidance and control of beams of charged particles whether they are the electrons that make the picture on the screen of old fashioned television sets or the beams of high-energy particles that circulate inside accelerators.

Magnetic fields are produced by moving charges. For example, a current I of charges moving in a long straight wire produces a magnetic field that surrounds the wire. For this special case the magnetic field has an intensity that drops off as the inverse of the distance R from the wire; the direction of the field is everywhere tangent to the circle of radius R:

$$B = \frac{\mu_0 I}{2\pi R}.$$ Magnetic field around a current in a long straight wire, p. 207

PROBLEMS

1. Suppose a particle made of equal amounts of negatively and positively charged matter with a total mass of $2\,\mu g$ is put initially at rest somewhere in space.

    a. You notice that although the negative and positive charges remain evenly mixed, this neutral particle begins to accelerate. What is likely to be the cause of such acceleration?

    b. Imagine you remove from the particle $1\,\mu C$ of negative charge. You observe that now when the particle is placed at rest at the same point in space as before, it begins to accelerate much more than before. What is likely to be the cause of this additional acceleration?

    c. You notice that once it is in motion, the acceleration of this charged particle remains essentially constant regardless of the direction in which the particle is moving. What does such behavior tell you about the presence of a magnetic field?

2. In Chap. 6, you learned that, even though it is negative charges that move in metals, the direction of current is conventionally chosen to be the direction in which positive charges would move to create the same current. To answer the question "Does this convention makes a difference to magnetic forces?" consider a wire with (positive) current flowing to the right and located in a uniform field pointing out of the page.

    a. Make a drawing of this situation.

    b. What would be the direction of the force on the wire if you assume that the current consists of the motion of positive charges?

    c. Remembering that the current actually consists of negative charges moving to the left, describe the direction of the actual force exerted by the magnetic field on the wire.

3. Draw a picture of two horizontal parallel wires placed 1 m apart and each carrying 1 ampere of current in the same direction.

    a. At the upper wire what is the direction of the B field caused by the lower wire? Hint: Consult Fig. 8.5 on p. 207.

    b. At the upper wire, what is the magnitude of the B field caused by the lower wire?

    c. What is the direction of the force on the upper wire exerted by the B-field from the lower wire?

    d. What is the average force on one electron in the wire? Assume the wires are 22-gauge copper (0.0644 cm diameter); the density of electrons is $8.43 \times 10^{22}\,\mathrm{cm}^{-3}$; and their average drift velocity is $2.27 \times 10^{-4}\,\mathrm{m\,s}^{-1}$ (as in Chap. 6).

    e. What is the force per meter on the wire?

    f. Put your result for the force into algebraic form. Instead of numbers use n (the number density of free electrons), v (the drift velocity of the electrons), I (the electric current), A (the cross sectional area of the wire), and μ_0 (the magnetic permeability of free space—the constant that appears in Eq. 3, etc.

    Notice that your final result depends only on the current I and the distance between the wires, not on the electron density n or the area A of the wire. This fact allows the official definition of the ampere to be the amount of current that produces a force of $2 \times 10^{-7}\,\mathrm{N\,m}^{-1}$ between two long parallel wires 1 m apart and

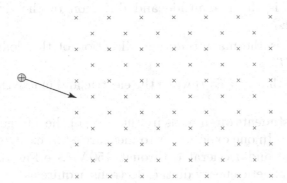

FIGURE 8.10 A positive charge enters a uniform magnetic field (Problem 4).

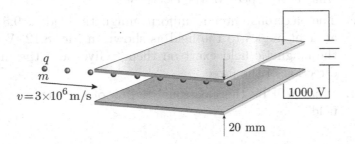

FIGURE 8.11 Arrangement for electric deflection in Problem 5.

carrying the same current. (The coulomb is then officially defined to be the amount of charge carried past a point in 1 s by a current of 1 ampere.)

4. A proton is accelerated from rest through a potential difference of 0.5 MV.

   a. What is its kinetic energy after this acceleration?

   b. What is its velocity?

   c. If it enters, as shown in Fig. 8.10, a uniform magnetic field of 0.1 T, pointing into the plane of the paper, does the proton's path curve toward the top edge of the page or toward the bottom edge?

   d. What is the radius of curvature of its path in the magnetic field?

5. An electric potential of 1000 V is placed across two plates separated by 20 mm, as shown in Fig. 8.11. A beam of particles each with charge q and mass m travels at $v = 3 \times 10^6$ m/s between the plates and is deflected downward. A uniform magnetic field applied to the region between the plates brings the beam back to zero deflection.

   a. What is the magnitude and direction of the deflecting electric field E?

   b. What is the magnitude and direction of the deflecting magnetic field B?

   c. Prove that $v = E/B$ when the electric and magnetic forces balance.

6. Physics students often measure the ratio of the charge e of an electron to its mass m. In one experiment to measure e/m, electrons are boiled off a hot filament and accelerated through 150 V (See Fig. 8.12).

   a. What kinetic energy do the electrons acquire as a result? Give your answer both in joules and in electron volts.

   b. After they have passed through a potential difference of 150 V, what is the speed of the electrons?

   c. The electrons enter a uniform magnetic field of 0.5 mT traveling at right angles to the field as shown in Fig. 8.12. What force will the magnetic field exert on them? Give both the magnitude and direction of the force.

   d. What will be the electrons' radius of curvature in the magnetic field?

7. A uniform magnetic field of 45.6 mT points in the z direction, as shown in Fig. 8.13. At time $t = 0$ a particle with a charge of -1.60×10^{-19} C is moving along the positive x-axis, as indicated in the diagram. It has a speed of 4.4×10^5 m/s and moves in a circle of radius $r = 10$ cm.

   a. What is the direction of the force on the particle when it has the velocity shown in the figure?

   b. What is the numerical value of the force on the particle?

   c. What is the momentum of the particle?

   d. What is the particle's mass?

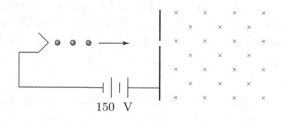

150 V

FIGURE 8.12 A beam of electrons accelerates through 150 V and then enters a uniform magnetic field of 0.5 mT pointing into the plane of the paper (Problem 6).

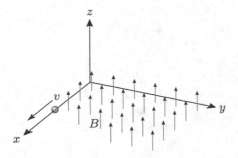

FIGURE 8.13 Direction of motion of a negatively charged particle in a uniform magnetic field at some instant of time (Problem 7).

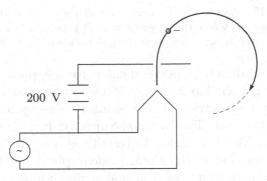

FIGURE 8.14 A beam of electrons accelerated through 200 V bends in a magnetic field (Problem 8).

8. Suppose you do an experiment in which electrons come off a cathode and accelerate through 200 V. Suppose these electrons enter a uniform magnetic field. They are moving perpendicular to it and bend as shown in Fig. 8.14.

    a. What is the kinetic energy of the electrons? Give your answer in electron volts.

    b. What is the direction of the force on them? Give your answer by drawing arrows on the diagram in several places.

    c. What must be the direction of the magnetic field to bend the electrons as shown?

    d. Explain how you figured out your answer to (c).

9. A cyclotron (Fig. 8.15) is an apparatus for accelerating charged particles to very high kinetic energies. Rapidly moving charged particles are coaxed into circular orbits by a strong uniform magnetic field that

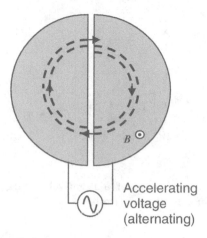

FIGURE 8.15 A diagram of a small part of the spiral path of particles accelerating inside a cyclotron where the magnetic field B is perpendicular to and out of the plane of the page. Acceleration occurs at the gap between the dees. (Problem 9).

points in a direction perpendicular to the plane of the particles' orbit. During each lap around the cyclotron, the particle is accelerated (twice) by an electric field, boosting its energy and also increasing the radius of its orbit. The National Superconducting Cyclotron Laboratory (NSCL) at Michigan State University is a world leader in the study of rare isotopes and fundamental nuclear physics. The K500 cyclotron at NSCL has a diameter of 3 m and a maximum magnetic field strength of 5 T. You can learn more (including the answers to this problem) at http://www.youtube.com/watch?v=xO4Dtz9vkiI.

    a. Singly charged argon ($m = 40\,\mathrm{u}$) ions are accelerated within the K500 cyclotron. What is the maximum speed to which they can be accelerated? (Express this as a fraction of the speed of light c.)

    b. Prove that the time needed for an $\mathrm{Ar^+}$ ion to make a full circle within the cyclotron is independent of its velocity. (This is true as long as $v \ll c$.) Calculate this time. Note that the electric field accelerating the charges must flip its polarity twice in this time.

10.   Figure 8.16 illustrates the operation of a mass spectrometer. For an accelerating voltage of 1000 V and a magnetic field of 0.1 T, helium ions ($\mathrm{He^+}$) have a circular orbit of radius 9 cm. (Note $m = 4\,\mathrm{u}$).

    a. Show that the mass-to-charge ratio of an ion, the accelerating voltage, the magnetic field, and the radius of the ion's orbit are related as follows:

$$\frac{m}{q} = \frac{B^2 R^2}{2 V_{\mathrm{acc}}}.$$

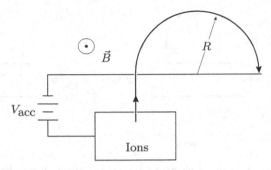

FIGURE 8.16 Schematic diagram of the mass spectrometer of Problem 10.

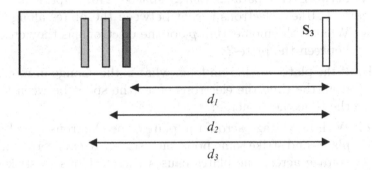

FIGURE 8.17 Representation of a developed photographic plate from a Bainbridge mass spectrometer (Problem 11).

b. If V_{acc} is increased to $4000\,\mathrm{V}$, what is the new orbital radius of He^+?

c. If V_{acc} is returned to $1000\,\mathrm{V}$, and B is doubled, find R.

d. Let B once again equal $0.1\,\mathrm{T}$. At what voltage would O_2^+ ions ($m = 32\,\mathrm{u}$) have a radius of $9\,\mathrm{cm}$?

e. What is the kinetic energy (in convenient units) of the helium and oxygen ions when they enter the magnetic field?

11. A representation of the photographic detector plate of a Bainbridge mass spectrometer is shown in Fig. 8.17. Three lines are visible where ions of a single element have struck the plate. S_3 is the location of the exit slit of the Wien velocity filter. The distances of the lines from S_3 are: $d_1 = 12.0\,\mathrm{cm}$, $d_2 = 12.5\,\mathrm{cm}$, $d_3 = 13.0\,\mathrm{cm}$. The line at d_1 is the darkest line. What is your best guess for the identity of the element? Explain your reasoning.

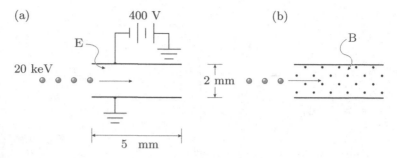

FIGURE 8.18 A beam of electrons passes (a) through an electric field (Problem 12) and then (b) through a magnetic field (Problem 13).

12. An electric field is produced by applying a potential difference of 400 V between two parallel plates spaced 2 mm apart (see Fig. 8.18). A beam of 20 keV electrons is sent between the plates along the x-axis.

    a. What is the momentum p_x of the electrons as they enter the space between the plates?

    b. If the plates are 0.5 cm long, what is the change in momentum Δp_y from the time the electrons enter the space between the plates to the time they emerge?

    c. With no voltage across the plates, the electrons travel on from the plates and strike some point on a screen 15 cm away. Turning on the voltage across the plates causes the electrons to strike a different point on the screen. What is the distance between the points that the electrons strike with voltage off and with voltage on?

13. Now the electric field in the previous problem is turned off and a uniform magnetic field is made to fill the space between the plates. The direction of B, the magnetic field, is out of the plane of Fig. 8.18.

    a. In what direction will the electrons be deflected?

    b. For what value of B will the magnitude of the change in momentum Δp_y of the electrons passing between the plates be the same as that produced by the electric field of the previous problem?

14. In a Wien velocity filter (page 210), perpendicular electric and magnetic fields are applied to moving charges so that only those charges with velocity $v = E/B$ pass through the fields without deflection. Figure 8.19 illustrates a complementary situation: charged particles moving through a conducting material (such as electrons moving through a copper wire) are confined by the walls of the conductor so that they move in straight lines parallel to the axis of the conductor. A magnetic field applied perpendicular to the motion causes the moving negative charges to separate

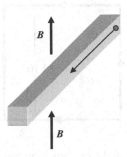

FIGURE 8.19 The Hall effect on charged particles moving through a conductor. (Problem 14).

from positive charge and produce an electric field within the conductor, perpendicular to the current, that offsets the magnetic force and keeps the charges moving in straight lines along the conductor. This induced electric field is called the Hall Effect. (Because blood is a conducting fluid, surgeons can measure blood flow rates using the Hall Effect.)

    a. The figure shows electrons moving along a long wire. What is the direction of the electric field E within the wire? (Make a sketch.)

    b. Suppose the wire has a square cross section with width and height equal to 1 cm. If the velocity of the charges is 5 $\mathrm{mm\,s^{-1}}$, and $B = 2.0\,\mathrm{T}$, what potential difference would appear across the wire? This potential difference can be measured with a voltmeter.

    c. If you measure current with an ammeter, you cannot distinguish between positive charges flowing through the meter in one direction, and negative charges flowing in the opposite sense. But the Hall Effect allows you to determine both the polarity and the direction of motion of the charges. Explain this: imagine positive charges moving opposite to the direction shown in the figure. What happens to the induced E field?

15. The very large magnetic field B needed for the cyclotron at the NSCL in Michigan is generated by passing a high current through a solenoid, or coil, containing thousands of loops of superconducting wire. The magnetic field contributed by each loop adds to create a large total field. But the electrons that flow through the solenoid wire to produce B feel a Lorentz force from the field they have generated. At high fields and currents, this force can be strong enough to destroy the coil.

    a. You know how to find the force on a single charge q moving with velocity v in a magnetic field B. A current I is made up of many

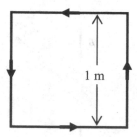

FIGURE 8.20   Geometry of the current loop for Problem 15.

such moving charges. Go back to Sect. 6.3 (p. 160) to find an expression for I in terms of the number of charges per unit volume, and then derive an expression for the force on a long straight wire of length L carrying a current I in a direction perpendicular to a uniform magnetic field B.

b. Consider a single loop of the solenoid, which for simplicity take to be square-shaped with side length 1 m, carrying a current of 100 A in the sense shown in Fig. 8.20. What is the direction of the field within the loop? What is the direction of the force on each of the four sides of the loop?

c. If $B = 5\,\mathrm{T}$, as in the NSCL K500 magnet, find the magnitude of the force on each side of the loop. Assume that the field is uniform. Compare this force to your weight. In magnets made with coils of superconducting wire great care must be used to assure that there are no small movements of coil wires because these can generate enough heat to turn off the superconductivity with the result that the energy stored in the magnetic field explosively dissipates in the coils. Designing a stable solenoid is not an easy task!

9

Electrical Atoms and the Electron

9.1 INTRODUCTION

One important result of physicists' increased understanding of electricity and magnetism was the recognition that atoms are electrical in nature. In the 1830s Michael Faraday's[1] studies of the flow of electricity through solutions contributed evidence that electricity is itself "atomic," i.e., made up of small indivisible units of electric charge. In 1894, G.J. Stoney proposed the word "electron" as the name for such a natural unit of charge. In 1897 the British physicist J.J. Thomson used electric and magnetic deflection to establish the existence of the tiny particle of electricity that we now call "the electron." His work also showed that the electron is a fundamental component of every atom and intimately related to its chemical properties. By the end of the first decade of the twentieth century the American physicist Robert A. Millikan had measured the electron's mass and charge to within one percent.

The experiments that led to the discovery of the electron and revealed its properties are good examples of how physicists use physics to discover more physics. These experiments illustrate the kinds of physical evidence that physicists find persuasive. They are also direct applications of the

---

[1] Michael Faraday, 1791–1867, born of a poor family near London, received only the equivalent of an elementary school education. Apprenticed to a bookbinder at thirteen, Faraday took to reading everything he could find, especially scientific books. Attendance at some lectures on chemistry by Sir Humphrey Davy led to his applying for and receiving a position as assistant to Davy. From this beginning, Faraday trained himself in science to the point where his discoveries in chemistry and electromagnetism, as well as his extremely popular public lectures, brought him renown and many honors.

C.H. Holbrow et al., *Modern Introductory Physics, Second Edition,*
DOI 10.1007/978-0-387-79080-0_9, © Springer Science+Business Media, LLC 1999, 2010

laws of electricity and magnetism laid out in the previous two chapters, and their results have led to pervasive, contemporary technologies that profoundly shape our lives today.

9.2   ELECTROLYSIS AND THE MOLE OF CHARGES

As early as 1800, the passage of electric current through a solution was known to decompose it into elemental components. For example, Davy decomposed water into hydrogen and oxygen gas in this way. In the 1830s, Michael Faraday studied the passage of electric current through fluids that are good conductors of electricity. A brine solution, sodium chloride dissolved in water, is a good example.[2] Acid and base solutions also are good conductors. From this work Faraday deduced that electrically charged atoms, "ions" as he named them, are the conductors of electricity in solutions. He established quantitative relationships between the amount of charge that passed through a solution and the amount of each element that was produced.[3]

Faraday called solutions that could be thus electrically decomposed "electrolytes." He invented several new words to describe the processes he observed. Some of these and their definitions are

electrolyte: A solution through which electricity can flow.

electrode: The (usually) metal pieces inserted into the electrolyte to pass the electrical current into and out of the solution.

electrolysis: The electrically induced breakup of molecules in the electrolyte when current flows through it.

anode and **cathode:** Respectively, the positively and negatively charged electrodes.

ions: The bodies, charged atoms or molecules, that constitute an electric current as they move through the electrolyte between the anode and the cathode.

anions and **cations:** Ions that move to the anode are called anions; they are negatively charged. Ions that move to the cathode are called cations and are positively charged.

---

[2]Ordinary tap water has enough dissolved material to be a pretty good electrical conductor, which is why you should not stand in a puddle of water during a lightning storm or become an electrolytic cell by touching a live wire.

[3]See *Great Experiments in Physics.* Morris H. Shamos, Ed., Holt and Co., New York, 1959, p. 128, for extensively annotated excerpts from the original publications of Faraday on electrolysis and electromagnetism, as well as from the works of other important physicists. Short biographies of the pioneers we are discussing in this chapter are also included.

Faraday varied the nature of the electrolyte and the size and shape of the electrodes. He tried different arrangements of electrodes to ensure that when gases were the product of decomposition all the evolved gas was collected. When sufficient care was taken, the universal result, now known as Faraday's law of electrolysis, was that *the measured amount of a substance decomposed was "in direct proportion to the absolute quantity of electricity* [i.e., total charge] *which passes."*

EXAMPLES

1. In one experiment with zinc and platinum electrodes immersed in water,[4] Faraday observed that hydrogen gas was generated from the platinum electrode while zinc oxidized and left the zinc electrode. Faraday collected the hydrogen gas and determined that it was released by the electrolysis of an amount of water that has a mass of 2.35 grains. Before electrolysis, the zinc electrode weighed 163.10 grains; after electrolysis it weighed 154.65 grains. This means that during the electrolysis of 2.35 grains of water, 8.45 grains of zinc were removed from the zinc electrode.[4]

If for every molecule of H_2O that is electrolyzed, one Zn atom is removed from the zinc electrode, then the ratio of the masses of the electrolyzed water and zinc should equal the ratio of their molecular weights. From Faraday's data

$$M_{Zn} = M_{H_2O} \frac{8.45 \, \text{grains of Zn}}{2.35 \, \text{grains of } H_2O} = 18 \times 3.60 = 64.7 \, \text{u},$$

which agrees well with today's accepted atomic mass of 65.4 u for Zn.

Notice that in this example it is not necessary to know the SI equivalent of a grain.

Although in his day there was no way to measure the amount of charge passed through an electrolyte, Faraday could make a current I that he knew was steady (constant). Then the amount of charge Q passed through an electrolyte is proportional to the elapsed time Δt because $Q = I \, \Delta t$. If he ran the same steady current I through different electrolytes for the same length of time Δt, the total amount of charge Q passed through each electrolyte would be the same. He weighed the amounts of different

---

[4]Pages 254–255 in Michael Faraday, *Experimental Researches in Electricity* (3 vols. bound as 2), Dover Publications, Inc., New York, 1965. The three volumes were originally published in 1839, 1844 and 1855 respectively.

TABLE 9.1 Relative masses of elements produced by electrolysis using the same amount of charge

Element	hydrogen	oxygen	chlorine	iodine	lead	tin	silver
Relative mass	1	8	36	125	104	58	108
Atomic weight	1	16	36	125	208	116	108

elements released by the passage of the same amounts of charge Q and discovered that the ratios of the weights were closely related to the ratios of the elements' atomic masses. You can see from Table 9.1 that for hydrogen, chlorine, and iodine the ratio of the weights collected is exactly the same as the ratio of their atomic masses. In other words, the amount of charge that releases one atomic weight (i.e., 1 mole) of hydrogen releases one atomic weight of chlorine, iodine, or silver, but the same amount of electricity releases only half a mole of oxygen, lead, or tin. Once again, simple integer ratios are strong evidence for atomicity. In Chap. 3, it was atoms of matter; here it is evidence for atoms of electric charge.

When it became possible to measure charges accurately, the amount of charge required to electrolyze exactly one mole of hydrogen, or silver, or chlorine, was found to be 96 485 coulombs. In other words, 96 485 C will cause the electrolytic deposition of one gram atomic weight of silver, i.e., 107.88 g, which is one mole of silver atoms.

Therefore, even without knowing the value of Avogadro's number N_A it appears quite reasonable to identify this quantity of charge, 96 485 C, as N_A basic charges. This suggests that there exists some basic, elementary charge and that 96 485 C is a mole of them. This quantity of electricity, the mole of elementary charges, is called "the faraday" and is often represented by the symbol F. If we designate this smallest possible quantity of electric charge by the symbol e (for elementary), then the faraday is

$$F = N_A e = 96\,485\,\text{C}. \qquad \text{the faraday} \qquad (1)$$

Figure 9.1 shows a typical setup for electrolysis along with a simplified modern interpretation of what goes on in the process.

EXAMPLES

2. Suppose that a cell arranged as in Fig. 9.1 carries a current of 15 amperes for 10 min. How much silver will plate out on the cathode?

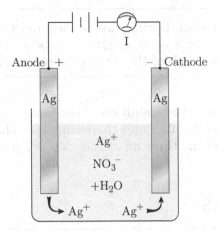

FIGURE 9.1 An electrolytic cell in which silver dissolves from a silver anode and plates onto the cathode.

The amount of charge, Q, delivered by this current during this amount of time is $(15\,\text{C/s})(10\,\text{min})(60\,\text{s/min}) = 9000\,\text{C}$. As a fraction of a faraday F this is

$$\frac{Q}{F} = \frac{9000}{96/,485} = 0.0933,$$

which from the definition of F must be the same fraction of silver's mole atomic weight of 107.88 g. Thus:

Deposited Wt. of Ag $= (107.88\,\text{g/mol})(0.0933\,\text{mol}) = 10.06\,\text{g}.$

Faraday's work strongly suggested that chemical reactions involve integer amounts of some basic quantity of electric charge. In conjunction with the chemical evidence for atoms, these results led to the idea of an integer number of these basic charges accompanying ions in solution. Faraday even was so bold as to suggest that it is the electrical charges that are the source of chemical "affinity," or bonding, as we would say now.

▪ EXERCISES

1. Faraday's results are shown in Table 9.1 along with the known atomic weights. Use these data to deduce the number of faradays required to produce one mole of each of the elements he studied. Assume the possibility of different numbers of elementary charges on different

ionic species (which Faraday apparently did not consider) and explain why a faraday of charge might not always generate a mole atomic weight. Explain each of the six results reported in the table.

Note that Eq. 1, the definition of the faraday F, intimately connects F with Avogadro's number and the elementary charge e. If you can determine any two of the three quantities N_A, F, or e, Eq. 1 will give you the third.

EXERCISES

2. In Chap. 5 you saw that a rough estimate of Avogadro's number ($\approx 6.6 \times 10^{23}$) could be obtained from kinetic theory and measurements of the viscosity of air. Using this result and the value of the faraday, estimate a value for the elementary charge e.

In electrolysis a given amount of charge releases a certain mass of an element. You can use this result to find the ratio of an ion's charge to its mass. For example, a charge $Q = Ne$ will release N atoms of hydrogen, and the total mass released will be Nm_H, where m_H is the mass of a hydrogen atom. The charge-to-mass ratio is found from the ratio of the total charge used to the total mass collected:

$$\frac{Q_{\text{total}}}{M_{\text{total}}} = \frac{Ne}{Nm_H} = \frac{e}{m_H}.$$

EXERCISES

3. You know that one faraday of charge releases one mole of hydrogen atoms. From that fact determine the value of the charge-to-mass ratio of the hydrogen ion.

4. Modify this calculation to find the charge-to-mass ratio of the oxygen ion with charge $-2e$.

5. Of all the possible singly charged ions that one can imagine forming from the periodic table of the elements, which one will have the largest value of its charge-to-mass ratio? The smallest?

Keep this result in mind for comparison with the charge-to-mass ratio of the elementary charge discussed in the next section.

9.3   CATHODE RAYS, e/m, AND THE ELECTRON

By the 1850s it had become possible to generate large electrical potentials that could supply sustained electrical currents. When a strong current was passed through a low-pressure gas in a glass container (tube), the gas would glow. Also rays were observed to stream from the cathode (the negative electrode) to the walls of the tube or to the anode. The study of these so-called "cathode rays" led to two revolutionary conclusions: atoms have parts; the parts are electrically charged. One of these electrically charged parts of the atom was the electron, an entirely new entity with a mass almost 2000 times less than the mass of hydrogen, the lightest atom.

The Electrical Nature of Cathode Rays

To prove the electrical nature of cathode rays was not easy. First, the idea of electrical particles was novel; second, if the rays were electrical they should be deflected both by magnetic fields and by electric fields. But electric deflection did not appear to occur. Although a magnetic field bent the rays as you would expect if they were electrical charges,[5] electric fields did not appear to affect them. In 1883 when the able German physicist Heinrich Hertz tried to deflect the rays by passing them through the electric field between two charged plates, he saw no effect and concluded that cathode rays were *not* electrical. Only later was it found that the lack of electrical deflection was caused by poor vacuum.[6]

In 1887 J.J. Thomson[7] began experiments that proved that cathode rays are electrical in nature. He also produced clear evidence that magnetic and electric fields bend cathode rays as though they have a well-defined, distinct ratio of charge e to mass m. This finding strongly

---

[5]Even when magnetic deflection was observed, it was not easy to understand. These were the first observations of particles with well-defined charges moving in magnetic fields, and most physicists of that time did not yet have the convenient tool of vector representations or all the "hand rules" that make it so easy (?) for you to predict the motion of charges in a magnetic field.

[6]Hertz's experiment used a vacuum at the edge of what was technically possible in his time, and it was not good enough. Ionization of residual gas in his cathode-ray tube produced enough conductivity to short circuit the applied voltage and reduce the electric field to a value too low to produce any deflection.

[7]Joseph J. Thomson, 1856–1940. See Shamos, *op. cit.*, p. 216. For an interesting account of Thomson and his work see http://www.aip.org/history/electron/jjhome.htm.

suggested that cathode rays are streams of distinct particles. We call these particles "electrons." For being the first to exhibit its existence and for determining some of its properties, Thomson is considered the discoverer (in 1897) of the electron.[8]

Using several different configurations of electrodes and tubes and bending the rays with a magnet, Thomson showed that the observed electrical charge was associated only with the rays. With improved vacuum technique Thomson showed that cathode rays are deflected by electric fields as they should be if they are electrically charged. He also showed that the heat generated in the anode when struck by cathode rays is equal to the electrical energy given to the beam by the voltage between the cathode and the anode.

Here is a simplified version of how to compare the heat energy collected in the anode with the kinetic energy in the beam. Build a metal cup of a shape that will collect all the rays coming from the cathode. Before you put this cup into the tube, measure how much heat energy raises the cup's temperature by 1 K. Say your result is 0.02 J.

Put the cup in a tube and pump out the air. Apply a potential difference of 400 V between the cathode and anode and observe that the cathode-ray beam current I is $2.0\,\mu$A at the anode. Collect this beam in the anode for 100 s, and observe that the temperature of the anode rises by 4.0 K. Because the temperature rises 1 K for every 0.02 J added, you calculate—assuming no serious losses by cooling—that the anode received 0.08 J of energy.

Now compare the energy heating the anode with the kinetic energy carried by your beam of cathode rays. In time Δt the beam delivers a total charge of $Q = I\Delta t$ that has been accelerated through a potential difference V and therefore has energy $K = QV$, making the total kinetic energy brought to the anode $K = VI\Delta t$. For $V = 400\,$V, $I = 2\,\mu$A, and $\Delta t = 100\,$s, $K = 400 \times 2 \times 10^{-6} \times 100 = .08\,$J.

The energy of heating equals the kinetic energy imparted electrically to the beam. This result supports the supposition that cathode rays are electrically charged.

Deflection

By Thomson's time pumps could produce vacuums good enough to permit electric deflection. Consequently, he was able to combine electric deflection

[8]For his discovery Thomson was awarded the 1906 Nobel Prize in Physics— http://nobelprize.org/nobel_prizes/physics/laureates/1906/index.html.

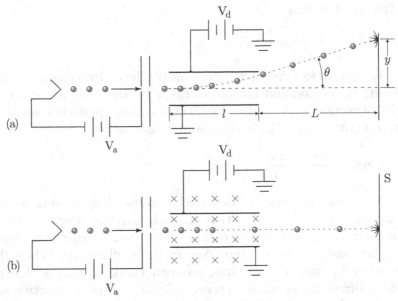

FIGURE 9.2   Thomson's arrangement of electric and magnetic fields to deflect cathode rays. The screen S on which the deflection was observed is at the extreme right.

with magnetic deflection and measure the charge-to-mass ratio e/m of cathode ray particles. This section explains how he did this and why he had to use both kinds of deflection.

Electric deflection is commonly achieved by passing a beam of charged particles between oppositely charged parallel plates as shown schematically in Fig. 9.2. You have seen that in the volume between such plates the electric field E is constant. While the beam is passing between the plates, the electric field exerts a constant force on the beam, upward in the case shown in Fig. 9.2. The deflection is usually measured by looking at the luminous spot made when the beam strikes a screen a distance L from the deflecting plates. When the field is off, the beam strikes one point on the screen; when the field is on, the spot is deflected some measurable distance y.

You can predict the amount of deflection by calculating the angle θ at which the beam emerges from between the plates when E is on. Assume the beam enters the region between the plates with a horizontal speed v_x but no upward motion. Because the field E exerts an upward force on them, the cathode rays leave the region between the plates with an upward speed of v_y. Because these particles still have their original horizontal speed v_x, they travel the distance L to the screen in a time t such that $v_x t = L$. In that same time t they travel vertically a distance $y = v_y t$.

The result is that

$$\frac{y}{L} = \frac{v_y t}{v_x t} = \frac{v_y}{v_x} = \tan \theta. \tag{2}$$

It is useful to think of deflection as produced by a change of momentum. Multiply the top and bottom of Eq. 2 by the mass m, and you see that the deflection angle is also the ratio of a beam particle's acquired upward momentum Δp_y to its horizontal momentum p_x.

$$\tan \theta = \frac{m v_y}{m v_x} = \frac{\Delta p_y}{p_x}.$$

The cathode rays are deflected because a force acts for some time interval Δt in the y direction on each particle. Each is traveling with momentum p_x in the x direction.[9] The force, acting over time, imparts an increment of momentum Δp_y in the y direction. When the force is applied for only a short time, you can think of it as a blow or swat in the y direction, as when a soccer ball rolls in the x direction across your path, and you kick it in the y direction. It doesn't go directly off in the y direction; rather, it is deflected and travels away from the x direction at some angle $\theta = \tan^{-1} \left(\frac{\Delta p_y}{p_x} \right)$ where Δp_y is the effect of your kick.

▪ EXERCISES

6. Why is the horizontal momentum p_x not affected by the electric field between the plates?

In Thomson's experiment the cathode rays pass quickly between the plates and experience a kick from the electric field E. Each charge e experiences a constant force $F_y = Ee$ during the time Δt it is in the region between the plates, and gets a kick of $\Delta p_y = F_y \Delta t$. If the length of these plates is ℓ and the speed of the electrons is v_x, then

$$\Delta t = \frac{\ell}{v_x}.$$

Because p_x is $m v_x$, the angle of deflection is

$$\theta = \frac{\Delta p_y}{p_x} = \frac{Ee\Delta t}{p_x} = \frac{eE\ell}{m v_x^2},$$

---

[9]This is a good description of what happens to the cathode rays, but in general the deflecting force need not be constant or at right angles to the initial path of the deflected particles.

where $\tan\theta$ has been replaced by θ because, as is often the case, $\Delta p_y \ll p_x$, and the small-angle approximation is valid. To find y, just multiply θ by the distance to the screen:

$$y = L\theta = \frac{eE\ell L}{mv_x^2}. \tag{3}$$

As expected, y depends on e, m, and v_x.

Electric deflection is a basic tool for controlling charged particles of all kinds. As Eq. 3 shows, electric deflection is inversely proportional to the kinetic energy of the particle being deflected; this is a general property of electric deflection.

The above calculation of displacement on the screen neglects y-displacement occurring while the cathode ray particles are between the plates. In cathode ray tubes, this is usually negligible compared to the amount of displacement that occurs after leaving the plates, but it is not hard to include it if you need to. (For example, see p. 261.)

Equation 3 was not enough for Thomson to find the value of e/m of cathode rays. He could measure y and E, but not v_x. You might think he could use the fact that he gave the rays their kinetic energy by accelerating them through a known voltage V_a (subscript "a" for acceleration). As a result

$$\frac{1}{2}mv_x^2 = eV_a,$$

but, as you can show, using this equation to eliminate v_x from Eq. 3 also removes e and m from the equation.

■ EXERCISES

7. Show that this last statement is true.

In contrast to electric deflection, magnetic deflection of a particle is inversely proportional to its velocity or, equivalently, its momentum. This property allowed Thomson to use magnetic deflection to determine v_x in Eq. 3 without eliminating e/m from the equation. First, he measured the deflection y caused by the electric field E; then he produced in the region between the plates a uniform B field with just the right strength to cancel the deflection due to E. The effects of the two fields cancel when

$$eE = ev_x B,$$

or

$$v_x = \frac{E}{B}. \tag{4}$$

(This is the condition for the Wien velocity filter discussed on p. 210.) Thomson could determine both B and E, which meant that he could replace v_x in Eq. 3 with its experimentally determined value. Or he could substitute Eq. 4 into Eq. 3 and solve to get

$$\frac{e}{m} = \frac{Ey}{B^2 \ell L},$$ (5)

where everything on the right side was measurable. These early measurements were not very precise; one value Thomson obtained was 0.71×10^{11} C/kg, about a factor of 2.5 less than the currently accepted value.

▪ EXERCISES

8. Derive Eq. 5.

9. If the separation between the plates was $d = 5$ mm and the deflecting voltage V_d (d for deflection) was 10^4 V, what was the value of the deflecting electric field?

10. You've seen that Thomson's experiments determined only the charge-to-mass ratio of cathode rays. How then did the mass spectrometry experiment described in Chap. 8 use a combination of electric and magnetic fields to determine ratios of the masses of charged ions? What additional assumption allows you to infer the mass ratios from mass spectrometry?

e/m by the Bainbridge Method

All methods of measuring e/m use both electric and magnetic fields. In the Bainbridge method shown in Fig. 9.3, a large tube is situated between two coils that produce a reasonably uniform magnetic field B. This field bends the cathode rays in a circle with a radius r that depends upon the particle momentum and the strength of the magnetic field. The beam of cathode rays is produced by a small "electron gun" oriented to inject electrons tangent to the circumference of the tube so that the beam can make as large a circle as possible. The gun gives the electrons kinetic energy by accelerating them across a potential difference of V_a volts. With a charge e and mass m, each electron gains a kinetic energy of

$$\frac{1}{2}mv^2 = eV_a.$$ (6)

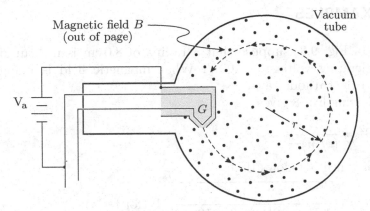

FIGURE 9.3 Diagram of an apparatus for measuring e/m. The source of electrons G directs a beam accelerated through a voltage V_a in a downward direction. An external magnetic field B bends the beam in the circle with a radius r as shown.

In modern versions of the Bainbridge apparatus the tube is evacuated during its manufacture and then refilled to a low pressure with an inert gas such as helium. When electrons pass through this gas, they cause it to glow, and the path of the rays becomes visible so that an observer can measure the radius of curvature r of the electron beam.

The external magnetic field B produces a force on the moving charges perpendicular to their path. This force supplies the centripetal force F_c that produces circular motion. Thus

$$F_c = \frac{mv^2}{r} = evB, \tag{7}$$

where r is the radius of the circle. Solve for v,

$$v = \frac{eBr}{m},$$

then square the result and substitute into Eq. 6 to get

$$\frac{2eV_a}{m} = v^2 = \frac{e^2 B^2 r^2}{m^2}$$

and

$$\frac{e}{m} = \frac{2V_a}{B^2 r^2}. \tag{8}$$

Every quantity on the right side of Eq. 8 is measurable, so e/m can be determined. The modern value is

$$e/m = 1.759 \times 10^{11} \, \text{C/kg}.$$

EXAMPLES

3. In Fig. 9.3, suppose that a radius of 8.0 cm is measured and the accelerating voltage is 400 V. What magnetic field is being used to bend the cathode rays?

Solving Eq. 8 gives

$$B^2 = \frac{2V}{(e/m)r^2}$$

from which it follows that

$$B = \sqrt{\frac{2 \times 400}{1.76 \times 10^{11} \times (8 \times 10^{-2})^2}} \text{ T (for tesla)}.$$

Evaluating this expression gives

$$B = \sqrt{7.10 \times 10^{-7}} = 8.43 \times 10^{-4} \text{ T}.$$

To check that the units are correct it may help if you use Eq. 4 to see that $1 \text{ T} = 1 \dfrac{\text{N/C}}{\text{m/s}}$.

The Significance of e/m

Thomson's value for e/m was remarkably large, or, to put it another way, the value of m/e was remarkably small. From experiments in electrolysis one finds that m/e for the hydrogen ion is about 10^{-8} kg/C (see Exercise 3). Thomson's value of m/e for electrons was more than a thousand times smaller, indicating that in terms of mass the electron was a very small part of the atom.

At least as important as its discovery was the recognition that the electron was a part of *every* atom. The electrons measured in this apparatus must come from the electrodes or the gas in the tube, but the same charge-to-mass ratio was measured when these materials were changed. This suggested that the electrons in every kind of atom are identical; that is, electrons are a fundamental building block of atoms.

Furthermore, a year earlier H.A. Lorentz had shown that he could explain a peculiar change that occurred in the frequency of light emitted by neutral atoms placed in a magnetic field if there was inside the atom a charged particle with a charge-to-mass ratio of about 2×10^{11} C/kg, the same value to within experimental uncertainty as the value of e/m measured by Thomson for a free electron. This was strong evidence that the electron is an internal part of every atom and intimately involved in atomic behavior.

9.4   THE ELECTRON'S CHARGE

Introduction and Overview

It is important to know the value of the electron's charge. It sets the scale of electric forces in atoms, and these forces determine atomic size. Moreover, knowledge of e yields N_A from the faraday, and the mass of the electron from e/m. Consequently, after Thomson's work, physicists sought to measure the value of the charge e directly.

The most conclusive of the direct measurements of e were made by the American physicist R.A. Millikan.[10] He obtained accurate results by studying the speed of motion in air of small electrically charged oil droplets moving under the influence of gravitational and electric fields. His work established that there is such a thing as an elementary indivisible unit of charge and that its value is $e = 1.6 \times 10^{-19}$ C.

Millikan's oil droplets were a few micrometers in diameter (a human hair is about $100\,\mu$m in diameter). When such a small droplet falls in air, it accelerates only briefly—until resistance from the viscosity of air balances the force of gravity—after which it falls at a slow, steady speed. These droplets are too small to be seen directly, but when illuminated with a bright light against a dark background, an individual droplet scatters enough light to make a visible glint. Millikan could then measure its speed of fall very accurately. The apparatus to apply the necessary fields and illuminate the oil drops is illustrated in Fig. 9.4.

He made the droplets by spraying oil through a tiny nozzle, and friction between the oil and the nozzle gave each droplet a tiny unknown charge. By measuring the speed v_g of a droplet falling under gravity, Millikan could determine its mass. By measuring the speed v_F of a droplet rising in an electric field, he could figure out how much electric charge the droplet carried. He made hundreds of measurements using different charges both on the same and on different droplets and found that all the charges on his droplets were integer multiples of a single, smallest amount e—the elementary charge.

Droplet Size from Terminal Velocity

To understand the effects of electric field on his charged oil droplets, Millikan had to know their masses. He determined the mass of a droplet by an ingenious application of the fact that when a small, spherical body falls

---

[10]Robert A. Millikan, 1865–1953. See Shamos, *op. cit.*, p. 238. The results of Millikan's experiment are reported in *Physical Review* **32**, 349 (1911).

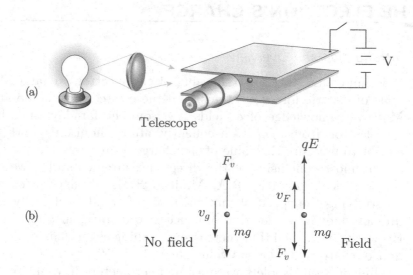

FIGURE 9.4 (a) Millikan's apparatus for determining the charge of an electron. (b) A charged droplet of oil can be observed through a telescope to fall at terminal velocity v_g between parallel plates in the absence of any electric field and to rise with terminal velocity v_F when an electric field E is present.

freely through a viscous medium, it quickly stops accelerating and falls at a steady rate v_g called its "terminal velocity." The terminal velocity v_g depends on the acceleration due to gravity, g, on the viscosity η of the medium, on the sphere's radius a, and on the density ρ of the oil of which the droplet was made. Millikan measured the terminal velocity of a falling droplet and from his knowledge of g, η, and ρ, found its radius a. Then he calculated the mass m of the spherical droplet from the fact that $m = \frac{4}{3}\pi a^3 \rho$.

EXAMPLES

4. If a spherical droplet of oil of density $\rho = 0.92 \times 10^3$ kg m^{-3} has a radius of $1\,\mu$m, what is its mass m? Because the mass of the sphere equals its volume times the density of the oil, you get

$$m = \frac{4}{3}\pi(1 \times 10^{-6})^3 (0.92 \times 10^3) = 3.9 \times 10^{-15}\,\text{kg} = 3.9\,\text{pg}.$$

How do you find the droplet's radius a from its speed of fall v_g? Think of an oil droplet falling freely through the air inside the chamber of Millikan's experimental apparatus (Fig. 9.4) with the electric field turned off. Two forces act on the droplet: gravitational attraction downward and viscous drag opposing the droplet's motion. The gravitational force is just mg. The retarding viscous force due to the air is proportional to the viscosity of air and to the droplet's velocity. The British physicist Sir George Stokes showed in the nineteenth century that the viscous force F_v acting on a small sphere of radius a falling with velocity v through a fluid with viscosity η is

$$F_v = 6\pi\eta a v. \tag{9}$$

Consider what happens when the droplet falls. It starts off with an acceleration of g, but as its velocity increases, the retarding viscous force increases, opposing the gravitational attraction and decreasing the droplet's acceleration. As the value of v continues to increase, the acceleration continues to decrease until the droplet reaches the velocity v_g at which the force F_v retarding the droplet just equals mg, the force of gravity downward. At this velocity v_g the opposing forces are balanced, the acceleration goes to zero, and the velocity stops changing. The velocity v_g at which this balance occurs is called the "terminal velocity."

You can find an expression for v_g in terms of the droplet radius a and density ρ. Begin by equating the viscous and gravitational forces, $F_v = mg$, and solving for v_g:

$$\begin{aligned} 6\pi\eta a v_g &= mg, \\ v_g &= \frac{mg}{6\pi\eta a}. \end{aligned} \tag{10}$$

Now express m in terms of a and the known density of the oil, ($\rho = 0.92\,\mathrm{g/cm^3}$ for the oil Millikan used), $m = \frac{4}{3}\pi a^3\rho$; substitute for m in Eq. 10 and do some algebra to get

$$v_g = \frac{2}{9}\frac{\rho g a^2}{\eta}. \tag{11}$$

Millikan measured v_g and used the known values of ρ, g, and η to find the radii and masses of droplets.

EXERCISES

11. Derive Eq. 11 from Eq. 10.

▽ EXAMPLES

5. How slowly does a droplet fall through air if its radius is only $1\,\mu$m, i.e., what is its terminal velocity? Assume that the density of the oil is $\rho = 0.92\,\mathrm{g\,cm^{-3}}$ and the air temperature is $20\,^\circ$C.

From Chap. 5 you know that for air $\eta = 18.3\,\mu$Pa s. Now use Eq. 11:

$$v_g = \frac{2}{9}\,\frac{(0.92 \times 10^3)(9.8)(1 \times 10^{-6})^2}{18.3 \times 10^{-6}} = 1.1 \times 10^{-4}\,\mathrm{m/s}.$$

A droplet $1\,\mu$m in radius will fall with a terminal velocity of $0.11\,$mm/s; it will take $9\,$s to fall $1\,$mm! For a small droplet in air this equilibrium speed is reached almost immediately, and an observer will see the droplet fall slowly at constant velocity.

▨ EXERCISES

12. What would be the terminal velocity of an oil droplet if it were $10\,\mu$m in radius?

To measure v_g Millikan watched a falling droplet through a telescope and measured the time t_g it took the droplet to fall a known, small height h marked by two reference lines in his telescope eyepiece. Then he calculated the terminal velocity as $v_g = h/t_g$. Knowing v_g, he used the inversion of Eq. 11 to find the droplet radius

$$a = \sqrt{\frac{9v_g\eta}{2\rho g}},$$

and then found the weight of the droplet:

$$mg = \frac{4\pi}{3}a^3\rho\,g = \frac{4\pi\rho}{3}\,g\left(\frac{9v_g\eta}{2\rho g}\right)^{3/2} = 9\pi\left(\frac{2\eta^3}{\rho g}\right)^{1/2}v_g^{3/2}. \tag{12}$$

Using his oil drop technique, Millikan could weigh a tiny oil droplet. Knowing the weight of the droplet, he could then determine how much electric charge q was on it.

▨ EXERCISES

13. Suppose a droplet has a terminal velocity of $0.43\,$mm/s. What is its radius and mass? Efficiency tip: Use the values worked out for a $1\,\mu$m radius drop and scaling arguments to avoid unnecessary calculation.

14. How big is the viscous force acting on the $1\,\mu$m-radius droplet when it is moving with terminal velocity? On the $10\,\mu$m droplet?

Finding the Charge on a Droplet

By applying an electric field Millikan could cause the charged droplets to rise. Turning the field repeatedly on and off, he could cause an individual droplet to rise and fall repeatedly as he observed its motion for hours. From the speeds of fall and rise that he measured, Millikan could determine the amount of charge on a droplet. He could also change the amount of charge on a given droplet, measure changes in its speed of rise, and calculate by how much the droplet's charge had changed. Here's how he did these things.

Millikan produced an electric field E by applying a potential difference V to a pair of metal plates separated by a small distance d. As noted earlier, this creates a constant electric field $E = V/d$ between the plates. When the field is on, the droplet experiences a force qE in addition to the downward force mg. The force qE could be either upward or downward, depending on the sign of q and the direction of E. However, Millikan was only able to experiment with particles that rose under the influence of the electric field; other particles would quickly fall out of the region he could observe in his telescope, so we'll work out the theory for this case. Call the terminal velocity v_F when the droplet is rising with the field turned on (F for field). When the electric field is off, the drop falls with a terminal velocity v_g (g for "gravity only").

For a given drop there is a simple proportion between its terminal velocity and the effects of gravity or the electric field acting on the droplet. As Eq. 10 shows

$$v_g \propto mg,$$

i.e., the downward terminal velocity is proportional to the downward force of gravity. When an applied electric field makes the droplet rise, the upward terminal velocity is proportional to $qE - mg$

$$v_F \propto qE - mg.$$

Divide the second expression by the first to get:

$$\frac{v_F}{v_g} = \frac{qE - mg}{mg}.$$

Solve for q:

$$q = \frac{mg}{E}\left(\frac{v_F}{v_g} + 1\right) = \frac{mg}{v_g E}(v_F + v_g). \tag{13}$$

Now use the expression for mg from Eq. 12 in Eq. 13 to find the charge on the droplet solely in terms of quantities that can be determined from measurements. You get

$$q = \frac{9\pi}{V/d}\left(\frac{2\eta^3}{\rho g}\right)^{1/2} v_g^{1/2}(v_F + v_g). \tag{14}$$

EXERCISES

15. Derive Eq. 14.

16. Show explicitly that the right-hand side of Eq. 14 has units of coulombs when all the quantities are measured in SI units.

EXAMPLES

6. For one particular droplet Millikan used a potential difference of 5085 V between two plates 16 mm apart. When the electric field was off, the droplet fell with a speed of 0.08584 cm/s; when the field was on, the droplet rose with a speed of 0.01274 cm/s. What was the charge on this droplet?

For $\rho = 0.92 \times 10^3 \, \mathrm{kg\,m^{-3}}$, $\eta = 18.2 \, \mu\mathrm{Pa\,s}$, and $g = 9.8 \, \mathrm{m\,s^{-2}}$, you find from Eq. 14 that

$$q = \frac{9\pi\, 0.016}{5085}\sqrt{\frac{2(18.2 \times 10^{-6})^3(8.584 \times 10^{-4})}{920 \times 9.8}} \tag{15}$$
$$\times (1.274 \times 10^{-4} + 8.584 \times 10^{-4})$$
$$= 2.97 \times 10^{-18} \, \mathrm{C}.$$

Notice that this is not really what we want to know. It is the charge on the given droplet; it is not the elementary charge. Only if a droplet had exactly one unneutralized elementary charge would the above equation give the value of the elementary charge. In fact a singly charged droplet could not be made to rise in the electric field of Millikan's apparatus, so he never observed such a droplet. To get around this problem Millikan changed the charge on a given droplet from time to time. He did this by slightly ionizing the air of the chamber with x-rays or radioactivity. A droplet would then pick up one or more ionized molecules and change its charge. As a result, the electric force on the droplet would change and so would the droplet's terminal velocity v_F. The velocity v_g stayed the same because the mass added to the oil drop by a few ions was utterly negligible compared to the mass of the droplet. By measuring the changes in v_F, Millikan could calculate by how much the charge had changed. He observed that this change was always an integer multiple of some smallest amount, 1.602×10^{-19} C by modern measurements. He concluded that this smallest change corresponded to the elementary charge we denote as e.

■ EXERCISES

17. How many elementary charges were there on the droplet referred to in Example 6?

18. Suppose the charge on a 1-μm radius droplet increased by $3e$ because it picked up three nitrogen ions. What would be the fractional increase in the droplet's mass?

Quantization of Electric Charge

It is interesting that Millikan did not need actually to measure the value of the elementary charge e in order to demonstrate experimentally that the electric charge on a droplet is always an integer multiple of e. His work is such a nice example of reasoning with ratios that it is worth looking at closely. Table 9.2 is a reproduction of data that Millikan published[11] to show the atomic nature of charge. (Other examples of Millikan's data can be found in Shamos's book.) Along with the data, Table 9.2 shows various parameters of his apparatus, such as voltage V, spacing between the plates d, the distance h of fall and rise of a droplet, etc. The data recorded by Millikan are the values of the quantities actually measured. These are the times of fall under gravity, t_g, and the times of rise when the field was on, t_F. You can see that when he caused the charge on a droplet to change, the time of rise with the field on changed because the electrical force qE changed.

There is enough information in the table so that from these times you (and he), using Eq. 14, could calculate the charge on every droplet. But

---

[11]R.A. Millikan, *Electrons (+ and −), Protons, Photons, Neutrons, Mesotrons and Cosmic Rays*, revised edition, University of Chicago Press, Chicago, 1947, p. 75.

TABLE 9.2   Data from Millikan's measurements on an oil drop[a,b]

t_g Sec.	t_F Sec.	$\dfrac{1}{t_F}$	$\left(\dfrac{1}{t'_F}-\dfrac{1}{t_F}\right)$	n'	$\dfrac{1}{n'}\left(\dfrac{1}{t'_F}-\dfrac{1}{t_F}\right)$	$\left(\dfrac{1}{t_g}+\dfrac{1}{t_F}\right)$	n	$\dfrac{1}{n}\left(\dfrac{1}{t_g}+\dfrac{1}{t_F}\right)$
11.848	80.708	.01236				.09655	18	005366
11.890	22.366	.04470	.03234	6	.005390			
11.908	22.390					.12887	24	.005371
11.904	22.308		.03751	7	.005358			
11.882	140.565	.007192				.09138	17	.005375
11.906	79.600	.01254	.005348	1	.005348	.09673	18	.005374
11.838	34.748	.02870	.01616	3	.005387			
11.816	34.762					.11289	21	.005376
11.776	34.846							
11.840	29.286	.03414				.11833	22	.005379
11.904	29.236		.026872	5	.005375			
11.870	137.308	.007263				.09146	17	.005380
11.952	34.638	.02884	.021572	4	.005393	.11303	21	.005382
11.860			.01623	3	.005410			
11.846	22.104	.04507				.12926	24	.005386
11.912	22.268		.04307	8	.005384			
11.910	500.1	.002000				.08619	16	.005387
11.918	19.704	.05079	.04879	9	.005421			
11.870	19.668		.03794	7	.005420	.13498	25	.005399
11.888	77.630	.01285				.09704	18	.005390
11.894	77.806		.01079	2	.005395	.10783	20	.005392
11.878	42.302	.02364						
11.880			Means		.005386			.005384

Duration of exp.	= 45 min.		Pressure	= 75.62 cm.
Plate distance	= 16 mm.		Oil density	= .9199
Fall distance	= 10.21 mm.		Air viscosity	= 1,824×10⁻⁷
Initial volts	= 5,083.8		Radius (a)	= .000276 cm.
Final volts	= 5,081.2		$\dfrac{l}{a}$	= .034
Temperature	= 22.82° C.		Speed of fall	= .08584 cm./sec.

[a] Taken with permission from p. 75 of R.A. Millikan, *Electrons (+ and −), Protons, Photons, Neutrons, Mesotrons and Cosmic Rays*, University of Chicago Press, ©1937 and 1947 by the University of Chicago

[b] The viscosity of air is given here in cgs units; the corresponding SI value is 18.24×10^{-6} Pas

this would be neither efficient nor very insightful. Also, as noted in the previous section, Millikan concentrated on the *changes* in charge. He did this because these changes were smaller multiples of e than was the total charge on a droplet.

From Eq. 14 you can see that the change in charge Δq is

$$\Delta q = \frac{9\pi}{V/d}\sqrt{\frac{2v_g\eta^3}{\rho g}}\left(v'_F - v_F\right), \tag{16}$$

where v'_F denotes the velocity of rise *after* the charge has changed and v_F denotes the initial velocity of rise (before the droplet charge changed).

Notice that for a given droplet, Eq. 16 means that

$$\Delta q \propto \left(v_F' - v_F\right),$$

or, since $v_F' = h/t_F'$ and $v_F = h/t_F$,

$$\Delta q \propto \left(\frac{1}{t_F'} - \frac{1}{t_F}\right),$$

because the distance h over which the fall and rise times were measured was the same for every droplet.

If the values of the various occurrences of Δq are multiples of some small, elementary value, then so must also be the values of the various occurrences of $\left(\frac{1}{t_F'} - \frac{1}{t_F}\right)$. In columns 4 and 5 of Table 9.2 you see Millikan's examination of this possibility. All the different values of $\left(\frac{1}{t_F'} - \frac{1}{t_F}\right)$ are convincingly close to being small integer multiples of the value $0.005386\,\mathrm{s}^{-1}$.

◼ EXERCISES

19. How close to being exact integer multiples of $0.005386\,\mathrm{s}^{-1}$ are the first three entries in column 4 of Table 9.2?

20. Explain why Millikan probably would have found the same result in column 6 of Table 9.2 even if there had not been an instance in which the charge changed by a single elementary unit.

Of course, if there is a basic "atom" of charge, then the total charge q on any droplet must be some integer multiple of it, and this quantity must be the same as the one of which any change in charge Δq is a multiple. From Eq. 14 it follows that

$$q \propto (v_g + v_F) \propto \left(\frac{1}{t_g} + \frac{1}{t_F}\right).$$

Consequently, the quantity of which $\left(\frac{1}{t_g} + \frac{1}{t_F}\right)$ is an integer multiple should be the same as the quantity of which $\left(\frac{1}{t_F'} - \frac{1}{t_F}\right)$ is an integer multiple. Columns 7, 8, and 9 of Table 9.2 show this to be the case—$0.005386\,\mathrm{s}^{-1}$.

■ **EXERCISES**

21. Explain why column 7 of Table 9.2 represents the total charge on a droplet (as distinct from the change in charge).

22. As well as showing the times of fall and rise, Table 9.2 lists the parameters of Millikan's experiment.
(a) Use these to show that the radius a is correctly calculated from his values of viscosity, speed of fall, and oil drop density.
(b) From these parameters and Millikan's conclusion that the elementary charge corresponds to $\left(\frac{1}{t'_F} - \frac{1}{t_F} \right) = 0.005386\,\text{s}^{-1}$ determine the value of the elementary charge in coulombs.

Millikan reported data on over 50 droplets altogether. He excluded measurements on any trial when he thought the charge on the droplet changed while he was measuring its speed in the electric field. In every case both the charge on a droplet and any change of that charge were integer multiples of a single value. The repeatability of these measurements is strong evidence that a unique atomic unit of charge exists. That charge e is called the "elementary charge." Often, somewhat carelessly ignoring its sign, people call e the "electron charge." The quantity e is also the magnitude of the charge on singly ionized atoms such as hydrogen ions and the like. The modern value for e is $1.602 \times 10^{-19}\,\text{C}$. The results you can obtain from Table 9.2 are slightly different from this value of e because they have not yet been corrected for inaccuracies that occur in Stokes's law when the size of the droplet is comparable to the mean free path of the molecules through which it is falling. Millikan made these corrections and obtained $e = 1.593 \times 10^{-19}\,\text{C}$.[12] Remember that the charge of an electron is negative, although the quantity e is usually given as a positive number. It is your job to remember the sign when it is needed.

Important Numbers Found from e

From accurate knowledge of e came accurate knowledge of Avogadro's number, the mass of the electron, and an order-of-magnitude estimate of

---

[12]Millikan claimed that his value for e was accurate to about a tenth of a percent. However, when other techniques became available to measure atomic spacings in solids to high precision, the value for Avogadro's number derived from these data disagreed with the value calculated from Millikan's e. The source of the discrepancy was ultimately traced to a slightly inaccurate value of the viscosity of air. Millikan used $18.240\,\mu\text{Pa s}$; the currently accepted value is $18.324\,\mu\text{Pa s}$. Scale Millikan's value for e by the ratio of these two numbers and see what you get for e.

the energy of interaction of an electron with its atom. Knowing a precise value of Avogadro's number, you can calculate the actual mass in kilograms of any atom or molecule.

■ **EXERCISES**

> **23.** Calculate Avogadro's number from e and the faraday F.
>
> **24.** Show how once Millikan had measured the elementary charge, he could determine that the mass of an electron is 9.11×10^{-31} kg.

Millikan's result permits you to estimate with moderate precision the energy of interaction of the electron with its atom. An electron a distance of $r_0 = 0.1$ nm—a reasonable number based on the sizes of atoms—from a singly charged ion will have a potential energy of $-ke^2/r_0 = -14.4$ eV. This number tells you roughly the amount of energy involved when atoms interact with each other by means of their electrons. In other words, this number tells you that the energy scale of chemistry is of the order of 10 eV.

9.5   SUMMARY

This chapter has examined three important contributions to our understanding of atoms. Faraday's studies of electrolysis and the regularities of electrolytic deposition showed that atoms have an electrical nature. The results hinted that electricity, like matter, has an atomic nature and that there exists such a thing as a mole of charge and, therefore, some sort of atom of electrical charge, some basic unit of electricity. Thomson discovered the electron by using electric and magnetic fields to show that in an electrical discharge in a gas the rays emanating from the cathode are particles with definite mass and charge. Because the cathode-ray particles were the same regardless of what kind of cathode material or residual gas was in the tube, his results supported other evidence that electrons are parts of all atoms. Millikan's precise measurements of the electronic charge confirmed the discrete nature of electricity and established the energy scale of the electron's interaction with an atom. His results yielded precise values of Avogadro's number and of the electron's mass as well as of its charge.

The indivisible "a-tom" had been cut, and two new, very important, questions arose at once. If the electron is in the atom, what does it do there? And if the electron is one part of an atom, what are the other parts? The answers to these questions led to a revolution in physics, but before you can understand this revolution, you need to study properties of light and other forms of electromagnetic radiation. This you do in the next chapter. Before taking up this new topic, however, you might like to see two interesting uses of electric fields to manipulate charged particles.

9.6   USES OF ELECTRIC DEFLECTION

A good idea can have many applications. You have seen how J.J. Thomson used electric deflection to determine that cathode rays are charged particles. Electric deflection of electrons has widespread application as the basis of the cathode ray tube that is the heart of the oscilloscope, arguably the most important instrument in all modern sciences. Electric deflection is also used for low-cost, high-speed, good-quality printing and to hunt for quarks.

The Inkjet Printer

The inkjet printer works by electric deflection of objects orders of magnitude greater in size than electrons. In the inkjet printer ink droplets are electrically charged and then steered by electric deflection so that the droplets form characters on paper. The essential features of an inkjet printer are shown in Fig. 9.5. About 10^5 droplets are sprayed onto the paper each second, with about 100 drops forming a single character.[13]

Each droplet has a diameter of 63 μm. Assuming that the droplets have the density of water, this diameter corresponds to a droplet mass of 1.31×10^{-10} kg. A controlled amount of electric charge between 80 and 550 fC can be put on each droplet by varying the voltage on the charge electrode. The droplets pass in a stream at a speed of 18 m/s between two small metal plates 1.3 cm long and 1.6 mm apart. These are the deflection plates.

How Much Can a Droplet Be Deflected?

If the maximum voltage that can be placed across the plates is 3.3 kV, what would be the maximum deflection of the ink drops on a piece of

---

[13]A quite detailed description of the considerations that went into the design of the inkjet printer developed by IBM for the IBM 46/40 Document Printer is given in "Application of Ink Jet Technology to a Word Processing Output Printer," W.L. Buehner, J.D. Hill, T.H. Williams, and J.W. Woods, *IBM J. Res. Develop.*, 1–9 (Jan. 1977), and in other articles in this issue of the *IBM Journal of Research and Development.*

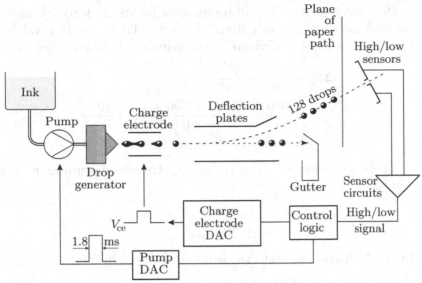

FIGURE 9.5 Inkjet printer showing droplet generator and charger along with deflection plates. ©1977 *International Business Machines Corporation. Reprinted with permission of The IBM Journal of Research and Development, Volume 21, Number 1.*

paper 1 cm away? You want to use a step-by-step approach to answer this question; stepwise reasoning is usually the best way to solve problems because it gives you more physical insight than blind use of an equation.

Notice that this printer is a scaled-up version of Thomson's method for deflecting electrons that was described in Sect. 9.3. A beam of charged particles with momentum p_x passes between a pair of charged plates and is deflected through some angle θ because the electric field between the plates imparts to the particles an increment of momentum Δp_y perpendicular to their path. By varying E you can vary Δp_y and control the deflection angle according to the relation

$$\tan\theta = \frac{\Delta p_y}{p_x} = \frac{y}{L},$$

where y is the vertical displacement a distance L from the deflecting plates.

■ EXERCISES

---

25. How should the apparatus be designed so it can move the particles to points other than on the y-axis?

How much is Δp_y? It will be the product of the force qE and the time interval Δt during which that force acts. This time interval is just the length of the plates ℓ divided by the velocity of the particles v. Therefore,

$$\Delta p_y = \frac{qE\ell}{v} = \frac{qV\ell}{vd}$$
$$= \frac{550 \times 10^{-15} \times 3300 \times 1.3 \times 10^{-2}}{18 \times 0.0016}$$
$$= 8.19 \times 10^{-10} \, \text{kg m s}^{-1}.$$

Then, to find θ you need to find p_x, the momentum along the x-axis. This is

$$p_x = mv_x = 1.31 \times 10^{-10} \times 18$$
$$= 2.36 \times 10^{-9} \, \text{kg m s}^{-1},$$

from which you see that Δp_y is about $\frac{1}{3}$ of p_x; in fact,

$$\tan\theta = \frac{\Delta p_y}{p_x} = \frac{8.19 \times 10^{-10}}{2.36 \times 10^{-9}} = 0.347.$$

Over a distance $L = 1\,\text{cm}$ from the deflection plates to the paper, this angle of deflection will produce a displacement y on the paper such that

$$\frac{y}{L} = \tan\theta = 0.347,$$

from which it follows that

$$y = 1\,\text{cm} \times 0.347 = 0.347 \, \text{cm} = 3.47\,\text{mm}.$$

With the distance to the screen $L = 1\,\text{cm}$ and the length of the deflecting plates $\ell = 1.3\,\text{cm}$, the IBM inkjet printer differs markedly from Thomson's arrangement in that L is not large compared to ℓ. As a consequence, the amount of displacement that takes place while the charged particle is between the plates is no longer negligible compared to the displacement that occurs as the particle travels to the paper.

It is not very difficult to calculate the vertical displacement that occurs while the particle is between the plates. Between the plates the displacement is just what occurs to a particle moving with a constant acceleration $a = qE/m$. That distance is the old, familiar (?) $y = \frac{1}{2}at^2$, where $t = \ell/v_x$. The value of the acceleration is

$$a = \frac{qE}{m} = \frac{qV}{md}$$
$$= \frac{5.5 \times 10^{-13} \, 3300}{1.31 \times 10^{-10} \, 0.0016}$$
$$= 8.66 \times 10^3 \, \text{m s}^{-2};$$

and the value of t is

$$t = \frac{\ell}{v_x} = \frac{1.3 \times 10^{-2}}{18}$$
$$= 7.22 \times 10^{-4} \text{ s.}$$

From these two numbers you can calculate that

$$y = \frac{1}{2}at^2 = \frac{1}{2}(8.66 \times 10^3) \times (7.22 \times 10^{-4})^2$$
$$= 2.26 \times 10^{-3} \text{ m,}$$

which is 2.26 mm and comparable in size to the 3.47 mm of displacement that occurs between the exit point and the paper. The total displacement of a droplet caused by electric deflection is therefore $2.26 + 3.47 = 5.73$ mm.

Noticing that this calculation shows that the droplets will be deflected 2.26 mm between two plates that are 1.6 mm apart, you may think that the droplets will hit the deflecting plates. This does not occur for two reasons. First, air resistance reduces the amount of deflection from what you calculate by the arguments given here. Second, as Fig. 9.5 shows, one of the plates is bent upwards to allow the maximally deflected droplets to escape without hitting a plate.

Quark Hunting

Electric deflection of droplets produced and charged like those in an inkjet printer has been used to look for quarks. Quarks are fundamental particles of which most other particles (but not electrons) are thought to be made. The proton is made of three quarks. One very unusual feature of quarks is their electrical charge: It is a *fraction*(!) of the elementary charge e. One kind of quark is thought to have a charge of $+2e/3$ and another to have $-e/3$. A burning question for over forty years has been, "Why has no one ever been able to see an isolated quark with these very unusual charges?"

It has not been for lack of searching. Quarks have been looked for in mine tailings, in seawater, in old stained-glass windows. They have been sought in niobium spheres and tungsten plates. They have been assiduously pursued with high-energy accelerators and experiments of great ingenuity. But no one has ever found one. They seem to occur only in groups inside other particles.

Separating Droplets by Deflection

One search for quarks used inkjet technology.[14] The idea was to generate a series of droplets with different numbers of charges on them and let them fall in a vacuum through a pair of electrically deflecting plates. If the number of charges on a droplet is an integer multiple of e, then deflections will occur in discrete amounts. For example, a droplet with a charge of $21e$ will undergo a certain deflection and land a discrete distance away from the landing points of droplets of charge $20e$ or $22e$. The droplets should fall in narrow bands with spaces between them. If a fractionally charged droplet came along, it would land in one of those spaces and so be observable. Figure 9.6 shows the apparatus. Droplets of tetraethylene glycol of mass $m = 16\,\mathrm{ng}$ and diameter $2r = 30.1\,\mu\mathrm{m}$ were injected at the top of a 22-m tall tower. The droplets were traveling at a speed of $v_1 = 8.6\,\mathrm{m/s}$ as they entered the 3.05-m long space between the deflecting plates where they were deflected by the electric field produced by a potential difference of $20\,\mathrm{kV}$ between the plates, which were $5\,\mathrm{cm}$ apart. After exiting from between the deflection plates, the droplets fell about $15\,\mathrm{m}$ to a detector that measured the amount of their sideways displacement.

Does this work? How much will be the separation between droplets when they are detected? Will there be enough space between the landing points of droplets that differ by one unit of charge so that you can distinguish any fractionally charged particle if one happens to be present? To answer these questions you need to calculate the sideways displacement of a droplet carrying a charge of $q = 20e$ for a deflecting voltage of $20\,\mathrm{kV}$.

Because the tetraethylene glycol droplets are accelerating in the direction of their motion under the force of gravity both as they pass between the deflecting plates and afterwards, it is not adequate to find the angle at which the particles emerge from between the plates as was done in the cases of a beam of cathode rays or a stream of charged droplets of ink. Instead, you must find the sideways displacement x_1 and the sideways velocity v_x at the time t_1 the droplet emerges from between the plates and also calculate the time t_2 that it takes the droplet to fall to the detector $15\,\mathrm{m}$ below. The sideways displacement after exiting from between the plates will be $x_2 = v_x t_2$, and the total sideways displacement will be $x_1 + x_2$.

To find the sideways velocity v_x as the droplets exit from between the plates, you need to use the sideways acceleration $a_x = qE/m$ produced by the electric field E during the time t_1 that the droplet is in the field. Then $v_x = a_x t_1$. But how do you find t_1?

---

[14] "Search for fractional charges using droplet-jet techniques," J. Van Polen, R.T. Hagstrom, and G. Hirsch, *Phys. Rev.* **D36**, 1983–1989 (1987).

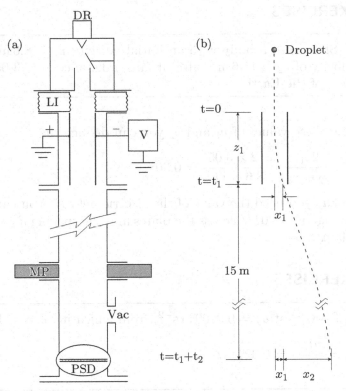

FIGURE 9.6 (a) Schematic diagram of a 22-m high evacuated tube inside of which charged droplets fall and are electrically deflected: DR is the droplet source; LI is a vibration isolator; V is the deflection voltage supply; MP is a mounting plate; Vac is the outlet to the vacuum pump; and PSD is the position-sensitive detector that records the fallen drops. (b) Variables and quantities for analyzing the deflection of a droplet; they are described in the text. *Taken with permission from J. Van Polen, R.T. Hagstrom, and G. Hirsch, Phys. Rev. Vol. D36, 1983–1989 (1987),* ©*1987 American Physical Society.*

To find t_1, use the fact that under constant acceleration the average velocity over some time interval t is $(v_2 + v_1)/2$, where v_1 is the velocity at the start of the interval and v_2 is the velocity at the end of the interval. Then the time t_1 to fall a vertical distance z_1 between the plates is

$$t_1 = \frac{2z_1}{v_2 + v_1},$$

where $v_1 = 8.6\,\text{m/s}$ is the velocity of the droplet as it enters the electric field between the plates and v_2 is its velocity as it exits. Of course, this means you must find v_2, but you can do this using the conservation of energy as was illustrated in Sects. 2.7 and 7.2.

▨ EXERCISES

26. Show that a body with an initial velocity $v_1 = 8.6\,\mathrm{m/s}$ will have a velocity of $v_2 = 11.6\,\mathrm{m/s}$ after it falls a distance $z_1 = 3.05\,\mathrm{m}$ near the surface of the Earth.

From these values of v_1 and v_2 you can deduce that

$$t_1 = \frac{2z_1}{v_2 + v_1} = \frac{2 \times 3.05}{8.6 + 11.6} = 0.30\,\mathrm{s}.$$

To find a_x you need the value of the electric field E. You can find E from the voltage $V = 20\,\mathrm{kV}$ across the plates and the spacing $d = 5\,\mathrm{cm}$ between the plates.

▨ EXERCISES

27. Show that $a_x = 0.0800\,\mathrm{m\,s^{-2}}$, from which it follows that

$$v_x = \frac{qE}{m}t_1 = 0.024\,\mathrm{m/s}.$$

At this point in the calculation you can find the displacement x_1 of the droplet during its passage between the plates. It will be

$$x_1 = \frac{v_{x0} + v_{x1}}{2}t_1 = \frac{0 + 0.024}{2}\,0.30 = 0.0036\,\mathrm{m} = 3.6\,\mathrm{mm}.$$

To find the distance x_2 that the drop moves sideways while falling under the acceleration of gravity through the remaining distance $z_2 = 15\,\mathrm{m}$, it is necessary to find its time of free fall, t_2. The approach is the same as for finding the time of passage between the plates.

▨ EXERCISES

28. Show that $t_2 = 0.93\,\mathrm{s}$.

Now knowing the fall time t_2 and the sideways velocity $v_x = 0.024\,\mathrm{m/s}$, you can calculate the sideways displacement x_2 of the droplet after it has left the plates.

$$x_2 = v_x\, t_2$$
$$= 0.024 \times 0.93 = 0.022\,\mathrm{m}$$
$$= 2.2\ \mathrm{cm}.$$

In the electric deflection of cathode rays the amount of displacement that occurred between the plates was often negligible compared to the amount that occurred afterwards on the trip to the screen. In this search for quarks, the displacement occurring between the plates is not negligible; it is $3.6\,\mathrm{mm}$ compared to $22\,\mathrm{mm}$. To find the total deflection you must add the two parts; this gives a total displacement of

$$x_1 + x_2 = 25.6\,\mathrm{mm}.$$

Although it is good to understand the step-by-step argument that has brought you to this answer, a complete algebraic expression is also useful.

■ EXERCISES

---

29. Show that a droplet of mass m and charge q that has passed through an electric field E in the quark-hunting apparatus will undergo a total displacement of

$$x_1 + x_2 = \frac{1}{2}\frac{qE}{m}t_1^2 + \frac{qE}{m}t_1 t_2. \qquad (17)$$

This expression tells you two very important things about the displacement: (1) It is proportional to the charge q on the droplet; (2) it is proportional to the strength of the electric field E across the deflecting plates.

Such an expression has several uses. For example, to check that their apparatus was functioning properly, the experimenters ran a separate set of experiments with $24.2\,\mathrm{kV}$ across the plates instead of $20\,\mathrm{kV}$. What would be the displacement produced by a deflecting voltage of $24.2\,\mathrm{kV}$?

Equation 17 shows that you can use a simple proportion to find the answer:

$$x' = \frac{24.2\,\mathrm{kV}}{20\,\mathrm{kV}}\cdot 2.56\,\mathrm{cm} = 3.1\,\mathrm{cm}.$$

More to the point, you can now answer the basic question about this experiment: With $20\,\mathrm{kV}$ across the plates, by how much will the displacement change if a charge of $20\,e$ changes by $\pm e$? The answer is

$$\frac{1}{20}\,2.56 = 0.128\,\mathrm{cm} = 1.28\,\mathrm{mm}.$$

■ EXERCISES

30. Explain the reasoning by which this answer was obtained.

This result means that a stream of droplets charged with different multiples of e will be spread out into a fan of streams, like fingers of a hand, each stream separated from the next by 1.28 mm. The authors say that the individual streams are only about 0.060 mm wide, so the gaps between the streams were pronounced. They looked with their detector to see whether there were any streams or occasional droplets that fell with a separation $\frac{2}{3}$ or $\frac{1}{3}$ of 1.28 mm. These might correspond to droplets carrying a single quark as part of its charge. The authors report that they saw no evidence of quarks.

PROBLEMS

1. In performing the experiment described in Example 9.1 Faraday collected H_2 in a volume of 12.5 cubic inches. His work was done at room temperature and at a pressure of 29.2 in Hg. He corrected this volume to 12.2 cubic inches at mean pressure (1 atm = 29.9 in Hg) and at "mean temperature" = 50 °F and determined that the mass of the collected gas was 2.35 grains (1 grain = 0.0648 g). What was the room temperature at which Faraday performed this experiment? Compare that to your room temperature.

2. In another electrolysis experiment, Faraday measured the amount of hydrogen and oxygen gas evolved from water at one electrode and the amount of tin deposited on the other (tin) electrode. He writes, "The negative electrode weighed at first 20 grains; after the experiment it...weighed 23.2 grains....The quantity of oxygen and hydrogen collected...= 3.85 cubic inches." [This is the volume corrected to 1 atm and to 50 °F.]

Use these data and the atomic weights of H_2 and O_2 to find the atomic weight of tin (Sn). (Remember that Sn is divalent—see Table 9.1.)

3. On page 164 you saw how to estimate the velocity with which electrons move when they carry a current through a wire. Using the following information, estimate the speed with which ions move between two electrodes in an electrolytic cell.

Assume that the cell is filled with 0.01 N sulfuric acid (pH $\approx$ 2.1, which means about 6×10^{18} H$^+$ ions in each cm^3) and that the electrodes have surface areas of 0.25 cm^2 and are 4 cm apart.

If the cell is run with a current of 0.2 A, what is the average speed with which ions travel from the anode to the cathode? How does your answer compare with the estimated speed of electrons in a current-carrying wire?

4. Millikan used Stokes's law, which says that a small sphere of radius a falling with velocity v in a homogeneous medium of viscosity $\eta = 18.6 \times 10^{-6}$ Pa s is subject to a retarding force

$$F = 6\pi\eta av.$$

He observed that a drop of oil of density $\rho = 0.92$ g cm^{-3} would fall 0.522 cm in 13.6 s.

    a. What was the radius of the drop?

    b. What was its mass?

5. If a charged oil droplet of density $\rho = 0.92$ g cm^{-3} and radius $a = 2\,\mu$m is just balanced against gravity's pull by an electric field of 1.9×10^6 V/m, what is the charge on the droplet? Give your answer in multiples of the elementary charge e.

6. A bowling ball of mass $M = 6$ kg is rolling down a hallway at 1.5 m/s. (see Fig. 9.7) If it rolled straight, it would hit an associate dean of students D standing at the end of the hall. You are standing at point U, 10 m from the end of the hall, and you intervene heroically by giving the ball a sideways kick as it passes you so that the ball rolls into the corner C instead of hitting D.

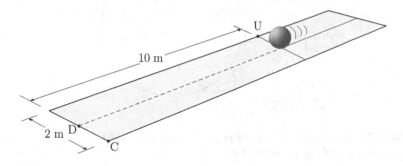

FIGURE 9.7 Ball rolling down a hallway towards D is deflected at point U so as to hit C instead (Problem 6).

    a. What must be the angle of deflection θ at point U if the ball is to hit C instead of D?

    b. How much sideways momentum Δp_y must you give the ball to produce the desired deflection?

    c. Explain the parallels between this problem and one of the experiments described in this chapter.

7. What was the first evidence that atoms are electrical in nature?

8. Faraday's work led to the realization that a mole of electric charges is about 96 500 C and that this amount of charge releases 1.008 g of hydrogen in electrolysis. From these results find the charge-to-mass ratio of an individual hydrogen ion, H^+, and explain your work.

9. Section 9.3 describes a method for measuring e/m of cathode rays. Suppose that you applied this method and observed a beam of mystery rays emerging from the source and that all that you knew about them was that they were charged particles, each carrying one elementary charge.

    Now you accelerate them through a potential difference of $V = 500\,\text{V}$. Then, as shown in Fig. 9.8, they emerge from the source and enter a uniform magnetic field of $B = 65\,\text{mT}$ (pointing into the plane of the page) and bend in a circle of radius 5 cm.

    a. Are these particles cathode rays? How do you know?

    b. What are some of the important conclusions drawn from Thomson's measurements of the charge-to-mass ratio of cathode rays?

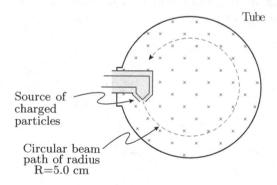

FIGURE 9.8   A beam of charged particles moves in a circle in a magnetic field pointing into the plane of the page (Problem 9).

10. In an electrolysis experiment a current of 0.5 amperes flowing for 20 min (1200 s) liberates 69.6 mL of gas at the cathode and 23.3 mL at the anode. You may assume that both gases are at STP ($T = 272.3$ K, $P = 1.013 \times 10^5$ Pa).

    a. How many moles of gas molecules are produced at the cathode?

    b. How many electrons were needed to form one molecule of gas at the cathode?

    c. Which of the following chemical reactions is consistent with the information given above?

$$2H_2O \rightarrow 2H_2 + O_2$$
$$2CO_2 \rightarrow 2CO + O_2$$
$$2N_2O_3 \rightarrow 2N_2 + 3O_2$$
$$2NH_3 \rightarrow N_2 + 3H_2$$

11. A droplet 10 μm in radius will fall in air at a steady velocity of 1.1 cm/s because its weight mg (m is its mass, g is the acceleration due to Earth's gravity) is balanced by the viscous force $6\pi\eta a v$ (η is the viscosity of air, 18.2 μPa s; a is the radius of the droplet; and v is the velocity of fall).

    a. Use these data to find the radius of a droplet that falls with a velocity of 8.59 $\times 10^{-4}$ m/s.

    b. When an electric field of 317.5 kV/m is turned on, the droplet in (a) is observed to rise at a steady speed of 2.41×10^{-4} m/s.

    c. Show that the upward electrical force on the droplet is 1.28 times the downward gravitational force mg.

    d. To within a couple of electrons, how many excess electrons are there on this droplet?

    e. What important conclusions do we draw from Millikan's measurements?

12. Faraday's work led to the realization that 96 500 C of charge will cause 1 mole of hydrogen or of silver or of chlorine to evolve from electrolysis. Thomson's work led to the measurement of $q/m = 1.76 \times 10^{11}$ C/kg for cathode rays. Millikan measured the elementary charge.

    a. What did Faraday's laws of electrolysis imply about atoms and electricity?

    b. What did Thomson's work reveal about cathode rays?

  c. Combining Faraday's result and Thomson's, find the value of the ratio of the mass of the hydrogen atom to that of the electron.

  d. What did Millikan's result reveal about electric charge?

13. a. Explain the significance of the faraday of charge $Q = 96\,500\,\text{C}$.

  b. Explain the significance of the discovery that all cathode rays have a definite charge-to-mass ratio of $1.7 \times 10^{11}\,\text{C/kg}$.

  c. Show by direct calculation how Millikan's experimental measurement of the elementary charge made it possible to determine

  i. Avogadro's constant.

  ii. The mass of the electron.

  iii. The actual masses of atoms.

  What fundamental property of electric charge did Millikan's work confirm?

14. In electrolysis $96\,500\,\text{C}$ of charge releases 1 mole of hydrogen atoms.

  a. From this information show how to determine the charge-to-mass ratio of the proton. Clearly state any assumptions you make.

  b. Suppose in measuring e/m of the electron you observed that electrons accelerated through a voltage acquire velocity of $v = 0.03\,c$ (where c is the speed of light in vacuum) and bend in the circle of radius $r = 3\,\text{cm}$ in a magnetic field of $B = 2\,\text{mT}$. What would you determine e/m to be? (Your answer should be different from the accepted value.)

  c. Explain why results like your answers to (a) and (b) implied to J.J. Thomson the existence of a new particle. What was it, and in what ways is it important?

15. Use a spreadsheet least squares routine to derive a best value for e from Millikan's data in Table 9.2. Do this in the following way: Plot a graph of $1/t'_F - 1/t_F$ vs n' (columns 4 and 5 in the table). Take the slope of this graph and multiply it by the factor (which you must determine from the information Millikan provides) that converts this slope into e. Explain why this calculation gives e. (Millikan corrected his results for inaccuracies in Stokes's law that occur when droplets are very small.)

16. The diagram in Fig. 9.9 shows a pair of parallel plates $60\,\text{mm}$ long and separated by $12\,\text{mm}$ to which a power supply (not shown) is connected. An electron beam enters the region between the plates from the left as shown.

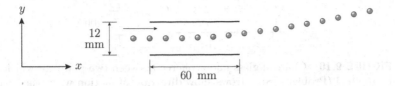

FIGURE 9.9   Electron beam passing between parallel plates (Problem 16).

  a. Assuming that the electric field between the plates is uniform, find the magnitude of the force exerted on any electron by the field given that the potential difference between the plates is 48 V.

  b. If the speed of the electrons is 5×10^7 m/s, find the electron's change of momentum in the y direction that results from passing between the plates.

  c. Suppose that the beam of electrons is replaced by a beam of negative hydrogen ions of the same speed (a hydrogen ion, H^-, has the same charge as an electron and a mass equal to 1836 electron masses). Which, if any, of your answers to (a) and (b) above would have to be changed? Why?

17. Give brief, compact answers to each of the following. Use specific facts, equations, or diagrams to support your answers.

  a. Explain why we think that there is such a particle as the "electron."

  b. Explain how Millikan's measured value of e can be used to find both Avogadro's number and the mass of the electron.

  c. What is the evidence for thinking that atoms are electrical in nature?

18. Olive Oyal, a promising young physicist, has decided to recreate Millikan's famous oil-drop experiment using modern materials. Rather than employing oil, Olive chooses precision glass microspheres of diameter 1.0 μm and density 2.5 g cm^{-3}. Figure 9.10 illustrates her experimental apparatus and three spheres A, B, and C. Olive observes that
(1) before the switch is closed, all three spheres fall with the same speed v_f, and
(2) after the switch is closed, sphere A rises with speed v_f, sphere B *falls* with speed v_f, and sphere C comes to rest.

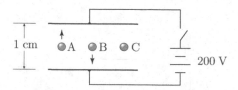

FIGURE 9.10 Charged glass microspheres between two plates to which a voltage can be applied (Problem 18). Arrows show direction of motion when the switch is closed.

 a. Based on the information given above, specify whether the charge on each of the three drops is +, −, 0, or indeterminate.

 b. Prove that the mass of each sphere is equal to 1.31 pg.

 c. Now calculate the charge on each of the three spheres. From these results what might Olive conclude is the basic, elementary, unit of charge?

19. The magnitude of the charge on an electron is 1.6×10^{-19} C; it is negatively charged.

 a. Could Millikan have done his experiment with the air pumped out of the volume of space where he made an electric field, i.e., between the plates? Why?

 b. Find the mass of a droplet of oil of density $0.92\,\mathrm{g\,cm^{-3}}$ if the radius of a droplet is $3\,\mu\mathrm{m}$.

 c. An electric potential difference of 5000 V is put across two large conducting plates separated by 2 cm. What is the value of the electric field in the space between the plates?

 d. If an electric field of $300\,\mathrm{kV/m}$ exerts a force of 4.8×10^{-14} N on an electrically charged drop of oil, how much is the electric charge of the oil drop?

20. We measure electrical charge in units of coulombs; a hundred years ago it was common to measure electrical charge in units called esu. The conversion is 3×10^{9} esu = 1 C.

 a. What is the charge of the electron measured in esu?

 b. Thomson measured e/m to have a value of 2.3×10^{17} esu/g. Convert this to C/kg and compare the result with the currently accepted value.

 c. Why was Thomson the first to see cathode rays be deflected by an electric field?

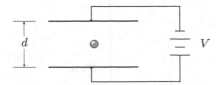

FIGURE 9.11   An oil drop suspended in an electric field. (Problem 22).

21.   J.J. Thomson measured e/m for cathode rays.

    a. Who was J.J. Thomson? Give a few biographical facts about the man.

    b. What are cathode rays?

    c. Thomson noted two significant properties of the rays. What were they and why were they important?

22.   A charged oil droplet is suspended motionless between two parallel plates ($d = 0.01$ m) that are held at a potential difference V as shown in Fig. 9.11. Periodically, the charge on the droplet changes, as in Millikan's original experiment. Each time the charge changes, V is adjusted so that the droplet remains motionless. Here is a table of recorded values of the voltage V:

    i. 350.0 volts.

    ii. 408.3 volts.

    iii. 490.0 volts.

    iv. 612.5 volts.

    a. In which case is the charge on the droplet the greatest?

    b. From the data above, determine the charge on the droplet for case (i) above. What assumptions do you need to make?
(Hint: the ratio of voltages = ?)

    c. Find the magnitude and direction of the electric field in case (i) above.

    d. What is the mass of the drop?

23.   From Faraday's experiment you find that during the electrolysis of water, 0.01 g of H_2 gas is liberated by the passage of 965 C of charge. Find the charge-to-mass ratio of a proton from these data.

24.   Particles each with mass m and charge q accelerate through a potential difference of $V = 100$ volts and enter a region of uniform magnetic

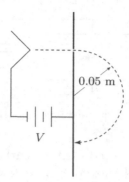

FIGURE 9.12 Accelerated charged particles enter a uniform magnetic field. (Problem 24).

field B, where their paths are circular with radius $R = 0.05$ m as shown in Fig. 9.12. The magnitude of the field is $B = 6.75 \times 10^{-4}$ T.

   a. Indicate the direction of B on a drawing.

   b. Derive the following relationship between the radius R and q, m, V, and B:

$$R^2 = \frac{m}{q}\frac{2V}{B^2}.$$

   c. Find the value of the charge-to-mass ratio (q/m) of these particles and identify them. Justify your answer!

25.   In a laboratory electrolysis experiment, 40 mL of N_2 gas and 60 mL of O_2 gas are liberated at the cathode and anode, respectively. The room pressure and temperature are 730 Torr and 25 °C.

   a. Which of the following overall reactions has taken place?

$$2N_2O \rightarrow 2N_2 + O_2$$
$$2NO_2 \rightarrow N_2 + 2O_2$$
$$2N_2O_3 \rightarrow 2N_2 + 3O_2$$
$$2N_2O_5 \rightarrow 2N_2 + 5O_2$$

   b. Figure 9.13 shows a plot of the volume of N_2 evolved at the cathode vs. the time in minutes. Which of the following statements most likely explains the "kink" in the graph?

     i. The current was increased after 15 min.

     ii. A gas leak developed above the cathode.

     iii. A gas leak developed above the anode.

     iv. Data points were collected more frequently after the first 15 min.

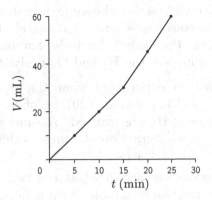

FIGURE 9.13 Rate of electrolytic evolution of N_2 (Problem 25).

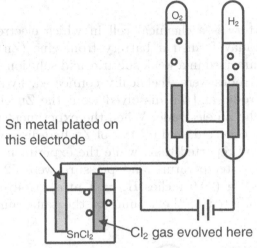

FIGURE 9.14 An arrangement for studying the electrolysis of $SnCl_2$. (Problem 26).

   c. At the temperature and pressure given above, what volume would 1 mole of N_2 gas occupy?

   d. What is the total mass of N_2 gas that has evolved after 25 min?

26.   Michael Faraday was a consummate experimentalist. In his earliest studies of electrolysis, he used the volume of gas evolved by the decomposition of water as a measure of the amount of electricity (total charge) passing through his electrolysis cells. Figure 9.14 illustrates an experiment to study the electrolysis of tin chloride ($SnCl_2$). Two platinum electrodes are immersed in a molten bath of $SnCl_2$ and connected in series with a second cell containing a weak solution of sulfuric acid in water. (Because the cells are connected in series, the same current flows through each cell.) In the reaction, gaseous Cl_2 is evolved at one electrode, and solid (metallic) Sn is plated onto the second electrode. Faraday measured the mass of the Sn and the volumes of the H_2 and O_2 evolved in the water cell.

Faraday stopped the experiment when the total volume of H_2 and O_2 equaled 63.1 cm^3, and found that 0.207 gm of Sn had been deposited.

    a. What volume of H_2 was evolved? Assuming standard temperature and pressure, what fraction of a mole is this? (The overall reaction is $2H_2O \rightarrow 2H_2 + O_2$.)

    b. What reactions are occurring at the two electrodes in the $SnCl_2$ cell? What fraction of a mole of Sn is deposited?

    c. From Faraday's numbers, find the atomic weight of Sn.

    d. If the Cl_2 gas had been collected, what would its volume have been?

27.   A battery is a chemical cell in which electrolysis occurs spontaneously. Faraday formed a battery from zinc (Zn) and platinum (Pt) electrodes immersed in a weak sulfuric acid solution (see Fig. 9.15). When the two electrodes were electrically connected, hydrogen gas evolved at the Pt electrode, and Zn dissolved from the Zn electrode. No gas was released at the Zn electrode. When the experiment was done, 205 cm^3 of H_2 had been collected, and 0.548 g of Zn had dissolved.

    a. Faraday reported that, while the experiment was in progress, the ambient temperature and pressure were 52 °F (11 °C) and 29.2 inches Hg (29.9 inches Hg = 1 atm). Convert the volume of H_2 gas collected to the volume of the same amount of gas at STP.

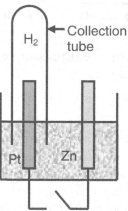

FIGURE 9.15 In this chemical cell electrolysis occurs spontaneously when the switch is closed; it is a battery. (Problem 27).

(Hint: you don't need to convert the pressure from inches Hg to Pa or Torr if you set up a ratio of volumes.)

b. Find, from the numbers above, the atomic weight of Zn.

28. In a modern version of Millikan's famous experiment, a charged radioactive sphere of mass $1\,\text{ng}$ $(=?\,\text{kg})$ is suspended motionless between two parallel conducting plates spaced by $1.0\,\text{mm}$.

The apparatus (plates and sphere) is in vacuum. Initially, a potential difference of $15.3\,\text{kV}$ is required to counteract the gravitational force.

a. Determine the magnitude of the following forces:

    i. Gravitational force.

    ii. Electric force.

    iii. Viscous force.

b. What is the charge on the sphere?

c. By radioactive decay, the sphere periodically emits a charged particle, and the voltage required to balance the gravitational force changes suddenly. You observe the following:

    i. The sphere is balanced by a voltage of $15.3\,\text{kV}$ for a short time, whereupon....

    ii. the sphere begins to fall, and the voltage must be increased to $30.6\,\text{kV}$ in order once again to levitate the sphere, and then....

    iii. the sphere falls and is unaffected by the magnitude or sign of the potential difference applied to the plates.

Explain these observations and determine the charge of the particle emitted in each radioactive decay.

29. In a lab experiment, an unknown species of charged particle is produced and accelerated through a potential difference of $200\,\text{V}$. The particles then enter a region of constant magnetic field $B = 1.0\,\text{mT}$, where they trace out a circular orbit of radius $r = 4.8\,\text{cm}$. The apparatus, particle trajectory, and magnetic field are as shown in Fig. 9.16.

a. Starting from the expressions for the magnetic force, the centripetal force, and the kinetic energy vs. accelerating voltage, prove that

$$\frac{m}{q} = \frac{B^2 r^2}{2V}.$$

b. From the information given, identify the particle.

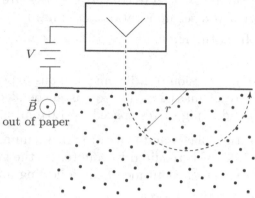

FIGURE 9.16 Charged particles accelerate through a voltage V and then travel in a region of uniform magnetic field $\vec{B}$ (Problem 29).

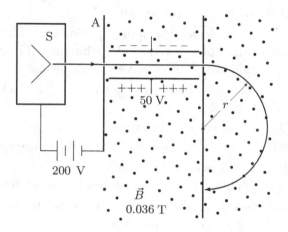

FIGURE 9.17 Experimental setup for Problem 30. Note that the region to the right of the anode A is filled with a magnetic field $B = 0.036\,\text{T}$ pointing out of the paper.

  c. Suppose the above particles were replaced by "negative muons," which have the same charge as an electron and a mass 207 times larger than an electron mass. What magnetic field $\vec{B}$ (direction as well as magnitude) is required if the muons (accelerated through the same potential difference) are to follow the same circular path as the original particles?

  d. Explain the significance of the discovery that all cathode rays have a definite mass-to-charge ratio of $0.57 \times 10^{-11}\,\text{kg/C}$.

30. As shown in Fig. 9.17, charged particles are produced in a source S and accelerated through a potential difference $V_{\text{acc}} = 200\,\text{V}$. The particles enter a region of constant magnetic field $B = 0.036\,\mathrm{T}$ (directed out of the paper) and pass undeflected through two charged parallel plates spaced by $1.0\,\mathrm{cm}$. The voltage across the plates is $50\,\mathrm{V}$.

    a. Study the figure carefully. Are the particles positively or negatively charged? How do you know?

    b. Prove that the particle velocity $v = 1.39 \times 10^5\,\mathrm{m/s}$.

    c. Find the mass-to-charge ratio of the particles, and from this identify the particle, choosing from the following list:

        i. Electron.

        ii. Alpha particle.

        iii. Proton.

        iv. None of the above.

31. From what you know about J.J. Thomson's discovery of the electron:

    a. Explain how one can use a combination of electric and magnetic fields to determine the ratio of charge to mass of a particle when you don't know either.

    b. Tell what there is about such experiments that leads you to conclude that cathode rays are particles?

    c. Describe the ways in which Thomson's experiments suggest that atoms contain electrons, i.e., that electrons are parts of atoms?

SPREADSHEET EXERCISE: THE MILLIKAN OIL DROP EXPERIMENT

Millikan's oil drop experiment had two important outcomes: it found the value of the charge on the electron, and, equally important, it provided convincing evidence that all electrical charge is some integer multiple of an elementary charge e. This computer exercise illustrates how to use a computer spreadsheet to analyze Millikan's data and show both of these outcomes. The data to be analyzed are shown in Table 9.3. The original table is in Millikan's paper "The isolation of an ion," *Phys. Rev.* **32**, 356–358 (1911); in the version given here the values of the charges and viscosity have been converted to SI units.

TABLE 9.3 Millikan's Oil-Drop Data in Modern Units

Distance between crosshairs $d_c = 1.010\,\text{cm}$
Distance between plates $d_p = 1.600\,\text{cm}$
Temperature $t = 24.6\,°\text{C}$
Density of oil at $25\,°\text{C}$ $\sigma = 0.8960\,\text{g cm}^{-3}$
Viscosity of air at $25.2\,°\text{C}$ $\eta = 18.36\,\mu\text{Pa·s}$

	G (sec)	F (sec)	n	e_n $(10^{-19}\,\text{C})$	e_1 $(10^{-19}\,\text{C})$
	22.8	29.0	7	11.50	1.642
	22.0	21.8	8	13.16	1.645
	22.3	17.2			
$G = 22.28$	22.4	— —			
$V = 7950\,\text{V}$	22.0	17.3	9	14.82	1.647
	22.0	17.3			
	22.7	14.2	10	16.48	1.648
	22.9	21.5	8	13.16	
	22.4	11.0	12	19.72	1.644
	22.8	17.4	9	14.82	
	22.8	14.3	10	16.48	
$V = 7920\,\text{V}$	22.8	12.2	11	17.99	1.635
$G = 22.80$	23.0	12.3			
	22.8	14.2			
	— —	— —	10	16.48	1.648
$F = 14.17$	22.8	14.0			
	22.8	17.0			
$F = 17.13$	— —	17.2	9	14.82	1.647
	22.9	17.2			
$F = 10.73$	22.8	10.9			
	22.8	10.9	12	19.72	1.644
$V = 7900\,\text{V}$	22.8	10.6			
$G = 22.82$	22.8	12.2	11	17.99	1.635
$F = 6.7$	22.8	8.7	14	22.90	1.636
	22.7	6.8	17	27.76	1.633
	22.9	6.6			
	22.8	7.2			
$F = 7.25$	— —	7.2			
	— —	7.3			
	— —	7.2	16	26.13	1.634
	23.0	7.4			
$F = 8.65$	— —	7.3			
	— —	7.2			

TABLE 9.3 Millikan's Oil-Drop Data *(cont.)*

	G (sec)	F (sec)	n	e_n $(10^{-19}$ C)	e_1 $(10^{-19}$ C)
	22.8	8.6	14	22.90	1.636
	23.1	8.7			
	23.2	9.8	13	21.24	1.635
F = 10.65	––	9.8			
	23.5	10.7	12	19.72	1.644
	23.4	10.6			
	23.2	9.6			
	23.0	9.6			
V = 7820	23.0	9.6			
F = 9.57	23.2	9.5			
G = 23.14	23.0	9.6	13	21.24	1.635
	––	9.4			
	22.9	9.6			
	––	9.6			
F = 8.65	22.9	9.6			
	––	10.6	12	19.72	1.644
	––	8.7	14	22.90	1.636
	23.4	8.6			
F = 12.25	23.0	12.3			
	23.3	12.2	11	17.99	1.635
	––	12.1			
	23.3	12.4			

Change forced with radium

	G (sec)	F (sec)	n	e_n $(10^{-19}$ C)	e_1 $(10^{-19}$ C)
	23.4	72.4			
	22.9	72.4			
F = 72.10	23.2	72.2	5	8.207	1.641
	23.5	71.8			
	23.0	71.7			
	23.0	39.2	6		
V = 7800	23.2	39.2			
G = 23.22	––	27.4	7	11.50	
	––	20.7	8	13.14	1.642
	––	26.9	7	11.50	1.642
	––	27.2			
	23.3	39.5			
F = 39.20	23.3	39.2	6	9.881	1.647
	23.4	39.0			
	23.3	39.1			

TABLE 9.3   Millikan's Oil-Drop Data *(cont.)*

	G (sec)	F (sec)	n	e_n $(10^{-19}$ C$)$	e_1 $(10^{-19}$ C$)$
	23.2	71.8	5	8.207	1.641
	23.4	382.5	4		
	23.2	374.0			
	23.4	71.0	5	8.207	1.641
	23.8	70.6			
$V = 7760$	23.4	38.5	6		
$G = 23.43$	23.1	39.2			
	23.5	70.3			
	23.4	70.5			
	23.6	71.2	5	8.207	1.641
	23.4	71.4			
	23.6	71.0			
	23.4	71.4			
	23.5	380.6			
	23.4	384.6			
$F = 379.6$	23.2	380.0			
	23.4	375.4	4	6.559	1.640
	23.6	380.4			
	23.3	374.0			
	23.4	383.6			
	—				
$G = 23.46$	23.5	39.2			
$F = 39.18$	23.5	39.2	6	9.881	1.647
$V = 7730$	23.4	39.0			
	—	39.6			
$F = 70.65$	—	70.8			
	—	70.4	5	8.207	1.641
	—	70.6			
	23.6	378.0	4	6.559	0

Saw it, here, at end of 305 sec, pick up two negatives

	G (sec)	F (sec)	n	e_n $(10^{-19}$ C$)$	e_1 $(10^{-19}$ C$)$
	23.6	39.4	6	9.881	1.647
	23.6	70.8	5	8.207	1.641

Mean of all e_1s = 1.640

Differences
$8.207 - 6.559 = 1.65$
$9.881 - 8.207 = 1.67$
$11.50 - 9.881 = 1.62$
$13.14 - 11.50 = 1.64$
Mean dif. = 1.64

THEORY

Millikan measured the terminal velocity v_1 of a charged droplet of mass m falling in air in a gravitational field g; he then applied an electric field F to the droplet and measured its terminal velocity as it rose against gravity under the effect of the field. He showed that for this situation the electric charge q on the droplet is given by

$$q = \frac{mg}{F}\left(\frac{v_1 + v_2}{v_1}\right).$$
(18)

His experiment was designed to work with the same oil droplet for hours at a time. In column G of Table 9.3 the values of t_1, the time the drop falls between the crosshairs when the electric field is turned off, are essentially constant because they are for the same droplet. Using radiation from a radioactive source, Millikan changed the charge on the droplet from time to time. Changes in charge changed the time t_2 that it took the droplet to rise in the applied electric field. These changes show up clearly in the variations of the times recorded in column F of Table 9.3. The change in charge, Δq, can be found from Eq. 18:

$$\Delta q = \frac{mg}{Fv_1}\left(v_2' - v_2\right) = \frac{mgt_1}{F}\left(\frac{1}{t_2'} - \frac{1}{t_2}\right).$$
(19)

The droplet's mass m is found from the time of fall t_1 by connecting the mass of the droplet to its size and density and then using Stokes's law to find its size from its measured terminal rate of fall. The mass m of a spherical droplet is

$$m = \frac{4}{3}\pi a^3 \sigma,$$
(20)

where σ is the density of oil and we neglect Millikan's correction for buoyancy due to ρ, the density of air. The drop's radius a is obtained from Stokes's law for the terminal velocity of a small sphere falling in a viscous medium,

$$a = \sqrt{\frac{9\eta v_1}{2g\sigma}},$$
(21)

where η is the viscosity of air. If you replace a in Eq. 20 with Eq. 21, you get

$$m = \frac{4}{3}\pi \left(\frac{9\eta v_1}{2g}\right)^{\frac{3}{2}} \left(\frac{1}{\sigma}\right)^{\frac{1}{2}}.$$
(22)

Dividing the measured value of t_1 into the distance between the cross hairs d_c gives the value of v_1, and this substituted into Eq. 22 gives the value of m.

The values of the constants needed for Eqs. 22 and 19 are:

$$g = 9.8\,\text{m/s}^2,$$
$$\sigma = 896\,\text{kg/m}^3,$$
$$\rho = 1.28\,\text{kg/m}^3,$$
$$\eta = 1.836 \times 10^{-5}\,\text{N s/m}^2,$$
$$d_p = 0.016\,\text{m},$$
$$d_c = 0.0101\,\text{m},$$
$$V = 7800\,\text{V}.$$

PROCEDURE

This procedure has two goals: First, to enable you to show that Δq is always an integer multiple of some basic quantity, and, second, to enable you to find the value of that quantity. The step-by-step instructions given below are for Microsoft®Office Excel.[15]

a. In an Excel spreadsheet, type into cell A1 your name; into cell A2 the title of this exercise; into cell A3 the date. Then, in A4 type 'quantity'; in B4 type 'value'; in C4 type 'units'. Next, in cells A5 to A11 enter the names of the quantities given in the above table. Enter their values into cells B5 to B11; enter their units into cells C5 to C11. (For computers you enter a number like 1.836×10^{-5} by typing 1.836E-5; the computer uses "E-5" instead of 10^{-5}.) In cell A14 enter 't1 (sec)'; in cell A15 enter 23.4, the average value of t1 in seconds.

b. In Cells A18 and A19 type 't2' and '(sec)'. This will be a header for the column of numbers that you will type underneath. Starting with cell A20, enter the values of t_2. Take them from column F in Millikan's data in Table 9.3 and use only the values that appear below the words "Change forced with radium." This will be 45

---

[15]Keep in mind that like all instructions having to do with computers, these only make sense when you already know how to do what they are telling you how to do. It is almost always easier to abase yourself and ask the person next to you how to do what you want to do.

values in all, starting with 72.4 and ending with 70.8. If you get ambitious, you can type in all 78 values.

You could use the times t_2 and d_c to calculate the velocities v_2 as Millikan did, but since the velocities are all proportional to $1/t_2$, it is simpler to calculate just the reciprocal times by setting up column B to contain $1/t_2$. To do this, type the header '1/t2' in B18 and '(1/sec)' in cell B19.

Now to calculate the reciprocals of t_2, type into cell B20 '= 1/A20'. The '=' sign tells Excel that you want it to compute something. After you type in '= 1/A20', copy this cell to the cells B21 to B65. Properly done the copy will put the value of '1/A21' into B21; the value of '1/A22' into B22; etc. Here is one way to do the copy: Put the cursor on cell B20 and click your mouse, i.e. select cell B20. The selected cell will have a little black square at its bottom right hand corner. Move the cursor to that little black square and a solid black cross will appear. Now hold down the click button and drag the B20 cell down the B column to cell B64; release the click button. Cells B20 to B64 should now contain the values of $1/t_2$ corresponding to the values in the A column.

c. You now have enough data in your spreadsheet to see whether there is anything to this idea of charges being integer multiples of some smallest quantity of charge.

A revealing way to see differences between the values of $1/t_2$ is to plot your data in a 'column chart', as Excel calls a bar graph. (Excel calls all graphs 'charts.') To make such a plot, first click on 'Insert' in the menu bar at the top of your computer screen; then click on 'chart'; choose 'column' for 'Chart Type' (it may be the default); click on 'Next' and you will get a panel asking for 'Chart Source Data'. The panel will have a box 'Data range'; click in it. Move your cursor to B20 to select the data to be plotted. With the cursor at B20 you will see a hollow cross; hold down the mouse button and drag the hollow cross to B64; release; press 'enter'. You should now see a graph of vertical bars, each of which represents the charge on the oil droplet; click on 'finish'. You can click and drag the graph to some convenient place on your spreadsheet. If you look at your graph carefully, you can see that the differences between the heights of the bars seem to be multiples of some smallest difference. And that is the point; the differences are equal.

Not convinced? Use your spreadsheet to make the evidence clearer. Sort the data in ascending order, and make another column chart to show that the successive changes have the same size, as they must if they are changing by some smallest, elementary amount.

To do this, make a copy of cells from B20 to B64 and put the copy in column D. (This is so you don't destroy your data and have to re-enter them.) To copy, select columns B20 to B64; then click the right-hand button of your mouse to get a pop-up menu; left click on 'copy' in the pop-up menu. Now move your cursor to, say, D20; right click the mouse; left click on 'paste special'; click on the radio button for 'values'; click 'OK'. The numbers in column D should be the same as those in column B.

To sort the numbers in column D, click on 'Data' in the menu bar at the top of the screen; click on 'Sort'. You may need to tell it to sort on column D, but if you have already selected a cell in column D, it will probably do the sort without further instruction. Make a column chart of the sorted numbers in column D. You should get a nice staircase with steps of equal height, strong evidence of the "quantization" of charge.

d. Now find the value of the amount of charge Δq that corresponds to the step height. You can do this from the sorted values by averaging each of two different groups of similar values and then subtracting the smaller average from the larger. For example, find the average of the values from D30 to D47 and then find the average of the values from D21 to D29. Then subtract the two averages to get a number corresponding to the smallest Δq.

This is a good opportunity to learn to use Excel functions. Select cell F20; this is where your average will appear. Just above the spreadsheet there is an entry box. At its left corner is f_x. Click on f_x to get the 'Insert function' panel. Type in 'average' in place of 'enter a brief description' in the search box; click on 'average' in the select box; click on OK; if you get a box asking for 'function arguments', select D30 to D47; press the 'enter' key. Do this again for cell F21 and cells D21 to D29. In cell F24, type '='; then select F20; type '-' (minus sign); select F21; press 'enter'. Voila! The difference you are after appears in cell F24. It is proportional to the value of the elementary charge Δq, so you must multiply it by the constants specified by Eq. 19. It's easiest to do this in steps.

A. Calculate a using Eq. 21. Type into a convenient cell, say H24, '=sqrt(9*[click on B8]*[click on B10]/[click on A16]/(2*[click on B5]*[click on B6])) and press 'enter'. If you made your entries correctly, the value of a, of the order of micrometers, will appear in H24.

B. Find the value of m. In cell H25 type '=4*pi()/3*[click on H24]^ 3*[click on B6]; press the 'enter' key. Your answer will appear. It should be of the order of tens of picograms.

Notice that to type in an exponent you use the ^ symbol. For example, to raise the contents of cell F17 to the power 3/2, you type 'F17 ^ (3/2)'.

C. Calculate the electric field strength F (which is Millikan's notation for what we usually call E). In H26 type '= [click on B11]/[click on B9]'; press 'enter' to get the value of F. Millikan's data show that the voltage varied slowly as he did his experiment, but the variation is not significant, and an average value of 7800 V will do very well.

D. Calculate Δq using Eq. 19. Type into H28 '= [click on H24]*[click on B6]*[click on A26]/[click on H25]*[click on F23]'; press 'enter'. You get an answer slightly larger than the accepted value and larger than Millikan's actual result because no corrections have been made for small deviations from Stokes's law.

Finally, spend countless hours fiddling with your spreadsheet, titling your graphs, labeling their axes, providing labels for your calculated quantities, and making the spreadsheet clear and informative. Hand in a printout of your data sheet, graphs, and the numerical value of Δq.

10.

Waves
and Light

10.1   INTRODUCTION

The preceding chapters have illustrated the power of the atomic hypothesis. From the observations of Dalton and Gay-Lussac you saw how to deduce the chemical composition of molecules and determine ratios of atomic masses. You also saw that by modeling gas atoms and molecules as featureless hard spheres, you could develop a kinetic theory of gases that allowed you to interpret physical quantities such as temperature and pressure in terms of the more fundamental concepts of kinetic energy and momentum. Also you saw how the kinetic theory can be used to connect the concept of mean free path to measured values of the viscosity of gases to obtain an estimate of Avogadro's number and, consequently, the size and mass of single atoms. Later, the experiments of Faraday, Thomson, and Millikan proved that atoms have internal structure, i.e., they are themselves composed of smaller, more fundamental particles. One of these particles is the electron, and it is removable, replaceable, and interchangeable. What other particles are contained in atoms? How are they assembled, and what holds them together?

Some answers to these questions were found by analyzing the light emitted and absorbed by atoms; other information came from probing atoms with electromagnetic radiation not visible to the human eye. The tools and techniques for such analysis and such probing could only be developed as physicists learned more about light and other forms of electromagnetic radiation. Because the wave properties of light are central to understanding how it can be used to reveal the inner secrets of the atom, you should now spend some time understanding waves and the wave behavior of light.

C.H. Holbrow et al., *Modern Introductory Physics, Second Edition,*
DOI 10.1007/978-0-387-79080-0_10, © Springer Science+Business Media, LLC 1999, 2010

10.2   THE NATURE OF WAVES

After a brief description of the general nature of waves, we will look closely at a special class of periodic waves, those with a pure sinusoidal waveform, the so-called "harmonic," or "sine," waves. Such waves represent simple physical situations: A sinusoidal sound wave is a pure tone; a sinusoidal light wave is a pure color.

Harmonic waves also have the virtue of being mathematically fairly simple. A few parameters are enough to completely describe a harmonic wave, and it is a major goal of this section to define and describe these: "amplitude," "wavelength," "frequency," "velocity," and "phase."

The manner in which waves combine when two or more come together at the same place is a fundamental property of waves, and it plays a major role in exploring matter in general and atoms in particular. Therefore, a large portion of this chapter is devoted to studying the combination, or "interference," of waves.

A Traveling Disturbance

What is a wave? A wave is a traveling disturbance without any transport of matter. For example, when you snap a jump rope, a pattern of deformation passes from one end of the rope to the other, but the parts of the rope stay put. Or again, consider a long line of upright dominoes. You can start a wave of falling dominoes by striking the end domino so that it topples against its neighbor, which then falls and strikes its neighbor, and so on. The disturbance—falling dominoes—propagates from one end of the line to the other, yet no domino moves far from its initial position.

The same behavior holds for other kinds of waves. Imagine tossing a pebble into a quiet pond. The resulting circular ripples—surface water waves—travel outward for several meters before disappearing. The individual water molecules, however, do not move more than about 1 cm. In fact, they return to their original positions after the waves die out. The sound waves reaching your ear during a physics lecture are a further example of this property. They have traveled a distance of 10 m or so, yet the individual air molecules have not moved more than a few microns in response to the passing wave. Consider the alternative: If molecules actually traveled from speaker to listener, then you not only would hear the lecture, but you would also smell the lecturer's most recent meal. "This lecture stinks!" would take on a new meaning.

Light waves are somewhat more abstract than the examples of the previous paragraph. For one thing, they can travel in a vacuum: They do not need a "medium" such as air or rope or dominoes or a water surface.

For another, the disturbance is in the form of a mixture of changing electric and magnetic fields. Nevertheless, it turns out that the description of wave properties that works well for sound waves in air or deformations traveling through solids works equally well for light.

Velocity, Wavelength, and Frequency

All waves travel with a finite velocity. This is certainly as true for light as it is for each of the other examples discussed above. The succession of observations and experiments over the last three centuries that have led to the determination of the speed of light ($\approx 3 \times 10^8 \, \mathrm{m \, s^{-1}}$) makes a fascinating story.

■ EXERCISES

> **1.** Anyone who has witnessed a fireworks display has direct evidence that light travels much faster than sound. It takes about 5 s for sound to travel 1 mile (1.6 km). How long does it take light to travel the same distance? If you hear an explosion 3 s after you see its flash, how far away did it occur? Write an equation relating distance to the time delay. Amaze your friends next July 4!

Consider again the example of circular water ripples. Figure 10.1 contains two cross-sectional views of the water surface, one at time t_1 and the other at a later time t_2. Each graph shows the height of the surface as a function of the radial distance r from the center, i.e., each is a plot of the waveform. The wave velocity is given by:

$$v = \frac{r_2 - r_1}{t_2 - t_1}. \tag{1}$$

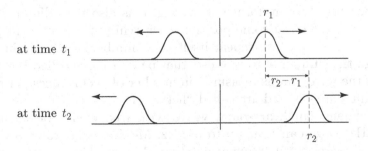

FIGURE 10.1 Views at two different times of water waves traveling across a surface.

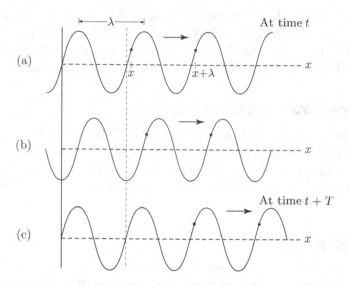

FIGURE 10.2 Views at three different times of a sinusoidal wave with period T traveling to the right: (a) at time t; (b) at time $t + \frac{T}{4}$; (c) at time $t + T$.

Now study Fig. 10.2, which illustrates a different waveform called, for obvious reasons, a sinusoid or sine wave. The graphs might represent the surface of water, the shape of a stretched string, the pressure (or density) fluctuations associated with a sound wave, or the variation in electric field in a light wave. The sinusoidal waveform is a particularly useful one for you to study, and much of the following discussion is devoted to it.

The sine wave in Fig. 10.2 is an example of a wave that is periodic. It is periodic in space (position) because the pattern of the wave repeats regularly as a function of position. As Fig. 10.2 shows, the form of the wave at any arbitrary point x is the same at distances $\pm\lambda$, $\pm 2\lambda$, $\pm 3\lambda$, $\pm 4\lambda$, ... from x. The shortest repeat distance λ is called the "wavelength." The Greek letter λ, i.e., lambda, is customarily used to represent the wavelength.

Because the waveform is traveling, it is also periodic in time. An observer stationed at some position x_0 watching the waveform pass by will see the basic pattern repeat itself. The number of repeats in a unit time (e. g., the number of wave crests that pass in 1 s) is called the "frequency" f of the wave, and is measured in number of occurrences per second. The SI units are s^{-1} and are called "hertz" (Hz).

Humans can hear sound waves with frequencies between 20 Hz and 20 kHz. Dogs can hear up to 35 kHz. Medical sonograms are taken with sound waves with frequencies above 1 MHz. The range of electromagnetic wave frequencies is enormous, and the uses are numerous and

TABLE 10.1   Some different kinds of electromagnetic waves

Commonly used name	Frequency		Wavelength	
AM radio waves	540–1500	kHz	–	
FM radio waves	88–108	MHz	–	
TV channels[a] 2–6	54–88	MHz	–	
TV channels[a] 7–13	174–216	MHz	–	
TV channels[a] 14–83	524–890	MHz	–	
Cell phone channels 128–249	824–849	MHz	–	
Cell phone channels 512–810	1850–1910	MHz	–	
Microwaves	–		30–0.03	cm
Infrared	–		300–0.7	μm
Visible	–		700–400	nm
Ultraviolet	–		400–30	nm
X-rays	–		30–0.03	nm
Gamma rays	–		<0.03	nm

[a] The FCC officially assigns these channel numbers to these frequencies. With the arrival of digital TV, some TV stations keep their brand name familiar channel numbers while actually using different channels. For example, WABC-TV Channel 7 in New York City broadcasts on channel 49.

vary depending upon the frequency. For example, the AM radio broadcast band uses frequencies of electromagnetic radiation between 540 and 1500 kHz. The FM band lies between 88 and 108 MHz. Your microwave oven cooks food with 1450 MHz electromagnetic radiation. Visible light has frequencies between 430 THz and 750 THz (that's nearly 10^{15} Hz). Other examples of electromagnetic waves are given in Table 10.1.

Since the distance between successive crests is λ, and the number of crests passing per second is f, then the wave velocity must be given by

$$v = \lambda f. \tag{2}$$

The situation is exactly analogous to a passing railroad train. If the length of each car is L meters, and N cars pass per second, then the velocity of the train is LN meters/sec.

The time between arrival of successive crests is called the period T of the wave. If f crests pass per second, then the time between crests must be $1/f$, i.e., the period is

$$T = \frac{1}{f}. \tag{3}$$

EXERCISES

2. "Concert A" is a pure sinusoidal wave having a frequency of 440 Hz. The speed of sound in air at room temperature is about 340 m/s. How far apart are the pressure maxima in the sound wave? The pressure minima? What is the period of the wave?

3. Complete Table 10.1 by calculating the missing entry (frequency or wavelength) in each row. What wavelength radiation is being transmitted by your favorite radio station?

Amplitude

If you were asked what the amplitude of a wave is, you could probably give a reasonable answer based on your intuition. It certainly is a measure of the magnitude of the disturbance, but for a sine wave like that shown in Fig. 10.2, amplitude usually refers to the *maximum* displacement from the undisturbed state. Thus in Fig. 10.2 the amplitude is the maximum height of the sine curve above the x-axis.

Represent the amplitude by A. Then you can write an equation for the wave in Fig. 10.2a as

$$y = A \sin\left(2\pi\frac{x}{\lambda}\right). \tag{4}$$

Notice that in Eq. 4 the argument of the sine must be an angle. For reasons of convenience (believe it) the angle is usually given—as here—in *radians*. Because this periodic function is to repeat every wavelength—that's the definition of a wavelength, the sine's argument must increase by 2π at one wavelength, by 4π at $x = 2\lambda$, by 6π at $x = 3\lambda$, and so on. That is exactly what Eq. 4 accomplishes.

EXERCISES

4. Suppose Eq. 4 describes a wave on a long string. Let $A = 5$ cm; $\lambda = 10$ cm. What would be the displacement y of the string at $x = 2.5$ cm? $x = 5$ cm? $x = 7.5$ cm? $x = 10$ cm? Graph the displacement y of the string vs. its length.

5. In the previous exercise, what would be the amplitude of the wave at each of the four x locations?

If for Exercise 5 you answered that the amplitude is 5 cm for each x, pat yourself on the back. Strictly defined, "amplitude" means the value of A, and this does not change with x. On the other hand, be aware that many physicists are sloppy in their terminology and may say "amplitude" when they really mean "displacement." There is not much you can do about that except be alert to the context.

▪ EXERCISES

6. Suppose Eq. 4 describes a wave traveling down a gas-filled pipe. Let $A = 5\,\mathrm{Pa}$; $\lambda = 10\,\mathrm{cm}$. What would be the extra pressure in the pipe at $x = 2.5\,\mathrm{cm}$? At $x = 7.5\,\mathrm{cm}$?

7. Suppose you are observing a light wave traveling down a thin transparent fiber. Let $A = 5\,\mathrm{V/m}$; let $\lambda = 10\,\mu\mathrm{m}$. What would be the electric field in the fiber at $x = 2.5\,\mu\mathrm{m}$? At $x = 5\,\mu\mathrm{m}$? At $x = 7.5\,\mu\mathrm{m}$?

Exercises 4, 6, and 7 show you how a single mathematical form, Eq. 4, represents different kinds of waves.

Phase

The argument of the sine wave in Eq. 4 is called its "phase." Referring to Fig. 10.2a, you can see that at $x = 0$ the phase of the wave is 0. At $x = \lambda/8$ its phase is $\pi/4$ radians; at $x = \lambda/4$ its phase is $\pi/2$ radians.

▪ EXERCISES

8. Show that the phase difference, $\Delta\phi$, between two points x_1 and x_2 is

$$\Delta\phi = 2\pi\frac{x_2 - x_1}{\lambda}. \tag{5}$$

Notice that the three sine curves of Fig. 10.2 have identical shapes but different phases. The curve of Fig. 10.2b has the same shape as (a) but it is shifted to the right. Do you recall that a function $f(x)$ is shifted to the right a distance h just by writing it as $f(x - h)$? So you can write the equation of Fig. 10.2b simply by shifting Eq. 4 by a constant amount of phase. From the figure you can see that this amount is $\pi/2$ radians, and therefore

$$y = A \sin\left(2\pi\frac{x}{\lambda} - \frac{\pi}{2}\right)$$

is an equation that describes the curve in Fig. 10.2b.

There is always more than one way to write the phase. For one thing, there is no way to tell whether the curve of Fig. 10.2b was shifted to the right by $\pi/2$ or shifted to the left by $3\pi/2$ radians. It would have been just as correct to write

$$y = A \sin\left(2\pi\frac{x}{\lambda} + \frac{3\pi}{2}\right).$$

For another, changing the phase by any multiple of 2π does not change the value of the function, so if you had not been told that (c) has been shifted one period, you would not be able to know that the phase of curve (c) differs by 2π from the phase of (a). Curve (c) will look the same as curve (a) whenever their phases differ by any multiple of 2π. There is also the fact that $\sin(\theta - \pi/2)$ is the same as $-\cos(\theta)$, so that Fig. 10.2b could also have been written

$$y = -A \cos\left(2\pi\frac{x}{\lambda}\right).$$

You can also think of phase difference as arising from a shift of the entire waveform along the x axis. From this point of view the curve in Fig. 10.2b arises from a shift of (a) a distance $\lambda/4$ to the right. Then

$$y = A \sin\left(2\pi\frac{(x - \lambda/4)}{\lambda}\right),$$

which is the same as the equation for curve (b) in Fig. 10.2.

▪ EXERCISES

9. Think of some other equivalent forms of Eq. 4.

You can use the same argument to make any function $f(x)$ into a moving function. Let $h = vt$ where v is the speed and t is time, then $f(x - h)$ becomes $f(x - vt)$ and it moves to the right at the speed v. To represent a moving sinusoidal waveform, just replace x in Eq. 4 with $x - vt$:

$$y = A \sin\left(2\pi\frac{x - vt}{\lambda}\right).$$

TABLE 10.2   Parameters of periodic waves

y	Displacement
A	Amplitude
x	Position
λ	Wavelength
t	Time
T	Period
$f = \frac{1}{T}$	Frequency
ϕ	Phase constant
$2\pi\frac{x}{\lambda} - 2\pi\frac{t}{T} + \phi$	Phase
$kx - \omega t + \phi$	Phase

If you use the fact that $v = \lambda f = \lambda/T$ and add an arbitrary phase constant ϕ, you get as the general form of a one-dimensional traveling harmonic wave moving in the direction of increasing x:

$$y = A \sin\left(2\pi\frac{x}{\lambda} - 2\pi\frac{t}{T} + \phi\right). \qquad (6)$$

Table 10.2 lists the notation and terminology commonly used to describe periodic waves.

Transverse and Longitudinal Waves

Waves are also characterized by the direction of the displacement of the wave relative to its direction of propagation. In Fig. 10.3a, the wave is traveling on a taut string, and each element of the string moves up and down vertically as the wave moves in the positive x-direction. When the displacement associated with a wave is perpendicular to the direction of propagation as in Fig. 10.3a, the wave is said to be a "transverse" wave. If the displacement is parallel to the direction of propagation as in Fig. 10.3b, the wave is said to be a "longitudinal" wave. Sound waves in air are longitudinal waves; light waves in a vacuum are transverse waves.

▦ EXERCISES

10. Imagine that Fig. 10.3a is a snapshot of the wave traveling to the right along the string. At the exact moment that the photo is taken, what is the direction of motion of the string at point A? At point B? At

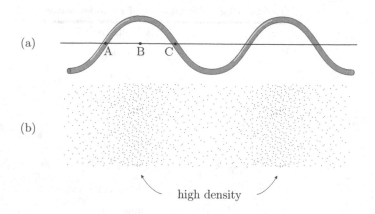

(a)

A B C

(b)

high density

FIGURE 10.3   (a) Transverse wave on a string. (b) Longitudinal sound wave in a gas.

point C? (Hint: Sketch the wave both at the time shown in the figure and at a short time afterwards.) How would your answers change if the wave were traveling to the left?

11.   Table 10.2 says that phase is sometimes given as $k\,x - \omega t + \phi$. The quantity k is called the "wave number" and ω is called the "angular frequency." Show that $k = \frac{2\pi}{\lambda}$ and $\omega = \frac{2\pi}{T} = 2\pi f$.

Intensity

Although waves do not themselves transport matter, they do carry momentum and energy from one point in space to another. It is by means of such transported momentum and energy that waves have effects. The warmth of sunshine tells you that light waves carry energy. The pain in your ears tells you that rock music transports momentum.

The specific definition of "intensity" of a wave in three dimensions is the amount of energy carried across a unit area in a unit time. Thus the SI units of intensity are $\mathrm{J\,m^{-2}\,s^{-1}}$ or, equivalently, $\mathrm{W\,m^{-2}}$.

▼ EXAMPLES

1.   The intensity of sunlight on the upper atmosphere of Earth is about $1.39\,\mathrm{kW\,m^{-2}}$. How much solar energy does Earth receive in $1\,\mathrm{s}$?

The Earth with its radius of about $R = 6400\,\mathrm{km}$ offers to the Sun a circular target with an area of $\pi R^2 = \pi(6.4 \times 10^6)^2 = 1.3 \times 10^{14}\,\mathrm{m}^2$. Therefore, in $1\,\mathrm{s}$ the Earth will receive

$$1.39 \times 10^3 \times 1.3 \times 10^{14} = 1.8 \times 10^{17}\,\mathrm{J}.$$

This means that the Sun delivers $180\,000\,\mathrm{TW}$ (terawatts) of power to Earth. About half reaches the surface. In 2005 the world-wide rate of human energy consumption was abut $16\,\mathrm{TW}$. Do you see why people would like to figure out how to make efficient use of solar energy?

It is important for you to know that *the energy carried by any wave is proportional to the square of its amplitude.* This should seem plausible to you if you know that the energy stored in a stretched spring is proportional to the square its stretch. Sound waves and waves along strings all stretch and compress the matter through which they travel just the way a spring can be stretched and compressed. It is not obvious that this rule should work for water waves and light waves, but it does.

10.3 INTERFERENCE OF WAVES

Imagine the following peculiar situation. After a long day's drive, you have just checked into the Bates Motel and are about to enjoy a warm, relaxing shower. When you enter the shower stall, you notice that it is equipped with two independent water nozzles, each of which operates normally and delivers a fine spray distributed uniformly over the entire stall. Imagine your surprise when, turning on both shower heads together, you notice that there are positions under them where you do not get wet at all, and other positions where you get four times as wet as with a single shower head! This does not really happen, of course, but it would if the nozzles emitted waves instead of particles of water.

Here is a more plausible example to illustrate the point. Suppose you and a dozen friends lug a CD player and two stereo speakers out onto a flat, grassy, open field. You set up the speakers a distance d apart, and then you and your friends form a line at a perpendicular distance D from the speakers as shown in Fig. 10.4. Now play a rather dull recording consisting of a single pure note, i.e., a sinusoidal wave, sustained for the entire duration of the CD. If only one speaker is plugged in, then you all will hear the note, with the observers near O_0 recording the highest

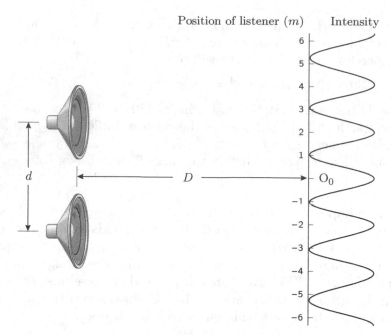

FIGURE 10.4 Acoustic interference experiment between two stereo speakers. The curve represents the intensity of the received sound as a function of position. Each tic mark signifies an attentive student listener.

intensity. When both speakers are activated, some of you will hear nothing at all, while O_0 will report an intensity four times greater than for the single-speaker case! This phenomenon is called "interference."

Interference is a unique and defining property of waves. If any kind of radiation exhibits interference, you know it is a wave. Thus, if the water from the shower actually behaved the way described above, you would know immediately that the water spray was wavelike. This idea that interference tells you when something is a wave is important later in this book. Interference is also very important for the study of atomic properties.

Interference Along a Line

Interference is a consequence of the property of superposition, a word that means that when two waves occupy the same points in space, their displacements just add together algebraically. To understand this principle consider a case in which two pulsed waves with the same shape approach each other on a long stretched string, as shown in Fig. 10.5. As a specific example, suppose that each pulse has a trapezoidal shape. What happens as the two pulses come together? As shown in Fig. 10.5, the string's shape

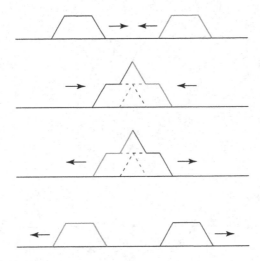

FIGURE 10.5 Two waves approaching on a string and combining. At each moment the resulting wave shape is the point-by-point sum of the two waveforms.

is exactly what you calculate if each pulse passes undistorted through the other and their heights simply add together at each point along the string. This simple additivity occurs for all small-amplitude waves and is called the "principle of superposition." It means that to find the waveform due to a combination of waves, you just add, point by point, the displacements of each wave.

▪ EXERCISES

12. To see that superposition can be quite interesting, consider what happens in Fig. 10.5 when the wave coming from the right has negative displacement, i.e., if the right-hand trapezoid is flipped over with respect to the axis.

Now look at Fig. 10.6. Imagine you are an observer standing on a line with two stereo speakers, one behind the other, emitting sine waves of the same frequency f and amplitude A toward you. The speakers are a distance L apart along the straight line running from the speakers to you.

With both speakers turned on, you hear the same frequency f as when only one speaker is on. However, you hear an intensity that varies dramatically when the distance L between the speakers is changed. As Fig. 10.6 shows, when $L = 0, \lambda, 2\lambda, \ldots$, the waves add together to form a larger wave with total amplitude $2A$. This is called "constructive interference."

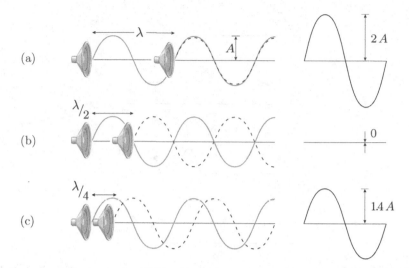

FIGURE 10.6 Constructive and destructive interference of sine waves. The output of the left speaker is the solid curve; the output of the right speaker is the dashed curve. One cycle of the sums of these two curves is shown at the extreme right.

If $L = \lambda/2, 3\lambda/2, \ldots$ the superposed waves cancel exactly, and you hear nothing even though both speakers are emitting sound! This is called "destructive interference." For other speaker separations, the resulting amplitude takes on values between 0 and $2A$. Since the sound-wave intensity is proportional to the energy carried by the wave, and since the energy is proportional to the square of the amplitude, the intensity for constructive interference is four times as large as it was for a single speaker.

Now think about what the sine curves sketched in Fig. 10.6 represent. They are "snapshots" of the wave. In this case all the snapshots were taken at the instant when the displacement of the wave emerging from each speaker was zero. The two speakers are "in phase"; they are both emitting a null at the same time, because the apparatus has been set up to make this happen. If, for example, the two speakers are driven by the same amplifier, then, because the signal from the amplifier travels quickly along wires to each speaker (almost at the speed of light), the sound waves will come out of the speakers with the same phases no matter where the speakers are located.

You could take your snapshot at any time. Suppose you took a snapshot of the waves just as the phases of the waves emerging from the speakers are both $\pi/2$. At this instant the wave displacements at the speakers would be maxima, but ϕ, the phase *difference* between the two waves, would still be zero, and they combine as before.

EXERCISES

13. Repeat the sketch shown in Fig. 10.6a at a time $T/4$ later, when the phase of the wave emerging from each speaker is $\pi/2$ so that the displacement is a maximum.

You see in Fig. 10.6 that at each instant of time the wave emerging from one speaker has the same phase as the wave emerging from the other a distance L away. But notice that this does not mean that the phases of the two waves are the same at the place where the waves overlap and superpose (add together). Since the waves are traveling to the right, it takes some time for the wave from the first speaker to reach the second speaker. As a result, the phase of the first wave as it arrives at the second speaker can not be the same as that of the wave just emerging from the second speaker. The snapshots in the figure show this. In the first snapshot (Fig. 10.6a), the speakers are separated by a distance λ. The wave emerging from the first speaker reaches the second speaker after a time $t = \lambda/v = 1/f = T$, one full period later. Therefore, its phase differs by exactly 2π radians from that of the wave just emerging from the second speaker. The two waves add together with this phase difference all along the line along which they are traveling to the detector, and because the phase difference is an integer multiple of 2π ($n = 1$) the interference is constructive. In snapshot (b) of the figure, the first wave requires a time $t = T/2$ to travel from the first to the second speaker, so all along the line of travel the two waves differ in phase by π radians, and there is complete destructive interference.

Keep in mind when looking at Fig. 10.6 that for a sound wave the sine curve describes small, longitudinal variations in pressure. The amplitude of the wave is then the maximum amount of increase in pressure produced by the source of the wave. (This maximum increase is usually quite small.) The sine curve shows you how the pressure in the sound wave swings up and down around the average pressure by this increment.

EXERCISES

14. Each speaker in Fig. 10.6 is emitting a $1\,\mathrm{kHz}$ sound wave with an amplitude A. Calculate three speaker separations L that lead to destructive interference, i.e., $A_{\mathrm{tot}} = 0$.

What happens when you superpose two waves differing in phase by an arbitrary angle? The two cases of interference you have already studied are for special values of phase difference. For *constructive* interference, the waves are in phase, i.e., their phases differ by an integer multiple of 2π radians, and they add in a simple, direct way. For *destructive* interference, the two waves are exactly out of phase, i.e., their phases differ by an odd integer multiple of π radians, and they completely cancel each other. But what if the phases of the two waves differ by some arbitrary angle ϕ? Can you see intuitively that there will be *partial* interference? You can show this by plotting on the same graph the two waves shifted in phase relative to each other by ϕ, and then adding them together point by point. (Go to p. 335 to see how to do this with a computer spreadsheet.)

There is a nice formula for the sum of two waves with equal amplitudes A that shows explicitly how their sum changes when you vary their phase difference ϕ. The formula is based on a (seldom remembered) trigonometric identity for combining two sines with arguments that differ in phase:

$$A \sin \alpha + A \sin(\alpha - \phi) = 2A \cos\left(\frac{\phi}{2}\right) \sin\left(\alpha - \frac{\phi}{2}\right). \tag{7}$$

This formula shows that when two waves of the same wavelength λ but differing in phase by an angle ϕ are superposed, they form a single wave of the same wavelength and with an amplitude $2A \cos\left(\frac{\phi}{2}\right)$. Notice that when $\phi = \pi$, the amplitude of the combined waves is zero, i.e., total destructive interference.

EXAMPLES

2. You can understand these statements better if you see how they apply to the waves emitted by the two speakers shown in Fig. 10.6. For these waves

$$\alpha = 2\pi \frac{x}{\lambda} \qquad \text{and} \qquad \phi = 2\pi \frac{L}{\lambda},$$

where the location of the speaker on the left is taken to be $x = 0$ and the location of the speaker on the right is $x = L$. When you replace α in Eq. 7 with $2\pi \frac{x}{\lambda}$, you find that the superposition of the two waves from the speakers in Fig. 10.6 is

$$y_{\text{left}} + y_{\text{right}} = 2A \cos\left(\frac{\pi L}{\lambda}\right) \sin\left(2\pi \frac{x}{\lambda} - \frac{\pi L}{\lambda}\right). \tag{8}$$

▨ EXERCISES

15. Show that for cases (a), (b), and (c) in Fig. 10.6 the amplitude of Eq. 8 gives the result shown in the figure.

16. What would be the amplitude of the superposed waves if the right hand speaker was at $L = \lambda/8$?

Example 2 is like most of the cases of interference that you will study in this book in that the phase difference is due to a difference in the length of the paths that each wave must travel to reach the detector. To analyze these cases you first find the difference in the distances traveled by two (or more) waves to reach a detector and then see how much phase difference results from that difference of path lengths.

In practice you usually measure intensity, not amplitude. Intensity (loudness of sound or brightness of light) is proportional to the square of the amplitude. This means that the intensity I of the wave arising from the superposition of two waves according to Eq. 8 is proportional to the square of $\cos\frac{\phi}{2}$:

$$I \propto 4A^2 \cos^2\left(\frac{\phi}{2}\right),\tag{9}$$

where, as before, $\phi = 2\pi L/\lambda$. The intensity of the sum of the two waves is equal to $4\cos^2(\phi/2)$ times the intensity of either wave:

$$\frac{I_{\text{sum}}}{I_{\text{individual}}} = 4\cos^2\left(\frac{\phi}{2}\right).\tag{10}$$

Notice that you do not need to know the constant of proportionality to get the result given by Eq. 10.

These results are summarized in Fig. 10.7. Part (a) shows the correspondence between the phase difference of the two waves (in this case, $\pi/2$) and the difference L in the distances from the wave sources to the detector (in this case, $\lambda/4$), where the relationship between the phase shift and the distance is $\phi = 2\pi\frac{L}{\lambda}$. The graph in Fig. 10.7b shows you that the amplitude and intensity are greatest when the phase difference is an integer multiple of 2π, and that they are zero when the phase difference is an odd integer multiple of π.

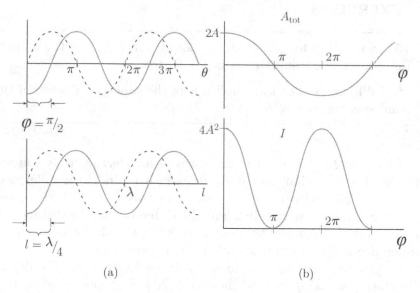

(a)   (b)

FIGURE 10.7 In (a) the upper graph shows a wave shifted in phase by $\pi/2$ radians relative to another; the lower graph shows that the phase shift arises because the sources are separated by a distance of $\lambda/4$. In (b) the upper graph shows how, at the detector, the amplitude of the two superposed waves varies as a function of their phase difference (produced by changing the separation of the speakers); the lower graph shows the intensity corresponding to the amplitude of the upper graph.

EXERCISES

17. Find a phase difference at which the sum of the two waves in Fig. 10.7a will have exactly the same amplitude and intensity as either wave individually.

18. What separation of the speakers will give this phase difference?

19. Suppose that for the two speakers one behind the other as shown in Fig. 10.6, total destructive interference first occurs at a separation of $L = 15\,\text{cm}$, and then at 45 cm, 75 cm, ...
(a) What is the frequency of these sinusoidal waves?
(b) If the amplitude of each wave is A, calculate the ratio of the intensity recorded by the listener at $L = 10\,\text{cm}$ to the intensity she would hear if one of the speakers was turned off.

Visualizing Waves in Three Dimensions—Wavefronts

The examples of two stereo speakers one behind the other and of two speakers side-by-side emphasize that phase difference is produced when waves generated in phase travel different distances to a detector. These examples are idealizations. Only under special circumstances do waves of sound or light travel in only one or only two dimensions. In general, sound and light waves travel in three dimensions.

Consider a mental-images-of-planes or MIP wave, invented to help you visualize waves in three dimensions. This is a sine wave that fills 3-D space and is traveling from left to right through a stack of large, thin, flat cards. As the wave passes along the stack of cards, they become colored. The color at each point on a card depends on the displacement of the wave at that point on the card, and the colors range from dark red through red to pink to white to light gray to dark gray to jet black. The shades of red stand for negative displacements; the shades of gray stand for positive displacements. Jet black is the maximum positive displacement, and so it corresponds to the wave's amplitude; similarly, dark red represents the negative amplitude. Because the wave's displacement at any point on any card in the stack depends on the value of the wave's phase at that point, each color also corresponds to a definite phase of the wave at each point.

Now comes the important idea of a *plane wave*. This wave has the property that at each card the phase of the wave is the same at every point on the card. If you suppose that at any instant of time you can freeze the effect of the MIP wave on the cards, the colors will be uniform over each card. The jet black card will be jet black everywhere; a pink card will be the same shade of pink everywhere; a white card will be white all over. The MIP wave is a plane wave because it has surfaces of constant phase that are planes perpendicular to the wave's direction of travel. At each instant of time the planar surfaces of constant phase coincide exactly with the planar cards. A surface of constant phase is often called a *wavefront*. Plane waves have planar wavefronts.

■ EXERCISES

20. Suppose you riffle through the stack of cards and observe that every 16th card is jet black. Do you see that this tells you that the wavelength is 16 card thicknesses?

21. If card 24 is jet black, where to its right will you find the next darkest red card?

22. Suppose you assign a phase of 0 radians to card 20. What then is the phase of the wave at card 24? Of card 36? Of card 4?

---

23.   After taking a snapshot of the stack, you wait 1 s and take another and find that in the second recording cards 27, 43, 59, ... are jet black. Do you see that the wave might be traveling to the right with a speed of 3 card thicknesses per second?

Suppose the MIP wave is a sinusoidal plane wave traveling from left to right. It has a wavelength of 16 card thicknesses, a velocity of 3 card thicknesses per second, and, by arbitrary choice, the phase of the wave at card 20 is 0 radians at time $t = 0$. To repeat, a wave is a plane wave when its phase is constant everywhere on any plane surface perpendicular to the direction of the wave's travel. The uniformity of color on the plane surface of each card represents the constancy of phase over a plane. Because you chose the phase of the wave at the all-white card 20 to be 0 radians, the phase of the wave at card 24 is $\pi/2$ radians, and the card is jet black. Remember, the cards don't move; they are the medium through which the wave moves.

Now apply these ideas to sound waves. A plane wave of sound is like the MIP wave except that instead of color at each card there is a compression or rarefaction of air pressure. Because these variations in air pressure occur along the line of travel of the wave, a sound wave is a longitudinal wave.

What about light? A sinusoidal plane wave of light is like the MIP wave except that at every point on each card, instead of color there exist both electric field E and magnetic field B. That's right; the two fields are present at every point on each card. For a plane wave the value of E is the same at every point on a given card; so is the value of B. You can think of a plane wave of light as a succession of sheets of electromagnetic field. The magnitudes of the two fields vary from sheet to sheet like the colors of the MIP wave as it passed along the stack of cards. Just as the density of coloration varies sinusoidally, so do the magnitudes of the electric and magnetic fields; the change of color from black to red in the MIP wave corresponds to the reversal of direction of the fields in an electromagnetic wave.

The E and the B fields point at right angles to each other and also at right angles to the direction of travel of the wave. Because the fields lie in the plane of the card and are perpendicular to the direction of travel, light waves are transverse waves. Transverse waves exhibit a phenomenon called polarization; it will come up later.

A plane wave is a special case.[1] Real waves are only ever approximately plane waves, but often you can arrange your apparatus to make waves that closely approximate them: If you are far from the source of light, any small portion of its wavefront will be flat enough to act like a plane wave for practical purposes; or you can use lenses to shape light waves to have flat phase planes over some useful area; or you can use light from lasers that produce beams of light with nearly flat wavefronts.

Interference in Terms of Wavefronts

You can use these ideas to analyze interference occurring in two dimensions. Look again at the example on p. 295 of two stereo speakers in an open field. As shown in Fig. 10.8, an observer at any position (D, y) with $y \neq 0$ is at different distances from the two speakers, i.e., $r_1 \neq r_2$. In the figure, the line $\overline{EC}$ is the difference between r_2 and r_1.

The type of interference (constructive, destructive, or partial) is determined by the difference between the two path lengths r_1 and r_2. Two waves emitted with the same phase from the speakers arrive at the observer (screen) with different phases if they travel different distances. The difference between their phases, often called their "relative phase," is

$$\phi = 2\pi \frac{r_2 - r_1}{\lambda}.$$

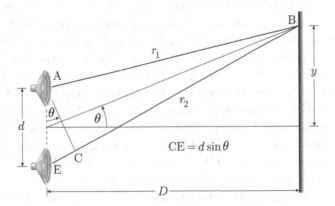

FIGURE 10.8 Geometry of the two-speaker interference experiment; the origin of the x-y coordinate system is at the midpoint of the line AE connecting the two speakers.

---

[1]There are other special-case wave forms. The phase of waves coming from a point source will be constant on spherical surfaces; such waves are called *spherical waves*. The phase of waves coming from a long narrow slit will be constant on cylindrical surfaces; these waves are called *cylindrical waves*. They are all idealization to a greater or lesser extent.

The relative phase of the two waves at any point B is fixed by the difference in the values of r_1 and r_2. This difference does not change in time even though every point of each wave and its corresponding phase move steadily (and rapidly) from source to detector (and beyond). As before, when $\overline{EC}$ is an integer multiple of λ then ϕ is an integer multiple of 2π, and there is constructive interference; if $\overline{EC}$ is an odd integer multiple of $\lambda/2$ then ϕ is an odd multiple of π radians, and there is destructive interference; for all other values of $\overline{EC}$ and the corresponding ϕ, there is partial interference.

▨ EXERCISES

24. Using the coordinates of Fig. 10.8, show that
$$r_1 = \sqrt{(y - d/2)^2 + D^2}$$
$$r_2 = \sqrt{(y + d/2)^2 + D^2}.$$

25. Assume the two speakers are a distance $d = 4\,\mathrm{m}$ apart and emitting a pure tone with $\lambda = 0.34\,\mathrm{m}$. For $D = 12\,\mathrm{m}$ at what y positions on the vertical line do observers hear the loudest sound? What frequency corresponds to this wavelength?

Often the distance D to the detectors is much greater than the separation d between the sources of the waves, i.e., $D \gg d$. For this case there is an approximation that you must learn to use. The idea is that when $D \gg d$ the lines $\overline{AB}$ and $\overline{EB}$ in Fig. 10.8 are essentially parallel. To understand this, imagine that the line of detectors is moved very far to the right so that D is in fact much greater than d. Can you see that then at the base of the isosceles triangle $\triangle ABC$ the two angles $\angle BAC$ and $\angle BCA$ become ever more nearly right angles? (This is the same as saying that the lines $\overline{AB}$ and $\overline{EB}$ are becoming nearly parallel.) As a result, the small triangle ACE becomes almost a right triangle, and you can find the length of $\overline{CE}$ from trigonometry: $\overline{CE} = d\sin\theta$. As you know, when this extra distance is equal to an integer number of wavelengths, there is constructive interference. Therefore, when $D \gg d$ the condition for constructive interference is quite accurately

$$d\sin\theta = n\lambda, \tag{11}$$

and the condition for destructive interference is

$$d\sin\theta = \left(n + \frac{1}{2}\right)\lambda, \tag{12}$$

where $n = 0, 1, 2, 3, \ldots$.

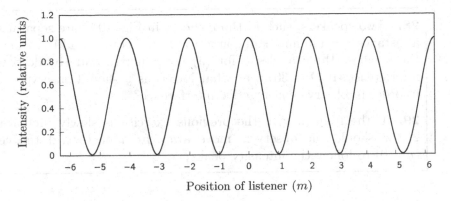

FIGURE 10.9 A listener moving along the line $x = D$ in Fig. 10.8 will hear the intensity from the two speakers vary as shown above.

Once again, observers at certain locations would hear almost nothing, while observers at other places would record an intensity four times that coming from a single speaker. Along the line of observers the sound intensity, which is proportional to the square of the amplitude, varies as a function of phase difference ϕ according to Eq. 9 on p. 301. Because observer position is directly related to ϕ, the difference between the phases of the waves arriving from the two speakers, a graph of intensity vs. observer position varies as the square of a cosine curve and looks like the graph in Fig. 10.9. This plot, the pattern of intensity vs. the position of the detector (the observer) along the line $x = D$, is an example of what is called an "interference pattern."

In general any combination of waves with a well defined relationship among their phases will form a three-dimensional pattern of intensity variations at every point in the space through which the waves are passing. On any plane surface in this space there will be a two-dimensional pattern of intensity variations. Such a pattern of spatial variation of intensity is the unmistakeable signature of interference. An interference pattern tells you that a wave is present, and the details of the pattern contain a great deal of information about the wave and its sources.

▪ EXERCISES

26. Use the equations in Exercise 24 to find an expression for $\sin \theta$ in terms of y and D. Simplify your answer for the case when $y \ll D$.

27. Compare the result you found in Exercise 25 to the result of Eq. 11. The requirement $D \gg d$ is not fulfilled in this case, but the result of Eq. 11 is surprisingly accurate.

28. Two speakers, such as those shown in Fig. 10.8, are separated by a distance $d = 1\,\mathrm{m}$ and emit sound waves of frequency $1\,\mathrm{kHz}$. Calculate the smallest three angles θ for (a) constructive and (b) destructive interference. If $D = 30\,\mathrm{m}$, at what locations y should observers stand in order to observe constructive interference?

29. If the frequency in the previous exercise is slowly increased to $2\,\mathrm{kHz}$, should the observers move away from or toward the center $(y = 0)$ to follow the intensity peaks?

10.4   LIGHT INTERFERES; IT'S A WAVE

What does all this have to do with light? In 1801 Thomas Young demonstrated that light forms interference patterns and, thereby, proved conclusively that *light is a wave.* Young's experiment was analogous to the sound-wave experiment described above. He directed a strong light source (sunlight) through a tiny aperture and then onto a card containing two closely spaced small holes. Light passing through the holes illuminated a distant screen and formed unmistakable intensity maxima and minima—certain evidence of interference. The bright maxima and dark minima are often referred to as fringes Interference fringes of light can be quite small and hard to see because the wavelength of visible light is much smaller than the dimensions of ordinary objects.

Wavelength of Light Is Color

The wavelength of visible light tells you its color. White light is a mixture of all colors of the rainbow. If white light passes through a glass prism, it spreads out into all its component colors—red, orange, yellow, green, blue, violet (roygbv). You will often need to know roughly what wavelengths of light correspond to which colors. A list is given in Table 10.3. *the longest visible wavelengths are red, and the shortest visible wavelengths are blue or violet.*

TABLE 10.3   Rough correspondence of wavelengths and colors of visible light

Color	Red	Orange	Yellow	Green	Blue	Violet
Wavelength (nm)	660	620	580	520	440	380

Analyzing Light: Interference of Light from Slits

Young's discovery that light is a wave was a major advance in human understanding of the physical world. The new understanding of light led to new research tools with which physicists and chemists learned more about atoms. They discovered that atoms emit and absorb light, and that different atoms emit and absorb different wavelengths. From highly precise measurements of these wavelengths and their intensities, they deduced the internal structure and behavior of atoms and molecules. In the early twentieth century their discoveries led to the development of quantum mechanics, a remarkable theory that revolutionized scientists' understanding of the behavior of atoms and molecules and their components.

The "diffraction grating" is one of the main tools that made possible precise measurements of wavelengths. A diffraction grating is a regular array of very narrow, closely spaced slits. When light passes through them, interference causes the different wavelengths present to separate out with a separation that is better than that obtained by passing light through a prism. The diffraction grating's importance as a tool for looking into the insides of atoms is reason enough to study its principles of operation, but there is the bonus that these principles will also help you understand basic ideas of quantum theory later in this book.

Double-Slit Interference

To learn how a diffraction grating works consider first the interference of light from just two narrow slits spaced close together. You can reproduce the essence of Young's famous experiment by shining a laser beam onto two closely spaced, narrow slits. As shown in Fig. 10.10a, a monochromatic[2] beam of light from a laser shines on the two identical slits S_1 and S_2.[3]

When illuminated by the plane wave radiation of the laser beam, the two slits become radiators themselves. From them emerge two cylindrical wave fronts centered on the slits. These cylindrical phase surfaces spread out at speed $c = 3 \times 10^8$ m/s, and the waves superpose everywhere in the space through which they travel. At any instant in time (snapshot), the electric field from a single slit is a maximum on the surfaces of concentric

---

[2]This word comes from the Greek roots "mono" (single) and "khromatos" (color). "Monochromatic" light is a single wavelength and frequency.

[3]Young did many of his experiments with a beam of sunlight formed with a pinhole. The results are much more evident when you use monochromatic light. To read excerpts from Young's papers on interference look in Morris H. Shamos, *Great Experiments in Physics*, Dover Publications, 1987, pp. 93–107.

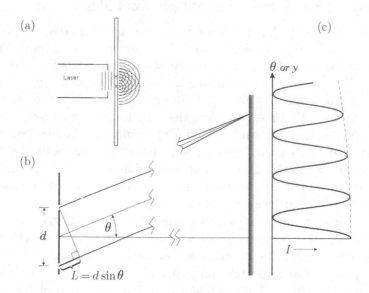

FIGURE 10.10 Geometry for a double-slit interference experiment. (a) Two side-by-side slits, with their long dimension perpendicular to the plane of the page, serve as in-phase sources of light; (b) variation of the path difference L with angle θ; (c) intensity of interference pattern on a screen far from the sources.

half cylinders spaced a distance λ apart. The circular arcs in Figs. 10.10 and 10.11, which are cross sectional views of the slits and wavefronts, represent the edges of such cylindrical phase surfaces. The intersection of two arcs corresponds to the intersection of cylindrical wave fronts from the two slits. Constructive interference occurs along the line of intersection of the two cylinders, i.e., the line running perpendicular into and out of the plane of the page.

By the time the cylindrical wave fronts have traveled a distance $D \gg d$, they are nearly flat. They are also nearly parallel, with only a very small angle between them. That small angle, however, means that if you insert a screen perpendicular to the wavefronts' direction of travel, you will see light and dark bands called fringes resulting from constructive and destructive interference. Examine Fig. 10.11b to see how intersecting plane waves produce constructive interference on a screen, but keep in mind that the angle between the planes is grossly exaggerated for purposes of illustration. Note also that the separation between planes is λ, so the spatial scale of parts of the diagram is vastly larger than what you observe in the laboratory. For light and two slits, just as for sound and two speakers

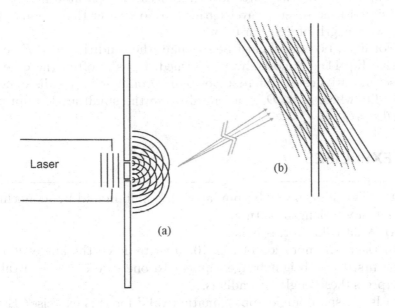

FIGURE 10.11   (a) Cylindrical wavefronts emerge from two slits illuminated by a laser beam. The lines along which the points of intersection lie are where there is constructive interference. (b) By the time these wavefronts reach the distant screen, they are nearly plane waves; the places where the points of intersection strike the screen are where intensity maxima occur.

(Fig. 10.8), the location of the bright fringes, the interference maxima, is given by

$$d\sin\theta = n\lambda, \qquad \text{double-slit interference } \mathbf{maxima} \qquad (11)$$

where n is any integer.

Observation of this interference pattern confirms that light is a wave. The pattern also yields reliable measurements of the wavelength of visible light. From measured values of d and θ, you find for visible light the values of λ given in Table 10.3 ($\lambda \sim 500\,\text{nm}$).

From Eq. 11 you can see that if d is too large, then $\sin\theta$ will be too small to measure, i.e., the fringes will be so crowded together you won't be able to measure their separation. On the other hand, if d is too small, that is if $d < \lambda$, then there is no value of θ that satisfies Eq. 11, because the geometry of the set up is such that the path difference between the waves from the two slits can never equal λ, and the phase difference can not become 2π, and, therefore, there can not be any fringes. In practice, to obtain an interference pattern with noticeably separated fringes, d should be greater than several wavelengths, but no greater than about 1000 times the wavelength of light, or 0.5 mm.

For light, because d must be so small, the condition $d \ll D$ required to derive Eq. 11 is almost always satisfied. If, as is often the case, you are observing fringes at different positions y on a screen a distance D from the slits where $y \ll D$, you can also use the small angle approximation $\sin \theta \approx y/D$.

▪ EXERCISES

30. Two slits spaced 0.5 mm apart are illuminated by monochromatic light of wavelength 650 nm.
(a) What color is this light?
(b) Using the notation of Fig. 10.10 write down the angles θ at which the first three bright fringes appear to one side of the central fringe. Express these angles in radians.
(c) Is the small-angle approximation valid for this exercise? Explain.
(d) If the interference pattern is projected onto a screen a distance $D = 2$ m from the slits, what is the distance between the fringes?

31. An electrical discharge in hydrogen gas emits a mixture of red, green, blue, and violet light. The source illuminates a pair of slits, causing an interference pattern to appear on a distant screen. Make a sketch showing the relative position of the first few bright fringes of each color. What color is the center of the pattern ($n = 0$)?

Single-Slit Diffraction

In Figs. 10.10 and 10.11 light is shown spreading out from the slits. This spreading is actually an interference effect called "diffraction" that you need to understand. A plane wave of light travels in a straight line, so you might expect that when such a wave passes through a rectangular slit, it will cast a bright rectangle with sharp edges on a distant screen. But if you look carefully, you will see the edges are blurred with faint lines of bright and dark where some light has spread into the shadow. This is diffraction.

The shape of the intensity pattern for the particular case of a rectangular slit of width b is shown in Fig. 10.12. (The slit is perpendicular to the page so Fig. 10.12 shows you its width, not its length.) If light were not a wave and there were no diffraction, the large central maximum would be constant (flat top) between $-\frac{b}{2}$ and $+\frac{b}{2}$ and zero everywhere else.

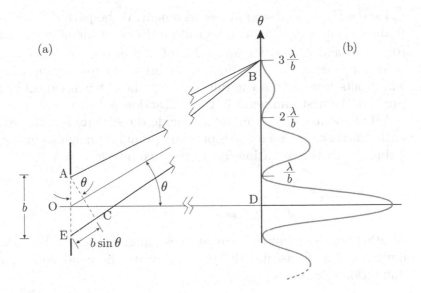

FIGURE 10.12 Single-slit diffraction: (a) geometry; (b) interference pattern.

Because of diffraction the light's intensity is not uniform on the screen, and it spreads into the shadow where it appears as secondary maxima. Notice that between bright places in the shadow there are dark places— intensity minima—where there is no light. This feature is not so surprising if you think of a single slit of finite width b as two slits of width $\frac{b}{2}$ with no separation between them. There might be interference between the light wavefronts coming from these two different halves of the slit.

By considering the slit of width b to be made up of narrower slits, you can calculate where the intensity minima of the diffraction pattern occur. To organize your thoughts about the arrangement, imagine a coordinate system on the diagram in Fig. 10.12. Put the origin in the middle of the slit; have the x-axis run from the slit to the screen; and have the y-axis run across the width of the slit. The edges of the slit are at $(0, \frac{b}{2})$ and $(0, -\frac{b}{2})$, and the long dimension of the slit runs along the z-axis, into and out of the plane of the diagram.

Now in Fig 10.12 use the same geometrical argument that you used in Fig. 10.10 for double-slit interference: Assume the distance from the slit to the screen is so large that wavefronts reaching it from different points of the slit are nearly parallel. Then the angle $\angle ECA$ is essentially a right angle. As a result, the path difference between the wave traveling from the origin O $(0,0)$ along the line $\overline{OB}$ and the wave traveling from A, i.e., $(0, \frac{b}{2})$, along $\overline{AB}$ is $\frac{b}{2} \sin \theta$. If this path difference equals $\lambda/2$ the two wave fronts will arrive out of phase by π radians and they will cancel each other; there will be complete destructive interference.

Here is the crucial step in the argument: If the path difference of waves from $(0, \frac{b}{2})$ and $(0, 0)$ is $\lambda/2$, so will be the path difference of waves from $(0, \frac{b}{2} - \delta)$ and $(0, -\delta)$ for any value of $0 < \delta < \frac{b}{2}$. In other words, there is an angle θ and a corresponding point B on the screen where all the wavefronts from across the full width of the slit will cancel in pairs and produce the first minimum in the diffraction pattern.

What about the other minima? Divide the slit into fourths, so that each path difference between corresponding points on neighboring segments is $\frac{b}{4} \sin \theta$. Then the condition for destructive interference is

$$\frac{b}{4} \sin \theta = \frac{\lambda}{2} \quad \text{or}$$
$$b \sin \theta = 2\lambda, \quad \text{etc.}$$

In other words, minima occur at those angles θ where $\overline{EC}$ is an integer multiple of λ. Consequently, you can write the condition for single-slit diffraction minima as

$$b \sin \theta = m\lambda, \qquad \text{single-slit diffraction \textbf{minima}} \qquad (13)$$

where $m = \pm 1, \pm 2, \pm 3, \ldots$. Given the slit width b and the distance D to the screen, you can use Eq. 13 to calculate the locations of the diffraction minima for any given wavelength λ; to find the locations of the maxima requires more discussion than is useful at this point.

Although Eqs. 11 and 13 look alike, they describe very different phenomena. Keep in mind that Eq. 13 tells you where the *minima* occur for diffraction from a single rectangular slit of width b, while Eq. 11 tells you where the *maxima* occur in the interference pattern arising from light passing through two very narrow slits spaced a distance d apart. For Eq. 11 you can use any values of $n = 0, \pm 1 \pm 2, \pm 3, \ldots$. For Eq. 13 you can use $m = \pm 1 \pm 2, \pm 3$, etc., but you can not use $m = 0$ because the diffraction pattern actually has a maximum at $\theta = 0$.

Combined Double-Slit and Single-Slit Patterns

In the discussion of the double-slit interference pattern there was no mention of the width of the slits. Only their separation d was taken into account. This is an oversimplification. Real slits have both finite width b and a separation d, and both slit width and slit separation affect the interference pattern. Together they result in an intensity pattern that is a double-slit pattern inside the envelope of a single-slit pattern as shown in Fig. 10.13c. A more complete argument would show you that mathematically the combined intensity pattern is the intensity of the idealized double-slit pattern multiplied by the intensity of the single-slit pattern.

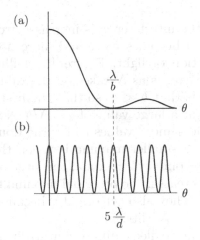

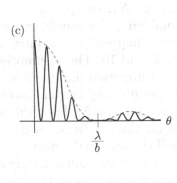

FIGURE 10.13 (a) Single-slit diffraction pattern; (b) double-slit pattern for $d = 5b$; (c) resultant interference pattern.

In other words, the intensity of the double slit interference given in Eq. 9 must be modified by the effect of the single-slit diffraction pattern. Figure 10.13 illustrates the resulting intensity distribution for the special case $d = 5b$. In general the peak intensities of a double-slit pattern decrease for large θ; they are reduced by the effects of single-slit diffraction occurring at each slit. Notice also in Fig. 10.13c the double-slit maximum corresponding to $n = 5$ in Eq. 11 is missing because the $m = 1$ minimum of the single-slit pattern Eq. 13 occurs at the same angle. When this happens, the single-slit pattern eliminates the double-slit maximum because it is multiplying it by zero.

Conversely, if you see a double-slit pattern where the 5th, 10th, . . . , maxima are absent, you know that the slit width b is 1/5 of the slit separation d.

▪ EXERCISES

32. A single slit of width $b = 0.2$ mm is illuminated with 633 nm light from a helium–neon laser.
(a) Calculate the angular positions of the first 3 minima.
(b) Is the small-angle approximation valid for these angles?
(c) If the interference pattern is projected on a screen 2 m away, how far apart are the dark fringes?

33. Sketch, as in Fig. 10.13, the interference pattern for $d = 3b$.

Multislit Interference Patterns

As the number of slits producing the interference is increased from 2 to many (N), the pattern changes and becomes more and more useful for analyzing the wavelength composition of light. Figure 10.14 illustrates what happens to the interference patterns as N takes on the values of 2, 3, 4, and 10. The assumption here is that $b \ll d$, so that the first single-slit diffraction minimum is located at a large value of $\theta \approx \lambda/b$. Note that large-intensity maxima occur at the same θ values as for the double-slit case, but for N slits there are $N - 2$ smaller maxima between the large ones. The smaller ones are called "secondary" maxima; the large ones are called "principal" maxima. As N increases, the principal maxima become narrower and more sharply defined. They also get brighter because of the larger amount of light collected by the N slits.

Such N-slit interference devices are called "diffraction gratings."[4] As N is made larger and d is made smaller, the principal maxima become sharper and separated by larger angles; the secondary maxima become smaller and negligible. The principal maxima occur at the angles such that

$$d \sin \theta = n\lambda. \tag{10}$$

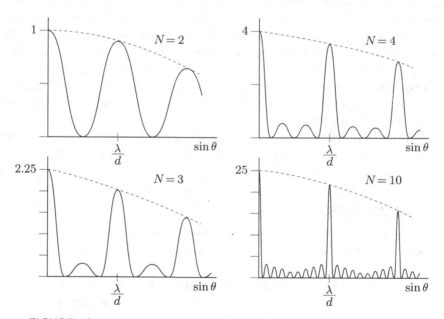

FIGURE 10.14 Multislit interference patterns for $N = 2, 3, 4, 10$.

---

[4]The word "grating" seems to have been chosen because the closely spaced slits reminded someone of spaced slits in iron grillwork of a fireplace grating.

These are the same angles as for maxima from double-slit interference, but now the maxima are extremely sharp and well defined. If θ and d are known accurately, then the wavelength λ can be determined with six-digit precision.

Equally important, a good diffraction grating will separate light consisting of two wavelengths that are nearly the same into two distinct principal maxima that can be located and compared with high precision. The larger you make N the better you can separate and distinguish maxima from two different wavelengths. It is useful to remember that the smallest difference in wavelength $\delta\lambda$ that can be separated is $\delta\lambda = \lambda/N$. The ratio $\delta\lambda/\lambda = 1/N$ is called the "resolution" of the grating.

▼ EXAMPLES

3. It is possible to prepare a glass plate with as many as 30 000 slits per centimeter. When a 1 cm wide piece of such a closely ruled grating is used to examine green light, $\lambda = 550\,\text{nm}$, the grating can distinguishably separate wavelengths that differ by as little as $550/(3\times10^4) = 0.018\,\text{nm}$.

Diffraction gratings can work by reflection as well as by transmission. You probably know that compact disks spread out reflected incident light into bands of colors. The reason is that the tracks on a compact disk are 1.6 μm wide and make the surface of the disk a nice diffraction grating.

Caution: In diffraction gratings, it's not unusual for the spacing d to be so small that the angle θ in $d\sin\theta = n\lambda$ is *not* a small angle. In this case, you cannot use the small-angle approximation.

▨ EXERCISES

34. If you look at the light reflected from a compact disk, what angular separation would you expect to see between the red light ($\lambda = 625\,\text{nm}$) that corresponds to $n = 1$ and that which corresponds to $n = 2$ in Eq. 11?

Spectra, Spectrometers, Spectroscopy

The collection of different wavelengths emitted by a source is called its electromagnetic "spectrum." The rainbow produced when water droplets spread out the visible wavelengths present in sunlight is the spectrum of sunlight. Because there is present some amount of almost every wavelength, the Sun's spectrum is said to be "continuous." Many sources, especially hot gases, emit light that is not continuous in its spectrum but is instead a mixture of quite distinct separate wavelengths. To observe a spectrum it is usual to define a beam of light from the source by passing the beam through a narrow slit; then when the beam is spread out by a prism or diffraction grating, the observer sees the separate colors in the light spread out into lines of light that are images of the slits. These images are called "spectral lines." Spectra composed of spectral lines are called "line spectra." The study of line and continuous spectra is called "spectroscopy," and those who perform such studies are called "spectroscopists." The study of the spectra of light from atoms is called "atomic spectroscopy."

Precision optical instruments that spread out light into its component wavelengths for the measurement of their values of λ are called "spectrometers." The principal components of a spectrometer that uses a diffraction grating are shown in Fig. 10.15. Modern optical spectrometers can measure wavelengths to a precision of 1 part in 10^5 and differences in wavelengths $\lambda_1 - \lambda_2$ to a precision of 1 part in 10^7. With modern techniques of laser-based technology, it is now possible to measure wavelengths to parts in 10^{15}–10^{18}. Powerful instruments such as these have been essential for exploring the structure of atoms and developing and testing the quantum theory.

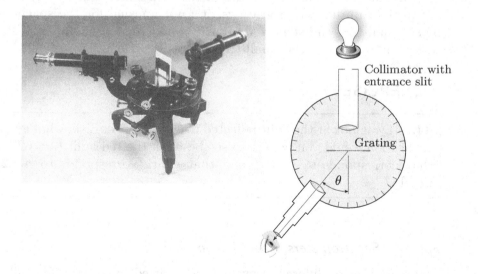

FIGURE 10.15 Components of a grating spectrometer.

10.5 ATOMIC SPECTROSCOPY

Helium, the lightest of the inert gases and the second-most-abundant element in the universe, was first discovered on the Sun in 1868. It was not found on Earth until 27 years later. Preposterous? Not at all. The discovery of helium illustrates the power of spectroscopic techniques.

Atomic spectroscopy reveals a most important experimental result: Isolated (i.e., gaseous) atoms gain and lose energy by absorbing and emitting light at certain precisely defined wavelengths (or frequencies); each atomic species has its own unique set of wavelengths. When light from a collection of these atoms comes through a slit to a diffraction grating, the output from the grating will be colored lines of light—the "spectral lines." The collection of lines is the spectrum of the atom. An atom's spectrum is its fingerprint. If the light emitted by a hot, glowing gas contains the spectral lines of, say, oxygen, then oxygen must be present in the gas. Wavelengths of spectral lines of several elements are listed in Table 10.4.

Helium was discovered when its spectral lines were seen in light coming from the Sun. Physicists attached a spectrometer to a telescope to study the spectrum of the hot gases of the solar corona when it was visible for several minutes during the total eclipse of 1868. The measured wavelengths were then compared to spectra obtained from laboratory samples of known elements. The spectral lines of hydrogen and sodium were easily

TABLE 10.4 Wavelengths (in nm) of some representative atomic spectral lines—only the strongest emission lines are shown

hydrogen—H	helium—He	neon—Ne
656.28	667.82	626.6
486.13	587.56	621.7
434.05	501.57	618.21
410.17	447.15	585.25
mercury—Hg	sodium—Na	argon—Ar
614.95	588.995	706.72
579.07	589.592	696.54
576.96	568.82	487.99
546.07	498.28	476.49
435.83		442.60
404.66		434.81

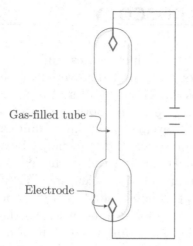

FIGURE 10.16   Discharge tube for producing visible spectral lines.

identified, but the solar corona also emitted a bright yellow line with a wavelength that did not match up with light from any known element. Physicists concluded that they had discovered a new element, and they named it helium from *helios*, the Greek word for the Sun. Twenty-seven years later, when the same yellow line was observed in the spectrum of gas released from a sample of uranium ore, researchers knew that they had discovered helium on Earth.

A common way to produce atomic spectra in the laboratory is a gas discharge tube like the one shown in Fig. 10.16. In this device electrons are accelerated through a low-pressure gas in a glass container, or "tube." The electrons collide with the gas atoms, transferring some of their kinetic energy to the atoms, which then shed their excess energy by radiating light at the wavelengths of their characteristic emission spectrum.

▉ EXERCISES

> **35.**   Light from a gas-filled discharge lamp is analyzed by a spectrometer using a diffraction grating with 10 000 slits per cm. Bright lines are recorded at the angles $\theta = 26°, 29°, 41°, 62°$, and $75°$. Using Table 10.4, identify the gas and specify the color of each line.

The spectroscopy of visible spectral lines has played a central role in the development of modern physics. When in the early twentieth century physicists sought to explain and understand the bewildering complexity of the spectra of even the simplest atoms, they concluded that some of Newton's basic ideas about motion do not apply to atomic-sized systems, and they developed a more fundamental description of the behavior of matter, now called "quantum theory" or "quantum mechanics." Spectroscopy has been a major tool for carefully testing the predictions of quantum theory and also for stimulating further theoretical advances.

In astronomy all of our information is conveyed by electromagnetic waves, especially visible light. Using spectroscopy, astronomers can measure the temperature and composition of the stars, their rotation rates, and also their motion relative to Earth.

10.6 PROBING MATTER WITH LIGHT

You can also use interference patterns to infer the geometrical structure of the objects that produce them and thus learn about the structure of bits of matter far too small to see. Figure 10.17a is a simple example from which you can extract information about the slits responsible for an interference pattern when neither N the number of slits nor d the spacing between them is known. There are no secondary maxima, so you know

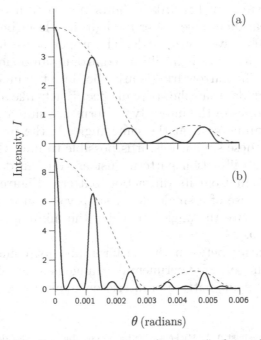

FIGURE 10.17 Interference patterns from two "unknown" slit structures. The dashed-line envelope of the fringes is shown to help you do Exercise 36.

it is a double slit pattern ($N = 2$). The first minimum ($m = 1$) due to diffraction falls at the third minimum due to double-slit interference. Therefore, the ratio of $d/b = 2.5$. If you know that $\lambda = 550\,\text{nm}$ (yellow-green), then you can determine d and b without ever examining the slits directly.

The ability to deduce things about a structure from the interference pattern it produces is important, because the structure causing the interference may be out of reach, as are stars, or perhaps too small to be measured with ordinary instruments, as are atoms.[5] The structures you can infer from interference patterns can be much more complicated than just collections of slits. For example, it is possible to reconstruct a crystal structure from the interference patterns produced when x-rays pass through the crystal. You will learn more about this in Chap. 14.

EXERCISES

---

36.   Given $\lambda = 550\,\text{nm}$, find d and b from Fig. 10.17a. Describe the slit structure (number, width, and separation) giving rise to the interference pattern in Fig. 10.17b.

The lower curve in Fig. 10.18 shows the pattern of intensity of light that will appear on a screen after the light in a laser beam passes through a single slit 100 wavelengths wide. The upper curve in Fig. 10.18 shows the intensity pattern of light diffracted around an opaque strip 100 wavelengths wide. The narrow fringes arise because the incident laser beam is only 16 times wider than the strip; in effect it acts like a double slit. Notice that the envelopes of the intensity patterns are similar. In particular, the diffraction minima occur at the same angles in the two cases.

Because of these similarities, it is possible to infer the width of a solid object from its diffraction pattern, just as you can deduce the width of a transparent slit from its diffraction pattern. Figure 10.18 shows that for the simple case of a single slit or strip you can use exactly the same technique: Measure the angles at which the minima occur and then use Eq. 13 to find b.

The similarities between the patterns are not coincidences. The fact that in the right sort of experimental arrangement an obstacle that blocks light forms the same pattern of interference minima as an opening of the

---

[5] The American physicist A.A. Michelson had a remarkable talent for devising instruments and techniques for precision measurements. For a fascinating and readable account of his exploits see A.A. Michelson, *Light and Its Uses*, University of Chicago Press, 1902.

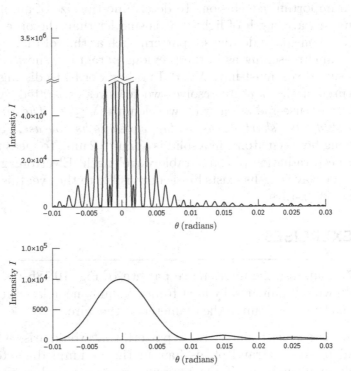

FIGURE 10.18 The lower curve shows the pattern of intensity from a single slit of width 100λ. The upper curve is the pattern of intensity of laser light diffracted around an opaque strip of width 100λ. The width of the laser beam is 16 times the width of the strip.

same size is related to "Babinet's principle." We mention the principle here only to emphasize that interference can be interpreted as arising from spaces between matter (e. g., slits), or, equivalently, from the matter itself (e. g., the screen material surrounding the slits).[6]

Regardless of the detailed shape of the object (slits, obstacles, lines, circles, whatever), there are fundamental limits on the information available from interference patterns. Consider once again a single-slit diffraction pattern such as that shown in Fig 10.18. To determine the size of the object, you must be able to locate the positions of the intensity minima or maxima. Because the positions of the minima are given by $b \sin\theta = m\lambda$ and because $\sin\theta \leq 1$, you must have $\lambda \leq b$ if you are to measure b. This

---

[6]The notable differences between the two intensity patterns in Fig. 10.18 arise largely because of the finite size of the illuminating laser beam, but there are other possible sources of differences that become significant when the width of the structure gets close to the size of the wavelength. For example, see R.G. Greenler, J.W. Hable, and P.O. Slane, "Diffraction around a fine wire: How good is the single-slit approximation?" *Am. J. Phys.* **58**, 330–331 (1990).

is an important conclusion: To determine the size of an object, you need to use a wavelength of light that is smaller than the object.

Now consider a double-slit pattern such as that of Fig. 10.13b on p. 315. The same reasoning as in the previous paragraph shows that in order to measure d you must have $\lambda \leq d$. In other words, to distinguish structural features of size x or to resolve two features separated by a distance x, you must use radiation with wavelength $\lambda \leq x$. *The tinier the objects you study, the shorter must be the wavelengths you use.* For example, the spacing between atoms in a solid is about 0.2 nm. To measure this spacing you need radiation of comparable wavelength. Electromagnetic radiation of these wavelengths exists but is not visible to the eye; this kind of "light" is called "x-rays."

■ EXERCISES

37. Suppose the interference pattern of Fig. 10.18b was created by a thin wire illuminated by light from a helium–neon laser at a wavelength of 633 nm. Determine the diameter of the wire.

38. Describe the interference pattern that would arise if Isaac Newton had placed a strand of his hair in the ~ 1 mm diameter beam of a helium–neon laser. If the pattern were projected onto a screen 2 m from the hair, how far from the center of the pattern would the first three minima fall? (If you can, borrow someone's laser pointer and try the experiment; look carefully.)

10.7   SUMMARY

A wave is a propagating disturbance. It travels with a finite speed, about 340 m s^{-1} for sound waves in air and 3×10^8 m s^{-1} for all electromagnetic waves in vacuum.

Sine waves can be used as the building blocks of all waveforms. A sine wave is characterized by its amplitude A, wavelength λ, frequency f (or, equivalently, its period T), and its phase. The sinusoidal variation of some property y, e. g., displacement, pressure, or electric field, traveling as a wave to the right is described by the equation

$$y = A \sin\left(2\pi \frac{x}{\lambda} - 2\pi \frac{t}{T} + \phi \right),$$

where $(2\pi \frac{x}{\lambda} - 2\pi \frac{t}{T} + \phi)$ is the phase, and ϕ is the phase constant.

If the variation of the property is along the line of travel of the wave, it is a longitudinal wave; if the variation is perpendicular to the direction of propagation of the wave, it is a transverse wave.

Both the energy carried by a sine wave and its intensity are proportional to the square of the wave's amplitude.

The velocity of propagation of a sinusoidal wave is given by

$$v = \lambda f.$$

This relationship holds for any periodic waveform.

Waves interfere with one another. Any phenomenon exhibiting interference must have a wave nature. Since light forms interference patterns, then light must be a wave.

Two important interference patterns are single-slit diffraction and double-slit interference. The *minima* of the single-slit diffraction pattern occur at angles θ given by the expression

$$b \sin \theta = m\lambda,$$

where b is the width of the slit and m is any positive or negative integer $\neq 0$.

The *maxima* of the double-slit interference pattern occur at angles θ given by the expression

$$d \sin \theta = n\lambda,$$

where d is the separation between the slits and n is any integer $0, \pm 1 \pm 2, \ldots$.

An array of N equally spaced slits ($N \geq 2$) has principal maxima satisfying the double-slit equation above. For $N \gg 2$, the array is called a diffraction grating. The integer n is called the "order" of the corresponding intensity maximum.

Diffraction gratings are used to measure accurately and precisely the wavelengths of light emitted from or absorbed by atoms. Each atomic species has a unique spectrum that identifies the atom and provides clues to its internal structure. Visible-light, or optical, spectroscopy has played an extremely important role in the development of modern physics and especially atomic physics.

If λ is known, an interference pattern can be used to study the object that generates it. Interference can be used to determine the sizes and structures of objects over an enormous range, from smaller than atoms to larger than stars. To distinguish structural features of size x with waves requires waves with wavelength λ of the order of x or smaller.

Light waves are remarkable in that their speed c in a vacuum is constant regardless of how fast an observer is moving toward or away from their source. In Newtonian physics this is impossible, and physics had to be completely restructured to take this property of light into account. The next two chapters will discuss this restructuring, Einstein's special theory of relativity, and some of its surprising consequences.

PROBLEMS

1. The diagram in Fig. 10.19 shows two waves labeled #1 and #2 as they appear at time $t = 0$.

    a. Write an equation for wave #1 in terms of its wavelength at $t = 0$.

    b. Write an equation for wave #2 at $t = 0$. Be particularly sure to get the phase constant correct.

2. In North America electricity is sent over wires in waves that have a frequency of 60 Hz. If these waves travel over wires at the speed of light, 3×10^8 m/s, what is their wavelength? Compare this to the wavelength of visible light.

3. A TV picture is generated by an electron beam that makes 512 parallel passes from left to right across the screen 30 times each second. In the days before cable TV, "ghosts," or secondary images, would appear due to reflections of the signal from nearby buildings. (See Fig. 10.20.) If a ghost appears 1 inch to the right of the main image on the screen, and the screen is 20 inches wide, what is the extra distance L that the reflected wave travels to reach the TV antenna?

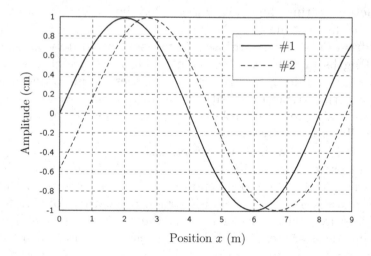

FIGURE 10.19   Two waves at time $t = 0$ (Problem 1).

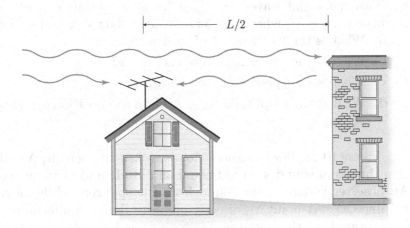

FIGURE 10.20 How reflections produce ghost TV signals (Problem 3).

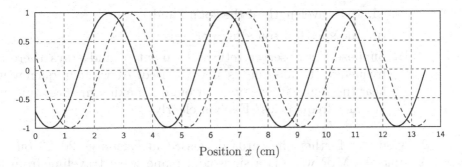

Position x (cm)

FIGURE 10.21 Two waves identical except for phase (Problem 0a).

4. Using a microscope a student finds that there are 50 slits per cm in a diffraction grating. She then takes light from a laser of unknown frequency and shines it through the grating onto a wall 6 m distant from the slits. On the wall she finds that the displacement of the first maximum from the center is 1.5 cm.

What is the wavelength of the laser light?

5. Consider the diagram in Fig. 10.21 showing a snapshot, taken at $t = 0$ s, of two waves.
  a. What is the phase difference between the two waves?
  b. What is the wavelength of the above waves?
  c. If snapshots taken 0.5 s apart show that the waves move 2 cm during that interval of time, what is the frequency of the waves?

6. Two sinusoidal waves are given by $y_1 = A\sin(kx - \omega t)$ and $y_2 = A\sin(kx - \omega t + \frac{\pi}{3})$, where $k = 5\pi\,\text{m}^{-1}$, $\omega = 800\pi\,\text{s}^{-1}$, and $A = 4.0\,\text{cm}$.

    a. What is the frequency f of each wave?

    b. What is the wavelength λ of each wave?

    c. What is the speed v of each wave?

    d. What is the amplitude A_{total} of the superposition of the two waves $y_1 + y_2$?

7. Figure 10.22 shows two small loudspeakers in open air, A and B, emitting sinusoidal sound waves of equal amplitude and of frequency 840 Hz. An observer listens to the sound from the speakers while moving along the line OO'. At point M, directly opposite the midpoint of the line AB, he is aware that the sound is the loudest; as he continues toward O', he notices that the intensity has dropped to zero at point Y. He measures the distances AY and BY to be 3.35 m and 3.15 m, respectively.

    a. From this information, find

        i. the wavelength of the sound and

        ii. the speed of sound in air.

    b. Suppose the observer returns to point Y and stays there as the frequency of the sound is slowly raised. At what frequency will he hear the sound intensity go through a maximum? Can this happen at other frequencies? Explain clearly.

8. Consider further the cards discussed in Exercises 20–23 on p. 303. Assume the MIP wave is a sinusoidal plane wave traveling from left to right. It has a wavelength of 16 card thicknesses, a velocity of 3 card thicknesses per second, and, by arbitrary choice, the phase of the wave at card 20 is 0 radians at time $t = 0$.

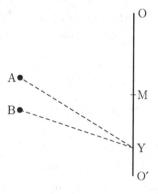

FIGURE 10.22 Speakers and listener for Problem 7.

a. Show that to the left of the jet black card 24 card 20 will be white. What are some possible phases corresponding to this card?

b. If you hadn't already been told that the wave is moving from left to right, why in Exercise 5 are the two snapshots consistent with the wave traveling to the left with a speed of 13 card-thicknesses per second?

c. If you didn't already know the direction of the wave, it could have many other speeds consistent with these observations from two snapshots. Give two of them in each direction.

d. Suppose you had not been told the wave is moving left to right. What measurements might you make to decide whether the wave was traveling to the right or to the left?

e. Suppose the wave is traveling to the right. Suppose also you can watch the color of card 24 without removing it from the stack. Describe what happens to the color of card 24 as you watch.

f. If you take a snapshot 3 s after the one you took at time $t = 0$, you will find that card 29 is white. What is the phase of the wave at card 29? What is the phase of the wave at card 13?

9. A laser emitting light of wavelength 600 nm illuminates a long, thin wire 20 cm from a screen and parallel to it. The illumination produces an interference pattern as shown in Fig. 10.23.

a. At what angle does the first minimum of the intensity pattern occur? The second minimum?

b. What is the diameter of the wire?

c. If the wire used above were replaced by one having twice the diameter, what would be the new positions on the screen of the first and second intensity minima?

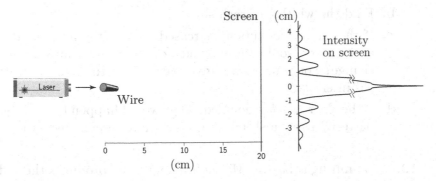

FIGURE 10.23 Intensity pattern produced when a laser illuminates a thin wire (Problem 9).

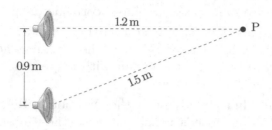

FIGURE 10.24 Two speakers emitting the same wavelengths in phase with each other—(Problem 10).

    d. How thin could a wire become and still produce at least one intensity minimum on the screen? What does this imply about using light to measure small objects?

10. Two identical small loudspeakers emit sound waves of wavelength λ, and the waves are in phase. The loudspeakers are separated by a distance of 0.9 m, as shown in Fig. 10.24.

    a. If $\lambda = 30$ cm, will the interference at point P be constructive or destructive? Explain.

    b. What is the frequency of the waves?

    c. If the frequency is halved, what kind of interference occurs at point P? Explain.

11. Light of wavelength 600 nm illuminates a double slit apparatus and produces the interference pattern shown in Fig. 10.25. Note that θ is expressed in radians.

    a. Find the slit spacing d.

    b. Find the width b of each slit.

    c. If the number of slits is increased to 3, with the same d and b as above, how would the intensity of the central maximum ($\theta = 0$) change? For your answer give the ratio of the new intensity to the old intensity.

    d. If the slit width b is doubled, what would happen to the interference pattern? Draw a sketch of the new interference pattern.

12. 600-nm light passing through a single slit produces the diffraction pattern shown by the solid line in Fig. 10.26.

    a. What is the slit width?

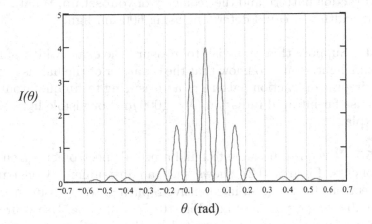

FIGURE 10.25 Interference pattern from the double slit arrangement in problem (11).

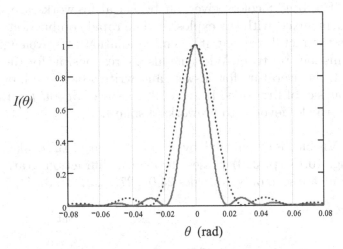

FIGURE 10.26 Single-slit diffraction patterns for problem 12. The solid line corresponds to light with wavelength $\lambda = 600$ nm.

    b. Light of a different wavelength produces the diffraction pattern shown by the dotted line. What is the wavelength of this light?

    c. Suppose, instead of changing the wavelength, you change the width of the slit to produce the dotted diffraction pattern. Did you widen or shrink the width of the slit? What is the new width?

13. If you know the wavelength of the light incident on a single slit, you can determine the slit's width from measurements of the observed diffraction pattern and the geometry of your set up. What is the minimum slit width you could determine using 600-nm light?

14. Suppose that you wish to measure the exact width of a human hair using light. You also know that the diameter of the hair is roughly 0.1 mm. Using the diffraction pattern as a measuring technique, would it be better to use far-infrared radiation ($\lambda = 1000\,\mu$m) or visible light ($\lambda = 600$ nm)? Explain.

15. It is quite possible to build a useful spectrometer from simple components. Figure 10.27 shows such an instrument. Construct your own spectrometer using an inexpensive replica grating (your instructor should be able to get you one) and Fig. 10.27 as a guide. Use your spectrometer to measure the spectrum emitted by the streetlamps on your campus or in your town. Identify the gas emitting the light. Table 10.4 (p. 319) may be of help.

16. The bright colors given off by aerial fireworks are due to metallic powders mixed with the explosive. The rapid combustion of the explosive heats the metal, causing it to emit radiation at wavelengths included in its emission spectrum. What metals are responsible for the various colors? Search the literature for answers and write a paragraph or two describing the physics of fireworks. Hint: A 1991 issue of *Scientific American* carried a full article devoted to fireworks displays.

17. Visible light emitted by a gas discharge tube like the one shown in Fig. 10.16 (p. 320) passes through a diffraction grating. Interference maxima are seen only at angles 24.0°, 27.1°, 32.3°, 37.4°, 54.4°, and 65.8°.

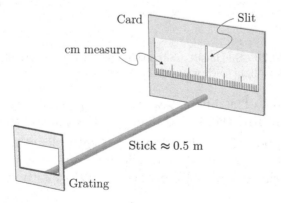

FIGURE 10.27   Basic components of a diffraction spectrometer (Problem 15).

FIGURE 10.28 A diffraction grating made of alternating strips of transparent and opaque material (Problem 18).

   a. Which angle belongs to the shortest wavelength light?

   b. The vapor within the discharge tube is one of those shown in Table 10.4 (p. 319). Which one is it? Show how you found your answer.

   c. Find the distance d between the slits of the diffraction grating.

18. A diffraction grating is made of alternating strips of transparent and opaque material, each of the same width b, as shown in Fig. 10.28. When monochromatic visible light of wavelength λ illuminates the grating, the first interference maximum appears at $\theta = \pm 15°$.

   a. At what other angles do interference maxima appear?

   b. Estimate the largest and smallest values of b that will produce a diffraction grating useful for working with visible light.

19. The amplitude A_{tot} of the sum of two waves of equal amplitude A and frequency f (and thus wavelength λ) that are out of phase by $\Delta\phi$ is

$$A_{\text{tot}} = 2A \cos\left(\frac{\Delta\phi}{2}\right).$$

This result can be represented geometrically by the sum of two vectors (arrows) of the same length A attached end-to-end with an angle $\Delta\phi$ between them as shown Fig. 10.29.

Use the law of cosines and show that the geometric representation in Fig. 10.29 gives the correct result for A_{tot}.

20. Consider a double slit apparatus with slit width b much smaller than the slit separation d (see Fig. 10.30a). When monochromatic light

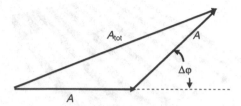

FIGURE 10.29   Amplitudes of waves that differ in phase by an angle $\Delta\phi$ add like vectors with the angle $\Delta\phi$ between them (Problem 19).

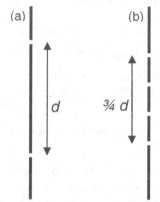

FIGURE 10.30   Two different arrangements of slits for Problem 20.

of wavelength λ passes through the apparatus, the first interference maximum falls at an angle θ and has an intensity I. Now imagine that two more slits are added in between the original two slits such that the four slits are equally spaced (Fig. 10.30b), and that the slit assembly is then shrunk to 75% of its original length. At the same angle θ, what is the new intensity? (Hint: use the geometric construction from Problem 19.)

21.   The interference pattern shown in Fig. 10.31 is observed when monochromatic light of wavelength 600 nm passes through an array of several evenly spaced slits. The angle θ is given in radians.

   a. How many slits are there?

   b. What are their spacing and width?

   c. The interference pattern appears on a screen 15 m away from the slits. How far apart on the screen are the points P and P'?

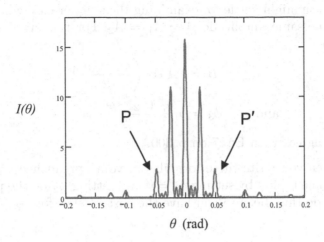

$I(\theta)$

θ (rad)

FIGURE 10.31 Interference pattern produced by 600-nm light passing through an array of evenly spaced slits (Problem 21).

SPREADSHEET EXERCISE: ADDING WAVES—INTERFERENCE

In this exercise you use a spreadsheet such as Microsoft Excel® to generate two sine waveforms with the same wavelengths λ but different amplitudes A_1 and A_2 and different phase constants ϕ_1 and ϕ_2. Then you add the waves together and see how their phase difference $\phi_2 - \phi_1$ determines their interference. A good way to make the interference apparent is to use the spreadsheet to graph the waveforms and their sum.

You can also show that the general formula for the sum of two waves

$$B \sin\left(\frac{2\pi x}{\lambda} + \phi_3\right),$$

where

$$B = \sqrt{A_1^2 + A_2^2 + 2A_1 A_2 \cos(\phi_2 - \phi_1)}$$

$$\text{and} \quad \phi_3 = \phi_1 \pm \cos^{-1}\left(\frac{A_1 + A_2 \cos(\phi_2 - \phi_1)}{B}\right),$$

gives the same result as you get when you numerically add the waves together point by point.

Or you might settle for examining the simpler case when the two waves have the same amplitude $A = A_1 = A_2$. For this special case B and ϕ_3 become

$$B = 2\,A\,\cos\left(\frac{\phi_2 - \phi_1}{2}\right)$$

and
$$\phi_3 = \phi_1 \pm \frac{(\phi_2 - \phi_1)}{2},$$

which agrees with Eq. 7 on p. 300.

When you write these formulas in your spreadsheet you may want to save some typing by setting $\phi_1 = 0$ and letting ϕ_2 be the phase difference. If you do this, be sure you have the correct sign for ϕ_2.

INSTRUCTIONS

The following step-by-step instructions are for Microsoft Excel®. They are supplied for two purposes. First, they are to help you through what may be unfamiliar uses of Excel and to show you how to fix up the graphs to be useful. Second, and more important, the instructions show you how to set up a spreadsheet with parameters that you can vary. If you follow these instructions, the spreadsheet will be set up so that you can change the phase difference by typing in a single number. Such an arrangement is very useful for doing a succession of calculations with different parameters. In this case you will be able to vary the phase difference and show how adding the waves can give constructive, destructive, or partial interference depending on the difference in phase between the two waves.

In what follows, the material on the left side of the arrows is what you should type; the address to the right of the arrow tells you in what cell of the spreadsheet to type it.

Begin by setting up some headings and putting in some initial values for the wavelength and the phases of the waves.

Phase   →A5

Radians→B5

Degrees →C5

phi-1   →A6

0   →B6

phi-2   →A7

=pi()/3   →B7

=degrees(B7) →C7

4 [wavelength]→B8

x   →A10

Next fill up 101 cells with successive values of x from 0 to 10 in steps of 0.1:

Enter 0 into A11; enter =A11+0.1 into A12; click on A12 and grab the little square in the lower right hand corner and drag it to A111.

Now put in some more headings.

Wave1 →B10

Wave2 →C10

Sum   →D10

Theory→E10

Now load some formulas, one for wave1 and another for wave2:[7]
=sin(2*PI()*$A11/$B$8 + $B$6) → B11
=sin(2*PI()*$A11/$B$8 + $B$7) → C11.

Then put the sum of these two waves into D11:
=$B11 + $C11 → D11.

Copy these formulas into the rest of the cells of their columns by clicking on B11 and dragging to B111; on C11 and dragging to C111; on D11 and dragging to D111.

You should now have the two waves in the B and C columns and their sum in the D column. Notice that if you want to change the phase difference, all you need to do is change the entry in cell B7. Try entering =PI() into B7. What happens to the sum?

Notice that this program is set up to have you enter the phase angle in radians into cell B7. When you do that, it automatically calculates the phase angle in degrees in cell C7. Perhaps you would like to do it the other way around.

---

[7]The dollar signs are important in the cell addresses. They freeze the reference to the cell so that if you move the cells containing the reference the address does not change. A dollar sign in front of the letter freezes reference to the corresponding column but allows the row to change; a dollar sign in front of both letter and the number freezes the reference to the particular cell. We want to do this when we have stored in a cell a number that we will use in many other places in the spreadsheet.

MAKING A NICE GRAPH

Try to set up your graph with a nice grid and uncluttered lines.
Begin by clicking on the Chart Wizard icon. Choose chart type "XY (Scatter)" and choose a sub-type ("smooth without data points" is good). Click "finish" and some sort of graph should appear.

To tell the graph what data to plot:
Right click on the grey area of your graph and choose "Source Data"; name your data "wave1"; click on "X values" and select A11 to A111 and click on the button at the end of the line containing this information; repeat for "Y values"; now click on "Add" and repeat for wave-2; again click on "Add" and repeat for "Sum."

Your graph should now contain three plots: wave1 wave2, and their sum.

To be quantitatively useful a graph should have a grid. Also its axes should be labeled and have reasonable scales. The various curves on your graph should be clearly labeled and easily distinguished. Right click on the grey area of your plot and choose "Chart Options." The various tabs available enable you to supply all the foregoing features.

Give each of the three curves on your graph a distinctive line style.

TASKS

a. Put some useful titles and identifying labels on your graphs.

b. Print two graphs, one for a phase difference of $\pi/3$ and another for a phase difference of 0.9π.

c. Add to the spreadsheet a column that calculates the sum of the waves according to the formula for adding waves of the same amplitude and wavelength. To take a square root, use the spreadsheet function =SQRT( ). Print one page of your spreadsheet, the first 35 lines or so, showing that direct addition of the waves and the addition formula give the same result.

d. Show the formula that you used above the column containing the evaluations of the formula.

Hand in your three printouts.

11.

Time and Length at High Speeds

11.1   INTRODUCTION

Electrons and light are two of the most important tools we have for learning about the structure of atoms. To use these tools correctly, you need to know how energy and momentum are transferred by objects moving with speeds approaching the speed of light. This means that you need to be able to use some of the ideas of Einstein's special theory of relativity.

The fundamental ideas of the special theory of relativity have to do with the nature of space and time. At high speeds it becomes apparent that these two concepts are interconnected in ways not suspected until Einstein proposed his theory. These connections mean that for objects moving with speeds approaching the speed of light, energy and momentum are related in ways quite different than experience with everyday speeds suggests. Although energy and momentum are the tools we need for studying atoms, the ideas of space and time that come from the special theory of relativity are so interesting in themselves that you should learn a little about them first.

It is fundamental to the special theory of relativity that the speed of light c is constant for all observers regardless of their relative motion. Equally fundamental is the idea that the laws of physics are the same in all frames of reference moving at constant velocity. This is called the "principle of relativity." In the following sections, after a look at some early experimental evidence for the constancy of c for all observers moving or at rest, we will see how these facts of nature lead us to expect moving clocks to run slow and moving lengths to become shorter. These curious predictions of the special theory of relativity have a surprising implication called the "relativity of simultaneity.": *Two events that occur*

C.H. Holbrow et al., *Modern Introductory Physics, Second Edition*,
DOI 10.1007/978-0-387-79080-0_11, © Springer Science+Business Media, LLC 1999, 2010

some distance apart and at the exact same time in your reference frame will occur at two different times in the reference frame of someone moving relative to you.

It is important for you to get a working understanding of what it means to say $E = mc^2$. You especially need to know how to describe kinetic energy at high speeds and how kinetic energy is connected to momentum. Therefore, the next chapter will deal with these topics and the related fact that mass depends on velocity.

The ideas presented in this chapter mean that Newton's description of matter and motion and time and space is wrong. In other words, most of the physics you have been taught so far is wrong! Yet Newton's physics works very well for the world in which you live because at low relative speeds it is the best approximation to the special theory of relativity. This statement has a precise mathematical sense: Newtonian physics is the limiting case of Einstein's theory of relativity for objects whose relative speeds v are much smaller than c. Thus Newton's physics is an excellent description of the world in which you live, sufficient to build skyscrapers or send astronauts to the Moon. You have not wasted your time learning Newtonian physics.

This idea of one formulation of physics as a limiting case of a more general theory is important. You will see this sort of thing happen often as you study more physics. To fully appreciate how one formulation can be obtained as the limiting case of another, you need a mathematical tool that permits you to create simple approximate equations from complicated exact ones. This tool is so important and so useful that we are going to introduce it to you before going any further.

11.2   APPROXIMATING A FUNCTION

Physics theories usually give us formulas, i.e., algebraic relations between some independent variables and some dependent variables. You have run into lots of these: $y = \frac{1}{2}g\,t^2$, $K = \frac{1}{2}m\,v^2$, $2\,a\,y = v^2 - v_0^2$, and so on. The algebraic equation is a particularly convenient representation of a simple function, but there are many other representations. For example, your math and physics teachers are always nagging you to use graphs. With graphs you get an overall picture of a function's behavior; with graphs you can represent complicated functions for which there are no formulas.

Figure 11.1 shows the graph of a smooth, single-valued function $y(t)$ vs. t. Perhaps it describes a mass falling and rising in some weird way that no formula can describe. But formulas are quite convenient, so here is how to *approximate* with a simple formula at least part of the behavior shown in the graph.

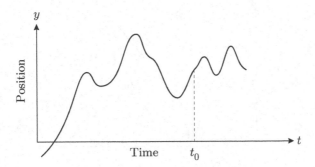

FIGURE 11.1 A graph of position as a function of time.

Straight-Line Approximations

To approximate the function look at such a small part of the curve that it is smooth and simple. The graph in Fig. 11.1 shows you that if you take a small enough part of $y(t)$ centered on some time t_0, the function will be almost a straight line. If you don't let t get very far from t_0, you can approximate $y(t)$ with a linear function, $y_{\text{line}}(t) = m\,t + b$, the equation of a straight line.

To write down the equation of the straight line that best approximates the curve in the graph around t_0, you need values for the slope m and the intercept b. You can see that the best straight-line approximation will be the line tangent to the curve $y(t)$ and touching it at the point $y(t_0)$. Figure 11.2 reminds you that for any point $y_{\text{line}}(t)$ on the line, the slope m is the constant ratio

$$m = \frac{y_{\text{line}}(t) - y(t_0)}{t - t_0},$$

so you can write the equation of the straight-line approximation at the point $(y(t_0), t_0)$ as

$$y_{\text{line}}(t) = y(t_0) + m\,(t - t_0). \tag{1}$$

■ EXERCISES

---

1. Consider the curve in Fig. 11.3. Find equations that give the best straight-line fit to the curve at points (a), (b), and (c).

You can use Eq. 1 to find a straight line to approximate a small segment of any curve, but it's particularly convenient to apply to it curves for which

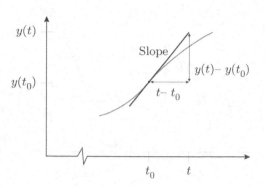

FIGURE 11.2 The best straight-line approximation to a curve near some point $(y(t_0), t_0)$ on the curve is the straight line tangent to the curve at that point.

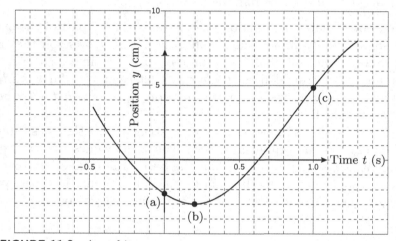

FIGURE 11.3 An arbitrary curve.

formulas—even quite complicated ones—exist. If you have a formula, you can use the fact that the slope of a curve at a particular point t_0 can be found by taking the derivative $y'(t)$ and evaluating it at that point; that is, $m = y'(t_0)$. Then you can write the approximation to the curve as

$$y(t) \approx y_{\text{line}}(t) = y(t_0) + (t - t_0)\, y'(t_0). \tag{2}$$

▼ EXAMPLES

1. Suppose the function is $y(t) = 5/t^2$. What would be the best straight-line approximation of this function at $t = 5$?

TABLE 11.1 Comparison of values of the exact and approximate versions of the function $5/t^2$

Time	Exact	Approx	Percent
t	$y(t)$	$y_{line}(t)$	difference
4	0.3125	0.2800	−10.4
4.4	0.2583	0.2480	−4.0
4.6	0.2363	0.2320	−1.8
4.8	0.2170	0.2160	−0.5
5.	0.2000	0.2000	0
5.2	0.1849	0.1840	−0.5
5.4	0.1715	0.1680	−2.0
5.6	0.1594	0.1520	−4.7

The derivative of $5/t^2$ is $-10/t^3$. The value of this derivative at $t = 5$ is -0.08. The value of the function at this point is $y(5) = 0.2$. Therefore, the equation of the straight-line approximation is

$$y_{line}(t) = 0.2 - (t - 5)\,0.08.$$

If you evaluate $y(t)$ and its approximation $y_{line}(t)$ at points in the vicinity of $t = 5$, you will get the results shown in Table 11.1. For any value of t between 4.8 and 5.2 the straight line approximation gives the same answer as the exact formula to within 0.5%. If that is precise enough, you can use the approximate formula.

■ EXERCISES

2. Use a spreadsheet to evaluate the function of Table 11.1 and its approximation. Then have the spreadsheet plot the two functions on the same graph. Choose your ranges of values and scale your plots to show clearly that the approximation works well around $t = 5$ and not so well for values more distant from 5.

3. Find the linear function that best approximates $5/t^2$ in the vicinity of $t = 10$. Plot both the exact function and the approximate one for values of t around $t = 10$.

Binomial Expansions

Functions of the form

$$y = (1 + x)^n$$

are common, and they deserve particular attention. Because the argument of the function has two terms, 1 and x, it is called a "binomial function." You will often want to approximate a binomial function around $x = 0$, so consider linear approximations for just this case.

To use Eq. 2 first find $y(0)$. For binomial functions of this form and for any value of the exponent n, $y(0) = 1$. Next find $y'(0)$. The derivative of the binomial function is

$$\frac{dy}{dx} = f'(x) = n(1 + x)^{n-1},$$

and when you evaluate this at $x = 0$ you get $f'(0) = n$ for any value of n. Inserting these values, you see that *around the origin* any binomial function of the form $y = (1+x)^n$ can be approximated by the straight-line equation

$$y = 1 + nx. \tag{3}$$

The quantity n here can have any value, integer or fractional, positive or negative. If you know Eq. 3, you then never have to do any differentiation. All you need to do is look at the exponent of the binomial. Another advantage is that it is easy to show that the approximation is quite good as long as x is much smaller than 1. If $x \leq 0.1$, the approximation will be accurate enough for most of our purposes. Equation 3 is an example of the first step of what is called the "binomial expansion."

▼ EXAMPLES

2. Find a linear approximation to the function $y = \sqrt{1 + x}$ at the origin. Recognize that $y = \sqrt{1 + x}$ is the same thing as $y = (1 + x)^{1/2}$ so that in Eq. 3 $n = \frac{1}{2}$, and the approximation is

$$y = 1 + \frac{x}{2}.$$

3. Here is the linear equation that best approximates the function $y(x) = 1/\sqrt{1 + x}$ near the origin:

$$y = \frac{1}{\sqrt{1 + x}} = (1 + x)^{-\frac{1}{2}} \approx 1 - \frac{1}{2}x$$

This example shows that n can be a negative fraction.

To use Eq. 3 you may need to manipulate a given binomial into the form $(1+x)^n$. For example, suppose the binomial is $y = \sqrt{4+8x}$. If you realize that you can factor the radical, you can rewrite it as $y = 2\sqrt{1+2x}$. Next imagine the substitution $z = 2x$, so that you get $y = 2\sqrt{1+z}$. Then apply Eq. 3 to get

$$y = 2\left(1 + \frac{1}{2}z\right) = 2 + 2x.$$

With practice you can do a lot of this in your head.

Fairly often, you may want to find an approximate function that is linear in some compound quantity. To see what this means, look at the following.

EXAMPLES

4. Suppose you have a function

$$K(p) = \sqrt{m^2 + p^2} - m,$$

where m is constant and p is the independent variable. Find an approximate formula for K when $p \ll m$. Then factor m out from the radical, and treat the compound quantity $\frac{p^2}{m^2}$ as a single entity. Because $p \ll m$, you know $\frac{p^2}{m^2}$ is small compared to 1, and you can write

$$K(p) = m\left(\sqrt{1 + \frac{p^2}{m^2}} - 1\right) \approx m\left(1 + \frac{1}{2}\frac{p^2}{m^2} - 1\right) = \frac{p^2}{2m}.$$

Not only does the above example show you an important trick for finding an approximate formula, it also illustrates one of those cases where the approximate formula is simpler than the exact one.

Amaze Your Friends!

You can use the binomial expansion to do mental arithmetic. Suppose you wish to know the square root of 1.03. You now know that this is the same as $(1 + 0.03)^{1/2}$, which has a binomial expansion of $1 + 0.03/2 = 1.015$. When you compare this to the 1.0149 you get with your calculator, you see that the approximation is very good.

▪ EXERCISES

4. What is the cube root of 1.05? Do it in your head.

5. What is the square root of 0.96? Do it in your head.

6. Use the binomial expansion to find the square root of 4.8. Do it in your head.

7. Evaluate the cube root of 0.00103. Do it in your head.

The Small-Angle Approximation

You have been using the small-angle approximation, $\sin\theta \approx \theta$ or $\tan\theta \approx \theta$ since you began taking physics (e. g., see Chap. 2). Now you can discover that it is just the best approximation of the sine function by a straight line.

It may help you to start your discovery by doing the following exercise.

▪ EXERCISES

8. Draw a graph of $y = \sin\theta$ for $-\pi/4 < \theta < \pi/4$. From the graph obtain the equation of the straight line that best approximates the graph in the vicinity of $\theta = 0$. Do your work in radians.

Now use $y(x) \approx y(t_0) + (t - t_0)\,y'(t_0)$ Eq. 2 to derive the small angle approximation.

▽ EXAMPLES

5. To apply Eq. 2 to $\sin\theta$ in the vicinity of $\theta_0 = 0$ make the following substitutions:

$$y(t_0) \longrightarrow y(\theta_0) = \sin 0 = 0,$$
$$y'(t_0) \longrightarrow y'(\theta_0) = \cos 0 = 1,$$
$$t - t_0 \longrightarrow \theta - \theta_0 = \theta - 0 = \theta.$$

So the best linear approximation to $\sin\theta$ around $\theta = 0$ is

$$\sin\theta \approx \theta,$$

which, voilà, is the small-angle approximation.

You can use this technique to produce approximations to any reasonable function.

■ EXERCISES

> **9.** Show that the best straight-line approximation to the function $y = \tan\theta$ around the point $\theta = 0$ is $y \approx \theta$.
>
> **10.** Show that the best straight-line approximation to the function $y = a\,e^x$ around the point $x = 0$ is $y \approx a(1 + x)$.
>
> **11.** Show that the best straight-line approximation to the function $y = \ln(1 + x)$ around the point $x = 0$ is $y \approx x$.

In discussing and listening to physics you will often need to know the linear approximations to $\sin\theta$, $\tan\theta$, e^x, $\ln(1 + x)$, and $(1 + x)^n$; learn them.

In this and the next chapter you will see how the approximation $(1 + x)^n = 1 + n\,x + \cdots$ shows that at low speeds Newtonian mechanics is the best approximation to the exactly correct special theory of relativity.

11.3 FRAME OF REFERENCE

The idea of a "frame of reference" is central to any discussion of motion. An object's frame of reference is the object and the collection of all things that are at rest relative to it. Thus, assuming that you are not doing anything very weird as you read this, your frame of reference right now includes yourself, your table, chair, lamp, room and its surrounding buildings and landscape.

Right away you can see that different objects can have different frames of reference. For example, if you are reading this while riding on a bus, then you, the vehicle, its seats, windows, aisle, and driver make up a reference frame. If there happens to be a car in the lane next to yours traveling at exactly the same speed, then it is also in your reference frame. But someone standing on the roadside as you drive past clearly is in a different reference frame, as are people in cars moving past your bus in either direction.

In important ways the appearance of the world depends on your choice of reference frame. If you are riding in a bus, the trees and houses move past you. If you are standing under a tree at the side of the road, you see the bus move by. If you are driving along in a car, you might see the bus approach from one direction and pass you while the tree approaches and passes you from the other direction. You can see that the velocities you measure depend on what frame of reference you choose.

Velocity Depends on Reference Frame

For everyday events we so often choose Earth as the frame of reference that we usually forget that a choice was made. If someone told you that the bus is going 60 mph, you probably would not ask what reference frame was being used. And you might chuckle if someone said that a telephone pole leaped at his car and dented the bumper. The statement is wry and ironic, but in the reference frame of the driver of the car it describes what happened.

The velocity describing an object's motion depends on what frame of reference is used.

■ EXERCISES

12. Imagine that you are standing out on the highway (see Fig. 11.4) as a bus drives by going east at 60 mph and overtakes a truck going east at 30 mph. There are three different reference frames involved here.

13. What reference frame has been used to describe this situation?

14. Describe the velocities using the reference frame of the bus.

15. What do the motions look like in the reference frame of the truck?

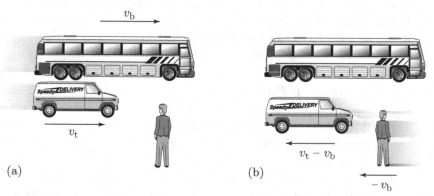

(a)   (b)

FIGURE 11.4 (a) Motion of bus and truck seen from observer's frame of reference S; (b) motion of observer and truck seen from bus's frame of reference S′.

> **16.** Simplicio says that momentum, mv, is an intrinsic property of a body like its color. What do you tell him? Give some numerical examples to illustrate your argument.

Does Physics Depend on Reference Frame?

A major historical achievement of physics is the description of how forces produce the velocities and paths followed by moving bodies. But if velocities and paths depend on the reference frame in which they are described, does this mean that the laws of physics are different in different reference frames?

You need to be careful about the word "mean" here. Clearly, the numerical values of physical quantities depend on the reference frame. If you take a 6 kg bowling ball at rest and set it rolling at 3 m/s down a bowling alley, it goes from having a kinetic energy of 0 J to having a kinetic energy of 27 J in your frame of reference. But if someone is traveling along a conveyor belt at 1 m/s beside the alley, then in her frame of reference the kinetic energy is initially 3 J, and then 12 J. In the first case the kinetic energy changes by 27 J; in the second case the change is 9 J, quite different numbers.

But a law of physics is not the same thing as the number it predicts in given circumstances. In this case the relevant law of physics is that a constant force F applied over some distance s changes the kinetic energy of a mass m by an amount equal to $F s$. The law $\Delta K = F s$ can be correct in both reference frames even if ΔK is different in the two reference frames, as long as the value of the product $F s$ also differs in the two reference frames.

Suppose the bowling ball was thrown in 0.5 s with a constant force of 36 N, so that the force was applied over a distance of 0.75 m ($s = \frac{1}{2}at^2$) in the reference frame of the bowling alley. Then in that frame of reference the change in kinetic energy would be 36 N$\times$0.75 m $=$ 27 J. In the reference frame of the conveyor belt, the force and the time of throw are the same,[1] but because the belt is moving away from the thrower at 1 m/s, the force is applied over a distance shorter by 0.5 s $\times$ 1 m/s $=$ 0.5 m. Therefore, in this frame of reference the force is applied over a distance of only 0.25 m and so produces a change in kinetic energy of 36 N $\times$ 0.25 m $=$ 9 J, which is the result obtained above. Thus, the equation $F s = \Delta K$ holds in the two different frames of reference although the actual values are different.

---

[1] Force and time behave this way in Newtonian physics, but not in Nature, which is why it is necessary to replace Newton's theory with Einstein's.

This idea that laws of physics are the same in different reference frames has led physicists to propound "the principle of relativity." The principle is most easily understood if we use only frames of reference that are in uniform, straight-line motion relative to one another. The bowling alley and the conveyor belt discussed above are good approximations of two such frames. (They are not moving in exactly straight lines, because they are all rotating around Earth's axis while going around the Sun in an ellipse.) Then experience leads us to say that the laws of physics are the same in all such frames of reference. This is the "principle of relativity."

Notice that if the laws of physics are identically the same in all uniformly moving frames of reference, then there is no experimental basis for thinking that any one frame is the "right" or "special" one. You choose a particular reference frame for describing some set of motions not because the physics will be correct in one and wrong in another, but because the chosen frame is convenient—maybe it makes calculations easier; maybe it makes interrelationships more evident.

The principle of relativity means that for physicists in enclosed laboratories moving with constant velocity relative to each other there is no experiment that can be done that will tell who is moving and who is not. *There is no special frame of reference,* no place in the universe that is absolutely at rest relative to everything else.

How Motion Described in One Frame is Described in Another

Given a description of motion in one reference frame, you often need to find its description in another. You have already transformed one description tFo another when you thought about the bus, the truck, and an observer (you). In the observer's frame of reference, call it the S frame (Fig. 11.4a), the bus was traveling with a speed of $v_b = 60$ mph, the truck with $v_t = 30$ mph, and the observer was at rest, $v_o = 0$ mph. However, in the bus's frame of reference, call it S'n (Fig. 11.4b), the velocities were $v'_b = 0$ mph, $v'_t = -30$ mph, and $v'_o = -60$ mph. You probably made the calculations without much thought about what you were doing. Now let's codify what you did.

First, notice that relative to the S frame, from which you are transforming, the S' frame, to which you are transforming, has a velocity $V = 60$ mph. (Use $+$ for eastward and $-$ for westward motion.) Then, to transform the velocities of objects in the S frame to their velocities in the S' frame, you do the following calculations:

$$v'_b = v_b - V,$$
$$v'_t = v_t - V,$$
$$v'_o = v_o - V.$$

The general rule to go from the S frame to the S' frame is to subtract the velocity of S' relative to S from the velocity of the object in S, i.e.,

$$v' = v - V. \tag{4}$$

Equation 4 is called a "transformation," because it transforms the velocity from the S frame to the S' frame.

■ EXERCISES

> **17.** Use Eq. 4 to find the velocities of the bus, truck, and observer in the frame of reference of the truck.

The most astonishing feature of Eq. 4 is that it is **WRONG**. It is an excellent approximation as long as all the velocities are small compared to the speed of light $c = 3 \times 10^8$ m/s, but *it is never exact.*

11.4   THE CONSTANCY OF c

The incorrectness of Eq. 4 follows from a most surprising fact: *The speed of light is the same in all reference frames regardless of their relative motion.*

You can see at once that this fact contradicts Eq. 4. Imagine a parked car that turns on its headlights. The light will come past you at a speed c. Now imagine that you are moving toward the car at 120 mph ($V = -120$ mph), and it turns on its lights. Equation 4 predicts that you will measure the speed of the light going past you to be greater by 120 mph. However, no change in the speed is observed experimentally, and so Eq. 4 must be wrong.

From early in the 1800s, when interference phenomena showed that light is a wave, physicists thought that light must be a disturbance in some medium and that light passed from one point to another like sound waves through a solid. As they learned more about electromagnetism, physicists deduced more properties of the medium and gave it a name. The hypothetical medium that carried light waves was called the "ether." If there were an ether and it carried light waves, then an observer moving relative to the ether would observe light moving faster or slower than c depending on whether the observer was moving toward or away from the source. For a number of reasons, particularly for logical consistency, Einstein concluded there was no ether and that c would be the same for all observers independent of the relative motions of the light source and the observers, and Eq. 4 must then be wrong. His conclusion was supported by the results of a remarkable experiment performed by the American physicists Albert A. Michelson and Edward W. Morley.

The Michelson–Morley Experiment

Although by the end of the nineteenth century most physicists thought there was some medium that supported the propagation of electromagnetic waves, there was no experimental evidence for its existence. Michelson set out to find experimental proof that the ether existed. He realized that in its orbit around the Sun, Earth must move through the ether and at some point during the year travel with at least its orbital velocity of 30 km/s relative to the ether. Under these conditions the speed of light measured along the line of Earth's motion through the ether would be slightly different from the measured speed of light traveling perpendicular to the direction of Earth's motion through the ether. Even using the high speed of Earth's motion in space, Michelson expected the effect to be small. Earth's speed relative to the Sun is only $(3 \times 10^4\,\text{m/s})/(3 \times 10^8\,\text{m/s}) = 10^{-4}$ of the speed of light, and, as you will see, the effect that could be measured was proportional to the square of this number, i.e., to 10^{-8}. Despite the smallness of the effect, Michelson expected to obtain a significant result because he had invented an ingenious instrument that could measure a difference in the travel times of two wavefronts to within a hundredth of a period of a light wave, i.e., to within 2×10^{-17} s.

Michelson Interferometer

In his work with interference of light waves, Michelson developed a device that he called the "interferometer." With it he could measure length differences as small as a few nanometers. The interferometer split light from an extended source into two wavefronts and sent them on round trips along two paths at right angles to each other and recombined them at the end of their trips. Two paths at right angles were exactly the right arrangement for testing the existence of the ether. He understood that when he measured a 6 nm difference between the lengths of the paths traveled by two wavefronts of light, he was also determining that the times for the two wavefronts to travel through the interferometer differed by $6 \times 10^{-9}/3 \times 10^8 = 2 \times 10^{-17}$ s. The capability of measuring a time interval so small meant that he could hope to measure the very small difference in the travel times expected for two wavefronts of light traveling along paths at right angles to each other in the ether.

Figure 11.5 shows a diagram of a Michelson interferometer. Its essential features are the two arms of equal length L at right angles to each other, the partially silvered mirror (G_1) that splits the incoming wavefront of light into two wavefronts, and the fully reflecting mirrors (M_1 and M_2) at

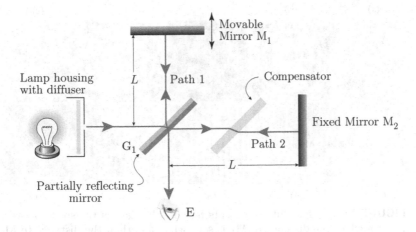

FIGURE 11.5 Schematic diagram of a Michelson interferometer.

the ends of the arms. To the left there is a source light that is arranged to produce a fairly flat wavefront (usually of a nearly pure single wavelength) that passes into the interferometer. G_1 produces a wavefront in each arm, one by reflecting half of the light into Path 1, and the other by transmitting half of the light along Path 2. The two wavefronts then go along their respective arms separately and are reflected by mirrors M_1 and M_2. When the wavefronts arrive back at G_1, half of each is reflected and half is transmitted. The result is that an observer E receives a superposition of two wavefronts, one that has made a round trip along Path 1 and one that has made a round trip along Path 2.

In an ideal case the two wavefronts would travel exactly the same distance and return to G_1, recombine exactly parallel to each other, and move on to the detector E. Under these circumstances the two wavefronts would be π radians out of phase over the whole field of view of the detector E, and a viewer would see a uniform darkness like that in Fig. 11.6c.[2] The ideal situation is hard to achieve, and usually the phase difference varies from constructive interference to destructive interference several times across the field of view. The result is a pattern of dark and light rings or bands called "fringes," like those shown in the other panels of Fig. 11.6.

---

[2]You might think that the two wavefronts should be exactly *in* phase, but a light wave traveling through air undergoes a phase change of about π radians when it reflects from a glass (or metal) surface; but when it is traveling through glass and reflects from a surface with air on the other side it undergoes very little phase change. In the Michelson interferometer one wave makes two reflections at air-to-glass interfaces, and the other wave makes one. As a result, the two wavefronts arrive at the detector essentially π radians out of phase when the two arms of the interferometer are exactly equal.

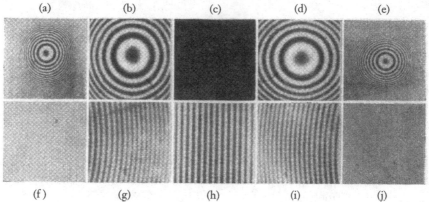

FIGURE 11.6   The white represents fringes; the dark is the space between them. (a) Fringes when the distance to M_1 is somewhat less than the distance to M_2; (b) when the distances are nearly the same (reflection from the beam splitter changes the phase by 180°, leading to destructive interference); (c) when the distances are exactly the same; (d) when the distance to M_1 is a bit more than the distance to M_2; (e) when the distance to M_1 is somewhat more than the distance to M_2; (f)–(j) are for the same mirror separations respectively as (a)–(e), except that now the mirrors are not parallel but slightly tilted relative to one another. *Taken with permission of the McGraw-Hill Companies from F.A. Jenkins and H.E. White,* Fundamentals of Optics *2nd edition, McGraw-Hill, 1950.*

Whatever the actual lengths of the two paths, light from the wavefronts from the two arms arriving in phase at the detector produces bright places in the interference pattern. If the length of the path through one arm is then changed by $\lambda/2$, the bright parts of the pattern will become dark and the dark parts will become bright—the pattern will shift. To change the path length by $\lambda/2$, you move a mirror half that distance to change L by $\lambda/4$, because the light makes a round trip between the beam splitter G_1 and the mirror. The interferometer is thus an extremely sensitive instrument, able to reveal length changes as small as 0.01 of the wavelength of light, i.e., five nanometers.[3]

A Moving Interferometer

If the speed of light c were constant relative to the ether—as the speed of sound is constant relative to the air through which it moves—Michelson knew that in an interferometer moving through the ether, the time t_1 for

---

[3]Since Michelson's time, the sensitivity of interferometers has been increased to the point that scientists searching for gravity waves have built an interferometer that can detect motion on the order of attometers (10^{-18} m). For more information read about LIGO on the web.

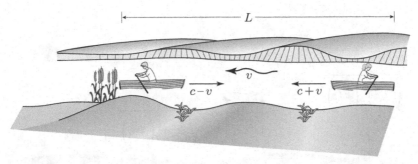

FIGURE 11.7 Rowing with a speed c relative to the water, your velocity upstream against a current moving with speed v will be $c - v$ relative to the bank; downstream it will be $c + v$.

light to complete a round trip along Path 1 would be different from the time t_2 to make a round trip along Path 2. To see why this would be so, consider the analogy of rowing a boat in a river. Figure 11.7 illustrates the situation. Assume you row your boat with a steady speed c relative to still water (the analog of the ether), and take v to be the speed of the river relative to its banks.

How long will it take you to row your boat upstream a distance L along the riverbank? Relative to the bank your speed will be $c - v$, so it will take you a time $t_u = \frac{L}{c-v}$ to travel upstream the distance L. Now turn around and row back. Your rowing speed relative to the water is still c, but now relative to the bank your speed is $c + v$ so it will take you only a time $t_d = \frac{L}{c+v}$ to get back downstream to your starting point. Your total travel time against the current and then with the current will be

$$t_2 = t_u + t_d = \frac{L}{c-v} + \frac{L}{c+v} = \frac{2Lc}{c^2 - v^2} = \frac{2L}{c}\frac{1}{\left(1 - \frac{v^2}{c^2}\right)}. \tag{5}$$

The factor $2L/c$ is the time it would take you to row the round trip in still water ($v = 0$).

Now suppose you row your boat a distance L directly across the stream and back. To reach a point directly across the stream, you must row your boat heading upstream. Suppose it takes you t seconds to row across. If, as you start out, there is a float in the water on the far side of the river and a distance vt upstream, and you row heading toward the float, you will reach the point opposite your starting point just at the same time the float does. This is the same as rowing into the current so that a component of your speed c cancels the v of the water. The result is that, although relative to the banks of the river you travel a distance L across the river, in the reference frame of the water you travel a distance greater than L. In Fig. 11.8 the dotted lines indicate the path of the boat relative

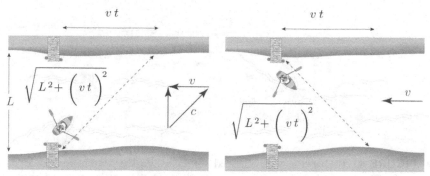

FIGURE 11.8 To row your boat to a point directly opposite on a stream moving with speed v, you must row upstream at an angle and row a distance relative to the water that is the hypotenuse of a triangle as shown.

to the water. Rowing to reach the other side, you will have to travel a distance vt across the current. This distance is at right angles to the L that you travel across the river, so the total distance you travel relative to the water is $\sqrt{L^2 + (vt)^2}$. Because your speed relative to the water is c, the time it takes you to travel the distance relative to the water is $t = \frac{\sqrt{L^2 + v^2 t^2}}{c}$. You can solve this equation for t and multiply by 2 to get the round trip time:

$$t_1 = 2t = \frac{2L}{\sqrt{c^2 - v^2}} = \frac{2L}{c} \frac{1}{\sqrt{1 - \frac{v^2}{c^2}}}. \tag{6}$$

■ EXERCISES

18. Show how Eq. 6 follows from the discussion in the preceding paragraph.

19. Show that if you direct your boat with speed c heading into the current just enough to cancel the effect of the flowing current v, then your velocity perpendicular to the banks will be $\sqrt{c^2 - v^2}$. This is a slightly different way to get the same result as Eq. 6.

Comparing Eqs. 5 and 6 you can see that the time t_1 to row a distance $2L$ across the river and back, perpendicular to the river bank, is different from the time t_2 to row the same distance upstream and return parallel to the river banks. If two boats start out together, one rowing across and back and the other rowing upstream and down, they do not return to their starting point at the same time even though they row at the same speed c relative to the water and travel the same distance $2L$ relative to the banks. If light travels at a speed c in some ether, the same would be true for wavefronts of light traveling through an ether stream. The difference in the two travel times is what Michelson expected to observe for wavefronts of light passing through his interferometer.

■ EXERCISES

---

20. Explain how Eqs. 5 and 6 show that it takes longer to go a distance L upstream and return than to go the same distance across the river and back.

In the rowboat analogy the boat traveling with speed c in still water stands for a wavefront of light traveling with speed c in stationary ether; the river banks represent the interferometer; and the water traveling with speed v past the river banks is the analog of the ether flowing through the interferometer with speed v because of Earth's motion through space. A boat moving up and downstream corresponds to a wavefront moving in one arm of the interferometer; a boat moving across the river is the analog of a wavefront moving in the other arm.

The beam splitter G_1 produces two wavefronts at the same instant of time. With the interferometer properly aligned with Earth's direction of motion, one wavefront moves upstream and downstream between the beam splitter G_1 and mirror M_2 along Path 2 in Fig. 11.5. The other moves along Path 1 perpendicular to the ether flow. Along the arm of Path 2 towards the mirror the speed of light relative to the interferometer should be $c - v$ Eq. 4. Coming back, the light is moving with the stream, and its velocity relative to the interferometer should be $c + v$. The total time t_2 for a round trip of a wavefront along Path 2 of the interferometer is then given by Eq. 5. The total time t_1 for a round trip of a wavefront along Path 1 perpendicular to the direction of Earth's motion through the ether is given by Eq. 6.

Δt, the difference between the times the wavefronts take to travel $2L$, is the difference between Eqs. 5 and 6:

$$\Delta t = t_2 - t_1 = \frac{2L}{c} \left\{ \frac{1}{1 - \frac{v^2}{c^2}} - \frac{1}{\sqrt{1 - \frac{v^2}{c^2}}} \right\}.$$

This expression for Δt gives a correct result, but an approximate expression obtained using the binomial approximation is more convenient and more informative:

$$\Delta t = \frac{2L}{c} \left\{ \left[1 - \left(\frac{v}{c}\right)^2 \right]^{-1} - \left[1 - \left(\frac{v}{c}\right)^2 \right]^{-1/2} \right\}$$

$$\approx \frac{2L}{c} \left\{ \left[1 + \left(\frac{v}{c}\right)^2 \right] - \left[1 + \frac{1}{2}\left(\frac{v}{c}\right)^2 \right] + \cdots \right\}$$

$$\approx \frac{2L}{c} \frac{1}{2} \left(\frac{v}{c}\right)^2. \tag{7}$$

The quantity $2L/c$ is the round trip time for light to travel through the interferometer when there is *no* motion relative to the ether.

When the two wavefronts reach the output of the interferometer, they form a pattern of interference fringes of the sort shown in Fig. 11.6. If one wavefront reaches the observer at a time slightly different from the other, the fringes in the pattern shift their positions because a difference in arrival times corresponds to a change in the phase difference between the wavefronts. For example, if the period of the light wave is T and if one wavefront is delayed by $\Delta t = T/4$ relative to the other, their phase difference will change by $1/4$ of 2π radians, and the pattern of fringes will shift in position by $1/4$ of the distance between two adjacent fringes. The amount of this shift of the fringes depends on the size of the difference in the round trip times of the wavefronts.

It is from Eq. 7 that Michelson knew, as was mentioned on pg. 352, that he was looking for an effect proportional to $v^2/c^2 \approx 10^{-8}$. This is small, and inserting some numbers makes the challenge of his experiment evident. In an early version of his apparatus, he used yellow light with $\lambda = 589$ nm; L was ~ 1.2 m; and v, the orbital speed of Earth, was 30×10^3 m/s. Thus

$$\Delta t = \frac{2.4}{3 \times 10^8} \frac{1}{2} \left(\frac{3 \times 10^4}{3 \times 10^8}\right)^2 = 0.40 \times 10^{-16} \text{ s}.$$

Light with $\lambda = 589$ nm has a period $T = \lambda/c = 19.6 \times 10^{-16}$ s, so a time difference of $\Delta t = 0.4 \times 10^{-16}$ s is ~ 0.02 of the period T and corresponds to a change of 0.02 of 2π radians in the phase difference between the two wavefronts. Consequently, Michelson expected to observe that the pattern of fringes had shifted by 0.02 of the distance between adjacent fringes from where the pattern would occur if there were no ether. He believed his apparatus was precise enough to reveal a shift as small as 0.01 of the distance separating adjacent fringes. This small shift corresponds to 0.01 of a period, i.e., 2.0×10^{-17} s or, equivalently, 0.01 of a wavelength, i.e., 6 nm.

Michelson did not need to know where the pattern of fringes would have been if it had not been shifted by motion relative to the ether, because he measured the fringe pattern and then rotated his entire apparatus through 90°. This rotation reversed the roles of the two arms. If in the initial orientation Path 2 was oriented parallel to the motion relative to the ether, then the travel time of a wavefront along Path 2 would be $\frac{2L}{c}\frac{1}{2}\frac{v^2}{c^2}$ more than along Path 1. After the interferometer was rotated, the longer travel time would be along Path 1, and the fringe pattern would shift by twice the amount predicted by Eq. 7. He expected that when the interferometer was rotated, the fringe pattern would shift by $0.04 = 1/25$ of the distance separating adjacent fringes. This is four times larger than the smallest shift that Michelson thought he could measure with his apparatus.

■ EXERCISES

> **21.** Suppose by an improved design Michelson could lengthen the arms of the interferometer from 1 m to 10 m. What then would be the predicted shift in fringes when the interferometer was rotated through 90°?

Michelson's Results

Michelson's first experiments did not show any detectable fringe shift. Although he was sure that his apparatus should have revealed the expected shift, he and Morley built a larger interferometer, shown in Fig. 11.9, that would produce a larger fringe shift. In the improved

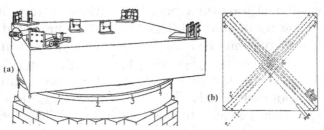

FIGURE 11.9 (a) The interferometer used by Michelson and Morley in 1886. The apparatus sits on a stone base floating in mercury. This makes it easy to rotate the interferometer through 90°. (b) A diagram of how the arm length L was lengthened by multiple reflections. *Taken from A.A. Michelson and E.W. Morley, Am. J. Sci. Vol. 34 No. 203, 333–345 (1887) as reproduced in* Selected Papers of Great American Physicists, *S.R. Weart, editor, published by the American Institute of Physics, ©1976 The American Physical Society.*

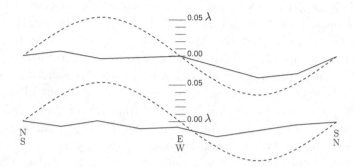

FIGURE 11.10 The solid lines are the observed fringe shifts vs. compass heading of the interferometer. The dotted curves are $\frac{1}{8}$ of the expected fringe shifts. *Taken from A.A. Michelson and E.W. Morley, Am. J. Sci. Vol. 34 No. 203, 333–345 (1887) as reproduced in* Selected Papers of Great American Physicists, *S.R. Weart, editor, published by the American Institute of Physics,* ©*1976 The American Physical Society.*

apparatus the mirrors were arranged so that the path of the light in each arm was lengthened by reflecting back and forth several times before the wavefronts combined. To permit easy rotation without mechanical distortion, they mounted the interferometer on a large stone base and floated it in mercury. With the effective arm length L increased to about $11\,\mathrm{m}$, the total fringe shift should have been 0.4 of a fringe. The experiment was carried out in 1886. Figure 11.10 shows their published results; there was no fringe shift.

"Now what?" he probably said to himself. Well, possibly the motion of the Sun through the Galaxy was canceling Earth's motion, producing zero velocity relative to the ether. If so, then wait half a year until Earth was on the other side of the Sun, and see double the effect. No luck there either. They tried many other possibilities, including checking that the arms did not change length by thermal expansion. In all cases the result of the experiment was still a "null." The fringe pattern did not shift.

This is the result that Einstein would have predicted nineteen years later. As already mentioned, for reasons of consistency and logic he concluded that there is no ether and that c is the same for all observers independent of how fast or slow they are moving relative to the light source. It is not clear whether Einstein was aware of Michelson's and Morley's result when he came to this conclusion.[4] Whatever the case, their result is strong evidence for Einstein's conclusion. Michelson and Morley

---

[4]Einstein later wrote that he first began thinking about this problem when he was twelve years old and wondered: What does a light wave look like to someone traveling with it at or close to the speed of light? What were you thinking about when you were twelve?

saw no fringe shift because the speed of light c is the same in all frames of reference. Consequently, it was c along each arm of the interferometer, and rotating the interferometer had no effect on the fringe pattern.

11.5   CONSEQUENCES OF CONSTANCY OF c

Einstein realized that the constancy of c and the principle of relativity could both hold only if space and time were interlinked in ways quite strange to Newtonian physics. He worked out a complete and consistent theory that accurately describes the motions and interactions of matter at high velocities as well as at low.

Moving Clocks Run Slow—Time Dilation

It follows from the constancy of c that the time interval between two events depends on the frame of reference from which they are observed. Einstein's theory correctly describes this behavior. To see that this behavior must occur, consider the "event generator"[5] shown in Fig. 11.11.

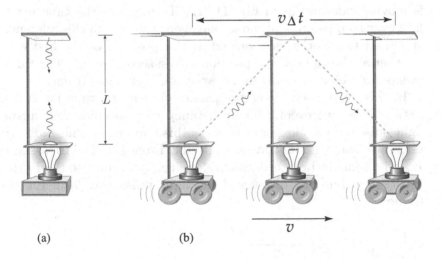

(a)   (b)

FIGURE 11.11   This event generator consists of two mirrors separated by a length L. (a) The system is stationary with respect to the textbook page; (b) the system is moving with velocity v with respect to the page.

---

[5]Sometimes called a "light clock."

This hypothetical device works by the emission of a pulse of light from near the bottom mirror (event 1). The pulse travels to the upper mirror, is reflected, and returns to the bottom one (event 2).

Descriptions of Einstein's theories often refer to "events." An event is any observable happening that occurs at a certain place and time. Each return of a light pulse to the bottom mirror is an event. (So is the arrival of the hand at some number on the face of a wall clock.) To measure the time of an event, it is convenient to imagine there are a clock and a recorder wherever the event occurs. A collection of such recorders and synchronized clocks at rest relative to each other is a reference frame.

Any recorder and clock at rest with respect to the event generator (Fig. 11.11a) will measure the elapsed time between events 1 and 2 to be:

$$\Delta t_{1\text{clock}} = \frac{2L}{c}$$

because the light pulse travels a distance $2L$. The label "1clock" is to remind you that this time interval is measured in the "one-clock" reference frame, the frame in which you need only one recorder and one clock to measure the time between events because both events occur at the same point in space.

Now imagine yourself in a reference frame in which the event generator is moving sideways as in Fig. 11.11b. To measure the time between the first event (emission) and the second event (return) in this reference frame, you need two clocks, one located at the position of the first event, and another at the right at the position of the second event. The times of the events are being recorded in a "two-clock" reference frame.

In this "two-clock" reference frame what is the time interval $\Delta t_{2\text{clock}}$ between the two events? The constancy of c specifies your answer. The light pulse travels a longer distance in this reference frame, but, as in every reference frame, it still travels at speed c. From Fig. 11.11b, assuming that L is the same in both frames of reference,[6] you can use the Pythagorean theorem to find the distance d the light pulse travels in the two-clock frame:

$$\frac{d}{2} = \sqrt{(L)^2 + \left(\frac{v\,\Delta t_{2\text{clock}}}{2}\right)^2},$$

and then you can use the fact that $d = c\,\Delta t_{2\text{clock}}$, and solve these equations and get:

---

[6]L is the same in both reference frames as long as it is perpendicular to the direction of their relative motion.

$$\Delta t_{2\text{clock}} = \frac{2L}{c} \frac{1}{\sqrt{1 - \frac{v^2}{c^2}}}. \tag{8}$$

The fraction $\frac{1}{\sqrt{1-v^2/c^2}}$ shows up so often in relativity that it is customarily given its own symbol, the lower case Greek letter gamma:

$$\gamma \equiv \frac{1}{\sqrt{1 - \frac{v^2}{c^2}}}.$$

You will often see Eq. 8 written as

$$\Delta t_{2\text{clock}} = \gamma \Delta t_{1\text{clock}}. \qquad \text{Time Dilation} \tag{9}$$

Notice that because v is always less than c, γ is always ≥ 1. The quantity $\gamma = 1$ when $v = 0$ and then increases toward ∞ as v approaches c. Consequently the time interval between two events is *always* longer in any two-clock frame than in the one-clock frame. This is a general result. For *any* pair of events, the time interval between them will be shortest in a reference frame in which the two events occur at the same place where their times can be measured with just one clock.

For any pair of events, there's an unlimited number of reference frames in which two clocks are necessary to record their times, but there is only a single one-clock frame. The time measured in this one-clock reference frame, $\Delta t_{1\text{clock}}$, is called "proper time."

Now you can understand what it means to say that "moving clocks run slow." A clock is an event generator; for instance, one event might be the second hand pointing at 12, and a second event, 10 s later, the second hand pointing at 2. If this clock is moving with respect to your frame of reference, you will need two recorders and two clocks in your reference frame to measure the time elapsed between the moving clock's two events.[7] As the sweep hand moves from 12 to 2 on the face of the moving clock, the sweep hands of your two clocks will go beyond 2. The measured time interval in your two-clock frame is always longer than the time interval that passes on the face of the moving clock; the moving clock is running behind the clocks in the two-clock frame. It is running slow in the two-clock frame. This is a general result.

---

[7]You may argue that, no, you only need one clock because you can sit in one place and still see both events on the moving clock. It turns out that the corrections you need to make to allow for the fact that light from at least one of the two events must travel some distance to your eyes (recorder) are equivalent to using two clocks located at the points in space where the events occur.

▼ EXAMPLES

6. Suppose a clock is moving past you at $v = \frac{\sqrt{3}}{2}c$. How many minutes would go by on clocks in your two-clock reference frame as 1 min goes by on the moving clock?

Find the value of $\gamma = \frac{1}{\sqrt{1-v^2/c^2}}$. For $v = \frac{\sqrt{3}}{2}c$,

$$\gamma = \frac{1}{\sqrt{1-\frac{3}{4}}} = 2.$$

Since $\Delta t_{2clock} = \gamma \Delta t_{1clock}$, 2 min will go by on the clocks in your reference frame while 1 min goes by on the moving clock.

The phenomenon of moving clocks running slow is called "time dilation." Time dilation is a property of time and space, not of clocks, and it has real physical consequences. The time intervals between all pairs of events—heart beats, the passages of a satellite overhead, creation and decay of radioactive atoms, the successive swings of a pendulum, the rotation of a wheel—are longer in two-clock frames of reference than in one-clock frames. For example, imagine a can of radioactive atoms such that half of them disintegrate every $1\,\mu$s. If you form the atoms into a beam and make them move past you at $v = \frac{\sqrt{3}}{2}c$, how long will it take for half of them to disintegrate? Because $\gamma = 2$, in your frame of reference you will find that it takes $2\,\mu$s for half of the particles in the moving beam to disintegrate. Their rate of decay is slower in your frame of reference.

▼ EXAMPLES

7. To see why no one noticed time dilation until the twentieth century, calculate the time dilation after an hour has elapsed on the clock in an automobile moving past you at 30 m/s (67 mph). How much would it be after a year?

To answer this question you must find γ, and you may face a difficulty that the binomial expansion can deal with. The ratio $v/c = 10^{-7}$, so $\gamma = 1/\sqrt{1-10^{-14}}$. If you try to calculate γ with your hand-held calculator, you are likely to get $\gamma = 1.000000000$. Fortunately, you know about the binomial expansion, so you can write down the answer without using any calculator:

$$\gamma = 1 + \frac{1}{2} \times 10^{-14},$$

The car's clock will lag behind your measuring clock by 5×10^{-15} times the elapsed time. This would be

$$(3600\,\text{s})(5 \times 10^{-15}) = 18 \times 10^{-12}\,\text{s}$$

after an hour. It would be $15.8 \times 10^{-8}\,\text{s}$ after a year. There are now clocks good enough to measure such small time differences, but they are not usually in automobiles.

■ EXERCISES

22. Suppose a clock moves past you at $v = 0.8\,c$. How far will it travel in 1 s (in your reference frame)? How much time will elapse on the traveling clock during that time?

23. What would be the time dilation of a clock moving at 30 km/s relative to Earth?

Time dilation may seem strange enough, but here is something else to think about. Imagine that you switch reference frames, so that you are in the frame of reference of the moving clock. Now the clocks of your former frame of reference are moving with velocity $-v$. What does the principle of relativity tell you about the behavior of those moving clocks when you measure their elapsed time using the clocks of your newly adopted reference frame?

The principle of relativity says—quite correctly—that observers in either frame will measure the clocks of the other frame running slow. How can that be? The key idea is that clocks that are synchronized in one reference frame are not synchronized in another reference frame. Take some time and think about it. Try doing Problems 6 and 8.

Moving Lengths Shrink—Lorentz Contraction

In relativity, length and time are intimately connected. Therefore, once you know that a moving clock runs more slowly than identical stationary clocks, it is not so surprising to find that moving lengths differ from stationary lengths. For example, a rod that is 1 m long at rest shortens in the direction of its motion; if it is moving at $\frac{\sqrt{3}}{2}\,c$, it will be half as long as when at rest.

This remarkable behavior follows from the constancy of the speed of light for all observers. To see that a moving rod must shorten, consider a

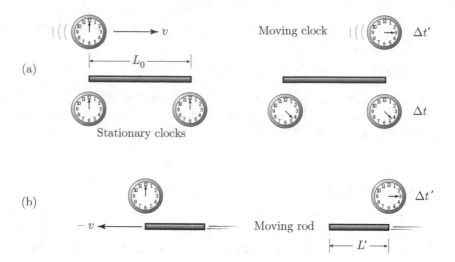

FIGURE 11.12 Measuring the length of a rod moving in the direction of its length. In (a) the length of a stationary rod is found by measuring the time $\Delta t_{2\text{clock}}$ it takes an object (here a clock) to travel from one end of the rod to the other at speed v. In (b) an observer attached to the moving clock sees the rod move past with speed $-v$ and finds its length by measuring the time $\Delta t_{1\text{clock}}$ the rod takes to pass his clock.

rod of length L_0 measured at rest and three recording clocks, two at rest relative to the rod and the other moving past it with a velocity v as shown in (a) and (b) of Fig. 11.12. How does each set of observers measure the length of the rod?

To measure the length of a rod at rest you can send an object from one end of the rod to the other at a constant speed, call it v, and measure the time $\Delta t_{2\text{clock}}$ that it takes to make the trip. The time interval $\Delta t_{2\text{clock}}$ is the time between event 1 when the object lines up with one end of the rod, and event 2 when it lines up with the other end of the rod. The length of the rod is then

$$L_0 = v\,\Delta t_{2\text{clock}},$$

where $\Delta t_{2\text{clock}}$ is the time interval measured by two clocks synchronized in the rod's reference frame and positioned one at each end of the rod. The subscript 0 indicates that this is the length of the rod in the (only) reference frame in which it is at rest. The measurement process is illustrated in Fig. 11.12a, where for the sake of the next step in this argument the object moving past the rod has been chosen to be a clock.

In the reference frame of this third clock, the rod is flying past with speed $-v$. This single clock is able to record the time of both events, since both occur at the position of the clock. Therefore, it will measure the length of the rod to be

$$L = v\,\Delta t_{1\text{clock}}.$$

From Eq. 9 you know that $\Delta t_{2\text{clock}} = \gamma \Delta t_{1\text{clock}}$. Combine this with the expressions for L and L_0 to get

$$L = \frac{L_0}{\gamma}. \qquad \text{Lorentz Contraction} \qquad (10)$$

This result says that a rod (or any object) is longest when measured in its own reference frame. This longest length is called its "proper length." When measured in any reference frame in which it is moving, the rod is shorter by $1/\gamma$. This shortening is called the "Lorentz contraction" after the physicist who first considered its possibility. Like all other predictions of Einstein's special theory of relativity, it is a consequence of the constancy of c for all observers and the principle of relativity.

Just as time dilation went unnoticed for so long, so too did Lorentz contraction, and for the same reason: the speed of light is so much larger than the speeds of objects familiar to humans that for us γ is usually very close to 1, and the effect is not observed.

▼ EXAMPLES

8. An example similar to Example 7 shows how small the Lorentz contraction is for everyday objects. What is the Lorentz contraction of a 5 m long automobile traveling at 30 m/s? First find the value of γ or just use the calculation in Example 7:

$$\gamma = \frac{1}{\sqrt{1 - (\frac{v}{c})^2}} \approx 1 + \frac{1}{2}\frac{v^2}{c^2} + \cdots$$

and use the fact that $v/c = 10^{-7}$, so $\gamma = 1 + 5 \times 10^{-15}$. From this it follows that the change in length of the speeding car is $(5 \times 5 \times 10^{-15}) = 25 \times 10^{-15}$ m, i.e., 25 fm. This is about the diameter of an atomic nucleus and is 10^7 times smaller than the wavelength of visible light. Clearly, the effect is not going to be very noticeable at speeds you experience in everyday life.

▦ EXERCISES

24. What would an observer find to be the Lorentz contraction of a 100 m long space ship passing by at 30 km/s?

The Doppler Effect

It is remarkable that any observer measures the speed of light from any source to be c regardless of how fast the source and observer are moving towards each other. But isn't there some effect of the relative motion? Yes, observers moving with different speeds relative to a light source will measure different frequencies (or wavelengths) of the same light. When a source and a detector are moving toward each other, the detector measures a higher frequency (shorter wavelength) than when they are at rest relative to each other; when they are moving away from each other, the detector measures a lower frequency (longer wavelength) than when they are at rest. This dependence of frequency on the relative motion of source and observer is called the "Doppler effect."

That such an effect must occur follows from time dilation. The derivation of the Doppler effect is a good way to practice your understanding of the ideas of the special theory of relativity. So consider the case in which the source and a detector are moving directly toward each other.

Assume that when a source and a detector are at rest relative to each other, as shown in Fig. 11.13a, the source produces a periodic wave of frequency f_0. The source then emits crests separated in time by $T_{1\mathrm{clock}} = 1/f_0$, and a detector stationary relative to the source records the arrival of crests separated in time by $T_{1\mathrm{clock}}$.

But what if the source and detector are moving toward each other at some speed v? The time interval between the arrival of two successive crests recorded by the detector will change for two reasons. First, there

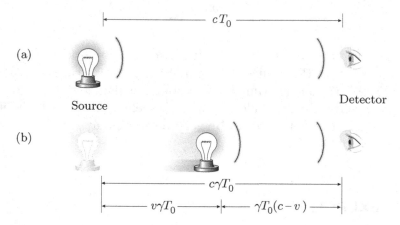

FIGURE 11.13 (a) The distance between crests of a wave moving between a stationary source and detector; (b) the distance between the crests when the source is moving toward the detector with speed v.

will be time dilation. If a time $T_{1\text{clock}}$ elapses on the clock of the source, a time $T_{2\text{clock}} = \gamma T_{1\text{clock}}$ will elapse on clocks in the detector's reference frame.

But $T_{2\text{clock}}$ is not the time interval T_{detector} between crests arriving at the detector, because each crest travels a shorter distance to reach the detector than the preceding crest, and it arrives sooner than if there was no relative motion between source and detector.

In the reference frame of the detector, the successive crests are emitted at time intervals of $\gamma T_{1\text{clock}}$, as shown in Fig. 11.13b. During this time the source moves closer to the detector by a distance $v\gamma T_{1\text{clock}}$, so the crests, instead of being separated by a distance $c\gamma T_{1\text{clock}}$, are separated by a distance $(c - v)\gamma T_{1\text{clock}}$. The time separation between these crests is this distance divided by c (because light waves travel at speed c in every reference frame):

$$T_{\text{detector}} = \frac{(c - v)\gamma T_{1\text{clock}}}{c} = \left(1 - \frac{v}{c}\right)\gamma T_{1\text{clock}}.$$

This expression can be simplified by using $\gamma = 1/\sqrt{1 - v^2/c^2} = 1/\sqrt{(1 - v/c)(1 + v/c)}$. Also, to make explicit that $T_{1\text{clock}}$ is the period of the emitted light in the source's rest frame, we relabel $T_{1\text{clock}}$ as T_{source}:

$$T_{\text{detector}} = \left(1 - \frac{v}{c}\right)\sqrt{\frac{1}{(1 - \frac{v}{c})(1 + \frac{v}{c})}}\, T_{\text{source}}$$

$$= \sqrt{\frac{(1 - \frac{v}{c})^2}{(1 - \frac{v}{c})(1 + \frac{v}{c})}}\, T_{\text{source}} = \sqrt{\frac{1 - \frac{v}{c}}{1 + \frac{v}{c}}}\, T_{\text{source}}.$$

To see what this means for frequencies use the fact that $f = 1/T$ and take the reciprocals of both sides of the equation. This gives

$$f_{\text{detector}} = \sqrt{\frac{1 + \frac{v}{c}}{1 - \frac{v}{c}}}\, f_{\text{source}}. \qquad \text{Doppler shift (approaching)} \qquad (11)$$

■ EXERCISES

25. Derive the relationship between f_{source} and f_{detector} if the source and observer are moving directly away from each other.

26. Distant galaxies are observed to emit spectra that have the same patterns as well-recognized atomic spectra but with their wavelengths all shifted toward the red. What does this tell you about the motion of distant galaxies relative to Earth?

How Do Velocities Transform?

This chapter began with a bus, a truck, and an observer moving in different frames of reference. Given the velocity v of an object in one frame, we transformed v to its velocity v' in another. The method of transforming velocities in one reference frame to another is summarized in Eq. 4. Although the method seems intuitively obvious, it must be wrong. It predicts that observers in different reference frames should see a given pulse of light travel with different speeds, but experiments show that light travels with the same speed for all observers regardless of their relative motions.

If Eq. 4 is wrong, what is correct? Einstein showed that if an object is moving with a speed v in one reference frame, then in a frame moving with a speed V relative to the first, the object will have a speed v' given by the relation

$$v' = \frac{v - V}{1 - \frac{vV}{c^2}}. \qquad (12)$$

This example, as all our examples so far, is for motion in one dimension.

Notice that this equation is like Eq. 4 except for the denominator. And notice that this denominator is not going to be significantly different from 1 unless v and V both begin to approach c.

■ EXERCISES

---

27. Suppose you measure the speed of a pulse of light. You find $v = c$. What would an observer in a frame moving with a speed $V = \frac{\sqrt{3}}{2} c$ find for the velocity v' of this light pulse? You know the answer, but show that Eq. 12 gives it.

28. When you calculated the speed of the bus in the reference frame of the truck (p. 351), you got $+30\,\text{mph}$. Calculate the difference between this answer and the relativistically correct answer.

Something to Think About

Let's conclude this section with a mindbender. Here is how to put a 10 m long pole into a barn that is only 5 m wide. Imagine a pole 10 m long when at rest moving with a speed of $\frac{\sqrt{3}}{2}\,c$ towards the open door of the 5 m wide barn, as shown in Fig. 11.14a. For this speed (as you should now begin to know by heart) $\gamma = 2$. This means that the pole is contracted to a length of 5 m in the frame of reference of the barn. Will it fit in the barn even ever so briefly before crashing into the back wall?

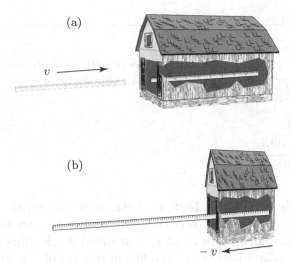

FIGURE 11.14 A 10 m pole and a 5 m barn with a relative speed such that $\gamma = 2$: (a) in the frame of reference of the barn; (b) in the frame of reference of the pole.

You can imagine a farmer standing by the barn door. Setting aside certain practicalities of reaction time, inertia of the door, air resistance, and how do you get a pole moving that fast in the first place, you can see that it is in principle possible to shut the door of the barn before the pole goes into the back wall. (If you want a little extra time for closing the door, have the pole move a little faster, so that it is a little shorter than 5 m.)

But to really bend your mind, imagine riding on the pole so that you are in the pole's frame of reference. Now the pole is 10 m long and the barn is moving toward you at $\frac{\sqrt{3}}{2} c$, so it is Lorentz contracted by $\gamma = 2$ to 2.5 m. This situation is shown in Fig. 11.14b. Will the pole fit in the barn? How do you reconcile these two quite different pictures of what is likely to happen? Be assured that the special theory of relativity says that these two different views of the pole entering the barn are correct and consistent.

The crux of the solution is that observers in the two different frames of reference will predict and observe different *sequences* of events: The farmer's clocks will show the barn door closed *before* his clocks show the rod crashed into the wall; the rider's clock will show that the door closed *after* the rod crashed into the wall. However, it would be too great a digression to explore how clocks synchronized in one reference frame are not synchronized in another, with the result that events simultaneous in one frame of reference are not simultaneous in another, and we leave the pole-in-the-barn problem for you to ponder on your own.

PROBLEMS

1. Evaluate each of the following without a calculator and using the kind of series approximation discussed in this chapter.
    a. $(1.02)^2$.
    b. $\frac{1}{0.96}$.
    c. $\sqrt[3]{1.09}$.
    d. $\sqrt{0.0408}$.

2. Michelson showed that if there were an ether, the hypothetical medium in which light was thought to travel, then light from a source moving with speed v through this medium would take a time $t_\parallel$ to travel a distance L and back along the line of motion of the source; and it would take a time $t_\perp$ to travel a distance L perpendicular to the line of motion, where

$$t_\parallel = \frac{2L}{c}\frac{1}{1 - \frac{v^2}{c^2}},$$

$$t_\perp = \frac{2L}{c}\frac{1}{\sqrt{1 - \frac{v^2}{c^2}}}.$$

Show, using the expansion technique, that $t_\parallel$ is greater than $t_\perp$.

3. The He–Ne laser commonly used in physics laboratories emits red light with a wavelength of $\lambda = 633\,\text{nm}$.
    a. What is the frequency of that light?
    b. Given that the time for light to travel a distance L parallel to Earth's motion through a hypothetical ether is

$$t_\parallel = \frac{L}{c - v} + \frac{L}{c + v} = \frac{2L}{c\left(1 - \frac{v^2}{c^2}\right)},$$

    while its time to travel perpendicular to Earth's motion is

$$t_\perp = \frac{2L}{c\sqrt{1 - \frac{v^2}{c^2}}},$$

    use the binomial expansion to find an approximate formula for $t_\parallel - t_\perp$.

   c. Find the value of $t_\parallel - t_\perp$ for Earth's motion of $v = 30\,\text{km/s}$ if $L = 1\,\text{m}$. Express your answer as a fraction of the period of the 633 nm red light.

   d. What did Michelson and Morley observe when they tried to measure the phase difference between the two arms, and why was their result important?

4. Evaluate without a calculator
   a. $1/1.05$.
   b. $(1.05)^2$.
   c. $(1.05)^{\frac{1}{2}}$.

5. A stick at rest has a length of 2 m. One observer, call her S′, moves with the stick at a speed of $0.6\,c$ past another observer, call him S. The length of the stick is parallel to her direction of motion.

   a. When S′ measures the length of the stick, what value does she obtain?

   b. When S measures the length of the stick, what value does he obtain?

   c. Which of these values is correct?

   d. How much time will elapse on the watch of S′ as the stick passes S?

   e. How much time will elapse on the watch of S as the stick passes him?

6. In the S frame illustrated in Fig. 11.15 Bob observes three clocks, e′, f′, and g′, moving at $v/c = \sqrt{3}/2$ past three stationary clocks, e, f, and g, 10 light seconds apart. (Clocks e′, f′, and g′ are synchronized in the S′ frame; e, f, and g are synchronized in the S frame.) Just as his

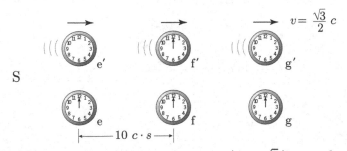

FIGURE 11.15 Three clocks moving at $v/c = \sqrt{3}/2$ past three stationary clocks (Problem 6).

stationary clocks all read 0, the three moving clocks line up exactly with the stationary clocks, and Bob observes that f′ also reads 0.

    a. How far apart are the moving clocks in their rest frame?

    b. What does e′ read in seconds when it is coincident with e?

    c. In S′ what does e read when e′ reads 0?

    d. In S′ what does g′ read when f′ reads 0?

    e. In S′ what does g read when f′ reads 0?

7. In the S frame of reference object 1 moves from the left with a velocity $v_1 = 0.8c$ while object 2 moves from the right with a velocity $v_2 = -0.8c$. What will be the velocity of object 2 in the rest frame of object 1?

8. This problem makes the point that what a clock reads and how much time elapses on it are not necessarily the same thing.

    Ann observes two clocks moving past her at $0.8c$, as shown in Fig. 11.16. She sees they are 16 light minutes apart and that as the right-hand clock comes by her, it reads 0 and her wristwatch reads 0.

    a. What is the proper length between the clocks? ("Proper length" means the length measured in the rest frame.)

    b. What does her wristwatch read when the left-hand clock passes her?

    c. If the two moving clocks have been synchronized in their rest frame, what does Ann find the left-hand clock to read when her wristwatch reads 0?

    d. How much time will have gone by on the right-hand clock when the left-hand one passes Ann? Explain.

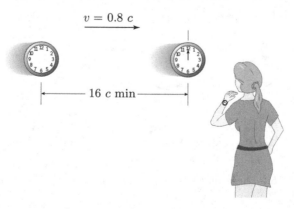

FIGURE 11.16 Two clocks moving past a stationary observer (Problem 8).

12.

Energy and Momentum at High Speeds

12.1   INTRODUCTION

Einstein's special theory of relativity modifies the Newtonian concepts of energy and momentum so that they correctly describe bodies moving at high speeds. The modifications lead to the best-known prediction of the theory of relativity: Energy has mass and vice versa,

$$E = mc^2,$$

and they also show that the relationship between kinetic energy and momentum that you have frequently used, $K = \frac{p^2}{2m} = \frac{1}{2}mv^2$, is only an approximation of the equations that are exact at all speeds. You now need to become familiar with the relativistically correct relationships and how they are used to extract information about atoms and the particles they are made of.

12.2   ENERGY HAS MASS

Einstein used the conservation of momentum and the fact that light exerts pressure to show that energy must have mass.

Light Exerts Pressure

Einstein knew, as did other physicists of his time, that light exerts a force on whatever it strikes. When a light wave is absorbed, it delivers an amount of energy ΔE to an object, and it imparts to the object a change

C.H. Holbrow et al., *Modern Introductory Physics, Second Edition*,
DOI 10.1007/978-0-387-79080-0_12, © Springer Science+Business Media, LLC 1999, 2010

in momentum $\Delta p = \Delta E/c$, where c is the speed of light. This means that light delivering a certain amount of power $\mathcal{P}$ is delivering momentum at a rate of

$$\frac{\Delta p}{\Delta t} = \frac{1}{c}\frac{\Delta E}{\Delta t} = \frac{\mathcal{P}}{c}.$$

But $\Delta p/\Delta t$ is force F, and, spread over a surface area A, it constitutes a pressure $P = F/A$.

It is common to specify intensity of light as the power delivered to a unit area. For example, the Sun delivers about $1.4\,\mathrm{kW\,m^{-2}}$ to the upper atmosphere of Earth. Dividing such a quantity by c gives the force per unit area; this is the pressure exerted by the light.

EXAMPLES

1. Sunlight absorbed by a square meter of collector above Earth's atmosphere results in a pressure of

$$P = \frac{1.4 \times 10^3}{3 \times 10^8} = 4.7 \times 10^{-6}\,\mathrm{Pa}.$$

This is quite small compared to atmospheric pressure of $10^5\,\mathrm{Pa}$, but it is enough to push a spacecraft to the outer reaches of the solar system.

EXERCISES

1. A laser beam 2 mm in diameter carries 0.5 mW of power. How much pressure does this beam exert when it is absorbed?

$E = mc^2$

The most famous result of Einstein's theory, the equivalence of mass and energy, is a straightforward prediction of his special theory of relativity; he also showed that $E = mc^2$ must hold if there is to be a relativistically correct law of conservation of momentum.

To remind yourself of an important consequence of the conservation of momentum, imagine a gun and a target mounted rigidly a distance L apart on a cart equipped with frictionless wheels. The situation is shown schematically in Fig. 12.1. When the gun fires a bullet of mass m to the right with speed u, the rest of the apparatus—the gun, target, wheels, etc.—with a total mass M, recoils to the left with velocity v.

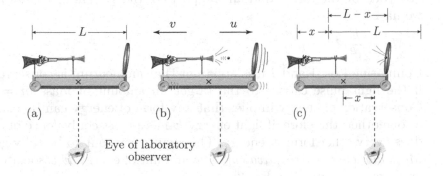

FIGURE 12.1 A gun and target are mounted on a frictionless cart. Before the gun fires, the total momentum of the system is zero. When the gun fires a bullet of mass m, the apparatus recoils until the bullet stops in the target.

The Newtonian form of conservation of momentum says that $Mv = mu$. However, by now you know enough relativity to be uneasy. There may be effects of the relative motion of M and m that make the Newtonian equation only approximate. You can evade this uncertainty by examining the situation of M and m after they come to rest. This will occur after the time Δt that it takes the bullet to reach the target. During that time the cart travels $x = v\,\Delta t$ to the left, and the bullet travels a distance $L - x = u\,\Delta t$ to the right.

Newtonian physics predicts that $Mv\,\Delta t = mu\,\Delta t$, so that after the transfer of the bullet we must have

$$Mx = m(L - x). \tag{1}$$

Since this result does not depend on the motion of anything, it must also be correct in the special theory of relativity as well as in Newtonian theory. Both theories require that a mass m be transferred a distance $L - x$ in order to conserve momentum.

Now consider the same setup with the gun emitting a pulse of light of energy E instead of a bullet of mass m. Because light carries momentum E/c, the gun, target, and cart must recoil with momentum $-E/c$. During the time Δt it takes the pulse of light to reach the target, the cart and attachments—still of mass M—recoil a distance $x = v\,\Delta t$, and the light travels a distance $L - x$ to reach the target. Since the light travels with speed c regardless of the recoil velocity, the pulse requires time $\Delta t = (L - x)/c$ to reach the target. In the Newtonian approximation, we have $Mv\,\Delta t = (E/c)\,\Delta t$, and therefore, after the pulse of light has been absorbed by the target and all the parts of the system are again at rest, we have

$$Mx = \frac{E}{c^2}(L - x). \tag{2}$$

Comparing Eqs. 1 and 2 it is apparent that momentum is conserved only if the light pulse carries to the target an amount of mass $m = E/c^2$. Conservation of energy implies that any form of energy can be converted into another; therefore, if light energy E has a mass equivalence of mc^2, so does every other form of energy. The result turns out to be fully general: *Mass and energy are equivalent.* The amount of energy E associated with any mass m is given by Eq. 3:

$$E = mc^2. \tag{3}$$

This famous equation also means that if the energy of an object increases, so does its mass, i.e.,

$$\Delta E = \Delta m \, c^2. \tag{4}$$

Suppose an object sitting on a flat frictionless surface is made to slide. It must have more energy moving than at rest because it has its kinetic energy in addition to any other forms of energy, and more energy means more mass. The equivalence of energy and mass means that an object will have more mass when it is moving than when it is at rest just because it *is* moving. Therefore, for a moving body the mass m appearing in Eq. 3 is greater than the mass m_0 of the body at rest.

It is helpful to distinguish between m and m_0, so m_0 is called "the rest mass" of a particle. This is always the mass that would be measured by an observer in the particle's rest frame.

Einstein's theory predicts that when a body with a rest mass m_0 moves with a speed v, it will have a mass m such that

$$m = \frac{1}{\sqrt{1 - \frac{v^2}{c^2}}} m_0 = \gamma \, m_0. \tag{5}$$

Therefore, Eq. 3 can be written as

$$E = mc^2 = \gamma m_0 c^2, \tag{6}$$

and Eq. 4 can be written as

$$\Delta E = \Delta mc^2 = (\gamma - 1)m_0 c^2. \tag{7}$$

From here on, unless the context tells you differently, m_0 will mean the rest mass and γm_0 or m will represent the relativistic mass. Notice also that Eq. 3 assigns a certain amount of energy to an object even when it is at rest. This is often called the particle's rest energy. For example, an electron has a rest mass of $m_0 = 9.11 \times 10^{-31}$ kg, so its rest energy is

$$m_0 c^2 = 9.11 \times 10^{-31} \times (3 \times 10^8)^2 = 8.20 \times 10^{-14} \text{ J}.$$

Almost never does one express the electron's rest energy in joules; the preferred units are electron volts. Then

$$m_0 c^2 = 5.11 \times 10^5 \text{ eV} = 511 \text{ keV}.$$

This is a fact you will use often.

Physicists frequently do not distinguish between rest energy and rest mass; after all, one is just the other multiplied by a constant. Although strictly speaking it is not correct, physicists often say: "The rest mass of the electron is 511 keV" when they mean $511 \text{ keV}/c^2$. (Note that keV/c^2 has dimensions of mass.)

■ EXERCISES

2. Given the validity of Eq. 5, derive Eqs. 6 and 7.

Experimental Evidence for $m = \gamma\, m_0$

To derive Eq. 5 would take us more deeply into Einstein's theory than we need to go. Instead, let's examine an experiment that shows that mass depends on velocity just as Einstein predicted.[1] The velocity dependence of mass is difficult to observe until v becomes comparable to c, as you can see by expanding γ:

$$\gamma = \left(1 - \frac{v^2}{c^2}\right)^{-1/2} = 1 + \frac{v^2}{2c^2} + \cdots.$$

The expansion shows that the fractional change in the mass due to its motion, $\Delta m/m = \gamma - 1$, will be less than 10^{-4} until $v/c = 0.014$. This corresponds to $v = 4243$ km/s, nearly 10 million mph and more than 140 times faster than Earth's motion in its orbit.

To find objects moving fast enough to show a measurable effect, Kaufmann used electrons emitted in the radioactive decay of atoms. This

---

[1] The experiment was done by Walter Kaufmann in 1901, (see pp. 502–512 in *The World of the Atom*, edited by H. A. Boorse and L. Motz, Basic Books, New York, 1966) four years before Einstein published his theory.

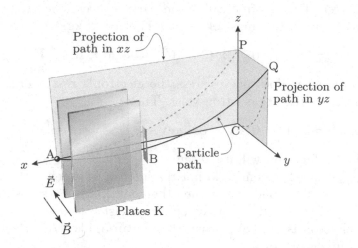

FIGURE 12.2 Kaufmann's arrangement for measuring e/m of electrons from radioactive atoms. Electrons from a point source of radioactive material at A pass between plates K across which is an electric field perpendicular to the plates; the electrons pass through a small hole B and strike a photographic plate in the z-y plane Q. A stack of permanent magnets provided the entire apparatus with a uniform magnetic field B parallel but opposite to E. The curve ABQ is the trajectory of one electron along a circular arc of radius ρ.

choice in 1901 is remarkable when you realize that radioactivity was only discovered in 1896 and the electron in 1897. As did J.J. Thomson, Kaufmann used a combination of electric and magnetic fields to measure e/m. But where Thomson had the E and B fields at right angles to each other, Kaufmann had them parallel.

Electrons from a tiny grain of radioactive material passed through this arrangement of fields and struck a photographic plate placed in the CPQ (y-z) plane in Fig. 12.2.

If the electrons all had the same velocity, they would follow a single path such as ABQ in Fig. 12.2 and they would all strike a single point on the photographic plate, e. g. the point Q for the trajectory ABQ. However, electrons from radioactive atoms have a wide range of velocities, and they will follow a correspondingly wide range of trajectories.

Kaufmann realized that the electric field E would deflect electrons in the y direction (Fig. 12.2) inversely proportional to the square of their velocity, i.e $\propto 1/v^2$, while a magnetic field B parallel to E would deflect the electrons in the z direction $\propto 1/v$. This was well known—see p. 239. As a result, in an ideal geometry the electrons arriving at the photographic plate would lie on a parabolic curve $y = kz^2$. The constant k would depend on the known geometry of his apparatus and on e/m the charge-to-mass ratio of the electron. In principle he could determine e/m for each point on the curve; that is, he could find e/m for electrons with quite different velocities. It's a clever idea.[2]

Of course, if m varies with v, then k will change with v, and the curve where the electrons strike the photographic plate will deviate from a parabola. By measuring the amount of deviation Kaufmann could determine by how much the mass changed as a function of v.

Kaufmann exposed a photographic plate for 48 h. He carefully measured the coordinates of five different points on the curve that appeared on the developed plate. From his measurements he determined the velocities and charge-to-mass ratios of electrons arriving at each of these points. The trajectory of one particular electron is shown in Fig. 12.2 along with a dashed line showing the curve formed by electrons with other velocities.

Data based on Kaufmann's measurements are shown in Table 12.1. The table shows velocities and corresponding values of e/m for his five data points. It is clear that e/m gets smaller as the velocity gets larger. This is just the effect that you would expect if m grows larger with speed. Also included in the table are values of $\gamma = \frac{1}{\sqrt{1-v^2/c^2}}$ corresponding to the measured values of v. If the observed variation in e/m arises because $m = \gamma\, m_0$, then multiplying any measured value of e/m by its corresponding γ should give e/m_0. The average of all the values of e/m_0 obtained from Kaufmann's data is 1.732×10^{11} C/kg, which agrees remarkably well with the currently accepted value of 1.759×10^{11} C/kg.

TABLE 12.1 e/m **for electrons of different speeds** v^a

Velocity v	e/m	e/m_0	γ	m/m_0
$(10^8$ m/s$)$	$(10^{11}$ C/kg$)$	$(10^{11}$ C/kg$)$		
2.815	0.620		2.891	
2.674	0.751		2.205	
2.526	0.927		1.854	
2.366	1.066		1.627	
2.230	1.175		1.495	

[a] These data are adapted from W. Kaufmann, "Die magnetische und electrische Ablenkbarkeit Becquerelstrahlen und die scheinbare Masse der Elektronen," *Nachrichten von der königl. Gesellschaft der Wissenschaften zu Göttingen* **2**, 143–155 (1902).

---

[2] It's also difficult to do and requires good vacuum, stable power supplies, and measurements accurate to a small fraction of a millimeter. In fact, although Kaufmann showed that mass increased with velocity, he was never able to use his measurements to convincingly confirm Einstein's theory compared to other plausible theories of the time.

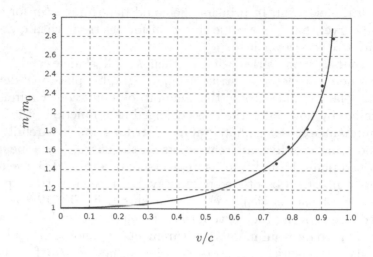

FIGURE 12.3   Comparison of results derived from Kaufmann's data (solid dots) with Einstein's prediction (solid line).

■ EXERCISES

3.   Use the above prescription to convert the values of e/m tabulated in Table 12.1 to e/m_0, and calculate the average value of e/m_0. Your results should show that Einstein's prediction that mass will vary as $m/m_0 = \gamma$ is supported by Kaufmann's experiment.

4.   Another way to show that Einstein's prediction is correct is to compute m/m_0 from Kaufmann's data and compare it to the values of γ calculated for the corresponding values of v. To find m/m_0 you need only divide the accepted value of e/m_0 by the measured value of e/m. Do this and put the values into column 5 of Table 12.1.

The graphical comparison of m/m_0 and v/c in Fig. 12.3 shows that Einstein's theory is consistent with Kaufmann's observations.

12.3   MOMENTUM AND ENERGY

If the mass of an object depends on its velocity, how does this affect our concepts of momentum and kinetic energy?

Relativistic Momentum

Let's begin by asking: What is the momentum of a particle in a frame where the particle is moving at velocity v? It turns out that for momentum to be conserved we must use the following specification for momentum:

$$p = mv = \gamma m_0 v.$$

This is the quantity that you must use to analyze collisions of moving particles, and it is the quantity p that you obtain from the relation $p = qBR$ (Eq. 2 on p. 205) when you deflect charged particles by a uniform magnetic field B.[3] For a particle moving with speed v, the combination γm_0 gives the increased mass. This increase is confirmed by measurements. In Kaufmann's e/m experiment, for example, the momentum determined from the curvature in a magnetic field yielded a larger value of m for a high velocity v and hence a smaller e/m than for a low velocity.

▼ EXAMPLES

2. Suppose that an electron has a velocity of $v = 0.25c$. What is its momentum? By what percent does its value of e/m differ from e/m_0, the low-energy result?

$$\gamma = \frac{1}{\sqrt{1 - (0.25c/c)^2}}$$

$$= \frac{1}{\sqrt{1 - 0.0625}} = 1.0328,$$

$$p = 1.0328 \times 9.11 \times 10^{-31} \times 0.25 \times 3 \times 10^8$$

$$= 7.05 \times 10^{-23} \text{ kg m/s}.$$

You can get the answer to the second part of the question by noting that if m is multiplied by 1.0328, then e/m must be divided by the same number. So

$$\frac{e}{m} = \frac{1.7587 \times 10^{11}}{1.0328} = 1.7028 \times 10^{11} \text{ C/kg},$$

or a 3.3% decrease.

---

[3] Buried in this statement and in the treatment of Kaufmann's data is a nontrivial assumption: Unlike mass, the electric charge is the same in all reference frames regardless of relative velocity. The success of the theory of electrodynamics, which deals with the transformations that determine how electric and magnetic fields appear in different reference frames, shows that this assumption is correct.

3. What is the radius of curvature of the path of an electron moving with $v = 0.25c$ in a magnetic field of $B = 1\,\mathrm{T}$?

You can answer this question using Eq. 2 and the value of the momentum p from the previous example.

$$R = \frac{p}{eB} = \frac{7.05 \times 10^{-23}}{1.602 \times 10^{-19} \times 1} = 4.4 \times 10^{-4}\,\mathrm{m} = 0.44\,\mathrm{mm}.$$

Relativistic Kinetic Energy

To handle kinetic energy correctly at high speeds you need to rethink the idea slightly. Kinetic energy is the *additional* energy that an object acquires because it is moving. Therefore, kinetic energy is the difference between the energy of a moving object, mc^2, and its energy at rest, m_0c^2:

$$K = mc^2 - m_0c^2 = (\gamma - 1)m_0c^2. \tag{8}$$

What would be the best approximation of this equation at low velocities, i.e., what should K be when $v/c \ll 1$? You know the answer, but you can use the binomial expansion to prove that it is the familiar formula for kinetic energy

$$K = m_0c^2 \left[\left(1 - \frac{v^2}{c^2} \right)^{-1/2} - 1 \right]$$

$$\approx m_0c^2 \left[1 + \frac{1}{2}\frac{v^2}{c^2} + \cdots - 1 \right] = m_0c^2 \frac{v^2}{2c^2}$$

$$= \frac{1}{2}m_0v^2.$$

Once again, Newtonian mechanics is the low-velocity approximation to Einsteins's theory; i.e., K is very accurately the old familiar quantity when speeds are small compared to c.

The equality of $E = mc^2$ is very general. It applies to all forms of energy, not just to kinetic energy. What effect does potential energy, for example, have on the mass? To be consistent with our principle of energy conservation, we have to conclude that these other forms of energy have mass also. Where would such mass be found? It can only be a part of the rest mass. But this will be the rest mass of the entire system, not just in a single part of the system, so it is not always clear where the new energy-associated mass is located in the system. Nevertheless, it must be the case that rest mass can include such things as thermal energy—the $\frac{3}{2}k_BT$ of average random-motion energy you learned about when studying the kinetic theory of gases—and potential energy. Does this mean that you are a little more massive at the top of a hill (where your potential energy is greater) than at the bottom?

▼ EXAMPLES

4. Find the change in mass corresponding to the change in gravitational potential energy of an 80 kg student after he climbs a 60 m high hill. Assume rather unreasonably that there is no loss of material through such things as respiration or perspiration. Is such a mass change detectable?

$$\Delta E = mg\Delta h = 80 \times 9.8 \times 60 = 47{,}040\,\text{J},$$

$$\Delta m = \frac{\Delta E}{c^2} = \frac{47{,}040}{9 \times 10^{16}} = 5.2 \times 10^{-13}\,\text{kg}.$$

It is not likely to be measured. Your gain in weight has nothing to do with climbing hills!

But wait! What about the energy that was used to do the work in climbing the hill? Since it had to come from chemical reactions in the body, all that has been done has been to convert chemical potential energy to gravitational potential energy. Both forms have equivalent mass, so there is no net change at all. Only if some outside agent (a roommate's car, perhaps) carried you up the hill would your mass change.

■ EXERCISES

5. Consider an electron and a point charge of $-1\,\mu\text{C}$. At first the two particles are at rest very far apart. What will be the change in the mass of this system if they are arranged to be at rest and 10 mm apart?

6. Calculate the change Δm in the combined mass m of a *positive* charge of $50\,e$ and an electron when both are at rest and a distance 1×10^{-12} m apart, relative to their combined mass when both are at rest and very far apart.

7. In x-ray machines electrons are accelerated through 100 kV or more. What is the speed of a 100 keV electron? What is its momentum? How much error do you make in your answers if you use the Newtonian equations instead of the relativistically correct ones?

Two things to remember:
It is common to refer to a particle by its kinetic energy. Thus when the previous problem told you that you have a 100 keV electron, you are expected to know that this refers to the electron's 100 keV of kinetic energy, not its total energy, which includes its rest energy and is $E = m_0c^2 + K = 611$ keV.

It is also usual to describe the momentum and energy of a moving particle from "our" rest frame. We call this reference frame the "laboratory," or "lab," frame of reference. For example, the curved path of a charged particle in a magnetic field tells you the particle's momentum in the lab frame. Unless they state otherwise, all the problems in this book ask for results in the lab frame.

Relation Between Energy and Momentum

It is more convenient to work with momenta and energies directly than it is to first calculate the velocity, find the relativistic factor γ, and then calculate kinetic energy or momentum. The momentum of a particle with a charge q accelerated to a high energy is often measured by bending it in a magnetic field, and its kinetic energy K is often known from the voltage V used to accelerate it, $K = qV$.

For such circumstances it is very helpful to have a relationship between momentum p and total energy E. You can find such a relationship from knowledge that $E = \gamma m_0c^2$ and $p = \gamma m_0v$. First square each of these equations to get

$$E^2 = \gamma^2 m_0^2 c^4 = \frac{m_0^2c^4}{1 - v^2/c^2},$$

$$p^2 = \gamma^2 m_0^2 v^2 = \frac{v^2/c^2}{1 - v^2/c^2}m_0^2c^2.$$

Multiply the second equation by c^2 to give it the same physical dimensions (energy squared) as the first and then subtract the second from the first. This will give

$$E^2 - p^2c^2 = \frac{1}{1 - v^2/c^2}m_0^2c^4 - \frac{v^2/c^2}{1 - v^2/c^2}m_0^2c^4 = \left(\frac{1 - v^2/c^2}{1 - v^2/c^2}\right)m_0^2c^4.$$

From this you can see that

$$E^2 = p^2c^2 + m_0^2c^4, \quad \text{or} \quad E = \sqrt{p^2c^2 + m_0^2c^4}, \tag{9}$$

which, as desired, enables you to find the total energy of a particle of known rest mass m_0 if you know its momentum, or to find the magnitude of its momentum if you know its total energy.

▼ **EXAMPLES**

5. Find the momentum of a 100 keV electron (Exercise 7. p. 385) without finding γ. Use SI units.

The basic approach is to solve Eq. 9 for p. This gives

$$p = \sqrt{\frac{E^2 - m_0^2 c^4}{c^2}}.$$

Then you need to determine E, and to do this you need to know m_0. From p. 635 of the appendix or from Exercise 17 in Chap. 9 you can find that $m_0 = 9.11 \times 10^{-31}$ kg for an electron. Then

$$E = K + m_0 c^2 = qV + m_0 c^2$$
$$= 10^5 \times 1.6 \times 10^{-19} + 9.11 \times 10^{-31} \times 9 \times 10^{16} = 9.80 \times 10^{-14}\,\text{J},$$

and

$$p = \sqrt{\frac{(9.8 \times 10^{-14})^2 - (8.20 \times 10^{-14})^2}{9 \times 10^{16}}} = 1.79 \times 10^{-22}\,\text{kg m s}^{-1}.$$

This is pretty messy. An easier and more common way to work with E and p is to abandon SI units and work with units based on the electron volt (eV).

12.4   MASSES IN eV/c^2; MOMENTA IN eV/c

Because units of electron volts are widely used in measurements of atomic particles, it is convenient to express their masses and momenta in terms of electron volts.

▼ **EXAMPLES**

6. To see how to use these units, consider a 100 eV electron. Suppose you want to find its velocity. It is of such low energy, i.e., $v \ll c$, that you can use

$$K = \frac{1}{2} m_0 v^2.$$

If you multiply the right-hand side by c^2/c^2, you will get

$$K = \frac{1}{2} m_0 c^2 \frac{v^2}{c^2}.$$

In this form the convenience of the units becomes more evident. Knowing that m_0c^2 for an electron is 511 keV, you get

$$100 = \frac{1}{2} \cdot 511 \times 10^3 \cdot \frac{v^2}{c^2},$$

which you can solve for v/c,

$$\frac{v}{c} = \sqrt{\frac{2 \times 100}{511 \times 10^3}} = 1.98 \times 10^{-2},$$

from which it follows that $v = 5.94 \times 10^6$ m/s.

The trick is to bundle together the right combination of m and c to get a quantity identifiable as an energy. Once you get used to it, using eV and suppressing the factor c^2 makes calculations easier and more informative.

EXERCISES

8. What is the velocity of the 200 eV electrons used for measuring e/m in many undergraduate laboratories?

EXAMPLES

7. You can also use this trick to find momentum. Suppose you want to know the momentum of a 100 eV electron. Because this is a low-energy electron, you can use the Newtonian approximation

$$K = \frac{p^2}{2m_0}.$$

Multiplying the right side by c^2/c^2 gives

$$K = \frac{p^2c^2}{2m_0c^2},$$

which can be solved for pc:

$$pc = \sqrt{2Km_0c^2}.$$

Now, if K is in eV and m_0c^2 is in eV, you will obtain pc in eV, so

$$pc = \sqrt{200 \times 511 \times 10^3} = 1.011 \times 10^4 \text{ eV},$$

and you get $p = 1.01 \times 10^4 \, \text{eV}/c$, or, more compactly, $p = 10.1 \, \text{keV}/c$. The units eV/c are units of momentum commonly used in nuclear and particle physics; they have dimensions M L T^{-1} like any other unit of momentum.

8. Find the momentum of a 1 MeV electron.

This electron has a kinetic energy twice its rest mass; therefore, it is a high-energy electron for which you must use the relativistically correct equations

$$E = K + m_0 c^2 = 1 + 0.511 \, \text{MeV} = \sqrt{(pc)^2 + (0.511)^2},$$

$$pc = \sqrt{(1.511)^2 - (0.511)^2} = 1.422 \, \text{MeV},$$

and thus $p = 1.422 \, \text{MeV}/c$.

Momentum is frequently found by measuring the radius of curvature R of a charged particle's trajectory perpendicular to a uniform magnetic field B.

Remember that a particle of charge e and momentum p will bend according to the relationship

$p = eBR$.   Eq. 2, p. 205

To use eV units in this equation requires a little ingenuity. Notice that if you multiply both sides by c, the dimensions of the equation become those of energy, i.e., pc has units of energy. At first glance it seems that you must evaluate $eBRc$ using consistent SI units and obtain an answer in joules. This will certainly work, but notice that if you convert the right side of the equation into electron volts, you divide by a conversion factor that has a numerical value just equal to e. This means that if you drop the factor of e and multiply B in tesla by R in meters and c in meters per second, your answer BRc will come out numerically in eV.

▽ EXAMPLES

9. What is the value of pc of an electron moving in a circle of radius 1 cm perpendicular to a uniform magnetic field of 0.5 T? The answer (in eV) is $BRc = 0.5 \times 0.01 \times 3 \times 10^8 = 1.5 \times 10^6 \, \text{eV} = 1.5 \, \text{MeV}$. So the momentum of this electron is 1.5 MeV/c.

10. What uniform magnetic field B will bend an electron having momentum 1.422 MeV/c in a circle of radius $R = 2.37$ cm?

Suppose you are unwilling to use eV units and cling to the SI. At least start with $pc = qRBc$. Then you need to convert the left side into joules:

$$1.422 \times 10^6 \text{ [eV]} \times 1.602 \times 10^{-19} \text{ [J/eV]}.$$

The right side will be

$$= 1.602 \times 10^{-19}\text{[C]} \ B\text{[T]} \ 0.0237\text{[m]} \ 3 \times 10^8\text{[m/s]}$$
$$= 1.139 \times 10^{-12}\text{[J/T]} \ B\text{[T]}.$$

Now solve for B.

But notice that the factor 1.602×10^{-19} appears on both sides of the equation. On the left it is the conversion factor with units of J/eV; on the right it is the elementary charge with units of C (coulombs). Although the units are different, the numerical factor is the same, and when you divide both sides by 1.602×10^{-19} J/eV, you get

$$1.422 \times 10^6 \text{ eV} = 0.0237 \times 3 \times 10^8 \ B \ \text{[C·m·m/s·T·eV/J} = \text{eV]}$$

where the equation $F = qvB$ shows you that C·m·m/s·T is J (joule).

So why not take this fact into account right from the start and skip the SI units?

$$B = \frac{pc}{Rc} = \frac{1.422 \times 10^6}{0.0237 \times 3 \times 10^8} = 0.20 \text{ T}.$$

11. Suppose you want to know the radius of curvature with which an electron with momentum of $10\,\text{keV}/c$ bends in a magnetic field of 10^{-3} T. Then

$$R = \frac{pc}{Bc} = \frac{10 \times 10^3}{10^{-3} \times 3 \times 10^8} = 3.37 \times 10^{-2}\,\text{m} = 3.37\,\text{cm}.$$

▨ EXERCISES

9. A proton, $(M_0c^2 = 938\,\text{MeV})$ bends with a radius of curvature of $1\,\text{m}$ in a magnetic field of $0.1\,\text{T}$. What are the momentum and the kinetic energy of that proton?

▼ EXAMPLES

12. Here is a repeat of Example 5 using units of electron volts. You are given that $K = 100\,\text{keV}$, and you *remember* that for an electron, $m_0c^2 = 511\,\text{keV}$. The total energy E is the sum of the kinetic energy K and the rest energy, $E = K + m_0c^2$, so $E = 100\,\text{keV} + 511\,\text{keV} = 611\,\text{keV}$.

Equation 9 tells you that $611\,\text{keV} = \sqrt{(pc)^2 + 511^2}$, which you can solve for pc:

$$611^2 - 511^2 = (pc)^2,$$

$$pc = 335\,\text{keV} \text{ and } p = 335\,\text{keV}/c.$$

▪ EXERCISES

10. Show that the momentum of $335\,\text{keV}/c$ obtained in Example 12 is the same as the value of $1.79 \times 10^{-22}\,\text{kg m s}^{-1}$ obtained in Example 5.

To work easily in eV you need to *know* the masses of the electron and the proton in eV/c^2. It may also be helpful to know the energy equivalent of one atomic mass unit, although for most purposes of this book you can use the proton or hydrogen-atom mass instead. Energy equivalents of these and other useful masses are summarized in Table 12.2. *From now on work all energy and momentum problems in units of eV (or keV or MeV, etc.).* The scale of these units is appropriate to atoms and smaller particles.

TABLE 12.2 Some masses in energy units

Entity	mc^2 (MeV)
electron	0.511
H atom	938.8
proton	938.3
neutron	939.6
1 u	931.5

12.5   WHEN CAN YOU APPROXIMATE?

In the preceding examples we told you when to use the Newtonian approximation and when not to. What if you have to decide on your own? How can you tell whether to use $K = p^2/(2m_0)$ or $K = E - m_0c^2$? Here we give you some guidance and some rules of thumb. Where the rules come from is left for later courses in physics.

Nonrelativistic Approximations

You will often have to solve one or the other of two related problems:
   a. Given the momentum p, find the particle's kinetic energy K.
   b. Given the kinetic energy K, find the particle's momentum p.

When v/c is small, you can solve these problems using the approximate Newtonian equations:

$$K = \frac{p^2c^2}{2m_0c^2},$$

$$\text{or} \quad pc = \sqrt{2Km_0c^2}.$$

As a measure of how good an approximation is, it is usual to use the "fractional difference" between the approximate and exact answers. (Fractional difference is often referred to as "fractional error.") Therefore, given an approximate value K_a and an exact value K_e, you want

$$\frac{K_a - K_e}{K_e} = \frac{\Delta K}{K_e}$$

to be small enough.

How small is "small enough"? It depends on what you are doing with your answer. If you are building a bridge or a particle accelerator, you want a smaller fractional error than if you are doing physics homework problems. For physics problems a fractional error of a few percent is often satisfactory. For the purposes of this book 2.5% is usually good enough.

▽ EXAMPLES

13. What is the fractional difference between 200 and 204? Between 200 and 220?

The answer for the first case is $(204 - 200)/200 = 0.02$, or, as one often says, 2%. For the second case the fractional difference is 0.1, or 10%.

The rest energy m_0c^2 is the key quantity for judging the accuracy of the approximate equations. The precision of results is governed by the ratio of K or pc to the rest energy m_0c^2. You need to know m_0c^2 to judge whether you can use approximate equations or if you must use the exact ones.

For example, if the ratio of the particle's kinetic energy to its rest energy, $K/(m_0c^2)$, is 0.1, a value of p calculated from $\sqrt{2Km_0}$ will deviate from the exact value by 2.5%. In other words, if you have a 51.1 keV electron and you calculate its momentum non-relativistically, your answer will differ from the exact answer by about 2.5%. For a proton the kinetic energy would have to be 94 MeV—10% of its rest energy—before the non-relativistic equations would be in error by 2.5%.

■ EXERCISES

11. What is the fractional error between the relativistic and non-relativistic calculations of momentum if the electron has a kinetic energy of 5.1 keV? 102 keV?

For a precision of 2.5% or better use the following rule of thumb:

- If $K/(m_0c^2) < 0.1$, you may use nonrelativistic equations to find momentum from kinetic energy.

■ EXERCISES

12. Suppose you had a 50 MeV proton. Would the Newtonian equations be good enough for calculating the proton's momentum? How much error would you make using the Newtonian equation?

If you know the momentum p and need to find the kinetic energy K, the nonrelativistic equation $K = p^2/(2m_0)$ will give an answer that has a fractional deviation from the relativistically exact answer that is $\leq 2.5\%$ as long as $pc/(m_0c^2) \lesssim 0.32$. This means that for an electron with momentum of 162 keV/c the nonrelativistic equation will give an answer that is 2.5% low.

So if 2.5% is good enough precision, you have the following rule:

- To find K from p you may use nonrelativistic equations when $pc/(m_0c^2) < 0.3$.

What if you know v and want to find K or p? When do you need to use the exact equations?

- To find K or p with an accuracy of 2.5% or better when you know v, you can use non-relativistic equations if $v/c \lesssim 0.2$.

If you would rather remember only one rule of thumb, just remember the most restrictive:

- If the ratio of $K/(m_0c^2)$ or $pc/(m_0c^2)$ or v/c is <0.1, you can use nonrelativistic equations with an error of less than 3%.

Ultrarelativistic Approximation

There is another extreme useful to know about. If the ratios of K or pc to the rest energy m_0c^2 are greater than ≈ 40, you can simply ignore the m_0c^2 terms in Eq. 9. The connections of E, K, and pc become

$K \approx E \approx pc$.

This is called the "ultrarelativistic" approximation, and it is good to better than 2.5%.

▾ EXAMPLES

14. What is the momentum of a $1\,\text{GeV}$ electron? In this case $K/(m_0c^2)$ is about 2000, so the particle is ultrarelativistic. Its momentum is $1\,\text{GeV}/c$.

12.6   SUMMARY

In the study of a moving particle the quantities of most interest are its kinetic energy K, its closely related total energy E, and its momentum $\vec{p}$.

At high speeds, i.e., when $v/c \gtrsim 0.1$, the values of these quantities differ appreciably from the predictions of Newtonian physics. This different behavior arises because inertial mass increases with a particle's speed:

$$m = \frac{m_0}{\sqrt{1 - v^2/c^2}} = \gamma m_0,$$

where v is the particle's speed, m_0 is its mass in its rest frame of reference, and

$$\gamma \equiv \frac{1}{\sqrt{1 - v^2/c^2}}.$$

The relativistically correct formulas connecting total energy E, kinetic energy K, momentum p, and mass m are

$$\vec{p} = m\vec{v}$$
$$E = mc^2$$
$$E = \sqrt{m_0^2 c^4 + p^2 c^2}$$
$$K = E - m_0 c^2.$$

The following relations are also useful:

$$\frac{v}{c} = \frac{pc}{E} = \sqrt{1 - \frac{1}{\gamma^2}}$$
$$\frac{pc}{m_0 c^2} = \sqrt{\gamma^2 - 1}$$
$$\frac{K}{m_0 c^2} = \gamma - 1.$$

In the limit of $v/c \lesssim 0.1$ the relativistically correct equations are very well approximated by the nonrelativistic Newtonian equations $\vec{p} = m_0 \vec{v}$, $K = p^2/(2m_0)$, and $E = K+$const. For many practical purposes when $v/c < 0.1$ or $K/(m_0 c^2) < 0.1$ or $pc/(m_0 c^2) < 0.3$, the Newtonian equations are sufficiently accurate and easier to use.

At speeds where $v/c > 0.9997$ or when $K/(m_0 c^2)$ or $pc/(m_0 c^2) > 40$, you can neglect the $m_0 c^2$ terms in the above equations. This is called the ultrarelativistic approximation. Then $K \approx E$ and $p \approx E/c$.

In modern physics the units of energy are electron volts (eV) and their SI multiples such as keV, MeV, GeV, TeV. They are used as units of kinetic energy and potential energy. By extension they are used as units of mass, eV/c^2, and units of momentum, eV/c. It is fairly common usage to say that the mass of a particle is so many eV, e. g., the mass of a proton is 938 MeV. You are supposed to know that this number includes the factor of c^2.

You need to know the values of the speed of light $c = 3 \times 10^8$ m/s and the rest masses of the electron ($511 \, keV/c^2$) and the proton ($938 \, MeV/c^2$).

PROBLEMS

1. A 9.38 MeV proton enters a magnetic field and is bent in a circle of radius $r = 0.2$ m.

   a. Is this a relativistic, nonrelativistic, or ultrarelativistic proton? How do you know?

   b. What is the extra mass of this proton arising from its motion? Give your answer in the appropriate multiple of eV.

2. Suppose you have an electron traveling with a velocity such that $\gamma = 3$.

   a. What is its kinetic energy?

   b. What is its momentum?

   c. What would be its radius of curvature in a magnetic field of $B = 0.1$ T?

   d. What is its velocity?

3. Suppose a stick and S′ are moving past S at a velocity such that $\gamma = 3$. If the stick has a mass of 1 kg when it is at rest,

   a. What does S′ measure its mass to be?

   b. What does S measure its mass to be?

4. What is the kinetic energy of the stick in the previous question

   a. as measured by S′?

   b. as measured by S?

5. a. We know that $E = mc^2 = \gamma m_0 c^2$. Use this relation to give a relativistically correct definition of kinetic energy K. Explain why this is a sensible definition.

   b. Show by a series expansion that for $v/c \ll 1$ your above definition is well approximated by the usual classical formula for kinetic energy.

   c. An electron is accelerated through a potential difference of 10 V. What is the relativistic increase in its mass? Explain.

6. A stationary particle having a rest mass energy of 1400 MeV disintegrates into two particles called "pions," a π^+ and a π^-, that travel in opposite directions, as shown schematically in Fig. 12.4. Each of the pions has a rest-mass energy of 140 MeV.

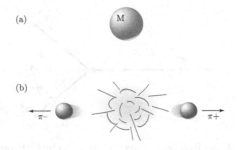

FIGURE 12.4 A particle disintegrates into two pions (Problem 6).

   a. Find the kinetic energy, in appropriate units, of each outgoing
particle.

   b. What is the value of γ ("gamma") for either pion?

   c. When the pions are at rest, they have a lifetime of 0.28 ns. According to an observer in the laboratory, what is the lifetime of the
moving pions (in ns)?

7. A positively charged K meson—a K^+ kaon—with a momentum $p =$
8.2 GeV/c enters a bubble chamber (a device in which charged particles
leave visible tracks) and decays into two pions: $K^+ \rightarrow \pi^+ + \pi^0$. The rest
mass of the kaon is 494 MeV/c^2 and that of each pion is 140 MeV/c^2.

   a. Calculate the total energy of the kaon. Do this exactly.

   b. What is the value of γ? Find the speed of the kaon in terms of c.
Is the ultra-relativistic approximation valid for this kaon?

   c. The K^+ has an average lifetime of 12.4 ns in its rest frame.
What lifetime would you measure in the rest frame of the bubble
chamber?

   d. In the rest frame of the bubble chamber how far (on average) would
a kaon travel before decaying? What distance would this be in the
kaon's frame of reference?

   e. Suppose that the two pions have the same kinetic energy. What
would that kinetic energy be? What is the value of γ for either
pion? What is the magnitude of the momentum of either pion?

   f. Assuming that the paths of the pions make equal angles θ, find θ
(Fig. 12.5).

8. The rest mass energies of a proton and an electron are 938 MeV and
0.511 MeV, respectively. Calculate the total energy of each of the following
particles:

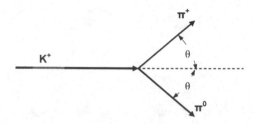

FIGURE 12.5 A positively charged kaon traveling through a bubble chamber disintegrates into two pions. This diagram shows their momenta before and after the disintegration (Problem 7(f)).

(a)   (b)

FIGURE 12.6 An electron and a positron collide and become two photons (Problem 9).

a. a 10 MeV proton.

b. an electron with $\gamma = 1000$.

c. a 500 MeV electron.

Are any of these particles "ultrarelativistic?" If so, which one(s)?

9.   There is a particle with the same mass as the electron but positively charged. It is called a "positron." The "Tristan" particle accelerator at Tsukuba, Japan, accelerates both electrons and positrons to energies of 25 GeV and then directs them into head-on collisions in which they annihilate one another; i.e., they turn into a pair of high-energy photons, as illustrated in Fig. 12.6. Give your answers to the following questions in eV.

a. Are these particles nonrelativistic, relativistic, or ultrarelativistic? Justify your answer briefly.

b. Find the momentum of a 25 GeV electron.

c. Suppose a single new particle is formed in the collision of a positron and an electron in this machine. Determine the momentum, kinetic energy, and rest mass of such a particle.

d. If instead of forming a new particle, all of the electron and positron energy is used to produce two photons of equal wavelength traveling in opposite directions (See Fig. 12.6b):

    i. Estimate the wavelength of each photon.

    ii. Why are two photons, rather than one, generated? (Hint: consider the conservation laws.)

10. What magnetic field will bend a 1.02 MeV electron in a circle with a radius of 1 m?

11. What magnetic field will bend a 1.02 MeV proton in a circle with a radius of 1 m?

12. What magnetic field is required to bend a 1 GeV electron in a circle 10 m in radius?

13. What magnetic field is required to bend a 1 GeV proton in a circle 10 m in radius?

14. What is the velocity of a 1.02 MeV electron? Express your answer as a fraction of the speed of light c.

15. An electron in a magnetic field of 1 kG (0.1 T) is bent in an arc with a radius of 1 m. What is the electron's kinetic energy?

16. For each of the preceding six questions, justify using the nonrelativistic approximation, the ultrarelativistic approximation, or the exact relativistic equations.

17. The rest mass energies of the electron and proton are respectively 511 keV and 938 MeV. For the cases i–iv below:
    a. Determine whether the particles are nonrelativistic or ultrarelativistic. Justify your answer.
    b. Calculate the kinetic energies (in eV) for each case, using the appropriate approximate formulas in each case *and also* the exact relativistic formulas.
        i. An electron with a momentum of 10 keV/c.
        ii. An electron with a momentum of 10 GeV/c.
        iii. A proton with $\gamma = 50$.
        iv. A proton with $v/c = 0.05$.

18. A proton and an antiproton each traveling with a momentum of 10 GeV/c collide head-on in a high-energy physics experiment. The proton and antiproton have equal rest masses $(938\,\text{MeV}/c^2)$, and charges of opposite sign but equal magnitude.

    a. What is the total energy before the collision?

    b. If they collide to form a new particle, determine its momentum, charge, and rest-mass energy.

    c. The new particle is unstable and decays into two pions, π^+ and π^-, both of which have rest masses of $140\,\text{MeV}/c^2$, and opposite but equal-magnitude charges. Find the resulting kinetic energy of each pion.

19. What is the speed, relative to the speed of light, of an electron with kinetic energy equal in value to its rest mass?

20. A typical nuclear reactor produces about $2.5 \times 10^9\,\text{J}$ of thermal energy in $1\,\text{s}$.

    a. What mass does this energy correspond to?

    b. If $200\,\text{MeV}$ of energy is released per fission, how many fissions occur per second in this nuclear reactor?

    c. What fraction of the mass of a ^{235}U nucleus is converted into thermal and radiant energy when it undergoes fission?

21. Find the energy released in the deuterium–deuterium fusion reaction

$$^2\text{H} + {}^2\text{H} \rightarrow {}^3\text{He} + {}^1\text{n}.$$

The rest masses of ^2H, ^3He, and the neutron ^1n are $2.014102\,\text{u}$, $3.016029\,\text{u}$, and $1.008665\,\text{u}$ respectively.

22. Consider the decay $^{55}\text{Cr} \rightarrow {}^{55}\text{Mn} + \text{e}^-$, where e is an electron. The nuclei ^{55}Cr and ^{55}Mn have masses of $54.9279\,\text{u}$ and $54.9244\,\text{u}$, respectively. Calculate the mass difference of the reaction. What is the maximum kinetic energy of the emitted electron?

23. In the electron–positron collider at Cornell University, electrons and positrons acquire oppositely directed momenta of $4.0\,\text{GeV}/c$ before they collide head-on.

    a. Suppose that after the two particles collide they unite to form a new particle *at rest*. What is the rest mass of the new particle?

    b. Suppose the new particle decays into a proton and an antiproton that move in opposite directions. What is the kinetic energy of either particle?

    c. What is the velocity relative to the speed of light, and what is the momentum of either particle?

13.

The Granularity of Light

13.1   INTRODUCTION

The discovery of the electron swiftly led to better understanding of the nature of matter. This in turn led to a revolution in the understanding of the nature of light. The most surprising outcome was the discovery that under many circumstances light and other forms of electromagnetic radiation behave like particles instead of waves. There are two outstanding examples of light behaving like particles. One example is called "the photoelectric effect" and the other "the Compton effect."

13.2   THE PHOTOELECTRIC EFFECT

Discovery of the Photoelectric Effect

In 1887 Heinrich Hertz showed that, as Maxwell had predicted, oscillating electric and magnetic fields produce radiation that travels with speed c. In effect, Hertz discovered electromagnetic radiation in the form of radio waves and provided convincing evidence that visible light is just a shorter wavelength form of such radiation. While doing this work, Hertz also discovered the photoelectric effect.

On one side of a room he generated radio waves by means of a high-frequency current sparking across a gap. The resulting waves crossed the room and induced sparking in a properly adjusted detecting apparatus. He observed that this induced spark was much larger when the metal tips of the spark gap were illuminated with light from the sparks of the generator across the room. When the light from the generating sparks was passed through glass before reaching the detecting spark gap, the spark

C.H. Holbrow et al., *Modern Introductory Physics, Second Edition,*
DOI 10.1007/978-0-387-79080-0_13, © Springer Science+Business Media, LLC 1999, 2010

became smaller, but when the light came through a quartz plate, the induced spark remained large. Hertz knew that glass absorbs ultraviolet light and that quartz does not, so he concluded that it was ultraviolet light causing the spark in the detector to become larger.[1]

In 1888, shortly after Hertz's observations, Hallwachs found that a negatively charged zinc plate would discharge when illuminated with ultraviolet light, while a positively charged plate would not.[2] This emission of negative electricity from the metal when illuminated with light of suitably short wavelength is called the "photoelectric effect." In 1889 Elster and Geitel showed that the photoelectric effect could be produced using *visible* light on metals that were amalgams of mercury with the alkali metals cesium, sodium, and potassium, as well as with zinc.

As you have seen, it was another decade before techniques were developed that gave reliable measurements on charged particles such as the electron. Understanding of the photoelectric effect did not progress much until it was possible to measure directly the actual charges released when light struck these metals.

Properties of the Effect

In 1899, two years after discovering the electron, J.J. Thomson, in England, and Philipp Lenard, in Germany, measured the charge-to-mass ratio q/m of the negative charges emitted in the photoelectric effect. They found q/m to be similar to the ratio already measured by Thomson for cathode rays, i.e., electrons, and concluded that the photoelectric charges must also be electrons.

Figure 13.1 is a diagram of the apparatus Lenard used to study the electrons emitted when light strikes a metal. Electrons produced using light are often called "photoelectrons." He used the apparatus both to measure the q/m ratio of the photoelectrons and to learn some things about their energy. The apparatus is in some ways like the one used to measure e/m. Here the cathode rays are produced when ultraviolet light from S shines through the window Q and strikes the cathode C, releasing charges. The flow of charges initiated by the light is often called the "photocurrent." The charges emitted at the cathode C have some initial energy K_i, and they acquire more kinetic energy by passing through a potential difference V maintained between C and A. The charges that pass through the hole in A proceed on with kinetic energy

$$\frac{1}{2}mv^2 = eV + K_i.$$

---

[1] Heinrich Hertz, *Electric Waves*, MacMillan & Co., London and New York, 1893.
[2] W. Hallwachs, *Ann. d. Phys.* **33**, 301 (1888).

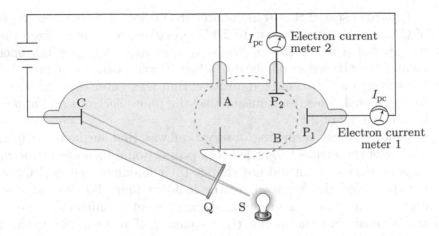

FIGURE 13.1 Lenard's apparatus for measuring q/m and $K_{\max}$ of photoelectrons. The dashed oval curve surrounds the region where a magnetic field is applied perpendicular to the plane of the page.

In Fig. 13.1 the dashed oval shows where a uniform magnetic field B can be applied directed out of the plane of the paper. When the field is off, charges coming through the hole A strike the anode P_1, and there is a photocurrent in meter 1; when the field is on and properly adjusted, the charged particles bend in a circle of radius R onto P_2. Then in meter 2 there is a photocurrent that becomes maximum for some value of B. R is measured from the geometry of the apparatus. If V is large enough so that $eV \gg K_i$, you can combine the energy equation with $mv = eBR$ and show that

$$\frac{e}{m} = \frac{2V}{B^2 R^2}. \tag{1}$$

Equation 1 is the same as Eq. 8 derived in Chap. 9 (p. 241). Lenard measured q/m to be 1.2×10^{11} C/kg and concluded that the emitted charges were electrons.

■ EXERCISES

1. How does this value compare with the value of e/m obtained by Thomson as described in Chap. 9? With the currently accepted value? Why do you suppose people measuring values of e/m that differed by $\approx 50\%$ or more all agreed that they were seeing the same particle?

Lenard observed that some photocurrent flowed even when the voltage of C was made positive—up to 2.1 V—relative to A. He realized that this meant that some of the electrons were emitted with enough energy to escape the attraction of plate C when it was positively charged. When the voltage on C became too positive, that is, greater than 2.1 V, no more charge flowed. This result meant that the photoelectrons had a *maximum* kinetic energy of 2.1 eV.

Lenard's most surprising observation was that an increase in the intensity of the incident light led to a proportional increase in the number of photoelectrons but did not change their maximum energy! Only when Lenard varied the frequency of the incident light did the maximum energy change. Later, more refined experiments confirmed these results and showed that the higher the frequency of incident light, the higher would be the maximum energy of the emitted electrons. In Fig. 13.2 schematic plots of photocurrent I_{pc} vs. the cathode voltage V show the effects of changing the intensity and also the frequency of the illuminating light.

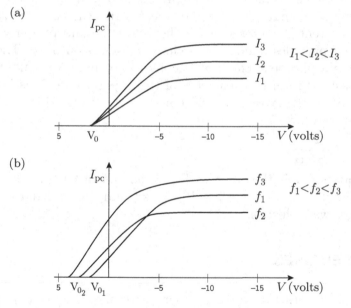

FIGURE 13.2 Photocurrent I_{pc} vs. cathode voltage V (a) for different intensities I_1, I_2, I_3 and (b) for different frequencies f_1, f_2, and f_3 of incident light. Increases in intensity increase the photocurrent but do not change the voltage V_0 corresponding to K_{max}; only increases in frequency increase K_{max}.

▪ EXERCISES

> **2.** When you shine ultraviolet light on a zinc plate connected to a charged electroscope, it discharges when negatively charged but not when positively charged. Use the discovery that the charges emitted in the photoelectric effect are electrons to explain these observations.

There was another surprising property of the photoelectrons. They began to flow as soon as the cathode was illuminated; there was no time delay between the arrival of the light and the emission of photoelectrons. This was unexpected for the same reason that it was a surprise to find that the maximum energy of the electrons did not depend on the intensity of light. Recall that the model of light as a wave implies that light emitted from a source spreads out through space and, like a water or sound wave, becomes ever weaker the farther it is from its source.

The fact that the intensity of light from a point source of light decreases as $1/r^2$ as you move a distance r from the source follows from the geometry of a sphere. As an example, consider light from the Sun, which emits a total of $W_0 = 3.9 \times 10^{26}$ J/s of electromagnetic radiation (mostly in the visible spectrum). By the time this radiation reaches Earth 1.5×10^{11} m from the Sun, its energy is spread over a sphere of area

$$4\pi r^2 = 4\pi (1.5 \times 10^{11})^2 = 2.8 \times 10^{23} \, \text{m}^2,$$

so that measurements of the solar energy incident at the top of Earth's atmosphere show that the amount of energy incident per second per unit area (the definition of intensity) is

$$\frac{3.9 \times 10^{26}}{2.8 \times 10^{23}} = 1390 \, \text{J s}^{-1} \, \text{m}^{-2}.$$

As the sunlight travels further and further, the area of the surrounding sphere grows as r^2, so the energy is spread over a larger area, and its intensity drops as $1/r^2$. Thus, at the planet Jupiter, which is 5.2 times further from the Sun than Earth, the intensity of sunlight striking the outer atmosphere of Jupiter is

$$\frac{1390}{(5.2)^2} = 51 \, \text{J s}^{-1} \, \text{m}^{-2}.$$

To see what is surprising about the fact that photoelectrons are emitted as soon as light strikes the metal, consider how quickly an atom could accumulate energy from a wave and get enough to eject the photoelectron. Assume that the photoelectric effect is to be produced with a narrow band of frequencies of sunlight, say between 700.0 THz and 700.3 THz (a deep blue light). This range contains about 7×10^{-5} of the total energy in the sunlight, and only half of that reaches the ground. Therefore, the intensity of the light for our experiment might be in the range of

$$3.5 \times 10^{-5} \times 1390 = 0.049 \,\mathrm{J\,s^{-1}\,m^{-2}}.$$

A beam of light with this intensity is very easy to see.

The research of Lenard and others showed that it takes different amounts of work to release an electron from different metals. Suppose you have a metal which requires about 2.9 eV to release an electron. How long do you expect it to take light with an intensity of 0.049 $\mathrm{J\,m^{-2}\,s^{-1}}$ to supply the necessary 2.9 eV to an atom? If the energy of light is spread smoothly over space, then the rate that energy is supplied to an atom is $W_a = I\pi r^2$, where πr^2 is the cross-sectional area of the atom of radius r and I is the intensity of the light. For an atom with a typical radius of 0.2 nm,

$$W_a = 0.049\pi(0.2 \times 10^{-9})^2 = 6.2 \times 10^{-21} \,\mathrm{J\,s^{-1}},$$

or

$$W_a = 0.038 \,\mathrm{eV\,s^{-1}}.$$

This result tells you that if this light were arriving as a wave spreading out from its source, it would take about 76 seconds before the incident light could supply the 2.9 eV the atom needs to release an electron.

■ EXERCISES

> **3.** Complete the calculation to show that it would take about 76 seconds, or well over a minute, to release one electron under the assumptions discussed above.

Einstein's Explanation: $E = hf$

The essentially instantaneous emission of an electron was an extraordinary result, totally at odds with classical physics. Sir William Bragg described the strangeness of the effect very well when he said:

> It is as if one dropped a plank into the sea from a height of 100 feet, and found that the spreading ripple was able, after travelling 1000 miles and becoming infinitesimal in comparison with its original amount, to act on a wooden ship in such a way that a plank of that ship flew out of its place to a height of 100 feet.

Think about this image. When you understand it, you will understand why the photoelectric effect was so remarkable. No matter how low the intensity of the light, energy was either delivered instantly and completely to the atom or not at all.

The idea of the continuous, smooth spreading of electromagnetic energy must simply be wrong. Einstein was the first to accept that possibility. He generalized an idea of Max Planck and asserted that light (both the visible part and all other forms of electromagnetic energy) carries energy in indivisible amounts, or "quanta." Each quantum of light of frequency f has an energy of hf, where h is a constant known as "Planck's constant" with the value 6.63×10^{-34} J s.

■ EXERCISES

> **4.** Find the value of h in eV s, units that will often be useful in this book.

Your answer $h = 4.14 \times 10^{-15}$ eV s means that light of frequency 700 THz consists of many packets, or quanta, each carrying an energy of $700 \times 10^{12} \times 4.14 \times 10^{-15} = 2.9$ eV.

If that is the case, said Einstein, then we can understand that each electron is emitted immediately following the absorption of a single quantum of light. It is then evident that reducing the intensity of the light just reduces the number of quanta, but not the time it takes a particular quantum to be absorbed and produce the emission of an electron.

Also, it is clear from conservation of energy that no electron can be ejected from the metal with more energy than the quantum of light brought to it in the first place. In fact, experimental evidence showed that even the electrons emitted with maximum energy did not have the full energy supplied by the "photon," as a quantum of light is often called. Einstein understood that it would take some energy to break the electron loose from the surface; this energy is called the "work function" of the metal and is often represented by the lowercase Greek letter phi, ϕ. Einstein summarized his ideas by a simple statement of the conservation of energy,

$$K_{\max} = \frac{1}{2} m v_{\max}^2 = hf - \phi, \tag{2}$$

which says that the maximum kinetic energy a photoelectron can have is the amount carried in by the photon less the energy used to break the electron loose.

Experimental Verification of Einstein's Equation

Precise experimental verification of Einstein's equation was difficult for practical reasons. The energies of emitted photoelectrons depend sensitively on the state of the metallic surface. Metal surfaces are always coated with oxide layers, and surfaces usually have thin films of oil or other contaminants. It is difficult to prepare surfaces so that from experiment to experiment the surface is in the same state. Millikan[3] partially solved this problem by installing a machine tool in an evacuated box and using it to shave off a thin layer of metal to make a surface in a well-defined state of cleanliness. Even then, it was hard to get reproducible results because a vacuum is never perfect, and a surface will quickly oxidize and pick up contaminants after it is cleaned. Millikan's results were convincing because he found an independent way to determine ϕ, and then he showed that there was a way to prepare the surface so that ϕ remained unchanged over several weeks. The surface was quite dirty, but it was reproducibly dirty. That was enough to make his measurements meaningful.[4]

Millikan measured the maximum kinetic energy of photoelectrons emitted from sodium metal illuminated with various wavelengths of light. His apparatus, shown schematically in Fig. 13.3, was entirely contained in vacuum. The wheel W would rotate one of the three metal cylinders—one of sodium metal, one of lithium, and one of potassium—under the knife K, which then shaved off a layer of metal to clean the surface. The cylinder was then rotated around to face a window O through which came a beam of monochromatic light.

The energy of the electrons was determined by measuring the photocurrent versus the electric potential difference between the sodium metal and the wire mesh B and collector C. This electric potential difference is the cathode voltage, V, and when it is positive, it impedes the flow of the electrons from the sodium cathode to the mesh and the anode. The cathode voltage was varied until the photocurrent vanished. The value of V_0 at which electrons just cease to flow corresponds to the maximum electron energy, $i.e.$, $K_{\max}$, because

$$\frac{1}{2}mv_{\max}^2 = eV_0$$

---

[3]R.A. Millikan, A direct photoelectric determination of Planck's "h," *Phys. Rev.* **7**, 355–388 (1916); Einstein's photoelectric equation and contact electromotive force, *Phys. Rev.* **7** (Second series), 18–32 (1916).

[4]This is another situation where inadequate vacuum resulting from the limits of the technology of the day caused much confusion that was cleared up only by ingenious design of the experiment and by carrying it out very carefully.

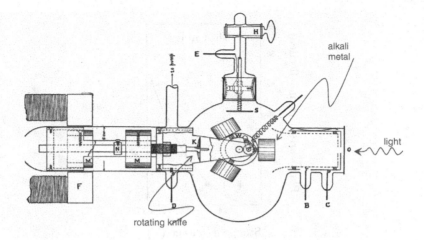

FIGURE 13.3 Diagram of Millikan's photoelectric effect apparatus. *Taken with permission from R.A. Millikan, Phys. Rev. Vol. 7, 355–388 (1916), ©1916 The American Physics Society.*

TABLE 13.1 **Experimental values of photocurrent I_{pc} vs. cathode voltage V for sodium illuminated with six different wavelengths of light**[a]

546.1 nm		433.9 nm		404.7 nm		365.0 nm		312.6 nm		253.5 nm	
V_0 (V)	I_{pc} (mm defl)	V_0 (V)	I_{pc} (mm defl)	V_0 (V)	I_{pc} (mm defl)	V_0 (V)	I_{pc} (mm defl)	V_0 (V)	I_{pc} (mm defl)	V_0 (V)	I_{pc} (mm defl)
0.253	28	0.829	44	0.934	82	1.353	67.5	1.929	52	2.452	68
0.305	14	0.889	20	0.986	55	1.405	36	1.981	29	2.568	38
0.358	7	0.934	10	1.039	36	1.458	19	2.034	12	2.672	26
0.410	3	0.986	4	1.091	24	1.510	11	2.086	5	2.777	16.5
				1.143	10	1.562	4			2.882	8
				1.196	3						
0.46 V		1.03 V		1.21 V		1.59 V		2.13 V		3.03 V	

[a] Adapted from Millikan, A direct photoelectric determination of Planck's "h," *Phys. Rev.* **7** 355–388 (1916). Three entries from his table have been corrected to agree with his graphs, and every entry has been corrected for 2.51 V of contact potential. The bottom row of the table contains the values of V_0 at which I_{pc} goes to zero as determined by Millikan's extrapolation. The current was read as millimeters of deflection (mm defl) on the scale of an electrometer.

(after correcting for the voltage produced by the battery action of the different metals of which the apparatus was made).

Millikan made careful measurements using several different wavelengths of light on lithium, sodium and potassium. His data for sodium are given in Table 13.1. These data illustrate an important difficulty of experi-

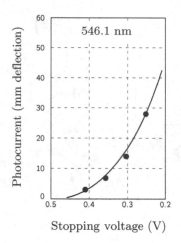

FIGURE 13.4 Use of extrapolation to determine the cathode voltage V_0 at which the photocurrent $I_{pc} \to 0$ for 546.1 nm light on sodium.

ments that try to determine the point at which a quantity becomes zero. Accurate location of this point is usually difficult because there are always a few electrons around from stray light or other sources, and they obscure the zero point. The answer is found by measuring V at points where the photocurrent is not zero and then determining V_0 by extrapolation, as shown in Fig. 13.4. There is always some guesswork in such extrapolation because the data points that will best inform you about where the zero point is are also the least reliable data points because they are small and more like the obscuring background. Millikan's extrapolated values are given in the last row of Table 13.1.

▮ EXERCISES

> **5.** Try some extrapolation yourself. Find the zero points for 433.9 and 404.7 nm light. How well do your values agree with Millikan's?

▽ EXAMPLES

> **1.** How can you find the work function from Millikan's data? For 546.1 nm light the cathode voltage at which the photocurrent becomes zero (found by extrapolation) is 0.46 V; therefore, the maximum energy K_{max} of electrons from the photoelectric effect induced with

TABLE 13.2 Work functions of some metals

Metal	Work function (eV)
silver	4.73
aluminum	4.20
calcium	2.7
cesium	1.9
potassium	1.76–2.25
sodium	1.90–2.46
rubidium	1.8–2.2
copper	4.1–4.5
iron	4.72
nickel	5.01
platinum	6.30

this light is 0.46 eV. Because the energy of a 546.1 nm photon is $1240/546.1 = 2.27$ eV, the work function must be $2.27 - 0.46 = 1.81$ eV.

EXERCISES

6. Calculate the work function using Millikan's data for $\lambda = 433.9$, 404.7, 365.0, 312.6, and 253.5 nm. How do your answers compare among themselves? With the value(s) given in Table 13.2?

7. Consider Eq. 2. In a plot of K_{max} of the photoelectrons vs. the frequency of the incident light, what functional dependence and shape of curve do you expect? What should be the slope?

8. Use Millikan's data to determine values of K_{max}. Plot them versus frequency. Do the slope and intercept agree with what you would expect from Eq. 2? Your results should show excellent agreement between Millikan's measurements and Einstein's predictions.

So Einstein was right. Electromagnetic radiation has its energy in packets. It is quantized. Light of frequency f has a smallest indivisible amount of energy hf. Light possesses a kind of atomicity; it consists of photons.

▼ EXAMPLES

2. What is the maximum velocity of photoelectrons emitted when 250 nm light strikes a clean aluminum plate?

Notice that in addition to answering the question, this example introduces you to an efficient way to calculate a photon energy $hc = 1240$ eV nm because

$$hc = 4.136 \times 10^{-15} \text{ eV s} \times 2.9979 \times 10^8 \text{ m s}^{-1} = 1240 \text{ eV nm}.$$

To find the maximum possible velocity, find hf and ϕ and then use conservation of energy Eq. 2 to calculate the maximum possible kinetic energy of an emitted electron. For Al, $\phi = 4.20$ eV (see Table 13.2). For 250 nm light,

$$hf = \frac{hc}{\lambda} = \frac{1240}{250} = 4.96 \text{ eV}.$$

(Learn to use this way to connect hf and λ; it permits you to use photon energies given in eV and wavelengths given in nm without having to convert them to SI units.)

From conservation of energy

$$\frac{1}{2}mv_{\text{max}}^2 = 4.96 - 4.20 = 0.76 \text{ eV}.$$

To find v_{max} rewrite the above to have mass in units of eV/c^2,

$$\frac{1}{2}mv_{\text{max}}^2 = \frac{1}{2}mc^2 \left(\frac{v_{\text{max}}}{c}\right)^2 = 0.76 \text{ eV},$$

which you can solve for v_{max}/c:

$$\frac{v_{\text{max}}}{c} = \sqrt{\frac{2 \times .76}{mc^2}} = \sqrt{\frac{1.52}{511 \times 10^3}} = 1.73 \times 10^{-3},$$

so

$$v_{\text{max}} = 1.73 \times 10^{-3} c = 5.2 \times 10^5 \text{ m s}^{-1}.$$

13.3 PHOTOMULTIPLIER TUBES: AN APPLICATION OF THE PHOTOELECTRIC EFFECT

Within five years of its discovery the photoelectric effect was used to measure ultraviolet radiation from the Sun. Since then, many instruments for detecting and measuring light have been based on the photoelectric effect. One of these, the photomultiplier tube, is widely used in industry and in research. This device can detect individual photons; it can directly demonstrate the granularity of light.

How the Photomultiplier Tube Works

The photomultiplier uses two different effects: the photoelectric effect and electron multiplication. When a photon strikes a surface coated with a material with low work function, an electron is sometimes emitted. This is just the photoelectric effect, and it represents a conversion of light energy into an electric current, i.e., into the motion of an electric charge.

A single electron charge is difficult to detect, and it is desirable to amplify it. This is done by placing nearby a second surface at a voltage positive with respect to the first. The electron emitted from the first surface is then accelerated toward the second surface. It gains enough energy so that when it strikes the second surface, it causes the emission of several electrons. This process is called "secondary emission," and it multiplies the electrons.

The multiplication can be repeated by placing a third surface nearby with a voltage positive relative to the second. Then if the surfaces are properly shaped and arranged, the electric fields associated with the voltage difference between them will direct all the electrons emitted from the second surface onto the third one. Each impacting electron causes the emission of several electrons, and the multiplication repeats, as shown in Fig. 13.5.

Parts of a Photomultiplier Tube

The surfaces that emit and collect electrons are called the "electrodes" of the photomultiplier. The first electrode, where the photoelectric effect occurs, is called the "photocathode." The last electrode, where the multiplied electrons emerge as a current, is called the "anode." (You can see that these names are adapted from Faraday's names for the parts of an electrolytic cell.) The electrodes between the photocathode and the anode are called "dynodes." It is at the dynodes that electron multiplication occurs.

An electron from the photocathode passing down a string of dynodes each 100 volts higher than the preceding one can give rise to an overall amplification from 10^3 to 10^8. Exactly what amplification occurs depends upon the voltage between the dynodes, the material of which the dynodes are made, the total number of dynodes in the string, and the particular geometric arrangement of the dynodes. Figure 13.5 shows schematically one possible arrangement of photomultiplier electrodes.

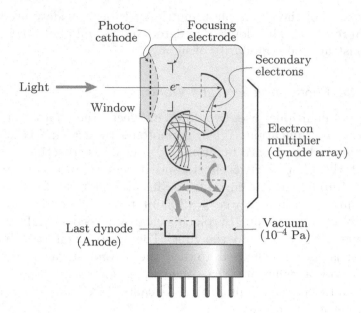

FIGURE 13.5 Schematic diagram of electrodes and electron multiplication in a photomultiplier tube.

▪ EXERCISES

> **9.** What is the overall amplification of a photomultiplier tube that has nine dynodes and an electron multiplication factor of six at each dynode?

Among the many practical considerations in the design of photomultiplier tubes is finding photocathode materials that can emit an electron for long wavelengths of light, i.e., materials with especially low work functions. Even with enough energy, not every photon causes the emission of an electron. Therefore, we look for materials in which production of a photoelectron is both energetically possible and maximally probable. Some alkali metals produce only one photoelectron for every 1000 incident photons, while some mixtures of alkalis, e.g., the multialkali Na K Sb Cs, can produce as many as 300 electrons for every 1000 photons. This ratio is very important and is called the "quantum efficiency." The two examples here illustrate quantum efficiencies of 0.1% and 30%. A quantum efficiency of 30% is about as large as you can get with visible light.

Scintillation Counting of Radioactivity: A Useful Application

When an energetic charged particle or a high-energy photon strikes a piece of matter, it may impart all or some of its energy to an electron in the matter. The electron's energy will often go to make a small flash of light, and if the crystal is transparent, as is, for example, sodium iodide doped with thallium, NaI(Tl), this flash can be detected. A small flash of light is called a "scintillation." By detecting and counting the scintillations, one detects and counts incident particles. Even better, the size of the light flash is proportional to the amount of energy the particle or incident photon leaves in the crystal. By comparing the amounts of light in different scintillations one compares the energies of incident particles. The scintillator converts the energy of an invisible charged particle or high-energy photon into a pulse of low-energy visible photons.

The photomultiplier tube is very effective for counting such scintillations. The tube (see Fig. 13.5) is arranged to view the crystal into which the incident particles are directed. Photons from a scintillation strike the photocathode of the photomultiplier, which emits a tiny pulse of electrons. These are then amplified, and there appears at the anode an electrical pulse signaling the presence of an incident particle; the pulse size is proportional to the energy deposited in the crystal by the detected particle.

13.4 SUMMARY

Electromagnetic radiation comes in discrete packets of energy called photons. For radiation of frequency f, the energy of a single photon is hf, where h is an experimentally determined constant called "Planck's constant." Its value is 4.14×10^{-15} eV s, but it is especially useful to *remember* that $hc = 1240$ eV nm.

Einstein's idea of the granularity of light explained the photoelectric effect. He showed that if light came in discrete quanta, then radiation of frequency f upon being absorbed into a surface could release electrons with a maximum kinetic energy

$$\frac{1}{2}mv_{\max}^2 = hf - \phi,$$

where ϕ is the work function of the surface. Millikan's experimental measurements verified Einstein's predictions.

PROBLEMS

1. You have been put in the basement of a building from which you can escape only by riding up to the ground floor on a coin operated elevator that requires $2. Someone passes you a $5 bill through the bars of the basement window. Fortunately, the elevator's money slot makes change. There is also in the elevator a video game that requires $0.50 per game.

    a. What is the maximum amount of money that you can have left when you reach the ground floor?

    b. What is the minimum amount of money that you can have left when you reach the ground floor?

    c. How is this situation analogous to the photoelectric effect?

2. Light is directed onto a metal surface for which the work function is 2 eV. If the light's frequency f is such that $hf = 5$ eV, what is the maximum energy with which an electron can be emitted from the surface?

3. Light with wavelength of $\lambda = 450$ nm shines on a cesium sample, and a photoelectric current flows. With a retarding voltage of 0.85 V the current goes to zero.

    a. What is the maximum kinetic energy of electrons emitted by the cesium?

    b. Find the work function for cesium.

4. A marvelous new metal, phonium, is found to have a work function of 1 eV.

    a. If a photon of 3 eV energy strikes phonium and causes the emission of an electron, what is the maximum kinetic energy that electron can have?

    b. Light of wavelength 620 nm strikes phonium. What is the maximum energy of electrons emitted by the photoelectric effect under these circumstances?

    c. What are two features of the photoelectric effect that support the idea that light energy comes as multiples of some smallest, indivisible packet?

    d. What is the highest energy photon that can be produced when electrons that have been accelerated through 40 kV stop by a sudden collision with a tungsten anode?

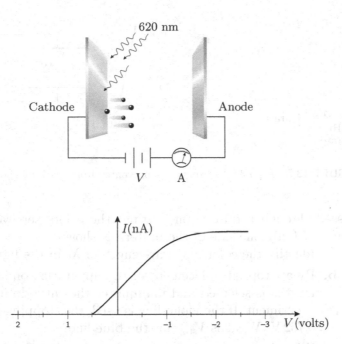

FIGURE 13.6 Photocurrent vs. cathode voltage (Problem 5).

5. The schematic diagram in Fig. 13.6 shows an apparatus for studying the photoelectric (PE) effect. When light of wavelength 620 nm shines on the cathode, electrons are emitted. The current I of electrons that reach the anode across a potential difference V is measured on the ammeter A. (In what follows neglect any contact potential.)

    a. What did Einstein conclude about the nature of light in order to explain the various features of the PE effect?

    b. As the voltage V is varied from negative to positive values, the current I changes as shown in the graph above.

        i. Why is the current of emitted electrons not zero when $V < 0$?

        ii. What is the maximum kinetic energy of an electron emitted when 620 nm light strikes the cathode?

    c. On the same graph, sketch how the current I varies as V is changed when the cathode is illuminated with 310 nm light.

    d. What is the work function (in eV) of the cathode?

6. Visible light from a mercury lamp is composed mainly of five wavelengths: 615 nm, 579 nm, 546 nm, 435 nm, and 405 nm. The lines are colored blue, violet, red, green, and yellow (in no particular order). When

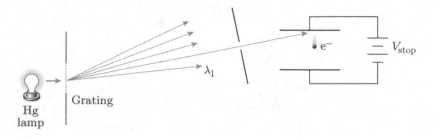

FIGURE 13.7   A photoelectron is just prevented from reaching the anode (Problem 6).

passed through a diffraction grating, the colors separate as shown in Fig. 13.7 (only one side of the pattern is shown).

   a. Identify the color and wavelength of λ_1 in the figure.

   b. By appropriate placement of the apparatus, one of the five wavelengths is selected and illuminates the cathode in a photoelectric experiment. If the violet line is used, the stopping potential $V_{stop} = 0.762\,\text{V}$. What is V_{stop} for the blue line?

   c. What is the stopping potential when red light from the mercury lamp is used? Explain your result carefully.

7. An electric discharge causes atomic hydrogen to emit photons with energies of 1.89, 2.55, and 2.86 eV. In an experiment these are passed through a 1 mm wide slit and then through a diffraction grating that has 5000 rulings per centimeter. The photons then go on and strike a screen 1 m away.

   a. What are the wavelengths of these photons?

   b. Describe what appears on the screen and where. Be quantitative. Also give colors and tell how you know them.

   c. Explain why these results suggest that light is a wave?

8. Millikan's photoelectric data for a lithium cathode are:

λ (nm)	433.9	404.7	365.0	312.5	253.5
V_0 (V)	0.55	0.73	1.09	1.67	2.57

Make a plot of the stopping potential versus frequency and find:

   a. The value of Planck's constant.

   b. The work function of lithium.

c. The cutoff frequency below which no electrons are emitted from the lithium cathode.

9. The stopping potential for photoelectrons emitted from a photocathode surface illuminated by light of wavelength 491 nm is 0.63 V. When the wavelength is changed to a new value, the stopping potential is found to be 1.43 V.

    a. What is the new wavelength?

    b. What is the work function of the surface?

    c. Use Table 13.2 to identify the material of which this photocathode might be made.

10. An experimental physicist tells you that when she studied photoemission from a certain material, she found the work function to be 2.35 eV and the threshold wavelength (i.e, the wavelength at which electrons just begin to be emitted) to be 438 nm. Would you believe these results or ask that she repeat her measurements? Justify your answer.

C H A P T E R

14.

X-Rays

14.1   INTRODUCTION

As the nineteenth century came to a close, three discoveries revolutionized physics, set the stage for remarkable technological developments, and
ushered in a new century and a new physics. In 1895 Roentgen discovered
x-rays; in 1896 Becquerel discovered radioactivity; in 1897 J.J. Thomson
discovered the electron. Each discovery became a new tool with which
physicists explored the atom and made the discoveries that have led to
the extraordinary technology that underpins our society today.

Each discovery resulted in a similar pattern of exploration. A new
particle or radiation was discovered. Its properties were determined; a
technology for its production was developed; and once its properties were
understood, it was used to explore matter further. For example, after
the discovery of x-rays, physicists learned to generate them, detect them,
measure their energy, produce beams of them, and use them as probes to
explore the way in which crystals are built up of atoms.

In this chapter we discuss the technology for producing and measuring
x-rays. A later chapter will discuss how x-rays are used to probe and
reveal basic features of atomic structure.

14.2   PROPERTIES OF X-RAYS

Roentgen discovered x-rays while studying cathode rays. He noticed that
certain materials outside and a little distance away from the cathode-ray
tube would fluoresce, i.e., give off light, when cathode rays were striking

C.H. Holbrow et al., *Modern Introductory Physics, Second Edition*,
DOI 10.1007/978-0-387-79080-0_14, © Springer Science+Business Media, LLC 1999, 2010

the wall of the tube. He realized that some new form of radiation was traveling from the tube to make the material fluoresce. He did not then know that it was short-wavelength electromagnetic radiation, so he called the mystery radiation "x-rays."

He quickly found that x-rays would cause fluorescence in many materials, that they would darken photographic plates, and that they would cause air to become electrically conducting. Fluorescence, photography, and ionization are still important ways to detect and measure x-rays.

14.3 PRODUCTION OF X-RAYS

You can produce x-rays by sufficiently large acceleration of any kind of charged particle. Electrons, because of their small mass, are the easiest and most practical to accelerate. The simplest way to give electrons a large acceleration is to bring them to an abrupt stop by colliding a beam of them with a metal target. The "Coolidge tube" shown in Fig. 14.1 works this way. Electrons are boiled off a hot cathode C, formed into a beam, and accelerated to a potential of some tens of thousands of volts— 10 kV to 100 kV. They then smash into a water-cooled metal target T, often made of copper or tungsten or molybdenum. The collision produces x-rays. The target is set at an angle so that x-rays can come off at a right angle to the tube without being absorbed by the target material.

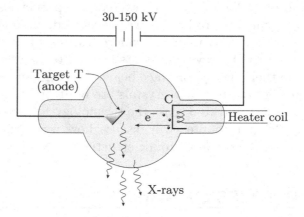

FIGURE 14.1 Diagram of a Coolidge tube for the production of x-rays. Electrons are boiled off from the cathode C, which is heated by the current produced by a power supply that is not shown. Electrons from C are accelerated toward target T by voltage from a high-voltage source. X-rays are produced when the electrons strike the target.

▽ EXAMPLES

1. You can see why water cooling might be needed. An electron current of $10\,mA$ at $50\,kV$ delivers $10 \times 10^{-3} \times 50 \times 10^3 = 500\,W$ of power to the target. Water cooling is a practical way to carry off this heat.

▨ EXERCISES

1. How much power would be delivered to the target by a $75\,kV$, $10\,mA$ beam of electrons?

14.4   X-RAYS ARE WAVES

The wave nature of x-rays remained hidden for more than 15 years after their discovery. As you know from Chap. 10, the test for wave nature is to look for interference. The results of early investigations were ambiguous. The shadow cast onto a photographic plate by x-rays passing around a sharp edge showed a faint fuzziness that might be diffraction. The problem was that if x-rays were waves, their wavelengths were so short that they did not exhibit much diffraction in the sizes of slits one could make in a machine shop and use in a laboratory. Indeed, the unsuccessful attempts at observing diffraction implied that the wavelength must be shorter than $0.1\,nm$, more than a thousand times less than the wavelength of visible light.

In 1912 von Laue suggested that nature provides slits or gratings with dimensions small enough to diffract waves as short as the x-rays might be.[1] From indirect evidence of shapes and sizes it appeared that crystalline matter was composed of atoms laid out in simple patterns with regular spacings of the order of a few tenths of nanometers. Von Laue realized that such arrays should act as three-dimensional diffraction gratings for x-rays, and he predicted that when a beam of x-rays of many different wavelengths passed through a crystal, interference would cause a single

---

[1] W. Friedrich, P. Knipping, and M. Laue, "Interferenzerscheinungen bei Röntgenstrahlen," (Interference phenomena with Roentgen rays), *Sitzungsberichte d. Bayer. Akademie der Wissenschaften*, 303–322 (1912); an abridged translation is given in *The World of the Atom*, edited by Henry Boorse and Lloyd Motz, Basic Books Publishers, New York, 1966, pp. 832–838.

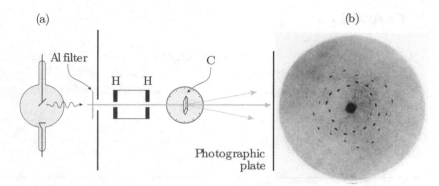

FIGURE 14.2 (a) Friedrich and Knipping's apparatus for producing and detecting x-ray diffraction: X-rays are collimated by apertures in the lead plates H and then diffracted by crystal C. (b) An example of the diffraction data they obtained: The black dots are places where diffraction concentrated the x-rays on the photographic plate. *By permission of Oxford University Press from M. Siegbahn,* The Spectroscopy of X-Rays, *p. 14, Oxford University Press, 1925.*

incident beam to emerge as a sheaf of beams the way a single beam of many wavelengths of light on a diffraction grating emerges as a fan of beams. He predicted that x-rays passed through a crystal would make a pattern of dots on a photograph placed to intercept the emerging sheaf of beams.

Acting on von Laue's idea, Friedrich and Knipping did the simple experiment diagrammed in Fig. 14.2a. They used a crystal of copper sulfate (because it was easy to obtain) and a beam of x-rays containing a broad range of wavelengths. The very first results confirmed von Laue's prediction. Figure 14.2b shows the pattern Friedrich and Knipping obtained. Each dot in the pattern corresponds to a diffracted beam caused by constructive interference from the regular array of atoms in the three-dimensional crystal. The experiment established that x-rays are waves and confirmed that crystals are three-dimensional ordered arrays of atoms.

Nowadays, the pattern of dots produced when a beam of x-rays containing a broad range of wavelengths diffracts from the crystal is called a "Laue pattern," and the dots are called "Laue spots." For his work von Laue received the Nobel Prize in physics in 1914.

14.5   THE BRAGG LAW OF CRYSTAL DIFFRACTION

The work of von Laue, Friedrich, and Knipping established the wave nature of x-rays, but it did not provide a quantitative description to guide

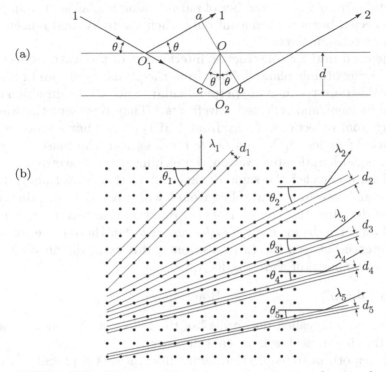

FIGURE 14.3 Bragg conditions for constructive interference of x-rays in a crystal. (a) Geometry for derivation of the Bragg law; (b) illustration of the presence of different sets of Bragg planes; in this illustration five different planes can satisfy the Bragg law for five different wavelengths and so produce five Laue spots.

further exploration. A quantitative theory was developed by W.H. Bragg and W.L. Bragg, father and son.[2]

They recognized that Laue spots could be described as arising from the constructive interference of x-rays reflected from parallel planes of atoms in the crystal. It is as though each atomic plane acts as a slightly reflective mirror, so that a small portion of the x-ray radiation of wavelength λ incident on an atomic plane at a grazing angle θ (see Fig. 14.3) is reflected—mirror-like—at the same angle θ. The rest of the radiation passes through this plane and a part of it is reflected at the same angle θ by the next plane immediately below, and so on, so that there emerges

[2]W.H. Bragg and W.L. Bragg, "Reflection of x-rays by crystals," *Proc. Roy. Soc.* (London), Series A, **88** (1913), 428–438; reprinted in part in *The World of the Atom*, pp. 845–852. See also *X-Rays and Crystal Structure*, W.H. Bragg and W.L. Bragg, G. Bell and Sons, Ltd., London, 1915.

from the crystal in the direction θ radiation made up of the combined reflections from a large number of parallel atomic planes. The intensity of the reflected beam is significant only when the individual reflected waves interfere constructively.

The condition for constructive interference of the waves reflected from neighboring atomic planes follows from the geometry shown in Fig. 14.3a. In the diagram, the lines are perpendicular to and in the direction of travel of the incident and reflected wavefronts.[3] They represent the wavefronts arising from reflections of wavefront 1 at two neighboring atomic planes. Because triangles O_1Oa and O_1Oc are identical, the lines O_1a and O_1c are of equal length, so wavefront 1 reaching a and wavefront 2 reaching c will have traveled the same distance. And then wavefront 1 traveling from a and wavefront 2 traveling from b will travel the same distance to a distant detector. But you see from the diagram that wavefront 2 must also travel an extra distance $cO_2 + O_2b = 2d\sin\theta$. For there to be constructive interference between the two wavefronts, this extra distance must be an integer number n of wavelengths λ, i.e.,

$$2d\sin\theta = n\lambda. \qquad \text{Bragg Law} \qquad (1)$$

The integer n is called the "order" of the reflection. Whenever the Bragg condition is satisfied, a Laue spot appears.

It is important to realize that many different sets of planes of atoms can be imagined in the same crystal. This point is illustrated by the different sets of lines connecting the dots in the diagram of Fig. 14.3b. Each set of planes can reflect x-rays. Of course, the spacings of these different sets of planes are different, and the angles of incidence of incoming x-rays vary with respect to the different sets of planes. The intensity of the diffracted beams drops off as the density of atoms in a plane decreases. The many spots in the Laue photograph (Fig. 14.2) arise because the incident beam contains a continuum of wavelengths, and the various planes of the crystal select out those wavelengths that satisfy Eq. 1 and reflect them as beams that are recorded on the film.

EXAMPLES

2. Suppose x-rays strike a crystal plane at a grazing angle of $20\,°$, and suppose the spacing d between the planes of the crystal is $0.2\,\text{nm}$. What

---

[3]Such lines are called "rays".

wavelength will emerge at $20°$ to the crystal? The first-order value of λ is just

$$\lambda = 2d \sin\theta = 2 \times 0.2 \sin 20° = 0.137\,\text{nm}.$$

There might also be a second-order or a third-order wavelength present.

▪ EXERCISES

2. Suppose the crystal in the above example is rotated to $30°$ with respect to the incident beam and the detector is rotated to $30°$ with respect to the crystal. What is the wavelength of the x-rays that are detected?

Powder Diffraction Patterns

If you use x-rays of a single wavelength, you can also obtain diffraction patterns from powdered samples. A powdered sample of crystalline material consists of many small crystals randomly oriented relative to each other; foils of hammered metal may also consist of many small randomly oriented crystals. When such a sample is irradiated with x-rays, the outgoing diffracted x-rays form a pattern of concentric rings like that shown in Fig. 14.4a and not the array of spots observed in the Laue experiments. Such diffraction from many small randomly oriented crystallites is called "Debye–Scherrer diffraction."

Each ring of the Debye–Scherrer diffraction pattern (Fig. 14.4) corresponds to a different spacing between planes in the crystal being irradiated. To see how a ring might arise, consider a single set of parallel atomic planes within a tiny crystal. If the crystal is tipped so that the incoming beam is incident at the angle θ satisfying the Bragg law, a beam will be diffracted and emerge at an angle 2θ relative to the incident beam and at an angle θ relative to the plane of the crystal, as illustrated schematically in Fig. 14.4b. For the particular spacing d there will be a few other orientations of the crystal relative to the beam that satisfy the Bragg law. These orientations correspond to the higher orders, i.e., $n = 2$, 3, etc.

If the crystal can be tipped only about an axis perpendicular to the plane of the page, as in Fig. 14.4b, the outgoing Bragg reflected beam will

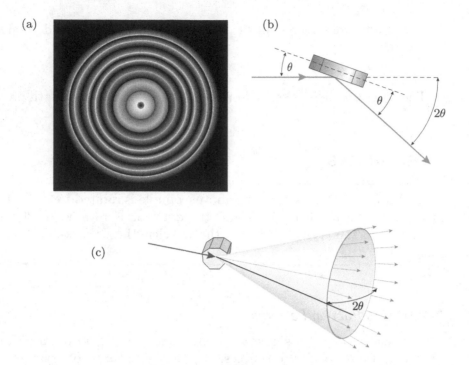

(a)

(b)

(c)

FIGURE 14.4 Why small randomly oriented crystallites produce ring-shaped x-ray diffraction patterns: (a) example of Debye–Scherrer diffraction; (b) schematic illustration of a crystallite satisfying the Bragg law for the incident x-ray wavelength; (c) randomly oriented crystallites will produce a cone of radiation at an angle 2θ to the axis of the incident beam; this forms the circular pattern on the photograph.

lie only in that plane. However, if the crystal were rotated about an axis parallel to the incident beam, the beam would go off at some angle out of the plane of the page. Therefore, from a sample consisting of a large number of tiny crystallites oriented every which way, the outgoing beams will lie along the surface of a cone around the incident beam, as shown in Fig. 14.4c. When these cones meet a photographic plate some distance away, they produce circles like those shown in Fig. 14.4a.

■ EXERCISES

3. Explain what pattern might appear on the photograph if the x-rays had a continuum of wavelengths. Why would this not be useful?

14.6   A DEVICE FOR MEASURING X-RAYS: THE CRYSTAL SPECTROMETER

The Braggs realized that diffraction from the planes of a single crystal spread x-rays out in space according to their wavelengths, and that this property could be the basis for constructing a spectrometer for x-rays. To make their x-ray spectrometer they arranged a crystal at an angle θ relative to the beam coming from an x-ray tube and placed a detector at an angle θ relative to the crystal, as shown in Fig. 14.5. According to Eq. 1, only those x-rays with wavelength $\lambda = 2d\sin\theta$ or, less likely, some simple fraction, e.g., $\lambda/2$, $\lambda/3$, will undergo Bragg reflection through the angle θ. Thus the device selects out from all the different wavelengths of incident x-rays the particular wavelengths that will produce a beam at angle θ. As θ is changed (by rotating the crystal), the wavelength of the Bragg reflected beam changes as determined by the Bragg equation. Then the detector is rotated to the new value of θ to intercept the outgoing beam of the new wavelength. By successive rotations of crystal and detector, through θ and 2θ respectively, you can measure the intensity of x-rays as a function of λ.

Determining the Spacing of Atoms in Crystals

Notice that to extract useful numbers you need to know either the spacing d between planes of atoms in the crystal or the wavelength λ. If you know the crystal spacing, you can use the crystal spectrometer to measure x-ray wavelengths. If you know the wavelength, then you can use those x-rays to measure the spacings of various crystals and so learn how crystalline

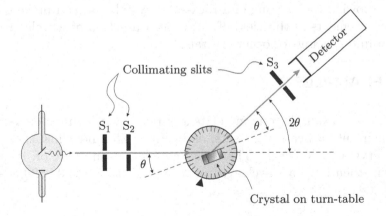

FIGURE 14.5   Schematic diagram of a Bragg spectrometer.

solids are put together. Physicists do both: They use crystals to study x-rays; they use x-rays to study crystals. But how do they get started? It is necessary to measure some crystal spacing or some x-ray wavelength. After that you can use one to measure others. Where do you get the first measurement?

From its Laue pattern the Braggs were able to infer that the Na and Cl atoms of a crystal of table salt sit on the corners of a cube equally spaced a distance d from one another. For such a simple structure the volume occupied by a single atom is d^3, and a numerical value of d^3 can be calculated from the molar weight M_M, the density ρ of salt, and Avogadro's number N_A.

A mole of NaCl has a mass of $M = 35.453 + 22.99 = 58.443\,\mathrm{g}$, and the density of salt is $\rho = 2.163\,\mathrm{g\,cm^{-3}}$. Therefore, the volume of a mole of NaCl is

$$\frac{58.443}{2.163} = 27.01\,\mathrm{cm^3}.$$

Because 1 mole of NaCl contains 2 moles, i.e., $2N_A$, of atoms, each atom occupies a volume

$$d^3 = \frac{27.01}{2 \times 6.022 \times 10^{23}}\,\mathrm{cm^3},$$

from which it follows that

$$d = 2.820 \times 10^{-8}\,\mathrm{cm} = 0.2820\,\mathrm{nm}.$$

It is common, especially in older books, to measure atomic-sized lengths in angstroms (Å) where $1\,\text{Å} = 10^{-8}\,\mathrm{cm} = 10^{-10}\,\mathrm{m} = 0.1\,\mathrm{nm}$. In these units the spacing between the planes of a rock salt crystal is $d = 2.82\,\text{Å}$. In modern textbooks, however, it is customary to use nanometers for atomic-sized dimensions, and we will do that.

Knowing the spacing of a rock-salt crystal, you can make quantitative measurements of the yield of x-rays as a function of wavelength, and you can find spacings of other crystals.

▽ EXAMPLES

3. For example, one might take x-rays from a tungsten target, form them into a beam with slits, as in Fig. 14.5, and allow them to strike a crystal of rock salt. The x-rays emitted at an angle of, let's say, 25° will then have a wavelength of $\lambda = 2d\sin\theta = 2 \times 0.2820 \times \sin 25° = 0.2384\,\mathrm{nm}$.

In this way the Bragg spectrometer is used to measure unknown x-ray wavelengths.

▼ **EXAMPLES**

4. The x-rays coming from the rock-salt crystal in Example 3 are allowed to strike a calcite crystal ($CaCO_3$). The calcite crystal is rotated until there is a strong reflection from it. This occurs when the angles of grazing incidence and reflection are each $23.1°$. From this measurement you can deduce that the spacing of the planes of atoms in the calcite crystal is

$$d = \frac{0.2384}{2\sin 23.1°} = 0.3036 \text{ nm}.$$

In this way a known x-ray wavelength is used to find an unknown spacing of atoms, i.e., the x-rays are used to explore the structure of crystalline matter.

■ **EXERCISES**

4. At what angle would such a beam of x-rays reflect strongly from a quartz crystal? From a mica crystal?

It is important to keep in mind that there are many other sets of planes within the crystals, and each set will give rise to an x-ray beam when the Bragg law is satisfied. The theory of these is elaborate, but the above simple examples suffice to show you how x-rays can be used to find new details of atomic structure.

The spacing of the calcite crystal is useful information because calcite was a crystal frequently used in early x-ray spectrometers. Its spacing and those of some other historically important crystals are given in Table 14.1.

TABLE 14.1 Spacing between planes of commonly used crystals at 18 °C

Crystal		d (nm)
Calcite	$CaCO_3$	0.3036
Rock salt	NaCl	0.2820
Quartz	SiO_2	0.4255
Mica	SiO_2	0.9963

14.7 CONTINUUM X-RAYS

Using known crystal spacings you can make a spectrometer from any convenient crystal and measure how much of each different wavelength is present in the x-rays. Ulrey used a Bragg spectrometer with a calcite crystal to measure the intensity of x-rays from a tungsten target as a function of the angle of orientation θ of the crystal. This angle is related to the wavelength by the Bragg law, Eq. 1.

Remember that in a Bragg spectrometer (Fig. 14.5) the detector is rotated to view an exit angle θ corresponding to the angle of grazing incidence θ. In other words, the detector must always be at an angle of 2θ relative to the axis of the incident beam of x-rays in order to register the intensity of the x-rays diffracted through the Bragg angle θ. With $V_{acc} = 50\,kV$ and setting the crystal and detector at many different angles, Ulrey obtained the data shown in Table 14.2: the intensity distribution of the x-rays produced when energetic electrons strike a tungsten (W) anode in an x-ray tube.

Ulrey's data[4] are plotted in Fig. 14.6 for several values of V_{acc}. They show two particularly interesting features: a continuous, smooth variation

TABLE 14.2 Intensity of X-rays vs. angle of grazing exit θ from calcite[a]

X-Ray intensity	θ	X-Ray intensity	θ
(relative units)	(degrees)	(relative units)	(degrees)
1.9	2.45	6.4	6.24
5.0	2.83	5.6	6.62
6.8	3.30	4.8	6.91
8.6	3.59	4.1	7.19
9.5	3.87	3.2	7.57
9.9	4.25	2.6	7.86
9.9	4.53	2.3	8.14
9.6	4.91	2.0	8.52
9.1	5.29	1.7	8.81
8.0	5.58	1.5	9.19
7.0	5.86		

[a] Taken from Ulrey, *Phys. Rev.* **11**, 401 (1918).

---

[4]C.T. Ulrey, "Energy in the continuous x-ray spectra of certain elements," *Phys. Rev.* **11**, 401–410 (1918).

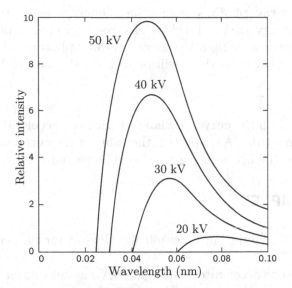

FIGURE 14.6 Relative intensity of continuum x-rays from electrons striking a tungsten target. Each curve is labeled with the voltage through which its corresponding electrons were accelerated. *Taken with permission from C.T. Ulrey, Phys. Rev. 11, 401–410 (1918)* ©*1918 The American Physical Society.*

of intensity from long wavelengths down to some short wavelength and a sudden termination, or "cutoff," of the curve at that short wavelength.

The continuum is to be expected from the simple picture of how x-rays are produced. Not all electrons striking the target undergo the same acceleration; some are accelerated suddenly, some less so. The distribution of wavelengths should follow the distribution of accelerations. The largest intensity corresponds to the acceleration that occurs with greatest probability.

But the short wavelength cutoff warns us that this simple picture is incomplete. Although we might expect extremely sudden decelerations to be less probable, there is no reason in classical physics why their probability should suddenly drop to zero. Classical physics leads us to expect that the continuum will peter out at short wavelengths, not stop suddenly.

14.8   X-RAY PHOTONS

The short-wavelength cutoff is explained by the granular nature of electromagnetic radiation. Just as there are photons of visible light, so also are there x-ray photons, and x-ray energy comes in packets $E = hf = hc/\lambda$.

Because x-ray photons cannot have energy larger than the maximum kinetic energy of the incident electrons, electrons of charge e accelerated through a voltage V cannot produce photons with energies greater than eV. Therefore, there will be a shortest possible λ, and it will be

$$\lambda_{\text{cutoff}} = \frac{hc}{eV}. \tag{2}$$

In Fig. 14.6 the curves are labeled with the accelerating voltage applied to the x-ray tube. As expected, the value of the cutoff wavelength becomes smaller as the accelerating voltage is increased.

EXAMPLES

5. What should be the cutoff wavelength for an accelerating voltage of 30 kV?

Electrons accelerated through 30 kV cannot deliver more than 30 keV of energy to the target. Therefore, the shortest possible x-ray wavelength is $hc/(eV) = 1240/30\,000 = 0.0413$ nm. This agrees well with Ulrey's data.

Table 14.3 shows cutoff wavelengths determined from Ulrey's data for various accelerating voltages.

EXERCISES

5. Use Eq. 2 to predict the cutoff wavelengths for the curves labeled 40 kV and 20 kV in Fig. 14.6. How do your answers compare with Ulrey's data?

TABLE 14.3   X-Ray tube voltage and cutoff wavelengths

Tube voltage (kV)	λ_{cutoff} (nm)
20	0.062
30	0.0413
40	0.031
50	0.0248

6. In an old fashioned color-television tube electrons pass through a potential difference of 30 kV before they strike the phosphorescent screen at the end of the tube. What is the maximum energy that an x-ray photon can have coming from this tube? What is the wavelength of that photon?

7. In the previous exercise what (approximately) is the most probable energy of photon that will be produced by the TV tube?

14.9   THE COMPTON EFFECT

Introduction

The existence of the short wavelength cutoff is a reminder that x-rays do not always behave like waves. Indeed, studies with the crystal spectrometer and other techniques for measuring x-rays show that in some circumstances they behave like hard, featureless particles. Photons can collide with electrons and bounce off them like tiny BBs.

Bragg scattering is an important special case of the interaction of x-rays with matter, but it is not the only kind of x-ray scattering. A beam of x-rays directed at a crystal will result in the weak emission of x-rays at all angles relative to the incident beam. In the early 1920s, the American physicist Arthur Holly Compton studied this kind of x-ray scattering. He scattered x-rays of a single well-defined wavelength from a small piece of carbon and observed that at scattering angles other than 0° some of the scattered radiation came out with its wavelength unchanged and some came out with its wavelength increased. He explained the change in wavelength as the result of a billiard-ball-like collision of an x-ray photon with a nearly free electron of a carbon atom. Such a change in wavelength (or energy) of a photon after it scatters from a charged particle is called the "Compton effect."

The discovery and explanation of the Compton effect decisively persuaded physicists of the granularity of light. This was important because the idea of corpuscular electromagnetic radiation, e.g., photons of light or of x-rays, was widely doubted even after Einstein explained the photoelectric effect in terms of the absorption of photons.

Compton Scattering

Figure 14.7 shows a diagram of the apparatus Compton used to exhibit the effect. X-rays of a well-defined wavelength[5] were generated by a specially designed x-ray tube and directed at a small block of graphite (carbon) place near the anode at the point R in the figure. The x-ray source and the graphite were surrounded by a lead box. The graphite scattered the x-rays in all directions, but collimating slits S_1 and S_2 allowed only x-rays scattered through a chosen angle θ to escape from the box. For the setup shown in Fig. 14.7 the scattering angle is $\theta = 90°$. These scattered x-rays are quite weak, so the graphite must be placed close to the anode to increase their intensity.

To measure the wavelengths of the scattered x-rays, Compton let them strike a calcite crystal that he used as an x-ray spectrometer. By rotating the calcite and the detector, as shown in Fig. 14.5, and using the Bragg Law, he could measure the intensity of the scattered x-rays as a function of their wavelength.

Figure 14.8 shows Compton's results for scattering angles $\theta = 45°$, $90°$, and $135°$. The y-axis in each figure is the intensity measured by the detector. The x-axis displays the Bragg angle θ_B of reflection from the calcite crystal in the spectrometer. To find the wavelength of each peak,

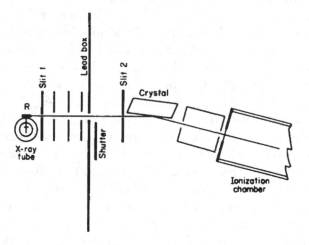

FIGURE 14.7 Diagram of Compton's apparatus. *Taken with permission from A.H. Compton, Phys. Rev. 22, 409–413 (1923) ©1923 The American Physical Society.*

---

[5] In addition to the spectrum of continuum radiation (Sect. 14.7), x-ray tubes also emit intense radiation at one or more well-defined wavelengths. These "characteristic" wavelengths become much more intense than the continuum radiation as the accelerating voltage is increased; you will encounter them again in Chap. 17.

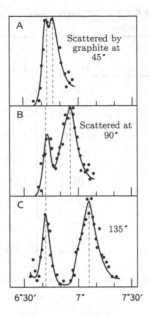

FIGURE 14.8 Compton's data showing the appearance of the longer-wavelength x-rays as the scattering angle θ is increased; the abscissa is the angle θ_B of the Bragg spectrometer. *Taken with permission from A.H. Compton*, Phys. Rev. *22, 409–413 (1923)* ©*1923 The American Physical Society.*

you apply Bragg's Law to θ_B. (Note that there are two different angles in this discussion: θ is the angle through which x-rays are scattered from the graphite, and θ_B is the angle of reflection from the calcite crystal of the spectrometer.)

▼ EXAMPLES

6. For calcite $d = 0.3036\,\text{nm}$. From the scale of Fig. 14.8 you can see that the primary peak occurs at $6°\,42' = 6.7°$. The wavelength of the x-rays making this peak must then be

$$\lambda = 2 \times 0.3036 \sin 6.7° = 0.0708\,\text{nm}.$$

The figure shows that a second peak emerges and moves toward larger Bragg angles as the scattering angle is increased. Larger Bragg angles correspond to longer wavelengths and thus to lower photon energies.

TABLE 14.4 Relative intensity of X-rays scattered from carbon

$\theta = 135°$ Intensity	Wavelength (nm)	$\theta = 135°$ Intensity	Wavelength (nm)
4.3	0.0688	13.8	0.0734
0.5	0.0689	20.0	0.0739
5.5	0.0697	25.8	0.0745
8.0	0.0703	25.6	0.0750
13.0	0.0707	18.9	0.0754
7.8	0.0713	10.8	0.0760
3.0	0.0719	4.8	0.0765
2.8	0.0725	2.0	0.0773
11.0	0.0729		

▮ EXERCISES

8. Use the data in Fig. 14.8 to find the wavelength of the lower energy photons when the scattering angle is 90°.

Table 14.4 shows the data Compton obtained when he scattered 0.0708 nm radiation, looked at x-rays coming out at 135°, and used the Bragg Law to find their wavelengths.

▮ EXERCISES

9. Make a graph of the data in Table 14.4 and draw a smooth curve though the points. Label the peak that has the same wavelength as the incident photons; label the peak that arises from Compton scattering. By how much is the wavelength shifted?

The occurrence of the shifted peak is the Compton effect. Compton explained the effect as due to elastic scattering, i.e., scattering like that of hard spheres bouncing off each other. One of the "spheres" was an electron in the carbon; the other was a packet of electromagnetic energy—the photon. Elastic scattering of a photon from an unbound charged particle is called "Compton scattering."

The fact that Compton's explanation worked so well led to general acceptance of the concept of the particle-like photon with energy hf, where h is Planck's constant and f is the frequency of the radiation.

Derivation of the Energy Change of a Compton Scattered Photon

Figure 14.9a is a diagram of the scattering process as Compton imagined it. A photon of energy hf enters from the left and strikes a stationary electron of rest mass m. After the collision, the electron recoils away with momentum p_e at an angle ϕ with respect to the line of motion of the incident photon, and a photon of energy hf' travels away at an angle of θ; the electron stops inside the target and is not measured. Because the electron has gained energy from the collision, the outgoing photon's energy of hf' will be less than that of the incoming photon hf.

Although the data in Table 14.4 are the wavelengths λ and λ', it is more convenient to discuss Compton scattering in terms of the corresponding frequencies f and f' and derive a relationship between f' and f and the scattering angle θ.

To do this use the relativistically correct expressions for conservation of energy and momentum. First notice that the relation of photon energy to photon momentum is

$$E = hf = \sqrt{0 + p^2 c^2}$$

because $m_0 = 0$ for a photon. Therefore, *for a photon,*

$$E = hf = pc.$$

■ EXERCISES

---
10.   Show from the above equation that for a photon, $p = h/\lambda$.

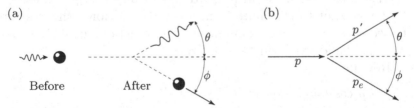

(a)   Before   After   (b)

FIGURE 14.9 (a) Schematic representation of a Compton scattering event. (b) Momentum diagram of Compton scattering.

Now use this relationship between photon energy and photon momentum to write down the total energy of the photon *and* electron before the collision and after. Before the collision, the energy is just the photon's energy plus the rest energy of the electron: $pc + m_e c^2$. After the collision, the total energy is the photon's somewhat lower energy $p'c$ plus the electron's larger energy, as shown below. Because energy is conserved, the total energy before the collision equals the total energy after:

$$pc + m_e c^2 = p'c + \sqrt{m_e^2 c^4 + p_e^2 c^2}, \tag{3}$$

where p_e is the momentum of the recoiling electron and m_e is its rest mass.

To find the relation of the scattering angle of the photon and its momentum after scattering, use conservation of momentum as well as conservation of energy. Conservation of momentum gives two more equations that you can use to eliminate p_e from Eq. 3 so that you can find the photon's momentum and thus its energy.

In Fig. 14.9a a photon travels toward positive x and scatters from an electron as shown. The electron recoils at an angle ϕ while the photon goes off at an angle θ relative to the x-axis. Figure 14.9b shows that there are two momentum equations, one for vertical, or y, momentum, and the other for horizontal, or x, momentum. In the y direction the total momentum is initially zero, so after the collision the amount of momentum up must equal the amount down, and

$$p' \sin \theta = p_e \sin \phi. \tag{4}$$

The amount of momentum in the x direction before the collision equals the total amount in the x direction after, so

$$p = p' \cos \theta + p_e \cos \phi. \tag{5}$$

Our aim is to find p' of the outgoing photon in terms of the incident p and the scattering angle θ. This means that we need to eliminate both ϕ and p_e from Eqs. 3, 4, and 5.

Whenever you have an angle α appearing in trigonometric functions that you want to eliminate from a set of equations, the first approach is to see whether you can combine the trig functions in such a way that you can use the identity $\sin^2 \alpha + \cos^2 \alpha = 1$.

Rewrite Eq. 5 as

$$p - p' \cos \theta = p_e \cos \phi \tag{6}$$

and square it to get

$$p^2 + p'^2 \cos^2 \theta - 2pp' \cos \theta = p_e^2 \cos^2 \phi.$$

Square Eq. 4 to get

$$p'^2 \sin^2 \theta = p_e^2 \sin^2 \phi.$$

Add the two equations and use the trig identity to eliminate ϕ:

$$p_e^2 = p^2 + p'^2 - 2pp' \cos \theta. \tag{7}$$

■ EXERCISES

11. Show that Eq. 7 is correct by deriving it as above but include any missing steps.

Replace p_e^2 in Eq. 3 with the expression given by Eq. 7, to eliminate the electron momentum. The result is an explicit expression for p' in terms of the incident p and the outgoing scattering angle θ:

$$(pc - p'c + m_ec^2)^2 = m_e^2c^4 + p_e^2c^2,$$

$$(pc - p'c + m_ec^2)^2 - m_e^2c^4 = p_e^2c^2 = p^2c^2 + p'^2c^2 - 2pp'c^2 \cos \theta,$$

$$pp'(1 - \cos \theta) = m_ec(p - p'), \tag{8}$$

$$p' = \frac{m_epc}{p(1 - \cos \theta) + m_ec}. \tag{9}$$

■ EXERCISES

12. Do this entire derivation from start to finish.

13. From Eq. 9 show that f' in terms of f and θ is

$$hf' = \frac{hf}{\frac{hf}{m_ec^2}(1 - \cos \theta) + 1}, \tag{10}$$

which is very convenient for calculating how the energy of a photon changes when it scatters.

It is often helpful to express these photon energies relative to the rest energy of the electron. Notice that if you define the ratio of the photon energy to the electron rest energy as

$$x = \frac{hf}{m_ec^2},$$

you can write Eq. 10 in the compact form

$$x' = \frac{x}{x(1 - \cos\theta) + 1}.$$ (11)

▽ EXAMPLES

7. Remembering that the rest mass energy of an electron is 511 keV, suppose a 256 keV photon undergoes Compton scattering through 180°. What will be the energy of the outgoing photon? What will be the energy of the recoiling electron?

For this case $x = \frac{1}{2}$. Since $\cos\theta = -1$, it follows from Eq. 11 that $x' = \frac{1}{4}$. This means that $hf' = 511/4 = 128$ keV. Because the electron must have the remaining part of the energy, its kinetic energy in this particular case will be 128 keV.

Another virtue of this version (Eq. 11) of the Compton scattering equation is apparent if you want to know about Compton scattering from particles other than an electron. Suppose you want to know what happens to a photon that scatters off a proton, $m_\text{p}c^2 = 938$ MeV.

▽ EXAMPLES

8. To see what happens when a 256 keV photon scatters 180° from a proton, calculate $x = 0.256/938 = 2.729 \times 10^{-4}$, from which it follows that $x' = 2.728 \times 10^{-4}$. There is hardly any effect.

▨ EXERCISES

14. If a 100 keV photon scatters from an electron through an angle of 90°, what will be the recoil energy of the electron?

15. What is the maximum energy that an electron can acquire from a 600 keV photon by means of Compton scattering?

The Compton effect is customarily described as a change in wavelength rather than as a change in frequency. To get it in that form divide both sides of Eq. 8 by pp' and also by m_e. This gives

$$\frac{1 - \cos\theta}{m_e} = c\left(\frac{1}{p'} - \frac{1}{p}\right) = c\left(\frac{\lambda'}{h} - \frac{\lambda}{h}\right),$$

where p and p' have been replaced by h/λ and h/λ' respectively, and it follows that

$$\lambda' - \lambda = \frac{h}{m_e c}(1 - \cos\theta). \tag{12}$$

When physicists refer to the "Compton scattering equation" or to the equation for "Compton wavelength shift," they mean Eq. 12, which gives the change in wavelength of the scattered photon as a function of its angle of scattering.

Equation 12 shows the curious fact that the change in wavelength due to Compton scattering does not depend upon the frequency (i.e., energy) of the incident photon.

EXERCISES

16. Is the frequency shift of the scattered photon independent of the frequency of the incident photon? Explain.

17. Table 14.4 shows actual Compton scattering data. What is the energy of the 0.07078 nm incident photon? Calculate the shift in wavelength you would expect when that photon scatters through 135°. How well does your answer compare with the data shown in Table 14.4?

Compton Scattering and the Detection of Photons

Radioactive materials often emit energetic photons called, for historical reasons, "gamma rays." Radioactive cesium emits a gamma ray with an energy of 662 keV. Obviously, this is not a visible photon.

EXERCISES

18. How do you know that this photon is not visible?

As previously noted, such photons, although not visible themselves, can cause other materials to emit visible light. We described in Chap. 13 how a photon can deposit energy in a crystal of sodium iodide [NaI(Tl)] and cause the crystal to give off a flash of light that can be detected with a photomultiplier tube. By measuring the relative intensity of the scintillations you can measure the relative amounts of energy left by the photons in the crystal.

A common measurement of photons is to count the number of flashes that appear in the crystal and sort the counts according to how bright the flashes are.

The graph in Fig. 14.10 was obtained after hundreds of thousands of photons passed through a NaI(Tl) crystal. Of these, somewhere between 1% and 10% produced flashes of light. The flashes were counted and sorted according to their brightness. The graph shows the number of flashes on the y-axis and the brightness along the x-axis. Because brightness is proportional to energy deposited in the crystal, it is convenient to label the brightness scale with the corresponding photon energy, and that is done here.

If flashes came only from complete absorption of incoming single-energy photons, there would be only one size of flash. In Fig. 14.10 you can see that many of the flashes correspond to photons of 662 keV. This group of photons is labeled "photopeak" on the graph. The energies of the emitted photons are very closely the same, but because the flashes of light that they produce vary somewhat in brightness, the peak in the graph is wide. We say that the "resolution" of the instrument is not perfect. The name "photopeak" always refers to the peak corresponding to the full energy of the incident photon.

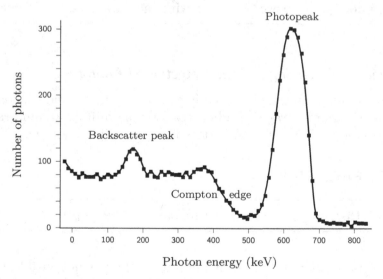

FIGURE 14.10 Distribution of energy deposited in a NaI(Tl) crystal by 662 keV photons emitted following the radioactive decay of ^{137}Cs.

Figure 14.10 shows that other things are going on besides absorption of 662 keV photons. The experimenter detected flashes of light over quite a range of brightnesses corresponding to the absorption of energy from 0 keV up to 800 keV. Two features show up particularly: the bump labeled "backscatter" and the sudden drop from a constant level of counts labeled "Compton edge."

It turns out that there are several ways for a photon to deposit energy in a crystal, and these depend on the energy of the photon. For example, between 250 keV and 1.02 MeV a photon is most likely to do either of two things. It may produce a photoelectron in the crystal and leave all of its energy in the crystal. Or it may Compton scatter off a nearly free electron of some atom and leave the crystal; the energy left in the crystal is then whatever was given to the electron by the Compton scattering. This will produce a smaller flash of light than total absorption because the energy left in the crystal is smaller. It is in this way that most of the lower-energy light pulses occur.

You can use your knowledge of Compton scattering to calculate the energies at which the backscatter peak and the Compton edge occur. The Compton edge corresponds to an event in which a photon enters the crystal and undergoes 180° Compton scattering from an electron of one of the crystal's atoms and then leaves the crystal with reduced energy. The electron remains behind with increased energy, which goes to produce a flash in the crystal. The electron can not acquire more energy than the amount imparted when the photon scatters through 180°, so there is a noticeable decline in the number of light pulses beyond this energy. This decline is called the "Compton edge."

▮ EXERCISES

> **19.** Under what circumstances will a photon that is Compton scattered out of the crystal leave the most possible energy with the electron from which it scattered?
>
> **20.** Calculate the maximum energy that a 662 keV photon will impart to an electron. How does your answer correspond to the energy on the axis of the graph in the region labeled "Compton edge"?
>
> **21.** To what do you attribute the counts observed at energies below the Compton edge?

The backscatter occurs because the NaI crystal is surrounded by dense material, and there is a fair probability that some photons will enter the crystal, pass through it unabsorbed, and then scatter off electrons of surrounding metal (usually aluminum, but it does not make much difference). These photons scatter back into the crystal (hence the name, backscatter) where they are absorbed. The crystal is usually arranged so that only those photons that bounce directly back are likely to reenter the crystal. Of course, photons that have scattered through 180° are reduced in energy, and the flashes of light that they can produce will not be as bright as those produced by the full-energy photons.

■ EXERCISES

> **22.**   What is the energy of a 662 keV photon after it Compton scatters 180° off Al? Off Si? Off Fe?
>
> **23.**   Compare your answer to the previous question with the energy corresponding to the "backscatter peak" in the graph. Why should they be the same?

14.10   SUMMARY

Useful Things to Know

X-rays are usually produced by sudden acceleration of charged particles. In practical circumstances x-rays are made when energetic electrons are brought to a sudden halt by collision with a metal electrode.

The existence of Laue diffraction spots and Debye–Scherrer diffraction rings shows that x-rays are waves with wavelengths a thousand times smaller than those of visible light.

Diffraction of x-rays by crystals obeys the Bragg law

$$2d \sin \theta = n\lambda, \qquad \text{Bragg Law}$$

where d is the spacing between successive planes of atoms in the crystal, θ is the angle of grazing incidence, λ is the wavelength of the x-rays, and n is an integer $n = 1, 2, \ldots$. In any given crystal there will many sets of planes with different spacings. The Bragg law holds for each set.

The Bragg law is the basis for the design and operation of the crystal spectrometer. With this device it is possible to measure x-ray wavelengths and map out the intensity of x-ray emissions as a function of wavelength.

X-rays also exhibit particle-like behavior. To understand the short-wavelength cutoff of the continuous x-ray spectrum, it is necessary to assume that x-rays are photons each with energy $E = hf$. Furthermore, the Compton effect, i.e., the decrease in energy of x-rays as the angle at which they are scattered from free or weakly bound electrons increases, can be understood as resulting from the scattering of one small structureless object by another.

Some Important Things to Keep in Mind

If the wavelengths of x-rays are thousands of times smaller than those of visible light, they can in principle be used to probe the structure of objects thousands of times smaller than the smallest things we can see with our eyes. But notice that x-ray photons are going to be thousands of times more energetic than visible ones; the energy scale shifts from eV to keV.

X-rays are a valuable probe that can yield interesting information about structures as small as 0.1 nm, that is, as small as atoms. By making x-rays of energies of 100 keV and higher, it is possible to probe even smaller structures.

Studies of atoms with x-rays supported two major advances of understanding about atoms. The works of von Laue and the Braggs showed that crystals are made of regular arrays of atoms. The Compton effect provided quite direct evidence that the electrons are constituent parts of every atom.

Be able to convert from energy to wavelength and vice versa without hesitation.

▼ EXAMPLES

9. For example, what is the wavelength of a 100 keV photon? Because $E = hf = hc/\lambda$, a 100 keV photon has a wavelength $\lambda = hc/E$. Never forget that $hc = 1240$ eV nm. Then if you know E in units of eV, you can immediately get that $\lambda = \frac{1240}{100 \times 10^3} = 0.0124$ nm.

▪ EXERCISES

24. What is the wavelength of a 12.4 keV photon?

FIGURE 14.11 Representation of a crystal lattice (Problem 1).

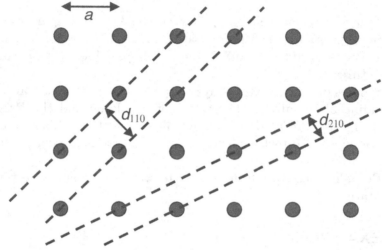

FIGURE 14.12 Schematic representation of a cubic crystal with interatomic spacing a (Problem 2).

PROBLEMS

1. The Bragg Law for crystal diffraction of x-rays is $2d \sin \theta = \lambda$. Use Fig. 14.11 to show:

    a. What the symbols d and θ refer to.

    b. What is meant by Bragg diffraction.

2. Figure 14.12 shows a cubic crystal with interatomic spacing a. Use trigonometry to find the interplanar spacings in terms of a for the atomic planes indicated by d_{110} and d_{210}.

3. Using the data in Table 14.2 on p. 432 and the fact that Ulrey's spectrometer used a calcite crystal:

    a. Calculate the wavelengths corresponding to angles of reflection, and then plot the intensity of the x-radiation vs. λ.

    b. What is the energy of the photon that corresponds to the wavelength at which the intensity is a maximum?

    c. By extrapolation of your graph determine the short-wavelength cutoff.

    d. What is the voltage difference through which these electrons passed before they struck the tungsten anode of the x-ray tube?

4. Electrons in an x-ray tube are accelerated through a potential difference of 50 kV.

    a. What is their kinetic energy?

    b. What is the speed of these electrons in terms of c?

    c. Are relativistic effects important in

        i. determining the kinetic energy?

        ii. determining the speed?

5. Ulrey observed that continuous x-ray spectra generated with electrons produced with the x-ray tube voltages shown in Table 14.3 (p. 434) exhibited sharp cutoffs at the wavelengths shown in the table.

    a. Predict what sort of a curve you will get if you plot f (or hf) vs. V. Explain your prediction.

    b. Using the insights you generated to make your prediction in the preceding, derive a formula relating the voltage V of an x-ray tube to the maximum frequency f_{max} of the emitted radiation.

    c. Why is this formula familiar to you?

6. Some teachers like to use general questions in exams such as:

    a. What is the Compton effect?

    b. How is the Compton effect important to our understanding of the nature of electromagnetic radiation?

7. A photon Compton-scatters from an electron through an angle of $90°$, as shown in Fig. 14.13. In doing so, it loses half its initial energy:

    a. What is the photon's initial wavelength?

    b. What are the initial and final energies of the photon?

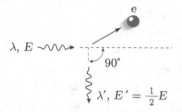

FIGURE 14.13   Compton scattering event (Problem 7).

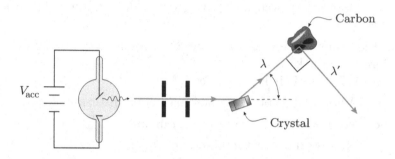

FIGURE 14.14   Compton scattering arrangement for Problem 8.

c. Suppose the photon scattered through 180° instead of 90°. What would be its final wavelength and energy (λ', E')?

d. When the photon scatters through 180°, what is the final energy of the *electron*?

8. Monochromatic x-rays of wavelength $\lambda = 0.243$ nm are produced in the following way: Electrons bombard a metal target in a standard x-ray tube, and the resulting radiation is reflected from a crystalline material, as shown in Fig. 14.14.

a. What is the purpose of the crystal?

b. What is the minimum accelerating voltage V_{acc} that could have produced x-rays of this λ?

c. The reflected x-rays go on to collide with a carbon block C, where their wavelength is shifted to λ' by the Compton effect. For the direction shown in the drawing, calculate the change in wavelength.

d. Does the x-ray gain or lose energy in this process? Where does the energy go? What is the most striking assumption made by Compton in his analysis of this phenomenon?

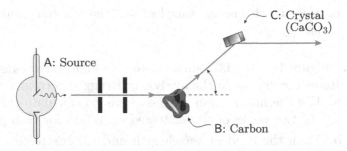

FIGURE 14.15   A Compton scattering arrangement in which the incident x-rays have a single well-defined wavelength (Problem 9).

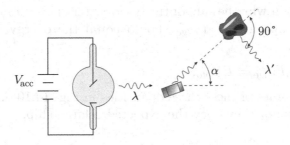

FIGURE 14.16   Compton scattering arrangement for Problem 10.

9. Write a short description of the experiment depicted in Fig. 14.15. Include the following:

    a. Identify and explain the *purpose* of items A, B, and C in Fig. 14.15.

    b. Discuss the significance of the experiment. Why were the experimental results, and the analysis of those results, important and surprising?

10. X-rays from an x-ray tube are Bragg reflected as shown in Fig. 14.16 from the planes of a calcite crystal spaced 0.3036 nm apart. Photons of wavelength $\lambda = 0.3654$ nm emerge from the crystal at an angle α with respect to their initial direction (see Fig. 14.16) and are Compton scattered through an angle of 90° as they pass through a second material.

    a. Find the angle α that the diffracted photon forms with its initial direction.

    b. What is the minimum accelerating voltage V_{acc} that would produce a photon of the given wavelength?

c. How much energy was lost by the photon, and where did the energy go?

11. Figure 14.8 (p. 437) shows three of Compton's scans obtained using calcite as the crystal in the analyzing Bragg spectrometer.

a. Use the information on the x-axis to calculate the wavelengths of the two peaks of the scattered radiation for each plot.

b. From the incident wavelength and the scattered angle use Compton's equation to calculate the wavelength of the radiation that scattered from electrons for each plot.

c. Compare the results of (a) and (b).

d. For each case calculate the energy gained by the scattered *electron*.

12. Explain why the sum of the backscatter energy E_{back} and the energy of the Compton edge E_{edge} should equal the energy of the photopeak E_{photo}, i.e.,

$$E_{back} + E_{edge} = E_{photo}.$$

Find the values of these three quantities in Fig. 14.10 on p. 444 and show whether or not they obey the expected relationship.

13.a. An electron tube (Fig. 14.17a) generates x-rays of intensity $I_1(\lambda)$, as shown in Fig. 14.17b. What is the voltage applied to the electron tube?

b. The x-rays from the tube are incident on a crystal. It is observed that the x-rays that undergo Bragg diffraction emerge at the angle shown in Fig. 14.18a. The intensity of the emerging beam $I_2(\lambda)$ is shown in Fig. 14.18b. What is the spacing d between the planes of the crystal from which the x-rays have diffracted?

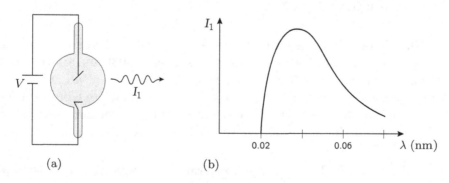

(a)   (b)

FIGURE 14.17   Intensity distribution of x-rays vs. λ for Problem 13(a).

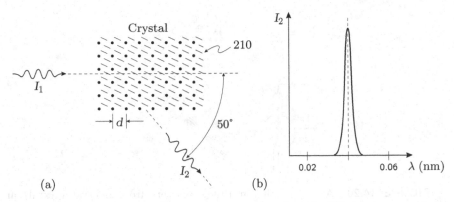

(a)   (b)

FIGURE 14.18 Schematic of Bragg crystal and diffracted x-ray intensity for Problem 13(b).

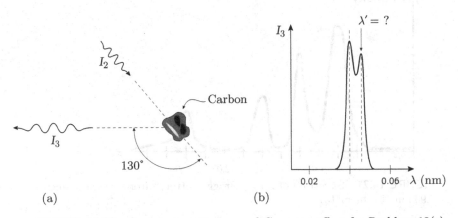

(a)   (b)

FIGURE 14.19 Compton scattering and Compton effect for Problem 13(c).

    c. The x-rays of part (b) are incident on a block of graphite, as shown in Fig. 14.19a. Photons Compton scattered through $130°$ are observed to have a distribution of intensity I_3 vs. wavelength, as shown in Fig. 14.19b. *Calculate* the wavelength λ' of the scattered x-rays.

    d. How much energy is gained by electrons that Compton scatter photons $130°$ as in part (c)?

14. X-rays generated in an x-ray tube are Bragg-reflected through $90°$ by a crystal (xtal in Fig. 14.20). The Bragg-reflected x-rays have a wavelength $\lambda = 0.4000$ nm.

    a. What is the spacing between the crystal planes causing this reflection?

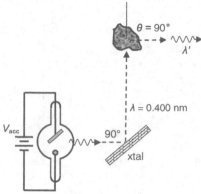

FIGURE 14.20 An x-ray photon Bragg scatters from a crystal (xtal) and then Compton scatters from carbon (Problem 14).

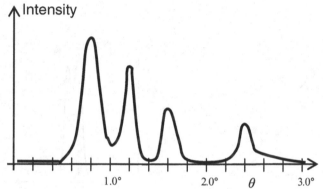

FIGURE 14.21   Spectrum of x-rays Bragg scattered from crystal planes separated by 0.94 nm (Problem 15).

   b. What is the wavelength λ' of the 0.4000 nm x-rays after they Compton scatter through 90° from an amorphous carbon block as shown in Fig. 14.20? Keep 4 significant figures in your answer.

   c. How much energy did each x-ray photon lose when Compton-scattered through 90°?

   d. It takes 11.2 eV to eject an electron from a carbon atom (i.e., to ionize the atom). What is the kinetic energy of the ejected electron after it has left the atom?

15.   An x-ray beam containing radiation of two distinct wavelengths scatters from a crystal, yielding the intensity pattern shown in Fig. 14.21. The spacing between the planes of the scattering crystal is 0.94 nm. Find the two wavelengths.

C H A P T E R

15.

Particles
as Waves

15.1 INTRODUCTION

You have seen that waves can act like particles. The energy of light waves is packaged in quanta called photons. Photons can scatter from electrons like tiny hard objects. If this behavior is not surprising or mysterious enough, it turns out that particles can act like waves. Electrons, protons, neutrons, and atoms all can exhibit interference and other forms of wave behavior.

15.2 THE DE BROGLIE WAVELENGTH

The first clue that particles might have wavelike behavior came when Louis de Broglie noted (as you saw in Chap. 14) that the momentum of a photon is

$$p = \frac{hf}{c} = \frac{h}{\lambda} \tag{1}$$

and suggested that perhaps particles might obey the same equation. In other words, he proposed that particles might behave as waves with a wavelength of

$$\lambda = \frac{h}{p}. \tag{2}$$

This wavelength of particles is called the "de Broglie wavelength."

What would be the size of such a wavelength? Some examples will show you how the wavelength of a particle depends on its kinetic energy K.

C.H. Holbrow et al., *Modern Introductory Physics, Second Edition*,
DOI 10.1007/978-0-387-79080-0_15, © Springer Science+Business Media, LLC 1999, 2010

▼ EXAMPLES

1. What is the de Broglie wavelength of a 10 keV electron? To answer this question, first find the momentum of the 10 keV electron. Because this energy is less than 10% of the electron's rest energy, you can use the nonrelativistic expression connecting momentum p and kinetic energy K:

$$\frac{p^2}{2m_e} = K.$$

To do your calculation in eV units, multiply and divide the left side by c^2 and solve for pc. This gives you

$$pc = \sqrt{2K m_e c^2} = \sqrt{2 \times 10 \times 511} = 101.1 \, \text{keV}.$$

From this calculated momentum, you can find the 10 keV electron's wavelength:

$$\lambda = \frac{h}{p} = \frac{hc}{pc} = \frac{1240}{101.1 \times 10^3} = 0.0123 \, \text{nm}.$$

2. What if the electron is relativistic? For example, suppose it has a kinetic energy of 1 MeV. You use the approach of the previous example, but now you must calculate the momentum using the relativistically correct formula

$$pc = \sqrt{2K \, m_0 c^2 + K^2} = \sqrt{2 \times 1 \times .511 + 1^2} = 1.422 \, \text{MeV},$$

from which it follows that

$$\lambda = \frac{hc}{pc} = \frac{1240}{1.422 \times 10^6} = 8.72 \times 10^{-4} \, \text{nm}.$$

This is very small.

3. At what energy will an electron have the same wavelength as a 10 keV x-ray? A 10 keV x-ray has a momentum of 10 keV/c and a wavelength of $\lambda = 1.240/10 = 0.124$ nm. An electron will have the same wavelength when it has the same momentum—10 keV/c. The kinetic energy K of an electron with that momentum is

$$K = \frac{p^2 c^2}{2m_e c^2} = \frac{10^2}{1022} = 98 \, \text{eV}.$$

The answers to Examples 1 and 3 highlight the fact that a photon and a particle with the same momenta (and, therefore, same wavelength) will usually have quite different energies. This is because the particle has rest mass and the photon does not. Conversely, a particle and a photon with the same energy will usually have very different wavelengths: A 10 keV photon has a momentum of 10 kev/c, about one-tenth the momentum of a 10 keV electron, so the photon's wavelength is

$$\lambda = \frac{hc}{pc} = \frac{1.24}{10} = 0.124 \, \text{nm},$$

about 10 times larger than the wavelength of the 10 keV electron.

■ EXERCISES

1. Find the wavelength of a 50 keV photon and compare it to the de Broglie wavelength of a 50 keV electron.

2. What energy of photon will have the same wavelength as a 50 keV electron?

15.3   EVIDENCE THAT PARTICLES ACT LIKE WAVES

De Broglie's idea that particles might have a wavelength has been strongly confirmed for all kinds of particles: electrons, neutrons, protons, atoms, subnuclear particles, etc. What is the evidence for such a wavelength? How can particles be made to exhibit wave-like behavior? The above examples suggest an approach: Create a stream of particles with an energy that corresponds to a wavelength for which a crystal is an appropriate grating, and see if you observe diffraction or other interference phenomena when the beam of particles strikes the crystal. Some version of this approach was used in the first experiments that exhibited wavelike behavior of particles—. G.P. Thomson's experiments, Davisson's and Germer's experiment, and Stern's experiment. The following section describes these experiments and also double-slit interference produced with electrons and the diffraction of neutrons by a crystal lattice. (The neutron is a neutral particle with mass a little larger than the mass of a proton.)

G.P. Thomson's Experiment

G.P. Thomson, the son of the J.J. Thomson who discovered the electron, took a very straightforward approach. If particles are waves, then a

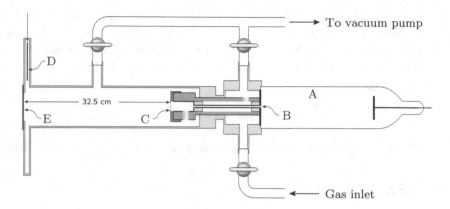

FIGURE 15.1 Schematic diagram of G.P. Thomson's apparatus for the study of electron diffraction.

beam of electrons of an appropriate momentum passing through a poly-crystalline foil should make a diffraction pattern of concentric rings like the pattern made by x-rays when they diffract through a polycrystalline sample, as described in Chap. 14. (See Fig. 14.4 on p. 428.)

To see whether electrons would produce such a pattern, Thomson used the apparatus shown schematically in Fig. 15.1. Electrons with kinetic energy of about 40 keV were produced in the tube A and then formed into a beam by passage through a tube B of bore 0.23 mm and length 6 cm. They struck and passed through a thin, polycrystalline gold foil at C. The resulting pattern of the electrons could be viewed on a phosphorescent screen E or recorded on a photographic plate D that could be lowered in front of the screen. The distance from the foil to the screen was 32.5 cm.

The electrons made the pattern shown in Fig. 15.2; it is repro-duced from Thomson's publications.[1] The rings are clearly analogous to those seen in the Debye–Scherrer patterns of x-ray diffraction from polycrystalline samples in Fig. 14.4.

Do these results agree with de Broglie's predictions? To answer this question, compare the pattern with the predictions of the Bragg law for x-ray diffraction in crystals:

$$2d \sin \theta = n\lambda. \tag{3}$$

First, recall from Chap. 14 that Debye–Scherrer diffraction of x-rays produces a pattern of rings, one for each order, $n = 1, 2, 3, \ldots$, for each dif-

[1]G.P. Thomson, "The diffraction of cathode rays by thin films of platinum," *Nature* **120**, 802 (1927); "Experiments on the diffraction of kathode rays," *Proc. Roy. Soc. London* **A117**, 600–609 (1928); *The Wave Mechanics of Free Electrons*, McGraw-Hill, New York, 1930.

FIGURE 15.2 The diffraction pattern from electrons on a polycrystalline gold foil observed by G.P. Thomson.

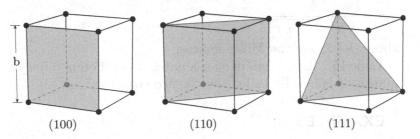

FIGURE 15.3 Miller indices of some important planes in a cubic crystal—the atoms are located at the corners of a cube.

ferent set of crystal planes. Therefore, the production of a similar pattern of rings by electrons strongly suggests that the electrons are undergoing Bragg diffraction like x-rays.

Second, using the various possible spacings d for a gold crystal, you can analyze the rings quantitatively. Figure 14.3 (p. 425) shows you that the value of d depends on which set of planes you are considering. For labeling the sets of planes there is a notation that we introduce here without much explanation.

Crystal planes are specified by a set of three small integers called Miller indices. They are usually written in parentheses in the form $(h\,k\,\ell)$. For example, the familiar, obvious parallel planes shown in Fig. 15.3a are denoted by (100). The planes that pass through a crystal at 45° as shown in Fig. 15.3b are labeled (110).

You can use the Miller indices of a crystal plane and the lengths of the sides of the smallest unit of the crystal to find the spacing

TABLE 15.1 Lattice constants for some common cubic crystals

Element	Lattice constant b (nm)
Al	0.404
Au	0.407
Ni	0.352
Cu	0.361

between adjacent planes. It is particularly simple to do this for the planes of cubic crystals. For cubic crystals the spacing d between the planes is related to b the shortest distance between neighboring atoms—this is the length of the edge of the elementary crystal cube—by the expression

$$d = \frac{b}{\sqrt{h^2 + k^2 + \ell^2}}, \tag{4}$$

where $(h\,k\,\ell)$ are the Miller indices.

Table 15.1 lists values of b for a number of different cubic crystals. This quantity b is usually called the "lattice constant."

▽ EXAMPLES

4. What is the spacing between adjacent (111) planes of a gold crystal? Table 15.1 gives the lattice constant for gold $b = 0.407$ nm. This means that the spacing between adjacent (111) planes is $0.407/\sqrt{3} = 0.235$ nm.

▨ EXERCISES

3. What is the spacing between the (110) planes of gold?

4. What is the spacing between the (331) planes of gold?

There is one more point to make about spacings between planes of a crystal. You will sometimes see reference to planes like $(h\,k\,\ell) = (200)$ which means that the spacing between such planes is $d = b/2$. But how can that be? If the spacing between planes is $d = b$, there should not be any atoms halfway between these two planes. And there aren't. Bragg reflection from (200) planes is the same thing as second-order Bragg reflection, i.e., $n = 2$, from the (100) planes.

To see that this is so, look at Eq. 3. For (200) planes, $d = b/2$, and the equation becomes

$$2\frac{b}{2}\sin\theta = \lambda \quad \text{or} \quad 2b\sin\theta = 2\lambda,$$

which is the same as Eq. 3 for (100) with $n = 2$; i.e., it is the second-order result. It is customary to deal with higher orders in the Bragg equation by using only $n = 1$ and then including the higher orders by appropriate choice of Miller indices and the corresponding value of d.

Now you can understand Thomson's experimental proof of the validity of de Broglie's idea. One of his tests was to observe how the diameter D of one ring—the one produced by the (200) planes—varied as he varied the voltage V through which the electrons were accelerated. He obtained the data shown in Table 15.2.

Note that $DV^{1/2}$, the product of the diameter of the ring and the square root of the accelerating voltage, is nearly constant. This is Thomson's proof that de Broglie was correct and $\lambda = h/p$.

Why is $DV^{1/2}$ constant if $\lambda = h/p$? First, the kinetic energy K of the electrons is proportional to the acceleration voltage V because $K = eV$, and for these non-relativistic electrons K is proportional to the square of the momentum because

$$\frac{p^2}{2m} = K = eV.$$

Therefore, the momentum p of the electrons is proportional to the square root of the accelerating voltage, i.e.,

$$p \propto V^{1/2}.$$

TABLE 15.2 Diameter of the (200) diffraction ring vs. electron voltage

Acceleration voltage V (kV)	Diameter D (cm)	$DV^{1/2}$ (cm keV)$^{1/2}$
24.6	2.50	12.4
31.8	2.15	
39.4	2.00	12.6
45.6	1.86	
54.3	1.63	12.0
61.2	1.61	12.6

Consequently, if de Broglie is correct and λ is proportional to $1/p$, it follows that $\lambda \propto 1/V^{1/2}$.

Then, if the electrons do behave like waves, the ring diameter D will follow from the Bragg law, $2d \sin \theta = \lambda$. If the scattering angle 2θ is small, $\sin 2\theta \approx r/L$, and $\sin \theta \approx r/2L$, where r is the radius of the diffraction ring and L is the distance from the gold foil to the photographic plate (see Fig. 14.4, p. 428, and Fig. 15.1, p. 458). But the radius r is proportional to the diameter D, so for fixed values of d and L,

$$D \propto \lambda,$$

from which it follows that

$$D \propto \frac{1}{V^{1/2}} \quad \text{or} \quad D\,V^{1/2} = \text{constant}$$

if de Broglie is correct.

You can see from the entries in the third column of Table 15.2 that $DV^{1/2}$ is nearly constant. It convinced most physicists that de Broglie was correct, and he was awarded the 1929 Nobel prize in physics.

Thomson further confirmed de Broglie's theory by analyzing the other diffraction rings shown in Fig. 15.2. He showed that they arose from Bragg diffraction from the (111), (200), (220), (113) + (222), and (331) + (420) crystal planes. (In the last two cases, rings from two different sets of indices fell too close to one another to be distinguished.)

■ EXERCISES

5. Calculate the values for $DV^{1/2}$ that are missing from Table 15.2.

6. Functional relationships are often verified by plotting measured quantities in some form that will give a straight-line graph if the expected relationships are correct. How would you plot the data of Table 15.2 to determine whether de Broglie was correct? Do it.

7. Show why it might be difficult to distinguish experimentally between diffraction from the (113) and (222) planes or between the (331) and (420) planes.

The Experiment of Davisson and Germer

The first experiment to show the wave nature of electrons was done by C.J. Davisson and L.H. Germer, working in the laboratories of the Western Electric Company, the manufacturing arm of the American Telephone and

Telegraph Company. Their original interest was to study the emission of secondary electrons from the electrodes of vacuum tubes (secondary electrons were produced when the electrons the tubes were intended to control struck the electrodes). When Davisson directed a beam of electrons at various samples, he observed that about 1% of the incident electrons were reflected back from them. He realized that the reflected electrons might be used to probe the structure of the atom just as E. Rutherford (about whom more later) used alpha particles to probe atomic structure.[2]

Davisson and Germer observed that when a beam of 54 eV electrons struck the (111) surface of a nickel crystal perpendicularly, the electrons diffracted just as light would diffract from a reflection grating.[3] The rows of atoms on the crystal surface acted as the lines of the grating, and the distance between these rows of atoms corresponded to the grating spacing D. A diffraction maximum in the intensity of reflected electrons occurred just where the grating equation $D \sin \theta = \lambda$ predicted for waves of wavelength $\lambda = h/p$.

For their experiment Davisson and colleagues used the apparatus shown schematically in Fig. 15.4. It was designed to find how the intensity of the outgoing electrons varies as a function of the angle between the incident beam and the scattered electrons. They could measure the intensity of

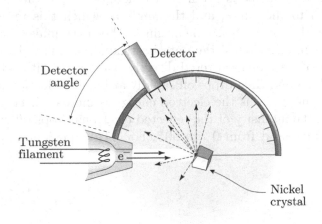

FIGURE 15.4 Schematic diagram of the apparatus of Davisson and Germer.

---

[2] For a short readable history of the Davisson–Germer experiment, read Richard K. Gehrenbeck, "Electron Diffraction Fifty Years Ago," *Physics Today* **31**(1), 34–41 (1978).

[3] Because of their electrical charge and low energy, the electrons penetrate no more than a few layers of atoms into the crystal surface, and therefore grating diffraction is a better description of what happens to such low-energy electrons than Bragg reflection from many planes in a three-dimensional crystal lattice.

electrons at any scattering angle they chose by moving the detector to that angle. They could also rotate the target and change the angle at which incident electrons hit the target's surface.

An electron gun made a beam of electrons with energy that could be varied from a few eV to around 200 eV. This beam was directed onto a target of single-crystal nickel, and electrons reflected from the target were detected and measured by catching them in a metal box.

The cut and orientation of the crystal determine which rows of atoms will be exposed to the incident electrons. The spacing between the rows of atoms affects the experimental results because it is the grating spacing. For the experiments of Davisson and Germer the face-centered cubic crystal was cut at right angles to the cube's diagonal, i.e., at 45° to each face exposing the (111) face as shown in Fig. 15.5b. Their electron beam came in normal to that face. For this orientation the rows of atoms on the (111) face are $D = 0.215$ nm apart, and this is the grating spacing.

Davisson and Germer looked at the variation of intensity of the scattered electrons as a function of scattering angle for different energies of the incident beam. Their results are shown in Fig. 15.6, where the electron intensity is plotted on polar graphs. Each point on these graphs shows the intensity of the scattered electrons and the angle at which it was measured. The intensity is given by the length of the radial distance from the origin to the curve, and the angle at which this intensity occurs is the angle that a line from the origin to the curve makes with the vertical axis. The progression of the data is striking. At all incident energies the intensity of electrons scattered directly backwards with a reflection angle of 0° is high and then drops off rapidly at larger reflection angles. But notice what happens as the electron energy is increased. For 40 eV incident electrons the intensity of the reflected electrons drops off smoothly as the angle is increased from 0° to 90°. For 44 eV incident electrons a bump

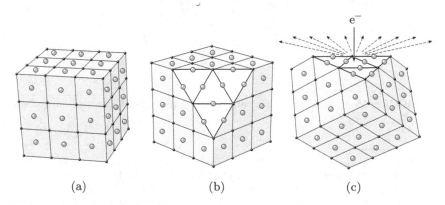

(a)   (b)   (c)

FIGURE 15.5 The arrangement of atoms of a Ni crystal. The (111) face is shown in (b). *Taken from J. Frank. Instit. Vol. 206, C.J. Davisson, "Are Electrons Waves?" (1928) with permission from Elsevier Science.*

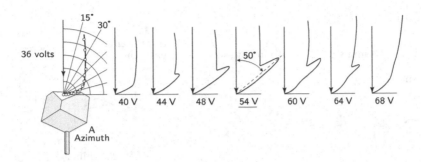

FIGURE 15.6 This succession of polar plots of the intensity of electrons reflecting off the surface of a nickel crystal shows that a diffraction maximum occurs at 50° when the electrons are accelerated through 54 V. *Taken from J. Frank. Instit. Vol. 206, C.J. Davisson, "Are Electrons Waves?" (1928) with permission from Elsevier Science.*

appears in the number of electrons reflected at 50°. The bump grows to a maximum as the energy is increased to 54 eV and then diminishes as the incident energy is increased further.

This bump is the expected interference maximum. According to the grating equation it should occur at 50° for a wavelength of

$$\lambda = 0.215 \sin 50° = 0.215 \times 0.766 = 0.165 \, \text{nm}.$$

How does this wavelength compare with de Broglie's prediction? An electron with kinetic energy $K = 54 \, \text{eV}$ has a momentum given by

$$pc = \sqrt{2m_e c^2 K} = \sqrt{1.02 \times 10^6 \times 54} = 7.43 \times 10^3 \, \text{eV}.$$

The de Broglie wavelength of such electrons is then

$$\lambda = \frac{hc}{pc} = \frac{1240}{7430} = 0.167 \, \text{nm}.$$

This degree of agreement is convincing. Many other measurements provide equally good agreement. In 1937 Davisson and Thomson shared a Nobel Prize for their independent experimental confirmation of the wave nature of particles.

"Double-Slit" Interference with Electrons

A striking example of wave behavior of electrons is the generation of a double-slit interference pattern by Möllenstedt and Düker.[4] Their apparatus is shown schematically in Fig. 15.7a. For two slits they used a thin metal-coated quartz fiber and applied to it a voltage of about 10 V.

---

[4]G. Möllenstedt and H.Düker, "Fresnel interference with a biprism for electrons," *Naturwiss.* **42**, 41 (1955); H. Düker, "Interference pattern of light intensity for electron waves using a biprism," Z. *Naturforsch.* **10a**, 256 (1955).

(a)

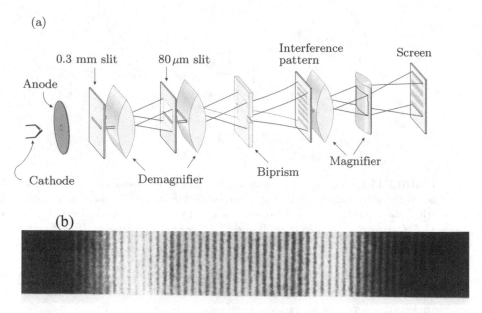

(b)

FIGURE 15.7   Double-slit interference of electrons: (a) apparatus; (b) interference pattern. *Taken from H. Düker, Z. Naturforsch, **10a**, 256 (1955) with permission of the publisher.*

This device slightly deflected the 50 nm wide beam of 19.4 keV electrons coming from the assembly of cathode, anode, and demagnifying electrodes in such a way as to make the electrons behave as though they had come from the two sources. (In optics this effect can be obtained with a device called a biprism, so that is the label given to the electrical equivalent in Fig. 15.7a.) The electrons then formed the interference pattern on a photographic plate as shown in Fig. 15.7b. Notice how strikingly it resembles the optical double-slit interference pattern described in Chap. 10. In Fig. 15.7b the bright lines are where the electron waves interfere constructively, and the black spaces represent destructive interference. *Electrons interfere just like photons.*[5]

---

[5]There is a certain amount of art that goes into a picture like Fig. 15.7b. If only the slits were producing the spacing of the electron fringes, they would be about 160 nm apart and very difficult to see. To make the pattern visible, a cylindrical electrical lens was used to magnify electrically the fringe-to-fringe dimension by a factor of 160. To suppress the effects of irregularities in the shape of the fiber, magnification of the dimension along the fringe was limited to between a factor of 5 and 10. Then the overall pattern was magnified optically a factor of 20 to make the picture with visible fringes.

■ EXERCISES

8. What is the wavelength of 19.4 keV electrons?

9. Suppose the effective separation of the sources of interfering 19.4 keV electrons was 6 μm and the effective distance from these sources to the photographic plate was 11 cm. What would be the distance between adjacent fringes on the photograph?

10. Suppose the fringes of the previous problem were magnified by a factor of 160 before they struck the plate. And then suppose the photograph was enlarged by a factor of 20 when it was printed. What then would be the spacing between the fringes? Compare your answer with the spacing of the fringes shown in Fig. 15.7b.

Waves of Atoms

The wavelike behavior of particles is quite general. In 1931 Estermann, Frisch, and Stern[6] showed that a beam of helium atoms would diffract from the surface of a crystal of LiF and that the wavelength of the atoms was what the de Broglie relationship predicted.

Since 1932, when the neutron was discovered, this neutral particle with a mass slightly larger than that of a proton has been widely used for diffraction studies of crystals and other matter. Figure 15.8 shows a comparison of diffraction data taken using the same wavelengths of x-rays and neutrons.[7] The target was powdered copper. The counter angle 2θ is twice the diffraction angle. The peaks occur at the angles where Debye–Scherrer maxima occur. In this case the diffraction patterns are not recorded on a flat screen, which would show the Debye–Scherrer circles, but with a detector that moves in an arc around the sample so that it crosses the circles and measures them as intensity maxima at their corresponding angles. The similarity of the two spectra is striking.

The energy of the neutrons was measured to be 0.07 eV. To understand the resulting diffraction patterns you need to know the neutrons' wavelength.

---

[6] I. Estermann, and O. Stern, "Diffraction of molecular rays," *Zeit. f. Physik* **61**, 95–125 (1930).
[7] C.G. Shull and E.O. Wollan, "X-ray, electron and neutron diffraction," *Science* **108**, 69–75 (1948).

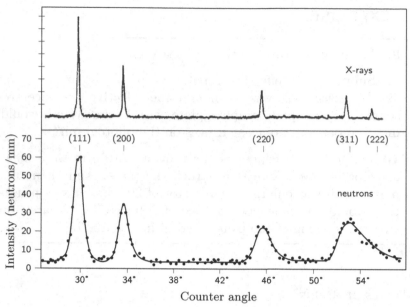

FIGURE 15.8 Bragg diffraction patterns for $\lambda = 0.1\,\text{nm}$ neutrons and x-rays in powdered copper. *Taken with permission from C.G. Shull and E.O. Wollan, "X-Ray, Electron, and Neutron Diffraction," Science 108, 2795, 69–75 (1948)* ©1948 *American Association for the Advancement of Science.*

EXAMPLES

5. To find the neutron wavelength from the de Broglie formula $\lambda = h/p$, first find the momentum p from the kinetic energy $K = 0.07\,\text{eV}$. For such a low energy you can find the product pc from the nonrelativistic relationship using the fact that for a neutron, $mc^2 = 939 \times 10^6\,\text{eV}$:

$$pc = \sqrt{2mc^2 K} = \sqrt{2 \times 939 \times 10^6 \times 0.07} = 1.146 \times 10^4\,\text{eV}.$$

This value of pc implies that the wavelength is

$$\lambda = \frac{hc}{pc} = \frac{1240}{1.146 \times 10^4} = 0.11\,\text{nm}.$$

This is a wavelength just about the size of an atom, and therefore just right for probing the structure of something made up of atoms, such as a crystal or a molecule. Moreover, because neutrons are electrically neutral they can, unlike electrons, pass through substantial amounts of matter without much attenuation; this property makes them exceptionally useful probes for studying crystalline structure.

▼ EXAMPLES

6. If these 0.07 eV neutrons are undergoing Bragg diffraction in the copper crystallites, at what angle should there be a maximum due to reflections from the (111) planes?

Use the Bragg law to answer this question, and solve $2d \sin \theta = \lambda$ for $\sin \theta$. First find d. From Table 15.1 note that the lattice constant for Cu is 0.361 nm, so the spacing d between the (111) planes is

$$d = \frac{0.361}{\sqrt{3}} = 0.208 \,\text{nm},$$

which yields

$$\sin \theta = \frac{\lambda}{2d} = 0.260,$$

so that $\theta = 15.1°$, and there should be a peak at the counter angle $2\theta = 30.2°$. There is!

■ EXERCISES

11. Calculate the counter angle at which you would expect to see a diffraction maximum from the reflection of 0.07 eV neutrons from the (200) plane of copper. Compare your result to the value given in Fig. 15.8.

12. Think of and prepare a nice graphical illustration to show that all the peaks in the neutron diffraction spectrum of Fig. 15.8 are consistent with Bragg diffraction.

15.4 SUMMARY AND CONCLUSIONS

All the objects you have learned to think of as particles—electrons, helium atoms, neutrons, protons—exhibit diffraction and interference just like waves. They act like waves with a wavelength λ given by the de Broglie relationship

$$\lambda = \frac{h}{p},$$

where p is the momentum of the particle and h is Planck's constant.

Some Useful Things to Know

Diffraction and interference are basic tools for working with particles. Consequently, you need to be able to find a particle's wavelength given its energy or momentum in order to predict how a beam of such particles will interact with an array of other particles, e. g., in a crystal lattice according to the Bragg law. Fairly often, this means that you must find the momentum p when you are given the kinetic energy K.

Before you can find the momentum of a particle from its kinetic energy, you must also be able to tell whether you can get by with the nonrelativistic relationship between momentum p and kinetic energy K of a particle of mass m,

$$pc = \sqrt{2mc^2 K},$$

or whether you need to use the relativistically correct relationship

$$pc = \sqrt{2Kmc^2 + K^2}.$$

The rule of thumb developed in Chap. 12 is that the nonrelativistic formula will be accurate to 2.5% if $K/(mc^2) \leq 0.1$. If $K/(mc^2) \geq 0.1$, it will be necessary to use the relativistically correct formula.

And of course, you must be able to solve all parts of this problem using only units of electron volts (eV). For this purpose it is helpful to remember that $hc = 1240\,\text{eV\,nm}$. Also remember that for an electron, $mc^2 = 511\,\text{keV}$; for a proton, $mc^2 = 938\,\text{MeV}$; for a neutron, $mc^2 = 939\,\text{MeV}$.

Waves, Energy, and Localization

Two important implications follow from the wave nature of particles. First, the smaller the object you study, the more energetic must be the probe. Second, particles can not be sharply localized in space.

Probing Small Objects Requires Large Energies

Remember from what you learned about waves that when you probe a structure you cannot learn about details that are smaller than the wavelength of your probe. Because the de Broglie wavelength is inversely proportional to momentum, the smaller you make the probe, the greater you must make its momentum.

Kinetic energy is proportional to the square of the momentum in the nonrelativistic case and proportional to momentum in the ultrarelativistic case. This means that the smaller is the thing you wish to probe, the larger must be the energy of the particles with which you do the probing. Since everything we know about tiny structures we learn by probing with something—photons, protons, electrons, or a number of other particles we have not yet talked about—the need for higher and higher energies to look at smaller and smaller objects is important to keep in mind.

Let's look at one consequence of this idea. Suppose you wish to study the internal structure of an atom with a beam of electrons. How energetic must the beam be?

▼ EXAMPLES

7. An atom is about 0.1–0.2 nm in size. If you want to see structure that is 1% of this size, you will want to have electrons that have a wavelength no larger than 0.001 nm. The momentum of such electrons is

$$pc = \frac{hc}{\lambda} = \frac{1240}{0.001} = 1.24 \times 10^6 \text{ eV}.$$

Remember the rule for going from momentum p to kinetic energy K: Use relativistically correct equations if $pc \geq 0.2mc^2$. Clearly, that is the case here, because $1.24 \times 10^6 \geq 0.2 \times 0.511 \times 10^6$ eV. Therefore,

$$K = \sqrt{m^2c^4 + p^2c^2} - mc^2 = \sqrt{0.511^2 + 1.24^2} - 0.511 = 0.83 \text{ MeV}.$$

In other words, to probe the structure of an atom to 1% requires you to use about 1 MeV electrons.

■ EXERCISES

13. The Thomas Jefferson National Accelerator Facility, in Newport News, Virginia, produces a beam of high-energy electrons for probing the interior of nuclei. It is designed to look at structures as small as 3×10^{-16} m. Roughly, what is the lowest electron energy that will achieve this design goal?

Particles are not Sharply Localized

The second important implication of the wave nature of particles (and the particle nature of waves) is that they can not be sharply localized; a wave by its very nature is spread out in space. This basic feature of a wave has forced physicists to conclude that until its position is actually measured an electron is located in more (sometimes many more) than one place at the same time. This strange idea also applies to momentum; i.e., until its momentum is measured, a particle can have a range of different momenta at the same time. The idea that an entity can simultaneously have more than one value of some definite physical property is essential to the successful description of the interactions and properties of atoms and subatomic particles. Chapter 19 discusses these matters further.

PROBLEMS

1. If a 2 keV electron has a wavelength of 0.0274 nm,
    a. what will be the wavelength of a 32 keV electron?
    b. what will be the wavelength of a 512 keV electron?

2. Measure directly from the circles in Fig. 15.2 on p. 459 and show that the progression of circle diameters is what you would expect for the diffraction of de Broglie waves.

3. Thomson says that the photographic plate that took the image shown in Fig. 15.2 on p. 459 was 32.5 cm from the gold foil. Use that number and the fact that the diameter of the (200) diffraction ring was 2.5 cm for 24.6 keV electrons to find the lattice constant of the gold foil. How does your result compare with the value given in Table 15.1? Warning: Remember that the angle of the diffraction ring relative to the incident beam is twice the diffraction angle θ.

4. Suppose you wanted to probe the structure of an atom using a beam of electrons. If you wanted to see structure on the order of the size of 0.1 of the radius of an atom, what energy of electrons would you need?

5. When you hold a wire or hair in front of a laser beam you get a diffraction pattern resembling what you get when you pass the laser beam through a slit of the same width.
    a. 2 eV photons diffracting around a wire as shown in Fig. 15.9 go on to strike a screen 1 m away. If the first minimum of the pattern occurs at 6.2 mm from the center of the pattern, what is the diameter of the wire?
    b. Suppose the wire is 50 μm in diameter and you wish to make the same diffraction pattern with electrons instead of with photons. What energy must the electrons have to produce a first minimum 6.2 mm from the center of the pattern?

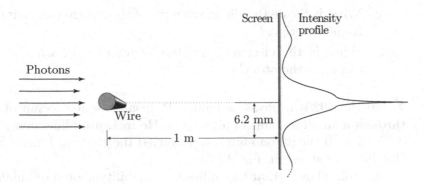

FIGURE 15.9   Light diffracting around a wire as in Problem 5.

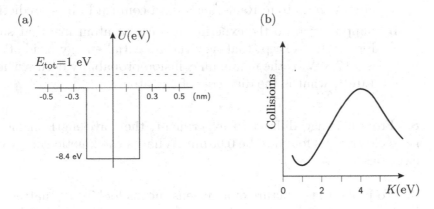

FIGURE 15.10   For Problems 6 and 7: (a) potential energy of an electron in an atom; (b) the probability of a collision between the electron and the atom as a function of the electron's kinetic energy.

   c. Suppose instead of a great, huge, thick wire, you had a wire of the diameter of a nucleus, i.e., $\approx 10^{-14}$ m. What energy would the electrons have to have in order to produce a diffraction minimum 6.2 mm away from the central maximum on a screen 1 m away?

   d. If the second maximum occurs at 9 mm, what is the angle through which the photons or electrons have been scattered to reach that point on the screen? Give your answer in radians.

6.   In Fig. 15.10a the solid line shows a simplified version of the potential energy of an electron in an argon atom. The dashed line shows the total energy of a particular electron.

   a. Is this electron bound in the argon atom? How do you know?

   b. What is the electron's kinetic energy when it is at 0.05 nm?

    c. What is the de Broglie wavelength of the electron when it is 0.25 nm from the origin?

    d. What is the electron's de Broglie wavelength when it is at the center of the atom?

7. In the 1920s a physicist named Ramsauer sent a beam of electrons through a small amount of argon gas. He measured how many electrons collided with the argon atoms as he varied the electron kinetic energy K. His data are shown in Fig. 15.10b.

    a. Bohr showed that the collision probability should be smallest when the de Broglie wavelength inside the atom is equal to the diameter of the argon atom. Show that if Fig. 15.10a is a good representation of an argon atom, Ramsauer's data confirm Bohr's prediction.

    b. Suppose you do the experiment with an atom like that shown in Fig. 15.10a, except that now the potential energy inside the atom is -17.8 eV. If the minimum collision probability occurs when $K = 1.0$ eV, what is the diameter of the atom?

8. From a Bragg diffraction experiment, the wavelength of monoenergetic electrons is found to be 0.06 nm. What is the kinetic energy of these electrons?

9. To probe the structure of a nucleus means looking at matter 10^{-15} m in size or smaller. What wavelength of probe would you want in order to study lengths this small?

10. What energy electron has a wavelength of 10^{-15} m?

11. What energy proton has a wavelength of 10^{-15} m?

12. The Davisson-Germer experiment (p. 462 et seq.) is another test of the de Broglie hypothesis $\lambda = h/p$. In this experiment an electron beam was incident perpendicular to the (111) face of a nickel crystal (see Fig. 15.12). Electrons interact strongly with the metal so that only the top few atomic planes are involved. The geometry of the incident and outgoing rays is shown in Fig. 15.12 for a single plane of atoms. Davisson and Germer found constructive interference at $\theta = 50°$ when the electrons were accelerated through 54 V.

    a. Using Fig. 15.11, prove that constructive interference occurs if $d \sin \theta = n\lambda$ where n is an integer.

    b. Find the electrons' de Broglie wavelength.

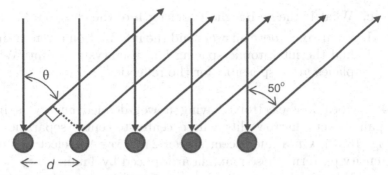

FIGURE 15.11   Schematic representation of the rays of an electron wave incident upon and reflected from a plane of nickel atoms (Problem 12).

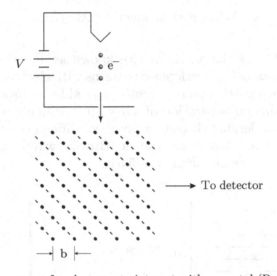

FIGURE 15.12   Arrangement for electrons to interact with a crystal (Problem 13).

   c. Find the interatomic distance d.
   d. What is the next higher voltage where an interference peak will occur at 50°?

13.   Electrons accelerated through a potential difference V pass through a narrow slit and strike the face of a cubic crystal, as shown in Fig. 15.12. When an electron detector (an ammeter) is placed in the position shown, a maximum current is recorded on the meter when $V = 50$ volts.
   a. What is the kinetic energy of the electrons after passing through the anode slit?
   b. Find the momentum of these electrons in units of eV/c. If an approximation can be used, then do so, but explain clearly why its use is justified.

   c. What is the de Broglie wavelength of the electrons?

   d. From your above answers and the information given in the drawing, find the interatomic spacing b in the crystal. (Hint: What atomic planes are responsible for the reflection?)

14. A beam of electrons having a well-defined energy is incident on a pair of very narrow slits whose center-to-center separation is $0.1\,\mu$m (Fig. 15.13). On a fluorescent screen 2 m away the electrons produce an intensity pattern whose maxima are spaced by 1 mm.

   a. What does the very existence of such an intensity pattern tell you about electrons?

   b. What is the momentum of these electrons?

   c. What is their kinetic energy?

15. A double-slit device known as an "electron biprism" is capable of producing interference patterns with electrons. With such a device, a team of experimenters recently was able to produce double-slit fringes with angular separation of 4.0×10^{-6} radians, using electrons that had been accelerated through a potential difference of 50 kV (Fig. 15.14.)

   a. Show that the de Broglie wavelength of the electrons in this experiment was 5.5 pm.

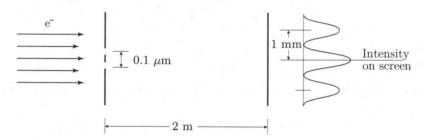

FIGURE 15.13 Electrons incident on a pair of slits (Problem 14).

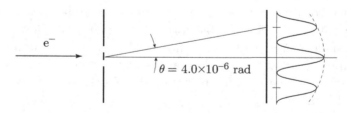

FIGURE 15.14 Experimental arrangement for producing electron-interference patterns (Problem 15).

b. What was the path difference between the interfering electron waves for the first minimum on either side of the central maximum?

c. What was the distance between the slits in this experiment?

d. If the accelerating voltage were increased from 50 kV to 200 kV, how would the angular separation between the fringes change? Explain briefly. A detailed calculation is not necessary.

16. Figure 15.15 shows a double-slit interference pattern for neutrons. Neutrons from a nuclear reactor were slowed and then directed into a double-slit apparatus with slit spacing of $d = 126\,\mu$m and slit width $b = 22\,\mu$m. When a neutron detector 5 m away from the slits was moved perpendicular to the neutron beam, it recorded the interference pattern shown in Fig. 15.15. Each data point is the number of neutrons detected in 125 min.

a. Using a rule and scale on the figure, find the distance between interference maxima.

b. Find the de Broglie wavelength of the neutrons.

c. What is the neutron velocity?

17. A gas of C_{60} molecules was created in an oven at temperature T. A beam of molecules escaped through a small hole in the oven, passed through the diffraction grating, and was detected at a distance of 1.2 m from the grating. The number of molecules detected as a function of the detector's distance from the center line is shown in Fig. 15.16.

a. From their data the experimenters determined the de Broglie wavelength of the C_{60} molecules was 4.6 pm. What was the distance between neighboring slits in the diffraction grating?

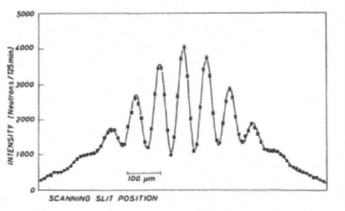

FIGURE 15.15 Interference pattern of slow neutrons obtained by Zeilinger et al. *Rev. Mod. Phys.* **60**, 1067 (1988) (Problem 16).

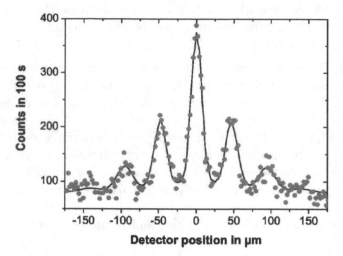

FIGURE 15.16 Interference pattern of C_{60} molecules (buckyballs) obtained by Nairz et al., "Quantum interference experiments with large molecules," *Am. J. Phys.* **71** 319–325 (2003) (Problem 17).

b. How fast were the C_{60} molecules moving?

c. Estimate the temperature T of the oven assuming that the molecular speed you found in part (b) is $\approx v_{rms}$ of the molecules in the oven.

d. Suppose that helium atoms (mass $= 4\,u$) were used instead of C_{60} molecules and that the same interference pattern was found. Find the ratio between the speed of the He atoms and the speed of the C_{60} molecules.

16.

Radioactivity and the Atomic Nucleus

In 1896 Henri Becquerel discovered that compounds containing uranium emit radiations that can penetrate opaque paper and even thin sheets of metal and cause photographic plates to darken. Like x-rays, these emissions ionized air and caused electroscopes to discharge, but unlike x-rays, they occurred without any external source of excitation. Becquerel's student, Marie Curie, named this spontaneous emission of ionizing radiation "radioactivity."

Research soon showed that radioactivity was not rare. Within a few years, Marie and Pierre Curie, working in France, discovered two previously unknown chemical elements, polonium and radium, that were radioactive. Over the next decade or so their work and the studies they inspired identified dozens of different radioactivities. Ernest Rutherford, at first in England, later with Frederick Soddy in Canada, and then again in England characterized and identified the radiations. They and their colleagues found that radioactive atoms emit either helium ions or electrons and as a result *change into other atoms*—a stunning overthrow of the idea of immutable, eternal chemical elements.

The helium ions and electrons emitted from radioactive atoms have energies typically 10^6 times greater than the energies characteristic of chemical bonding; moreover, in some cases they continue being emitted for billions of years (Gy). This was so puzzling at first that some physicists seriously considered that energy might not be conserved.

Such a drastic hypothesis became unnecessary after it was found that a given sample does not emit particles indefinitely; the radioactivity always runs down eventually. The large magnitude of the energies became understandable later when Rutherford discovered that every atom possesses a compact, extremely dense, positively charged core—the atomic nucleus. This discovery ultimately led to the recognition that nuclei are composed

C.H. Holbrow et al., *Modern Introductory Physics, Second Edition*,
DOI 10.1007/978-0-387-79080-0_16, © Springer Science+Business Media, LLC 1999, 2010

of two different kinds of particles, protons and neutrons, held together by a new force, different from the electromagnetic and gravitational forces, and much stronger. As you will see, the strength of the new force and the small dimensions of the nucleus can explain the large energies released in radioactivity.

Rutherford also was one of the first to realize that each species of radioactive atom has a well-defined, characteristic probability for undergoing spontaneous disintegration and transformation. This discovery was gradually understood to mean that the moment in time at which any individual atom will decay is purely random and unpredictable *in principle*, so that radioactive decay deeply contradicts Newtonian ideas of causality. This kind of fundamental randomness required the introduction of revolutionary new ideas into physics, ideas with implications that are still surprising and mystifying physicists.

16.1   QUALITATIVE RADIOACTIVITY

Becquerel Discovers Radioactivity

In the early weeks and months after Roentgen's discovery of x-rays there was intense activity in many laboratories directed towards discovering the source and nature of the rays. When the French physicist Henri Becquerel attempted to understand the production of x-rays, he accidentally discovered an entirely new phenomenon—radioactivity.

Many substances that have been bombarded by cathode rays or illuminated by beams of light continue to emit light after the incident radiation has been turned off. This delayed emission of light is called "fluorescence." Becquerel speculated that x-rays might be associated with fluorescence. To test his idea he wrapped a photographic plate in black paper so that no light would leak in. Then he coated the outside of the wrapper with uranium sulfate salts, which were known to fluoresce strongly when illuminated by sunlight. His idea was to let sunlight strike the salts and produce fluorescence. Then any x-rays that were produced would penetrate the opaque wrapper and expose the photographic plate inside. After letting the package sit all day in the sun to give the material plenty of time to produce the supposed penetrating radiation from the fluorescence, he opened the package in the dark and developed the plate. He found that the plate had darkened in just the way that photographic materials respond to x-rays. However, another plate showed the same degree of blackening even though it had been exposed to less sunlight and therefore less fluorescence. When uranium salts were not exposed to sunlight at all and so could not fluoresce, the wrapped photographic plates still

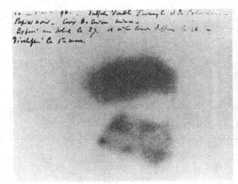

(a)   (b)

FIGURE 16.1   (a) The smudgy patches on Becquerel's photographic plate that signaled the existence of radioactivity. Notice the faint outline of a cross produced where a metal cross was placed to keep radiation from reaching the plate. (b) Some experiments were performed using uranium salts in ampoules like these. ©*Bibliotheque Centrale M.N.H.N. Paris, 1998.*

exhibited the same darkening. At this point, Becquerel concluded that he was dealing with something other than x-rays. Results of one of his earliest experiments are shown in Fig. 16.1. Only forty-nine years separate these smudges from the explosion of the first atomic bomb.

To show that it was the uranium in the salt that was responsible for the radiations that darkened photographic plates, Becquerel exposed the plates to a piece of uranium metal. If the radiations come from uranium, you would expect the effect on the plates to be stronger from the higher concentration of uranium in the metal than from the lower concentration in the salt. This is just what he observed.

Becquerel established that the radiations could cause the discharge of charged electroscopes, and he began to quantify how much radiation a sample of radiating material produced. He took the amount of darkening of photographic film, or the rate at which an electroscope discharged, to be a measure of what he called the "activity" of the sample. If one sample darkened film more or discharged an electroscope faster than another, he said that the sample had greater "activity." Quantitative measures of activity are now much more concrete and precise.

Becquerel found that the penetrating radiation that darkened his photographic plates could be bent by magnetic fields in the same way as cathode rays. He showed that the charged particles had a velocity of about 1.6×10^8 m s^{-1} (more than half the speed of light) and a charge-to-mass ratio of about 10^{11} C/kg. Remember that e/m for electrons is 1.76×10^{11} C/kg. In Chap. 12 you saw how in 1900 Kaufmann showed that one kind of radioactive ray had a charge-to-mass ratio of the order of 10^{11} C/kg. This and other evidence established that energetic electrons were being emitted.

■ EXERCISES

1. What is the energy of the electrons found by Becquerel?

The Curies Discover New Radioactive Elements

Becquerel's work motivated the start of two extraordinary research efforts, that of Marie and Pierre Curie, and that of Ernest Rutherford. Marie Sklodowska was a brilliant young student from Poland at the Sorbonne, in Paris, when she met and married Pierre Curie, a recent recipient of a doctoral degree. Becquerel's results became known just at the time when Marie Curie was casting about for a topic for her own doctoral research. She decided to look for other materials exhibiting radioactivity. In 1898 she found that thorium was also radioactive, but she was not alone in discovering it.

She did notice, however, that the mineral ore from which uranium was extracted was several times more active than uranium itself. She deduced that there had to be other unknown substances present that were much more active than uranium. That same year she succeeded in separating a chemically distinct radioactive material that was considerably more active than uranium or thorium. As it was apparent that she had discovered a new chemical element, she named it "polonium" in honor of her homeland.

A second radioactive substance also was observed to separate out along with barium from the ore. By heroic efforts Marie Curie extracted and purified about 0.1 g of this substance from a ton of ore. The new element was roughly a million times more active than an equivalent amount of uranium. She was able to determine its atomic weight, now known to be 226.0 u, and enough of its chemical properties to be sure that it was another new element. She named it "radium."

From the perspective of more than a century later you may find it hard to appreciate the great significance of the Curies' accomplishments. They showed that radioactivity could be used to identify new elements, elements that began to fill a large gap in the periodic table between bismuth ($Z = 83$) and thorium ($Z = 90$). Even more important, their discovery of new radioactive elements showed that radioactivity was more than just a peculiarity of one or two elements. It became clear that radioactivity is both widespread in nature and related in some basic way to the internal structure of individual atoms. Their work was a major contribution to the realization that radioactivity represented a new physical phenomenon profoundly different from any previously known.

The Curies also greatly advanced experimentation on radioactivity. By purifying radium, they made available a strong source of radiation that could easily be used for further research; it was no longer necessary to depend on the weakly active uranium samples that took a day to cause darkening of a photographic plate. They also began quantitative studies of the rate at which a gas is ionized by radium emissions; these studies led to the development of a standard for expressing the activity of a radioactive substance. The activity of 1 g of radium became the standard, and other activities were measured and described in equivalent grams of radium.

Alpha, Beta, and Gamma Rays

Many people had a hand in the early discoveries of radioactivity, but Ernest Rutherford was largely responsible for making sense of the phenomenon. One question that arose immediately after the discovery of radioactivity was whether the radiations were x-rays or something different. Rutherford, working in the laboratory of J. J. Thomson (of the e/m experiment) became interested in finding the answer. He soon identified two kinds of rays. The first were easily stopped by very thin foils; the others were much more penetrating. The easily stopped rays he named "α rays"; the more penetrating rays he named "β rays." A 0.02 mm thick foil of aluminum or a piece of ordinary paper would stop 95% of the radiations from uranium. These were alpha rays. Clearly, the more penetrating ionizing radiation—the other 5%—were beta rays and caused the darkening of Becquerel's photographic plates or the discharge of an electroscope.

It was natural to analyze the two kinds of rays by electric and magnetic fields (see Fig. 16.2). As already mentioned, e/m measurements showed beta rays to be very similar to the cathode rays of J. J. Thomson. The conclusion was that beta rays are energetic electrons.

Alpha radiations bend in a magnetic field as energetic, positively charged particles with the charge-to-mass ratio of fully ionized helium (He^{++}). The fact that helium gas, relatively rare on Earth, was often found in minerals that contained significant amounts of uranium or thorium suggested to Rutherford that alpha particles are helium ions.

At the University of Manchester, in England, Rutherford did experiments that directly showed that alpha particles are ionized helium atoms. He built an apparatus that collected the alpha particles as they stopped. Even with the much stronger sources of radioactivity that were by then available, he had to run his experiment for several months. At the end of that time he could show by optical spectroscopy that his previously

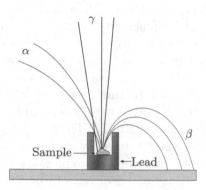

FIGURE 16.2 A schematic diagram of the behavior of different kinds of radioactive rays in a magnetic field: the α rays bend as positively charged particles with small q/m; the β rays bend as negatively charged particles with q/m of an electron; the γ rays are not affected by the magnetic field.

empty apparatus now contained measurable amounts of helium. This was the outcome to be expected if alpha particles were helium ions and formed neutral atoms as they stopped in the apparatus.

■ EXERCISES

2. What is the charge-to-mass ratio of an ionized hydrogen atom (proton)? Of doubly ionized helium? Would it be difficult to distinguish hydrogen from helium by their charge-to-mass ratios?

3. As part of measuring the charge-to-mass ratio of alpha rays from radium, Rutherford measured their velocity and got about 2.5×10^7 m s^{-1}. Later, more accurate measurements gave 1.53×10^7 m s^{-1}. Assume that alpha particles are helium ions, and use the later value of velocity to calculate the energy of these alphas.

By 1900 a third kind of ray was identified. They were named "γ rays." They are quite penetrating, but as Fig. 16.2 shows, they do not bend in a magnetic field and therefore cannot be charged particles. Later, after they were found to behave like x-rays, gamma rays were recognized to be energetic photons. They are emitted following the emission of alpha or beta rays and without radioactive transformation of the emitting atom.

▪ EXERCISES

4. What is the direction of the magnetic field that produces the deflections shown in Fig. 16.2?

Radioactive Atoms of One Element Change into Another

It soon became evident that in the radioactive process, atoms of uranium, thorium, and so on were actually changing into other kinds of atoms, a process called "transmutation." Rutherford realized that the emitted radiation was a result of these changes. But changing atoms? The phrase is a self-contradiction because the word atom means uncuttable, and generations of chemical studies had convinced people that the elements were just that—elemental, fundamental building blocks of matter. To suppose now that atoms could transform violated ideas built up over more than a century.

Let's consider some of the evidence for the occurrence of such changes. Uranium compounds emit both alpha and beta rays. Rutherford and Soddy[1] found that they could by chemical means separate from the uranium whatever was emitting beta rays, while leaving almost all the alpha activity behind with the uranium. They called the beta-emitting fraction uranium-x and denoted it at first by UX and later by UX1. They observed that the beta activity gradually built up again in the uranium sample and could then be separated again as more UX1. Significantly, the chemical techniques to separate the UX1 were the same as those required to separate thorium from uranium, and it is now known that UX1 is the mass-234 isotope of thorium.

Second, a buildup of an alpha-emitting, radioactive gas was observed in closed vessels containing samples of radium. Like the beta activity of UX1, this alpha-emitting substance could be drawn off a radium sample. This substance had the chemical characteristics of a noble gas. Initially called "emanation," this gas is now called "radon"—a household word in recent years.

Third, the observed radioactive elements were found to group into different families that we call "radioactive series" or "chains of radioactive decay." The members of each family have atomic masses differing by integer multiples of $4\,u$. Consequently, there are four possible different

---

[1]Frederick Soddy, 1877–1956, English physicist who worked with Rutherford at McGill University, in Canada. They proposed the disintegration theory of radioactivity. Soddy introduced the idea of "isotope" to explain how there could be several different radioactivities for a given chemical element. He received the 1921 Nobel Prize in chemistry.

chains. One has atomic masses that are integer multiples of 4, i.e., $4n$—where n is any integer; another has masses that are integer multiples of 4 plus 1, i.e., $4n + 1$; another has masses that are $4n + 2$; and a fourth chain has masses that are $4n + 3$. You can see that such families exist because only the emission of an alpha particle changes the mass of a radioactive atom appreciably, and it changes the mass by ~ 4 u. Emission of a beta ray changes the atomic number by one unit, but barely affects the atomic mass.

▽ EXAMPLES

1. Most naturally occurring uranium atoms have an atomic weight of 238 u. To what decay chain do they belong? Divide 238 by 4; you get 59 with a remainder of 2. Therefore, this uranium isotope belongs to the $4n + 2$ decay chain.

As Fig. 16.3 shows, an atom that emits an alpha ray transforms into an atom two elements lower down the periodic table; beta emitters trans-

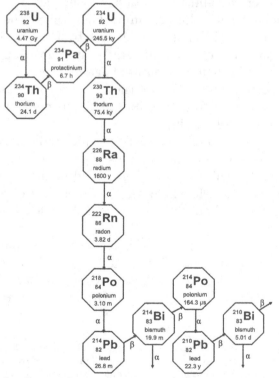

FIGURE 16.3   A map of a sequence of radioactive transformations and their half lives. The end of the sequence at mass-206 lead is not shown.

form to elements one position up the periodic table. Thus, the mass-238 uranium atom that emits an alpha particle becomes a mass-234 thorium atom. And when that thorium atom emits a beta ray it becomes a mass-234 protactinium atom, which a short while later emits a beta ray and becomes a mass-234 uranium atom. The occurrence of such radioactive "transformations" overthrew the fundamental, deeply held belief in the immutability of chemical elements.

Radioactivity also forced physicists and chemists to enlarge what is meant by an atom of a chemical element. In the course of the radioactive decays just described, the mass-238 uranium atom changed into a mass-234 uranium atom. To understand radioactive transmutations it was necessary to recognize that there can be atoms of the same chemical element that have substantially different masses; these different atoms of the same chemical element are called isotopes—another atomic surprise.

■ EXERCISES

5. To what radioactive series do radium atoms with atomic mass of 226 u belong?

6. Identify the decay chains of the elements listed in Table 16.1.

7. What will be the mass of the isotope of lead that is the end product of the radioactive decay of mass-238 uranium?

TABLE 16.1   Some of the first discovered radioactivities

Name	Symbol	Atomic number Z	Mass number A	Half-life
polonium	Po	84	210	138.4 d
radon, thoron, thorium emanation	Rn, Tn Em	86	220	55.6 s
radon	Rn	86	222	3.83 d
radium	Ra	88	226	1600 y
thorium	Th	90	232	14.1×10^9 y
thorium	Th, UX1	90	234	24.1 d
uranium	U	92	238	4.47×10^9 y

16.2   QUANTITATIVE PROPERTIES OF RADIOACTIVITY

Measures of Activity

After it was recognized that each instance of alpha or beta radioactivity corresponds to the disintegration and transformation of a single atom, activity was defined to be the number of disintegrations per second. In honor of the Curies, a commonly used unit of activity is the "curie," abbreviated Ci. By international agreement one curie (1 Ci) of activity is 3.7×10^{10} radioactive emissions per second. This is almost the activity of one gram of radium, so it is approximately correct to say that the activity of 1 g of Ra is 1 Ci.

▮ EXERCISES

> **8.**   Suppose a 100 μg sample of Ra causes an electroscope to discharge in 30 s, while 15 g of an unknown sample of radioactivity causes it to discharge in 100 s. What is the activity of the unknown?

The official SI unit of activity of radioactive substances is called the "becquerel" and is abbreviated Bq. A becquerel is defined to be one disintegration per second. Therefore, $1 \, \text{Ci} = 3.7 \times 10^{10} \, \text{Bq}$.

Radioactive Decay and Half-Life

The persistence of radioactivity posed a major mystery. Uranium and thorium samples emit very energetic particles at an apparently steady rate. Their activities seemed to stay constant. Could a sample really keep releasing such energetic particles indefinitely?

The answer to this question came when new and different radioactive substances were separated from uranium and thorium. For example, the UX1 that Rutherford and Soddy chemically separated from uranium and that carried all the beta activity showed measurable changes in activity over time. Rutherford observed that UX1's beta activity exhibited a gradual decay after it had been separated from uranium. He measured the ionizing current produced by UX1 and observed that it dropped off over time, as shown in Table 16.2.

In about 24 days, the activity of UX1 decreased to half of its initial value and kept decreasing by half in each succeeding 24-day period. The time for the activity to fall by one half is called the "half-life," or $T_{1/2}$. Every radioactive substance has been found to have its own unique half-life,

TABLE 16.2   Ionizing current caused by UX1 over several months

Ionizing current (arb. units)	Elapsed time (d)	Ionizing current (arb. units)	Elapsed time (d)	Ionizing current (arb. units)	Elapsed time (d)
124.6	0	55.4	28	23.9	56
114.3	2	53.6	30	23.3	58
111.2	4	48.7	32	22.0	60
109.1	6	47.8	34	21.4	62
96.2	8	42.3	36	19.2	64
93.0	10	42.1	38	18.2	66
85.6	12	39.9	40	16.5	68
81.3	14	38.2	42	17.1	70
80.2	16	36.1	44	15.4	72
72.2	18	34.3	46	15.1	74
68.3	20	30.8	48	13.5	76
65.9	22	29.2	50	14.1	78
59.5	24	26.4	52	11.8	80
57.4	26	25.1	54		

and therefore all radioactive materials do eventually run down. Changes in the activity of uranium were difficult to detect because for uranium $T_{1/2} = 4.5 \times 10^9$ y, so that in any reasonable period of time the relative change of activity was not measurable.

The data in Table 16.2 behave in a way familiar to students of growth and decay: the ionization current falls off exponentially. Using the ionizing current produced in an electroscope by a sample of UX1 as a measure of the sample's activity, Rutherford and others observed that relative to the initial current I_0, the ionizing current I being produced by the beta radiations from UX1 diminished exponentially with the passage of time,

$$I = I_0 \, e^{-\lambda t}, \tag{1}$$

where λ is called the "decay constant" or "disintegration constant."

This behavior of the ionizing current is just what you would expect if the activity of a sample is proportional to N, the number of atoms that have not yet decayed. To see why this is so, realize that each radioactive emission decreases N, and in a time Δt, N decreases by ΔN atoms. The activity—the number of disintegrations per second—is the ratio $\Delta N / \Delta t$.

If, as claimed, the activity is proportional to N then

$$\frac{\Delta N}{\Delta t} = -\lambda N, \tag{2}$$

where λ the constant of proportionality will turn out to be the decay constant in Eq. 1. The minus sign tells you that N is decreasing with time. For the normal case of large N and comparatively small ΔN, you can approximate $\Delta N / \Delta t$ by a derivative. (This means that you are treating the discrete decays and integer changes in numbers as smoothly varying functions.) In other words, you are replacing Eq. 2 with

$$\frac{dN}{dt} = -\lambda N. \tag{3}$$

Integrating this equation for N as a function of time, you get

$$N = N_0 \, e^{-\lambda t}, \tag{4}$$

where N_0 is the number of radioactive atoms at time $t = 0$. Equation 4 is the law of radioactive decay: *In the absence of any source that is producing them, the number of radioactive atoms decreases exponentially over time.*

Differentiation of Eq. 4 shows that if the equation is correct, then the activity of a UX1 sample should diminish exponentially as was observed. Differentiation with respect to time gives the activity A:

$$A = \left| \frac{dN}{dt} \right| = \lambda N_0 \, e^{-\lambda t} = A_0 \, e^{-\lambda t}, \tag{5}$$

where A_0 is the activity at time $t = 0$. The ionizing current is produced by a small but fixed fraction of the radiations emitted by UX1, so $I \propto A$ (and $I_0 \propto A_0$), and Eq. 2 follows from Eq. 5, i.e. the ionizing current that Rutherford measured will have the same exponential decay as the sample's activity.

EXAMPLES

2. A good test of whether radioactivity obeys Eqs. 4 and 5 is to plot the logarithm of the measured activity against time. If the result is a straight line, you know that the decay is exponential.

Figure 16.4 shows that Rutherford's UX1 data obey the relationship nicely.

The decay constant λ tells you how long a sample will last. Notice that if λ is large, a sample will decay quickly; if λ is small, the sample will be long-lived. Equivalently, for a given number of atoms, a large value of λ

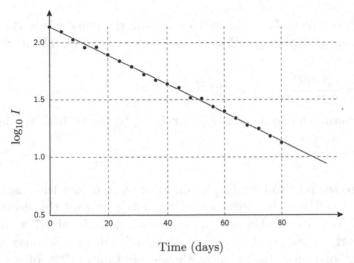

FIGURE 16.4 Plot of the logarithm (base 10) of the activity of UX1 vs. time. The straightness of the line shows that the decay of UX1 is exponential.

means a high activity. In other words, *the decay constant is a measure of the probability that an atom will decay.*

As long as the number of radioactive decays ΔN is small compared to the number of atoms present N, the probability that an atom will decay in the time interval Δt is the ratio of ΔN to N. Referring back to Eq. 2, you can see that the probability is $\lambda \, \Delta t$ that an atom will decay in a time interval Δt. *As long as this probability is much less than one,* you can say that $\lambda \, \Delta t$ is the probability that an atom will decay within the time interval Δt.

You now have two measures of the likelihood that an atom will decay, the half-life $T_{1/2}$ and the decay constant λ. These two quantities are connected through the decay law, Eq. 4.

▽ EXAMPLES

3. To see how to find λ if you know $T_{1/2}$, consider the case of UX1.

If you know the half-life, you can find λ without knowing either dN/dt or N. For UX1, $T_{1/2}$ is 24.1 days, or $24.1 \times 86\,400\,\mathrm{s}$. During this time the activity drops to one-half of its initial value, and because the activity dN/dt is proportional to N, the number of radioactive atoms N must have dropped to half of the initial value, i.e., $N/N_0 = \frac{1}{2}$. You can use this fact and Eq. 4 to write

$$\frac{1}{2} = e^{-\lambda T_{1/2}} = e^{-\lambda \, 24.1 \times 86\,400},$$

which connects $T_{1/2}$ and λ. To solve for the constant in the exponent, take the natural logarithm of both sides of the equation. Since $\ln 2 = 0.693$, you get

$$\lambda = \frac{0.693}{2.1 \times 10^6} = 3.3 \times 10^{-7}\,\mathrm{s}^{-1}.$$

In general, the decay constant is related to the half-life by the equation

$$\lambda = \frac{\ln 2}{T_{1/2}}. \tag{6}$$

4. To see how to find $T_{1/2}$ if you know λ, consider the case of radium.

You can use the definition of the curie to find the decay constant of radium. Remember that a curie is an activity of $3.7 \times 10^{10}\,\mathrm{s}^{-1}$ and is nearly equal to the activity of $1\,\mathrm{g}$ of radium. Because the atomic weight of radium is $226\,\mathrm{u}$, one gram contains $1/226$ of a mole of Ra atoms. Therefore, for a one-gram sample, N_0 is Avogadro's number divided by 226, and the activity of one gram of radium is, from Eq. 5,

$$-\frac{dN}{dt} = \lambda \frac{6.02 \times 10^{23}}{226} = 3.70 \times 10^{10}\,\mathrm{s}^{-1},$$

from which it follows that

$$\lambda = 1.39 \times 10^{-11}\,\mathrm{s}^{-1},$$

which with Eq. 6 gives

$$T_{1/2} = \frac{\ln 2}{\lambda} = 4.99 \times 10^{10}\,\mathrm{s},$$

which is 1580 years, slightly different from the currently accepted value of 1600 y for $T_{1/2}$ of Ra because modern measurements show that the activity of $1\,\mathrm{g}$ of Ra is $0.988\,\mathrm{Ci}$, 1.2% smaller than was thought when the curie was defined.

▪ EXERCISES

9. Derive the general relationship between $T_{1/2}$ and λ.

It can be shown that Eqs. 4 and 5 imply that radioactive decay is a purely random occurrence. Although you can know from measurement the probability of decay, there is no way to predict when a particular atom will disintegrate. It is as though the atoms are playing Russian roulette. Each atom has a revolver with a cylinder containing many empty chambers and one loaded one. In the course of a second each atom spins its cylinder, puts the gun to its head and pulls the trigger. λ is the probability that the gun will go off. During the next second all the remaining atoms spin the cylinders of their guns and play another round. The value of λ found in Example 16.4 shows that 139 Ra atoms out of every 10^{13} lose in each round of Russian roulette. The UX1 atoms play a much tougher game. In a 1-s round of Russian roulette, 33×10^5 UX1 atoms out of 10^{13} will lose.

The exponential decay law implies that in any given time interval the nucleus has the same chance of emitting a particle as in any other similar interval. This is strange. People do not age this way. Barring accidents, there is a relatively narrow range of ages in which people die. If they behaved like radioactive nuclei, half of the original population would die by a certain age, let's say forty. And half of the remaining population would die by the age of eighty, half again by one hundred and twenty and so on. Thus in a population of two million people, one-half million should be over one hundred and twenty, and nearly two thousand over four hundred years of age! We know that there are various aging processes and that people do not just die for no reason. Stars are the same. Stars like our Sun have fairly well-defined lifetimes. They do not just up and die at random. Why are nuclei so different from people and stars? How can one nucleus live for perhaps 1 s and another identical nucleus for ten billion years?

◾ EXERCISES

10. The most common naturally occurring uranium atoms have a half-life of 4.47×10^9 y. What is their decay constant? What are their odds to lose at Russian roulette if they spin the cylinder once?

11. A certain kind of rubidium atom has a decay constant of $7 \times 10^5 \, \text{s}^{-1}$. What is the half-life of these atoms?

12. In the previous exercise what difficulty occurs when you try to answer the question: "What is the probability that one of these special rubidium atoms will decay in a second?" How can you get around the difficulty?

We have touched the edges of a very profound problem here. Today, as they have for nearly a century, physicists hotly discuss and disagree about the meaning of the apparently causeless randomness at the atomic and nuclear levels. The issue will arise repeatedly as you study more physics.

16.3 DISCOVERY OF THE ATOM'S NUCLEUS

As soon as their energies, masses, and charges were known and reliable sources were developed, alpha rays were used to probe the atom's insides. The startling result was the discovery that atoms are mostly empty space containing a tenuous cloud of electrons around a dense, compact nucleus.

Alpha Particles as Probes of the Atom

In 1906 Rutherford observed that when a narrow beam of alpha particles passed through a sheet of mica, it made a slightly broader line on a photographic plate than when the mica was removed. He realized that the alphas were scattering from the atoms in the mica, but only through quite small-angles. In 1908 his student, Hans Geiger, examined this small-angle scattering in more detail. These first studies were interpreted using a model of an atom suggested by J. J. Thomson, who pictured the atom as an assemblage of electrons embedded in a sphere of smeared-out positive charge. Because the electrons were embedded within the positive charge like plums in a pudding, the model was called "the plum-pudding model." The Thomson idea seemed to explain some observations but not all of them.

Thomson's model implies that alpha particles passing near an atom will scatter *only* through small angles. You can see this is so by estimating an upper limit for the scattering of an alpha particle from an electron. Such a collision is like a bowling ball striking a ping pong ball, because the alpha particle is so much more massive than an electron. If you imagine a bowling ball just brushing the edge of a ping-pong ball, you may see that the electron is never going to scatter through an angle larger than 90°.

The mass of an electron is $0.511 \, \text{MeV}/c^2$; the mass of an alpha particle is about $4 \times 931.5 \, \text{MeV}/c^2$. The ratio of $M_\alpha/m_e = 7300$. Consequently, when a massive alpha particle moving with a velocity v_α and momentum $p_\alpha = M_\alpha v_\alpha$ collides head-on with an electron sitting at rest, the alpha imparts to the electron a forward velocity of $2 \, v_\alpha$. Thus the maximum possible change in momentum of the electron is

$$\Delta p = 2 m_e v_\alpha \ll p_\alpha. \tag{7}$$

If the momentum Δp were carried away from the alpha particle at a right angle, the alpha would be deflected through through an angle θ, as shown in Fig. 16.5, where

$$\theta \approx \tan \theta = \frac{\Delta p}{p_\alpha}.$$

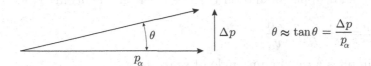

FIGURE 16.5 Upper limit of small-angle scattering produced by a small change in momentum.

Although the alpha cannot lose this much momentum at right angles, and the calculated deflection is greater than the largest possible scattering angle, the result serves as an upper limit of what is possible. Notice that the angle is just twice the ratio of the masses of the two particles:

$$\theta = 2\frac{m_e}{M_\alpha} = \frac{1}{3650} = 0.0003 \text{ rad} = 0.015°.$$

Clearly, the electrons in an atom would be unlikely to scatter the alpha very much even if there were many such collisions.

But might the positive charge itself produce appreciable scattering? Again a simple model is informative. Imagine that the atom is a ball of positive charge of radius R and some charge Ze, where e is the elementary charge and Z is some integer. Then the maximum force exerted on the alpha particle, which itself has a radius $r_\alpha \ll R$, by the atom's positive charge would be, from Coulomb's law,

$$F = k_c \frac{Z_\alpha Z e^2}{R^2},$$

where Z_α is the number of elementary charges on an alpha particle, i.e., 2, and k_c is the constant $9 \times 10^9 \text{ N m}^2 \text{C}^{-2}$ appearing in the Coulomb force law. You might think that the force would get larger if the alpha particle approached closer than the surface of the atom, but this is not so. Inside a ball of charge the force on the alpha particle is proportional to the alpha's distance from the ball's center, dropping to zero as the alpha particle moves to the center. This means that you can estimate the force by taking its value just at the surface of the sphere, where it will be largest.

To make it easy to estimate the change in momentum, use this maximum value of the force and assume that it acts constantly over a time interval equal to the time Δt for an alpha to travel a distance equal to the diameter of the atom:

$$\Delta t = \frac{2R}{v_\alpha}.$$

This gives an upper bound on the change in momentum of

$$F \, \Delta t = k_{\mathrm{c}} \frac{e^2 Z_\alpha Z}{R^2} \frac{2R}{v_\alpha},$$

which implies a maximum angle of scattering of

$$\theta = \frac{\Delta p}{p} = 2k_{\mathrm{c}} \frac{e^2 Z_\alpha Z}{R M_\alpha v_\alpha^2}. \tag{8}$$

The radius R of an atom is about $0.1\,\mathrm{nm}$; Z for a gold atom is 79; $\frac{1}{2}M_\alpha v_\alpha^2$ is the kinetic energy of an alpha particle, or about $5\,\mathrm{MeV}$; and, of course, $k_{\mathrm{c}} e^2 = 1.44\,\mathrm{eV\,nm}$.

■ EXERCISES

---

13. Show that the above estimate of the maximum angle of scattering gives about $0.026°$ for alphas scattering from gold atoms.

Discovery of the Atomic Nucleus

Rutherford had another student, Ernest Marsden, look for larger-angle scattering. The method of detection was interesting. When individual alpha particles hit a screen coated with zinc sulfide they make a flash of light, or "scintillation," that can be seen by the completely dark-adapted human eye. This was a very sensitive technique for observing rare events. For the reasons given above, Rutherford did not expect to see any large-angle scatterings. According to the Thomson model, scatterings through large angles would have to be the result of many successive small-angle scatterings. Since each would be random, the successive scattering directions would tend to average out, and only on very rare occasions would the events add up to a significant overall deviation.

The experiment was set up using an intense source of alphas in an evacuated chamber (Fig. 16.6). A narrow beam was formed by a pair of thin metal plates with small holes that served to collimate the stream of alpha particles. Gold foil was used as a target in order to get massive atoms of a material that could easily be made into sheets so thin that alphas could pass through. Marsden immediately began seeing significant numbers of alphas deflected through fairly large angles. After he modified the apparatus to allow observation of alphas scattered in any direction, he saw scintillations at even larger angles, some even in the straight backwards direction. He found that 1 out of every 8000 alphas scattered through an angle of $90°$ or more.

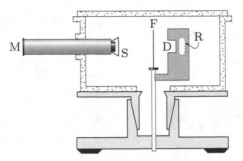

FIGURE 16.6   Apparatus used by Geiger and Marsden. Alpha particles from a source R pass through a collimator D and strike a foil F. Scintillations from scattered particles are produced on the screen S and are observed with a microscope M. The chamber is evacuated and can be rotated to different angles about the foil.

This result was a complete surprise. Because a solid is a collection of atoms in contact with one another, the alphas had to pass through hundreds of atoms to get through the foil. The experiment showed that the material was very porous, that most of the alphas passed through it like bullets through a rain shower. Astonishingly, however, the same material exerted very strong forces on the few alpha particles that were bounced back. As Rutherford said, "It was quite the most incredible event that has ever happened to me in my life. It was almost as incredible as if you fired a 15-inch shell at a piece of tissue paper, and it came back and hit you."

■ EXERCISES

14. Show that the simple model used above to estimate the small-angle scattering will produce larger angles of scattering if you imagine that the positive charge has a much smaller radius R than assumed above.

Rutherford realized that the scattering of the alpha particles through large angles could be explained if the positive charge of the atom is concentrated very compactly in the center of the atom. Indeed, the crude approximation of Eq. 8 suggests that if R, the radius of the positive ball of charge, is 10^{-3} to 10^{-4} smaller than the radius of the atom, then θ will be large. This makes sense, because near such a compact ball of charge the electric field is very large, and an alpha particle passing close to the compact core would experience a very large force. In the case of a head-on collision, an approaching alpha particle slows down and comes to a halt at the distance r from the charge where the alpha's initial kinetic energy is

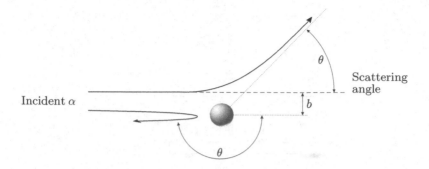

FIGURE 16.7 Trajectories of two scattered alpha particles. The upper alpha particle is scattered through an angle $\theta < 90°$; the other alpha particle collides nearly head on with the nucleus and scatters through $\theta \approx 180°$. The quantity b, the distance of closest approach if there were no electric force present, is called the scattering parameter.

completely converted to electrostatic potential energy; then it accelerates backwards away from the atom until it is far away and has regained its original speed. This is scattering through 180°, as shown by the trajectory at the bottom of Fig. 16.7. The upper trajectory in Fig. 16.7 shows the deflection through a smaller angle of an alpha particle that is not incident directly head-on to the atom. The amount of deflection is less the farther the alpha is from a head-on collision. This kind of scattering of one atom from another owing to the Coulomb force between their nuclei is called "Rutherford scattering" or "Coulomb scattering."

Assuming that the scattering center, i.e., the nucleus, acted as a point charge, Rutherford calculated that the number of alpha particles scattering through an angle θ will be proportional to the square of the charge Ze of the scattering nuclei and inversely proportional to the fourth power of the sine of half of the scattering angle, $\theta/2$. He also predicted that the number scattered would be proportional to the inverse of the square of the energy of the alpha particles, E_α. In short he predicted

$$\Delta N \propto \frac{Z^2}{E_\alpha^2 \sin^4\left(\frac{\theta}{2}\right)}, \tag{9}$$

where ΔN is the number counted.

Geiger's and Marsden's experimental data on the scattering as a function of angle could then be compared with Rutherford's theory. These data are shown in Table 16.3, which has several interesting features. For example, it shows convincingly that Rutherford's prediction that ΔN is proportional to $1/\sin^4\left(\frac{\theta}{2}\right)$ is correct. If such a proportionality holds, then the product of ΔN and $\sin^4\left(\frac{\theta}{2}\right)$ should be a constant. Column 4 shows that the product varies by no more than ±20%, while the number of counts varies by almost four orders of magnitude.

TABLE 16.3 Number of alpha particles scattered through various angles θ from gold

Scattering angle, θ (deg)	$\sin^4 \frac{\theta}{2}$	Number of scintillations ΔN	$\Delta N \sin^4 \frac{\theta}{2}$
150.	0.8705	33.1	28.8
135.	0.7286	43.0	31.3
120.	0.5625	51.9	29.2
105.	0.3962	69.5	27.5
75.	0.1373	211.	29.0
60.	0.0625	477.	
45.	0.02145	1435.	
37.5	0.01068	3300.	35.2
30.	0.004487	7800.	
22.5	0.001449	27300.	39.6
15.	0.0002903	132000.	38.3
30.	0.004487	3.1	
22.5	0.001449	8.4	
15.	0.0002903	48.2	
10.	5.77×10^{-5}	200.	
7.5	1.83×10^{-5}	607.	
5.	3.62×10^{-6}	3320.	

■ EXERCISES

15. Fill in the blanks of column 4 of Table 16.3.

16. Compare the variation of the product with the variation of ΔN for the bottom six entries of Table 16.3.

17. Plot the data of Table 16.3 in such a way that they should all lie on the same straight line if Rutherford's theory is correct.

Another feature of Table 16.3 deserves notice. Stop and think a moment about how these data were obtained. Imagine that you are one of two people sitting in a completely darkened room holding mechanical counters and counting the flashes of light from a tiny screen only a few cm^2 in area. It took you close to an hour of sitting in total darkness before your eyes became sufficiently adapted to be able to detect the tiny flashes adequately. Even then you could reliably count no more than 90 scintillations per minute and no fewer than 5. How do you suppose someone counted the 1 32 000 counts shown in Table 16.3? Or how did they get 33.1 counts?[2]

To deal with the limitations of these human detectors several tactics were used. The simplest was to put in front of the alpha source an aperture made small enough to limit the number of alphas until their frequency of scattering was within human capacity to count them and then correct the data to make them correspond to some standard size of aperture. Another approach was to start with a very active source, say 100 mCi, with a short half-life, e.g., ^{222}Rn with a half-life of 3.82 d. At the beginning of the experiment observers would make measurements at the large scattering angles where the yield is small; then as the days went by and the activity dropped they would make measurements at smaller angles, where the yield was larger. Of course, they had to correct their data for the fact that the decay of the source led to a diminished number of incident alphas.

■ EXERCISES

18. If the counts in the two sets of data in Table 16.3 differed only because of a change in aperture size, what ratio of aperture diameters would account for the observed differences?

19. Suppose that it took 10 days to do an experiment in which ^{222}Rn was the source of alpha particles. At any given angle what would be the ratio of the number of scattered alpha particles at the beginning of the experiment to the number scattered at the end of the experiment?

---

[2] J.B. Birks, *The Theory and Practice of Scintillation Counting*, Pergamon Press, Macmillan Co., 1964, p. 4-5, writes: "... at one famous laboratory during this period all intending research students were tested in the dark room for their ability to count scintillations accurately. Only those whose eyesight measured up to the standards required were accepted for nuclear research; the others were advised to take up alternative, less physically exacting, fields of study... Marsden, who counted hundreds of thousands of scintillations in his historic experiments on α-particle scattering, has recalled how on train journeys his colleague Geiger would urge him not to put his head out of the window, lest a chance smoke particle should impair his efficiency as a human scintillation counter. Truly the early nuclear physicists needed to be men of vision."

---

20. Geiger's and Marsden's alpha source was ^{222}Rn and they did their experiment 51 h after the 100 mCi source was prepared. By what factor should they multiply the initial source activity to get the activity at the time they ran their experiment?

Rutherford's successful explanation of these data established the validity of the assumption that each atom contains a very small, dense core of charge. Rutherford named this compact center of the atom the "atomic nucleus," often called just the "nucleus" for short.

Nuclear Size and Charge

Rutherford scattering provides a way to determine the size of a nucleus. The scattering formula Eq. 9 is valid only when the alpha and the nucleus do not come into contact. This assumption works because it is a general result that the electric field outside a sphere of charge (the nucleus in this case) is the same as that of a point charge. Therefore, an alpha particle will experience the field of a point charge as long as it does not go inside the surface of the nucleus. If the alpha does penetrate the nucleus, however, then the force will not be that of a point charge of magnitude Ze, and Eq. 9 will no longer be correct.

The easiest case to analyze is a head-on collision, where the incident alpha particle slows down and stops when all its kinetic energy becomes electrostatic potential energy. Then the distance of closest approach is the value of r that gives a value of the potential energy equal to the alpha's incident kinetic energy

$$\frac{1}{2} M_\alpha v_\alpha^2 = k_c \frac{e^2 2Z}{r}. \tag{10}$$

Here r is the distance between the center of the alpha particle, of charge $2e$, and the center of the nucleus, of charge Ze. The other symbols are as defined for Eq. 8. Suppose the energy of the alphas is 5 MeV. Then the distance of closest approach to a gold nucleus will be

$$r = \frac{1.44 \times 2 \times 79}{5 \times 10^6} = 4.55 \times 10^{-5} \, \text{nm} = 45.5 \, \text{fm}.$$

■ EXERCISES

---

21. What would be the distance of closest approach for 5 MeV alpha particles bombarding an aluminum foil?

Of course, when the distance of closest approach brings the alpha particle inside the surface of the nucleus, Eq. 10 will no longer be valid, because inside the nucleus the electric field of the charge Ze is no longer the electric field of a point charge. This means that to find the nuclear size all you need to do is keep increasing the energy of the incident alphas until the experimental results deviate from the predictions of Eq. 9. Experiments show that for gold, Rutherford's theory no longer correctly predicts the distance of closest approach when $r \leq 8.9 \, \text{fm}$; for aluminum the theory fails when $r \leq 5.5 \, \text{fm}$; for nitrogen it fails when $r \leq 4.8 \, \text{fm}$. We infer that the theory fails at these values of r because the outer edge of the alpha is starting to penetrate the nuclear surface, and we conclude that these values of r represent the radius of the nucleus plus the radius of the alpha particle. Many experiments have shown that nuclei have radii a few femtometers in size and, in general, the nucleus of an atom with an atomic mass A has a radius r_N that is reasonably well given by the expression

$$r_N = 1.2 \, \text{fm} \times A^{\frac{1}{3}}. \tag{11}$$

EXERCISES

22. From Eq. 11 what is the radius of an alpha particle?

23. What is the radius of an aluminum nucleus? Is your answer consistent with the observation that Rutherford's theory does not hold for a head-on closest approach of 5.5 fm?

24. What is the radius of a gold nucleus? Is your answer consistent with the observation that Rutherford's theory does not hold for a head-on closest approach of 8.9 fm?

25. What is the radius of a proton?

Rutherford also realized that alpha scattering could be used to determine the charge of a nucleus. Measurements made on several different metals showed that the nucleus has a number of elementary charge units approximately equal to half of its atomic mass. In 1913 Moseley's work on atomic x-rays, described in Sect. 17.4, further supported the nuclear model and confirmed the Dutch amateur scientist Van der Broek's proposal that the number of elementary charges in the nucleus of a chemical element is the same as the element's atomic number Z.

Rutherford's work established the existence of the atomic nucleus. For this reason Rutherford is recognized as the discoverer of the atomic nucleus and the founder of the nuclear model of the atom. In this model a positive charge Ze is concentrated along with most of the mass of the atom in a volume about 10 fm in diameter. This is only about 10^{-4} of an atomic diameter and means that *an atom is mostly empty space.* Because atoms are electrically neutral, there must be Z electrons filling the relatively large empty volume around the nucleus. Rutherford's discoveries produced the most profound change in the concept of the atom since Dalton's ideas a little over 100 years earlier.

▨ EXERCISES

---

26. Approximately what fraction of the volume of a gold atom is occupied by its nucleus?

27. Approximately what fraction of the mass of a gold atom is in its nucleus?

28. Estimate the density of nuclear matter. If all the people in your college or university were compressed into a sphere with the density of nuclear matter, how large would the sphere's diameter be?

29. What is the preferred plural of "nucleus"?

16.4 NUCLEAR ENERGIES

As noted earlier, the energies of the radiations emitted in radioactivity are astonishingly large. Individual alpha particles have energies on the order of 5 MeV. The average energy of emitted beta particles is typically a few MeV. The discovery of the nucleus made it possible to explain such large energies.

Energies of Alpha and Beta Particles

Measurements of alpha and beta particles show that they are more energetic than the highest-energy cathode rays produced. The velocities found when measuring the charge-to-mass ratios of beta rays and of alpha rays show that these particles have energies on the order of several MeV.

To measure such energies you can use some of the tools you learned about in Chap. 12 to analyze results obtained from magnetic deflection experiments.

EXAMPLES

5. For example, when the betas from UX1 enter a uniform magnetic field of 0.04 T, some of them bend into a circular arc with a radius of about 2.8 cm. What is the energy of these beta rays?

Recall from Chap. 12 that the momentum p of particles of mass m and charge $q = Ze$ can be directly obtained from the radius of curvature R of their path in a magnetic field B in units of eV, using

$$pc = ZBRc, \tag{12}$$

where Z is 1 for an electron or a proton, 2 for an alpha particle, and so on. (As usual, Eq. 12 has been divided by a number equal to the elementary charge in order to convert joules to eV; as a result, a factor e has been removed from the equation.)

If the particle is nonrelativistic, you can calculate its kinetic energy K from

$$K = \frac{p^2 c^2}{2mc^2}.$$

If the particle is relativistic, you *must* use

$$K = \sqrt{m^2 c^4 + p^2 c^2} - mc^2. \tag{13}$$

For our example,

$$pc = 1 \times 0.04 \times 0.028 \times 3 \times 10^8 = 3.35 \times 10^5 \, \text{eV}.$$

When you compare this number to the rest energy of an electron, you find that $pc/(mc^2) = 0.656$. This is not small compared to 1, so you should use the fully relativistic Eq. 13, from which it follows that

$$K = 0.511 \left(\sqrt{1 + 0.656^2} - 1 \right) = 0.511 \times 0.197 = 0.101 \, \text{MeV}.$$

EXERCISES

30.   Calculate the kinetic energy of alpha particles that bend with a radius of $R = 0.63$ m in a magnetic field of $B = 0.5$ T. Assume that the alpha particles are doubly charged helium atoms with atomic masses of 4 u; 1 u has a mass of $931.5 \, \text{MeV}/c^2$.

Your answer to Exercise 30 should be 4.8 MeV, which is the energy of an alpha particle emitted by radium. It is an extraordinary amount of energy. Because one gram of radium emits $1\,\text{Ci}$, or 3.7×10^{10} alphas per second, a $1\,\text{g}$ sample of radium will give off energy at a rate that is the energy per particle times the number of particles per second $=$ $4.8 \times 10^6 \times 1.6 \times 10^{-19} \times 3.7 \times 10^{10}\,\text{J/s} = 0.028\,\text{J/s}$.

To appreciate the significance of such a number, compare it with more familiar quantities such as the heat from burning one mole of hydrogen gas to make water, one of the more energetic chemical reactions known (recall the *Hindenburg*). This will produce about 2.9×10^5 joules of heat, as you can verify by looking up the heat of formation of one mole of water. A mole of radium, 226 grams, gives off $6.78\,\text{J/s}$. Although it will take a mole of radium nearly $12\,\text{h}$, or half a day, to produce the same energy as burning the equivalent number of hydrogen molecules, the radium keeps on producing energy for thousands of years! The energy that comes out in an alpha particle is more than a million times that released when two atoms of hydrogen combine with an atom of oxygen.

■ EXERCISES

31. Calculate the ratio of energy released by the radioactive decay of a mole of radium to the energy released when a mole of H_2 burns:

$$2H_2 + O_2 = 2H_2O + \text{heat.}$$

32. How long will it take a mole of radium to yield 90% of its radioactive energy?

It is simpler to compare radium and H_2O using electron volts.

▼ EXAMPLES

6. How many electron volts of energy are released in the formation of an H_2O molecule compared to the number released in the emission of an alpha particle?

We have that the formation of 6×10^{23} water molecules releases $290\,\text{kJ}$. Converting this amount of joules to electron volts you find that this is the same as releasing

$$\frac{2.9 \times 10^5}{1.6 \times 10^{-19}} = 1.8 \times 10^{24}\,\text{eV.}$$

The amount of eV per water molecule is this number divided by 6×10^{23}, which is $3.0\,\text{eV}$.

This result shows that while chemical reactions occur at energies of a few electron volts (eV) per atom, alpha decays involve millions of electron volts (MeV). Radioactivity was thus clearly revealed to be a new phenomenon profoundly different from anything previously known.

In this age of big numbers and super hype, of trillion dollar debts, nuclear bombs, and space probes (the last, of course, still depending on burning hydrogen with oxygen), it takes an effort of imagination to put oneself in the place of those early workers in radioactivity. But try to appreciate the impact of magnifying virtually overnight by a factor of a million the energies that people were used to.

Where could the energy possibly be coming from? What kind of mechanisms could convert such huge amounts of stored energy to kinetic energy? The presence of radioactive species with half-lives short enough to show detectable decreases in activity indicated that the energy had to be coming from some process drawing on a limited though large store of energy.

To see how a tiny nucleus of positively charged matter might store the high energies involved in radioactivity, consider the potential energy involved in the forces that hold the charge in place. Unlike the Thomson model, in which negative electrons were imagined to be sprinkled around, neutralizing small regions of the atom, the nuclear model has a lot of positive charge in a very small volume.

EXAMPLES

7. What is the electrostatic potential energy of an alpha particle in uranium?

For $Z = 92$, you can imagine 2 elementary charges being repelled by the other 90 at distances on the order of 7.4 fm. The potential energy will then be

$$\text{P.E.} = \frac{k_c e^2 \times 2 \times 90}{7.4} = 35\,\text{MeV},$$

where you should use the convenient fact that $k_c e^2 = 1.44\,\text{eV nm} = 1.44\,\text{MeV fm}$ so that you can use the nuclear radius directly in femtometers and get your answer directly in MeV.

This very large energy implies a strong repulsive force between the positively charged alpha and the positive charge of the rest of the nucleus. How do all these positively charged particles stay together when the forces pushing them apart are so large?

There must be very strong attractive forces in the nucleus, strong enough to overcome the large electrostatic repulsion between protons. These forces are something new. They hold the nucleus together, so we call them "nuclear forces." They produce a large negative potential energy; they bind the nucleus together; and they hold the alpha inside for the lifetime of the uranium atom before it decays.

■ EXERCISES

> **33.** Repeat the above calculation to estimate the minimum negative nuclear energy required to compensate for the positive electrostatic energy so that a gold nucleus can stay together. Imagine that the two halves of the gold charge, $Z/2$, are point charges 12 fm apart.

The scale of the energies involved makes it plausible that the nucleus is the region where most of the atom's mass is located and where most of the action occurs in the emission of radioactive particles.

16.5   THE NEUTRON

The next major advance in understanding the nucleus came in 1932 when Chadwick discovered the neutron. The identification of the neutron as a component of the nucleus on an equal footing with the proton completed the basic picture of the nucleus that we have today.

The neutron was the third basic particle to be discovered—the electron and the proton were discovered first. The neutron is electrically neutral and has a mass of 1.0086649 u, slightly larger than the mass of a proton. Although they are stable inside many nuclei, neutrons outside of a nucleus are radioactive and decay with a half-life of 10.4 min.

Every nucleus is made up of neutrons and protons. The number of protons is Z, the atomic number of the corresponding element. The number of neutrons is often designated N. The sum of these two numbers is called the "mass number" of the nucleus and is usually written as A. The mass number of an atom is the integer nearest to its atomic weight. Collectively, the neutron and the proton are referred to as "nucleons." Thus the mass number A is the number of nucleons in a nucleus,

$$Z + N = A.$$

Each combination of Z and N specifies a unique nucleus. Nuclei with the same value of Z and different values of N are called isotopes. You have met isotopes before, but now that you know about neutrons you can see how there might be a mass-one hydrogen and also a mass-two hydrogen atom. Mass-two hydrogen has a nucleus consisting of a proton and a neutron. There is also a radioactive isotope of hydrogen with $A = 3$. The nucleus of this isotope has two neutrons and one proton.

There is a standard notation for representing different nuclei. It uses the chemical symbol to indicate the value of Z (which means that fairly often you will need to look at a periodic table of the elements to find out Z) and has the mass number as a left-hand superscript. Thus the proton is just ^{1}H. The mass-2 isotope (called "deuterium") is ^{2}H. The mass-3 isotope (called "tritium") is ^{3}H. The alpha particle is just the helium nucleus ^{4}He. (There is another, very rare, stable isotope of helium, ^{3}He, which as you can see has one neutron fewer than ^{4}He.)

Sometimes the value of Z is supplied explicitly; then it is written as a left-hand subscript.[3] For instance, there are two naturally occurring nonradioactive isotopes of carbon: $^{13}_{6}$C and $^{12}_{6}$C. There is also a radioactive isotope of carbon that occurs in nature: $^{14}_{6}$C. Note that as must be the case, $Z = 6$ for all the carbon isotopes.

The general form of this notation is ^A_ZX$_N$, where X represents any chemical element symbol and A, Z, and N are respectively the nucleon, or mass number; the atomic number; and the neutron number. The entire assemblage of neutrons, protons, and electrons is called a "nuclide." There are 272 stable nuclides plus 55 radioactive nuclides that occur naturally on Earth. About two thousand radioactive nuclides have been made artificially.

Visualizing a nucleus as a collection of Z protons and N neutrons helps to understand the transformations that result from radioactive decay. Alpha decay removes from the nucleus two neutrons and two protons; it reduces A by 4 units and Z by 2 units. That is why uranium-238 ($^{238}_{92}$U) turns into thorium-234 ($^{234}_{90}$Th), and, in general, alpha decay can be described as

$$^A_Z\text{X} \rightarrow\, ^{A-4}_{Z-2}\text{Y} + \alpha.$$

Beta decay causes a neutron in the nucleus to become a proton. As a result, Z increases by 1 unit, but A does not change. There is a different kind of beta decay we have not discussed in which a *positively charged(!)* electron (called a "positron") is emitted and a proton becomes a neutron. For this so-called positron emission A does not change, but Z decreases by one unit.

---

[3]Occasionally, the neutron number is given also, as a right-hand subscript, e. g., $^{14}_{6}$C$_8$.

Gamma-ray emission does not change A, Z, or N. It does reduce the energy stored in a nucleus, and it changes some nuclear properties we have not talked about.

Because each nuclide is uniquely characterized by its values of Z and N, it is convenient to lay out the nuclides in a chart where Z is along one axis and N along the other. Then each nuclide can be represented as a box located at the coordinates (N,Z). A piece of such a chart is shown in Fig. 16.8; you can also find such charts on the World Wide Web.

■ EXERCISES

34. Get a chart of the nuclides on the World Wide Web at the URL http://www.nndc.bnl.gov/chart/reZoom.jsp?newZoom=1.

Use the chart to find and write down the chain of decays that connects the atoms of ^{235}U to the first nonradioactive nucleus in the $4n+3$ chain.

16.6   SUMMARY

Radioactivity led to the discovery of the atomic nucleus. This tiny core of the atom contains 99.98% of its mass in about 10^{-12} of the atom's volume. The nucleus consists of particles called nucleons of which there are two different kinds—the proton and the neutron. The number of protons in a nucleus is the atomic number Z of the atom, and it determines uniquely which chemical element an atom is. The neutron number N then determines which isotope of the element the atom is. The number of nucleons in an atom is called its mass number A, and $Z + N = A$. A nucleus with mass number A has a radius of about

$$1.2 \times A^{\frac{1}{3}} \times 10^{-15} \, \text{m}.$$

An isotope of element X is specified using the notation

$$^{A}_{Z}X_{N},$$

where the Z and N values are often omitted because they are redundant (if you know the atomic number of X).

Radioactivity revealed that some kinds of atoms can spontaneously transform into others. Some do this by emitting an α particle (a helium nucleus, i.e., ^{4}He), which decreases Z by 2 units and A by 4 units. Others emit a beta ray (electron), which decreases Z by 1 unit but leaves A unchanged.

Nuclei of a given kind will emit alpha rays with characteristic, well-defined energies on the order of 5 MeV. Gamma radiations also are emitted with well-defined, characteristic energies just like the visible line-spectra emitted from atoms, except that gamma-ray energies can be several MeV in magnitude rather than the few eV of visible light. By contrast, beta-decay electrons do not come out with well-defined energy. Usually there is nothing like a line spectrum of electrons. In a collection of identical nuclei that undergo beta decay, some will emit low-energy electrons and some high, with energies ranging from 0 up to some maximum energy on the order of a few MeV.

Each type of radioactivity obeys the law of radioactive decay

$$N = N_0 e^{-\lambda t},$$

which shows that the emitting nuclei decay away exponentially over time. The rate of decay is specified by the disintegration constant λ, which is unique to each species of nucleus and to each type of decay. In nuclear physics it is common to use the half-life $T_{1/2}$ instead of the disintegration constant, where

$$T_{1/2} = \frac{\ln 2}{\lambda}.$$

The energies associated with nuclear properties are 5 to 6 orders of magnitude greater than those observed in the atomic processes typical of chemical interactions. These large energies and the fundamentally random nature of radioactive decay can be taken into account only by new physics: the identification of a new force in nature, the so-called "strong," or "nuclear," force, and the ideas of quantum mechanics.

PROBLEMS

1. Marie Curie had to process a large amount of ore to extract 0.1 gram of radium. The reason was that the radium isotope she discovered ($T_{1/2} = 1600$ y) was a decay product of mass-238 uranium ($T_{1/2} = 4.5 \times 10^9$ y), and over geological spans of time would exist only in equilibrium with the uranium from which the radium originated.

    a. Explain why at equilibrium the activities of radium and uranium are equal and show that under these conditions

$$\lambda_U N_U = \lambda_{Ra} N_{Ra}.$$

    b. Under equilibrium conditions, what will be the ratio of the number of Ra atoms N_{Ra} to the number of U atoms N_U? Tests on old uranium deposits show that they contain this ratio of radium to uranium.

   c. Marie Curie discovered radium in tailings of pitchblende, the principal uranium-bearing ore. Assume that when originally mined her ore was 50% uranium by weight. How many kilograms of pitchblende tailings did she have to process to isolate 0.1 g of radium?

2. What would be the potential energy of a particle of charge $+2e$ a distance of 10 fm from a particle of charge $+90e$?

3. Two protons approach each other towards a head-on collision with kinetic energies of 100 keV each.

   a. What is the total momentum of the system? Explain your reasoning.

   b. Before the collision, when they are far away from each other, what are the system's total kinetic energy K_{tot} and its electrical potential energy U?

   c. What is K_{tot} at the point of closest approach? Find the separation r_0 between the two protons when they are at the point of closest approach.

   d. What is the force (magnitude and direction) that one proton exerts on the other when they are separated by a distance r_0?

   e. What did Rutherford conclude from his analysis of Geiger and Marsden's experiment with α particles?

4. If a gold nucleus has a radius of 7 fm, what is the maximum kinetic energy that an alpha particle can have and still not penetrate the surface of the nucleus during a head-on collision?

5. What is the closest distance that a 1-MeV proton can come to a gold nucleus?

6. How much error would you make if you worked Example 16.5 nonrelativistically? Considering other values you have seen for early measurements of e/m, do you think that Becquerel would have been bothered by such a discrepancy?

7. Table 16.4 shows data taken by Geiger and Marsden for a study of large-angle alpha scattering (H. Geiger and E. Marsden, "The laws of deflection of α-particles through large angles," *Phil. Mag.* **25**, 604–623 (1913)).

TABLE 16.4   Large-angle alpha scattering data (Problem 7)

θ (deg)	No target (cpm)	Target (cpm)	Subtract back- ground	$\Delta N \sin^4 \frac{\theta}{2}$
150	0.2	4.95		
135	2.6	8.3		
120	3.8	10.3		
105	0.6	10.6		
75	0.0	28.6		
60	0.3	69.2		

Some of the data in Table 16.4 were taken without any target in order to determine what background counts might be. Do the correction for background, and show that the corrected data obey the $1/\sin^4 \frac{\theta}{2}$ dependence expected for Rutherford scattering.

8.   Figure 16.8 on p. 513 is a small piece of a chart of the nuclides. From the chart find and write down in standard nuclide symbols:
   a. Three stable nuclides that are not isotopes of one another.
   b. Three stable isotopes of the same element; give their relative abundance in nature.
   c. Three radioactive nuclides; give their half-lives and their decay products.

9.   Which of the nuclides in the chart on p. 513 is naturally occurring and radioactive? Is it more likely to be a nuclide that was created at the time of the formation of Earth or one that is made by cosmic rays? Why?

10.   Of the stable nuclides shown in the chart on p. 513, how many individual nuclides are there that have both an odd number of protons and an odd number of neutrons?

11.   The chart on p. 513 shows that for very light nuclides the number of protons and number of neutrons is roughly equal. But as the mass number increases, in any given stable nucleus there get to be more neutrons than protons. Can you suggest why this might be?

12.   The mass of a deuterium atom is 2.01410177 u. How much energy is needed to break it into a hydrogen atom (^{1}H) and a neutron?

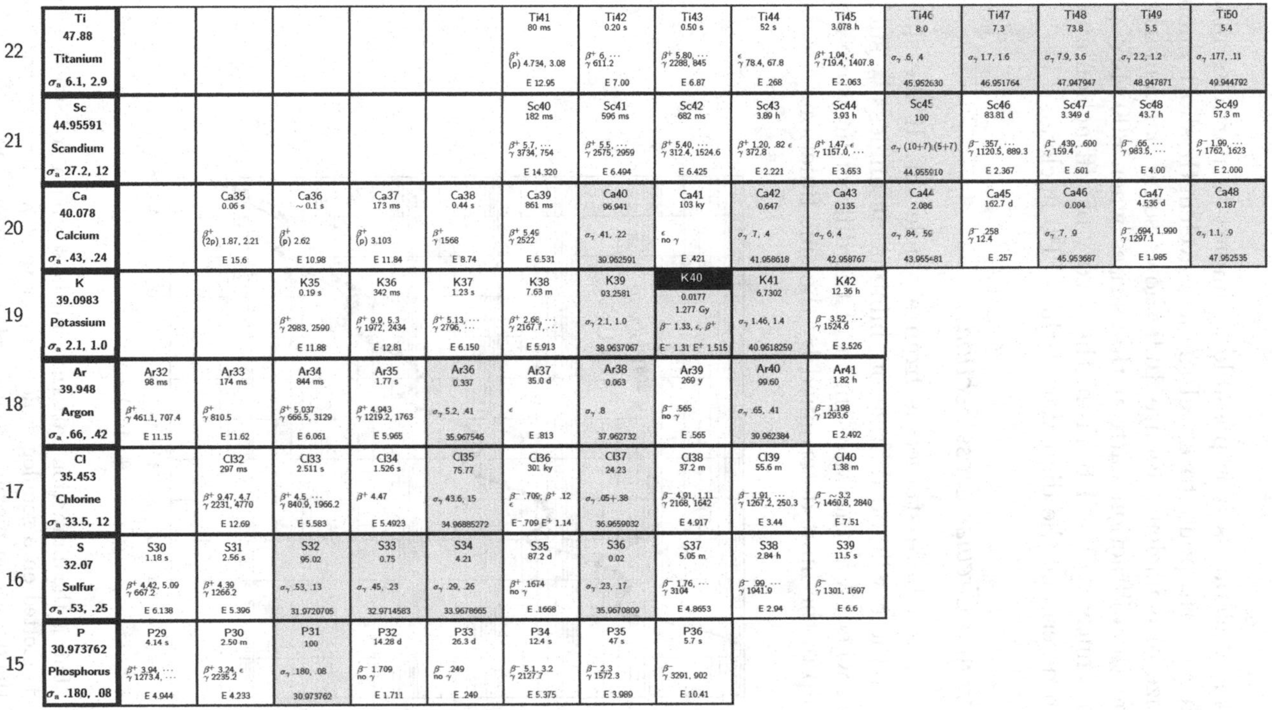

FIGURE 16.8 Each square is a distinct nuclide. The atomic number Z increases vertically from 15 to 22 here, and the neutron number N increases from left to right, going from 10 to 28 here. The column of squares at the left hand edge show the name, symbol, and *chemical* atomic weight of the element of the adjacent row of nuclides. Each grey square is a stable nuclide; its percentage natural abundance is given just below its symbolic name and mass number A; the square with a black bar is a naturally radioactive nuclide, here ^{40}K written as K40. The unshaded squares represent radioactive nuclides, and the number below the symbol is its half-life and the next lines show the principal kinds of decay.

FINDING THE RADIUS OF A NUCLEUS

Introduction

With modern accelerators it is possible to impart to particles momenta high enough to correspond to a deBroglie wavelength short enough to probe the size of a nucleus. Figure 16.9 shows some particularly good data showing the diffraction pattern that arises when 800 MeV protons scatter from a nucleus of ^{208}Pb. From the theory of diffraction and these data you can determine the diameter of the ^{208}Pb nucleus.

Diffraction from a Circular Cross Section

The diffraction you have studied is from a single slit. If the slit has a width b, then there will be diffraction minima at angles θ_n such that $b \sin \theta_n = n\lambda$ where λ is the wavelength of the diffracting wave and n is any integer up to the limit determined by the sine function.

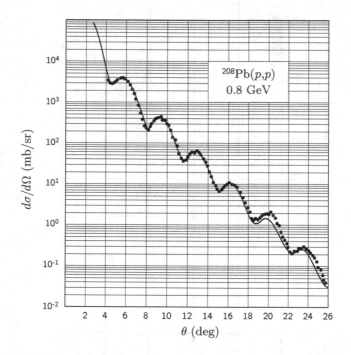

FIGURE 16.9 This is the diffraction pattern from the scattering of 800 MeV protons from ^{208}Pb. It is plotted on a semilog scale in order to show the very wide range of magnitudes of the diffraction maxima. *Taken with permission from G.F. Bertsch and E. Kashy, "Nuclear Scattering," 61, 859–859.* ©*1993 American Association of Physics Teachers.*

Diffraction from a circular aperture or from a circular cross section—do you remember Babinet's principle?—is slightly different from the case of a slit. The first diffraction minimum occurs when

$$b \sin \theta = 1.22\lambda.$$

The other minima occur very nearly according to the expression

$$b \sin \theta_n = 1.22\lambda + (n-1)\lambda, \tag{14}$$

where b is the diameter of the circular stop.

Consequently, for angles small enough for the small-angle approximation to hold, the angular separation of any two adjacent diffraction minima will be

$$\theta_{(n+1)} - \theta_n = \Delta\theta = \frac{\lambda}{b}. \tag{15}$$

Find the Nuclear Radius

   a. Find the wavelength λ of 800 MeV protons.

   b. Then use the data shown in Fig. 16.9 and determine the diameter of ^{208}Pb. Check to see whether the first minimum falls at 1.22 times $\Delta\theta$.

To check your answers and to see some more interesting information about nuclear radii, you may want to look at G.F. Bertsch and E. Kashy, "Nuclear scattering," *Am. J. Phys.* **61**, 858–859 (1993) from which this problem is adapted.

17.

Spectra and the Bohr Atom

17.1 INTRODUCTION

We come now to a new aspect of atoms: the existence of discrete energy states. Niels Bohr's idea that atoms can possess only certain well-defined amounts of energy was a major development in our understanding of atoms. In 1911 Bohr, a young Dane who had just received his Ph.D. in physics from the University in Copenhagen, came to England to visit for a year. He worked for a while in J.J. Thomson's laboratory in Cambridge, and then in early 1912 Bohr transferred to Manchester to work with Rutherford. Inspired by Rutherford's concept of the atomic nucleus, Bohr subsequently developed a nuclear model of the hydrogen atom that predicted the wavelengths emitted in the spectrum of atomic hydrogen. The agreement of his predictions with observations was startlingly good.

Bohr's nuclear model of the atom introduced two new ideas about the inner working of atoms that are basic to our present-day understanding of the atom: Atoms can exist only in special "stationary" states of well-defined energy; and an atom's angular momentum comes in integer multiples of $h/(2\pi)$ (i.e., $\hbar$). Bohr's model illustrates these ideas even though the model has been replaced by quantum mechanics. Despite its fundamental defects, Bohr's model continues to be of practical use because it often provides helpful insights into complicated problems more easily than does the full mathematical treatment of quantum mechanics. For these reasons we will examine Bohr's model—frequently called "the Bohr atom"—in some detail.

C.H. Holbrow et al., *Modern Introductory Physics, Second Edition*, DOI 10.1007/978-0-387-79080-0_17, © Springer Science+Business Media, LLC 1999, 2010

17.2   ATOMIC SPECTRA

You have already seen in Chap. 10 that each particular kind of atom emits light of well-defined characteristic wavelengths. Measurement, tabulation, and analysis of these wavelengths are the tasks of the subfield of physics called "spectroscopy." Quite soon after the discovery of the existence of these well-defined wavelengths, spectroscopists noted striking patterns and regularities among the observed wavelengths. Bohr showed that these patterns reveal much about the internal structure of the atom.

Wall Tapping and Bell Ringing

Historically, atoms have been difficult to study because they are so small. To get around this difficulty we have had to find ways to investigate objects too small to see, feel, or sense directly. One way is to collect a huge number (moles) of identical copies of the objects and look at their collective behavior. This is what you do when studying the pressure of a volume of gas or a chemical reaction. Another way is to whack the objects somehow and see whether interesting pieces break off. Cathode rays (electrons) and x-rays are some of the results of such whacking. Rutherford's alpha scattering from gold foils is another example of learning about atoms by whacking them quite hard. Although the results are interesting and informative, bashing objects lacks finesse and surely greatly modifies what you are studying.

You can also learn about an object by jiggling it gently and seeing what you can deduce from its subsequent wiggles. Have you ever tried to find a framing member of a house wall behind plaster or wallboard? A simple way is to go along the wall tapping with your finger. The sound will change when you reach the more solid area right over a support piece. The sound is a result of vibrations set up by your tapping, and the combination of vibrations changes character as the structure of the wall beneath the tapping changes. You use the quality of the sound to infer what structure lies beneath the point on which you tap.

You already use differences in the quality of sound to infer differences in structure. For example, you can easily tell the difference between a bell being struck and a piano string being hit. You can tell a banjo from a guitar, a saxophone from a tuba. The structure of each instrument determines its distinctive tone; the structure determines how much vibration of each frequency occurs whenever the object is excited in some way. In principle, it should be possible to learn something about the internal structure of the device by analyzing the combinations of frequencies present in any tone.

Something like this can be done with atoms using light rather than sound. Atoms made to vibrate will emit electromagnetic radiation, as was discovered in studies that began in the middle of the nineteenth century. The work of Kirchhoff, Fraunhofer, Balmer, Rydberg, and many others showed that each atomic element emits a distinctive pattern of light frequencies when properly stimulated. In Chap. 10 you learned that such a pattern is called the "spectrum" of the element.

Atomic Spectral Signatures

If each species of atom has a unique and distinctive spectrum, it is natural to think that spectra might tell us something about the internal structure of atoms. For testing this idea, hydrogen is the best choice because it has the simplest spectrum of all the atomic elements.

The simplicity is evident in Fig. 17.1, which shows a hydrogen-atom spectrum photographed by one of the authors. Notice the striking regularity with which the lines progress. They form a series of lines that get closer and closer together as the color goes from red (long wavelength) to violet (short wavelength) to the (invisible) ultraviolet. The series of lines approaches a well-defined limit, called the "series limit." (There are also lines in the picture that come from hydrogen molecules rather than hydrogen atoms; ignore these.)

The pattern's regularity is mathematically simple. In 1885 a Swiss school teacher named Johann Balmer devised a simple algebraic formula that accurately describes the sequence of observed wavelengths:

$$\lambda_n = 364.6 \, \frac{n^2}{n^2 - 4} \quad \text{nm}$$

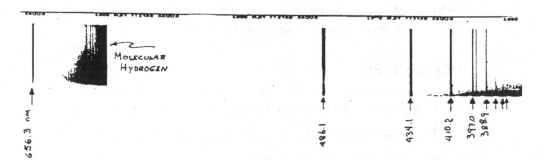

FIGURE 17.1 Photograph of the visible and near-ultraviolet spectrum of hydrogen.

where 364.6 nm is a constant chosen to match the formula to the observed wavelengths, and n is any integer greater than 2. A few years later Rydberg rearranged Balmer's formula into the form usually used today:

$$\frac{1}{\lambda_n} = R \left(\frac{1}{4} - \frac{1}{n^2} \right). \tag{1}$$

EXERCISES

1. Find the constant R from one of the wavelengths given in Fig. 17.1.

2. Plot a graph of $1/\lambda_n$ vs. $1/n^2$ to obtain a value for R from the slope. Check with the result of the previous problem.

3. Calculate the shortest wavelength, called the series limit, in the Balmer series.

Balmer's formula by itself was not earthshaking and did not immediately lead to any new insight into atomic structure. What it did do was to give hints and clues for discovering other series. For example, Rydberg rewrote the first term as $1/2^2$ and generalized it to $1/n'^2$. Then Eq. 1 became

$$\frac{1}{\lambda_n} = R \left(\frac{1}{n'^2} - \frac{1}{n^2} \right), \tag{2}$$

and when he assumed other integer values of n' he predicted entirely new series with different sets of wavelengths and different series limits. The series predicted for $n' = 1$ was later found by Lyman in the far ultraviolet spectrum of atomic hydrogen; another series predicted for $n' = 3$ was found in the near infrared spectrum by Paschen.

EXERCISES

4. Calculate the series limits for the Lyman and Paschen series.

5. Find the first three lines (longest wavelengths) for the Lyman series.

6. Find the first three lines of the Paschen series.

7. Using the same scale of wavelengths, draw the positions of the first three lines and the series limits of all three series described above.

These regularities in the spectra of atomic hydrogen provided Bohr with important clues to the inner workings of the atom. He devised a model of the hydrogen atom that exactly predicted its spectrum. Let's see how he did that.

17.3   THE BOHR ATOM

Need for a Model

J. J. Thomson's "plum pudding" model of the atom (see Sect. 16.3) could not predict the details of the observed spectrum of hydrogen or any other element. Initially, the nuclear model was also unpromising. In the first place, classical physics showed that electrons bound to a positively charged nucleus must revolve in orbits. But why would they have only those particular orbits that represented the special set of frequencies observed in the spectrum of hydrogen? Even more of a problem was how they could stay in orbit. Any orbiting charged particle must radiate electrical energy due to the acceleration it undergoes as it is bent into the circular, or perhaps elliptical, orbital path. Thus orbiting electrons would constantly lose energy and spiral in to the nucleus, just as Earth-orbiting satellites that are low enough to encounter some atmosphere gradually lose energy and spiral in to Earth. In the case of atomic electrons, though, the time to decay would be microseconds or less, not months or years! Classical physics had no way to explain why we have atoms at all or why they have the sizes they do.

Bohr's Ideas

Bohr was successful by being able to unstick himself from the accepted rules about how things should work. That in itself, though, is not necessarily remarkable. There is many a crank around doing the same thing. What was remarkable was Bohr's ability to invent new rules that worked and could be generalized to predict new results. We will not try to reproduce his actual steps, but will trace a similar path that is easier to follow.

Bohr decided to ignore classical physics' inability to account for stable atoms. He gave up trying to relate the hydrogen spectra directly to internal motions, and he took the nuclear model at its face value. If there had to be stable orbits, he reasoned, one should make stability a basic property of the model rather than worrying about how orbits couldn't be stable.

Turning first to the spectrum problem, you can get new insights into Rydberg's version of the Balmer formula by using the photon idea to connect wavelength to energy. Since c/λ is frequency, and the Planck constant, h, times frequency is energy, you can multiply both sides of Rydberg's equation by hc to get

$$E = \frac{hc}{\lambda} = hf = hcR \left[\frac{1}{4} - \frac{1}{n^2} \right],$$

which is Eq. 1. Substituting the experimentally observed value of $1.097 \times 10^{-2}\,\mathrm{nm}^{-1}$ for R and $1240\,\mathrm{eV\,nm}$ for hc gives

$$hf = 13.6 \left[\frac{1}{4} - \frac{1}{n^2} \right] \mathrm{eV}. \tag{3}$$

In terms of photons it does not make sense to think of the hydrogen atom as "ringing" in some complicated way like a musical instrument. Any one hydrogen atom must emit a photon with essentially only one frequency for any one event. This suggests that a single photon is emitted whenever a single hydrogen atom changes its energy. This photon must have one of the wavelengths observed by spectroscopists (656 nm, 486 nm, . . .). If the atom had an energy E_i before and E_f after emission of the photon, the photon energy would be:

$$hf = E_i - E_f, \tag{4}$$

where we assume $E_i > E_f$ because an atom would have to lose energy to produce a photon. Comparing Eq. 3 with Eq. 4, it is natural to think of the two terms on the right-hand side of Eq. 3 as separate energies. These different energies might be energies of different configurations of the atom, what we call "energy states" of the atom.

The important idea here is the assumption that *the atom exists only in particular definite energy states*. This is quite different from classical physics, where the energy of a system can vary continuously. It is like saying that a baseball thrown near the surface of the Earth can have 2 J of energy or 3 J of energy, but nothing in between.

Quantizing the Hydrogen Atom's Energies

Using classical mechanics Bohr derived an expression for the energy of the electron in the atom as a function of the distance r between the electron and the nucleus (proton). Then he took a giant step beyond classical physics: He invented a rule that permitted only certain values for the atoms's energy E. In this way he was able to explain Eq. 3. Today we say that Bohr "quantized the energy" of the atom.

Bohr assumed that the electron is kept in a circular path around the nucleus by the attractive Coulomb force between the electron and the proton that is the nucleus of the hydrogen atom. Why its energy did not leak away as classical physics predicted he did not know. It obviously didn't, so he set the question aside to be dealt with at some future time. The proton is so much more massive than the electron that it remains nearly stationary while the electron moves around it. Then since the Coulomb force (p. 159) is supplying the centripetal force, it follows that

$$\frac{mv^2}{r} = \frac{k_c e^2}{r^2}. \tag{5}$$

(Because the proton charge is $+e$ and the electron charge is $-e$, the electrical force is attractive.) Rewriting Eq. 5 in terms of momentum,

$$\frac{p^2}{mr} = \frac{k_c e^2}{r^2},$$

you get two expressions, one for momentum p and one for orbital kinetic energy K:

$$p = \sqrt{\frac{m k_c e^2}{r}} \tag{6}$$

and

$$K = \frac{p^2}{2m} = \frac{k_c e^2}{2r}. \tag{7}$$

Because electrostatic potential energy U is

$$U = \frac{-k_c e^2}{r},$$

the total energy, i.e., the sum of the kinetic and potential energies, is

$$E_{\text{tot}} = K + U = -\frac{k_c e^2}{2r}. \tag{8}$$

Notice that the total energy depends only on r. Therefore, for the total energy to take on only certain values, there must be some rule that correspondingly constrains r. Here Bohr took a bold step. In effect he said that the rule is that the product of the radial position r and the momentum p must always be some integer multiple n of the Planck constant divided by 2π, i.e.,

$$rp = n\frac{h}{2\pi} \equiv n\hbar. \tag{9}$$

There was no precedent for this rule, although there surely had to be some connection to the Planck constant. After all, something had to take on specific values, had to be quantized, and h was already associated with the quantum of light, the photon. There are several equivalent ways to rationalize Eq. 9. One is to recognize that rp is the so-called angular momentum of the system. Equation 9 then states that the allowed values of the angular momentum are integer multiples of $\hbar$, i.e., *the angular momentum is quantized in units of* $\hbar$. This assertion is validated by modern quantum theory. Because we have not discussed angular momentum in this book, let's consider an argument that makes use of the wave nature of matter introduced in Chap. 15. Although not logically sound, it is a convenient mnemonic.

Consider the implications of the electron having wave properties. In the presumed "orbit" such a wave must come back on itself. If after each circling of the nucleus the wave's phase has changed by an exact multiple of 2π, the succession of waves will reinforce constructively. If the phase difference is anything else, there will be destructive interference over time. In other words, only those circular orbits will exist for which the circumference is an integer number of wavelengths $2\pi r_n = n\lambda$. (It is stretching things to use the idea of a wave along the circumference of a circle and yet assume an exact radius.) Using the de Broglie relation between wavelength and momentum,

$$p = \frac{h}{\lambda} = \frac{hn}{2\pi r_n},$$

and substituting for p from Eq. 6 yields a special set of values of r, which we denote as r_n:

$$r_n = \frac{n^2 h^2}{4\pi^2 m k_c e^2}. \tag{10}$$

We can use these values of r_n to find the allowed energies of the hydrogen atom. Replacing r_n in the total energy equation, Eq. 8, by Eq. 10 gives a set of discrete energies E_n one for each integer value of $n = 1, 2, 3 \ldots$:

$$E_n = -\frac{2\pi^2 m k_c^2 e^4}{n^2 h^2}. \tag{11}$$

These special values of E that can occur are called "energy states" or "energy levels". The lowest energy E_1, i.e., $n = 1$, is often referred to as the "ground state" of the atom. Because of this association of the energy states of the hydrogen atom with the integers n, it is customary to label the energy states with the index n. Because the restriction of n to integer values forces there to be a finite difference, a *quantum* of difference, between energies, the index is called a "quantum number." For atoms more complicated than hydrogen, n continues to be very important, but other such indices are needed as well, and they are all called quantum numbers. To distinguish the n quantum number from the others, it is called the "principal quantum number," or, sometimes, the "radial quantum number."

The difference in energy between a lower-energy state of quantum number n' and a higher-energy state of quantum number n, is just

$$E_n - E_{n'} = \frac{2\pi^2 k_c^2 e^4 m}{h^2} \left[\frac{1}{n'^2} - \frac{1}{n^2} \right].$$

Therefore, if, as Eq. 4 proposes, a photon of energy hf is emitted when the atom changes from state E_n to state $E_{n'}$, the photon's energy would be

$$hf = E_n - E_{n'} = \frac{2\pi^2 k_c^2 e^4 m}{h^2} \left[\frac{1}{n'^2} - \frac{1}{n^2} \right],$$

which has exactly the same form as Eq. 3 when $n' = 2$. Putting in values for the constants to find the factor multiplying the bracketed terms gives

$$\frac{2\pi^2 k_c^2 e^4 m c^2}{(hc)^2} = \frac{2\pi^2 1.44^2 \, 511 \times 10^3}{1240^2} = 13.6 \, \text{eV},$$

exactly as observed experimentally.

This is a spectacular result. Until Bohr, Eq. 1 was only an empirical guess, and the value of R was an experimentally determined number. Bohr's model of the internal structure of an atom yielded an expression for R in terms of fundamental constants from several areas of physics that when evaluated numerically is in exceptionally good agreement with the experimental value. Clearly, Bohr's result cannot be just a fluke. Therefore, although the model with its ad hoc assumptions has serious flaws, some of its elements must be correct and must play a role in a full theory of the atom.

Although the wave nature of the electron makes the concept of an orbit with a well-defined value of r questionable, the calculated values of r_n from Eq. 10 do correspond roughly to the atom's size. The smallest one, r_1, is often given a special symbol, a_0,

$$a_0 = \frac{h^2}{4\pi^2 m k_c e^2} = 0.0528 \, \text{nm},$$

and is called the "first Bohr-orbit radius." All other possible Bohr-model radii are thus

$$r_n = n^2 a_0 = 0.0528 \, n^2 \, \text{nm},$$

where the quantum number n can take on any integer value. In Chap. 19 you will see that the value of a_0 is what is obtained when you use Heisenberg's uncertainty principle to estimate the minimum energy of the hydrogen atom. This means that the Bohr model is consistent with this fundamental principle.

Do not attribute too much significance to these values of r_n. Do not think that the electrons move in well-defined orbits in the atom. The uncertainty principle emphatically denies the possibility of such orbits. The electron in an atom is much more wavelike than particle-like. It does not and cannot have a well-defined location. Thus, although the value of r_n gives an indication of the spatial extent of a hydrogen atom that has an amount of energy E_n, r_n does not label the path of an electron inside the atom the way a planetary radius labels the planet's orbit around the Sun.

Energy-Level Diagrams

It is possible to describe *any* atom in terms of its energy states. An important graphical aid for representing the energies of an atom is the "energy-level diagram." As an example, consider the hydrogen atom. You can calculate the energies E_n of its quantum states from the relation

$$E_n = -\frac{13.6}{n^2} \; \text{eV}$$

and arrange them vertically on a scale, as shown in Fig. 17.2. The lines represent the "energy levels" of the atom. In the diagram shown here, the zero of the scale is the energy for n equal to infinity. Note that from Eq. 10, this condition corresponds to an infinite separation between the electron and the nucleus, i.e., the hydrogen atom would be ionized. This means that 13.6 eV of energy is required to free an electron from the hydrogen atom in its ground state.

The energy-level diagram looks like a ladder with "rungs" getting closer together as they approach the top. *A photon is emitted by an atom when it drops from one energy level to a lower one.* The energy of the emitted photon is equal to the difference in energy between the initial and final energy states of the atom. We say that the atom has undergone a "transition" from one energy state to another.

You can use this picture to understand the various spectral series that are emitted by gaseous atomic hydrogen. A hydrogen atom can undergo a transition from any higher energy state to any lower one. Different transitions have different probabilities of happening. If a hydrogen atom is ionized, an unbound electron, i.e., an electron outside of the atom, can make a transition to any energy level of the atom.

In the laboratory, when an electric current is passed through hydrogen gas in a thin glass tube, the resulting electric discharge ionizes some hydrogen atoms and excites the electrons in others to higher energy states. After excitation an atom loses its energy by a cascade of successive jumps. Which jumps occur differ from one atom to the next, and there is no guarantee that any given atom will go from any particular energy state

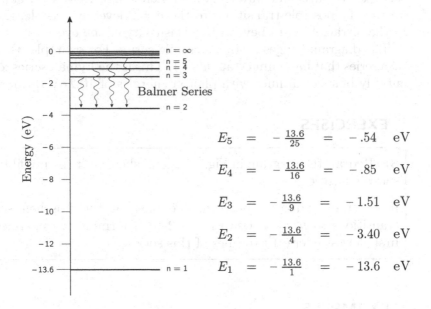

$$E_5 = -\frac{13.6}{25} = -.54 \quad \text{eV}$$

$$E_4 = -\frac{13.6}{16} = -.85 \quad \text{eV}$$

$$E_3 = -\frac{13.6}{9} = -1.51 \quad \text{eV}$$

$$E_2 = -\frac{13.6}{4} = -3.40 \quad \text{eV}$$

$$E_1 = -\frac{13.6}{1} = -13.6 \quad \text{eV}$$

FIGURE 17.2 Energy-level diagram of the hydrogen atom.

to another. The process is quite random, and the atom can de-excite by skipping steps. (Modern quantum theory can predict the probabilities of these transitions, but the Bohr model cannot.) The result is that when a collection of atoms de-excites, photons of many wavelengths will be emitted by different atoms. For any given atom, the whole sequence of photon emissions down to the ground state usually takes a very short time, from microseconds to nanoseconds. An observed spectrum is made up of photons emitted from a variety of different transitions occurring in many different hydrogen atoms.

The transitions that produce the Balmer series of spectral lines are shown in Fig. 17.2 by the vertical wavy arrows connecting higher energy states with the $n = 2$ energy level. Note that the final state of the Balmer line transitions is not the lowest possible energy state of the atom. After a Balmer transition, the hydrogen atom will change into its lowest, most stable, energy state—its ground state. In other words, it will make a transition from $n = 2$ to $n = 1$, and it will emit a photon that has a wavelength too short (energy too high) to be visible to the eye.

The Balmer series is apparent in the laboratory because it is the only series that consists of wavelengths of visible light. Even so, only three or four of the possible transitions to the $n = 2$ level are visible; the rest are in the ultraviolet and beyond the sensitivity of the eye.

The diagram suggests other possible series. For example, there should be a series that has a final state $n' = 1$. This is the Lyman series, consisting entirely of spectral lines with photon energies of 10.2 eV or more.

▪ EXERCISES

8. Redraw the diagram in Fig. 17.2, and show on it the possible Lyman series transitions.

9. In the hydrogen spectrum the lines of the Paschen series are transitions to $n' = 3$. Use Fig. 17.2 to determine the energies of the first (lowest energy) two lines of this series.

▾ EXAMPLES

1. What photon energies would be produced from atoms excited to $n = 3$?

The electron could go to the ground state by two routes. One would be to jump directly to $n = 1$, emitting the second line of the Lyman series, or it could go from 3 to 2 and then to 1. These transitions would be the first lines of the Balmer and Lyman series, respectively. Thus there are three possible photon energies:

$$E_3 - E_1 = -1.51 + 13.6 = 12.1 \, \text{eV},$$
$$E_3 - E_2 = -1.51 + 3.4 = 1.9 \, \text{eV},$$
$$E_2 - E_1 = -3.4 + 13.6 = 10.2 \, \text{eV}.$$

▪ EXERCISES

10. Eventually, two more hydrogen series in the infrared were identified: the Brackett series with $n' = 4$, and the Pfund series with $n' = 5$. Find the photon energies of the first two lines in each series.

> **11.** What are the wavelengths of the lines calculated in the previous problem?
>
> **12.** Suppose a group of hydrogen atoms are all excited to $n = 5$. How many different photon energies are emitted as the assemblage decays?
>
> **13.** Find the energies of all the photons that could be emitted from hydrogen atoms excited to $n = 4$.

17.4   CONFIRMATIONS AND APPLICATIONS

Bohr's model produced some remarkably accurate results. But in important ways it was *ad hoc*, introducing some strange new concepts just to explain a limited set of data. For the new ideas to become credible, there had to be further predictions that could be confirmed by experiment.

Energy Levels

Existing only in one well-defined state of energy or another, an atom can lose energy only in discrete amounts. But if energy is emitted only in certain well-defined amounts, energy also can be absorbed only in discrete amounts. If a particular atomic transition emits a photon of a certain energy, the same energy must be absorbed in order to reverse the transition and raise the atom from a lower to a higher energy level.

You can add energy to atoms just by shining light on them. If a photon in a beam of light meets an atom and if the photon's energy hf is *exactly equal* to the difference between the atom's present energy state and some higher energy state, the photon can be absorbed by the atom, and the atom will be excited to the higher energy level. The absorbed photon is then lost from the light beam. If nothing is in the path of a beam of light traveling from a hot body such as a lamp filament and passing through a slit and a diffraction grating, the grating spreads the beam out into a continuous spectrum of colored light ranging from red to violet. If there is a gas between the source and the spectrometer slit and if that gas absorbs light at certain well defined wavelengths, the intensity of light will be reduced in the spectrum at the positions corresponding to the absorbing wavelengths. As a result there appear in the spectrum thin strips of reduced intensity that are dark compared to the brighter parts of the spectrum. These dark strips are called "absorption lines."

It is also possible to make a beam of light of a single wavelength that can be varied. If as the wavelength is changed it takes on a value corresponding to the photon energy that matches the energy of a transition, the beam will be sharply absorbed and its passage through the gas will be much diminished.

Suppose you were to send a beam of visible light through ordinary hydrogen gas in a transparent container. Would photons at the wavelengths of the Balmer series be absorbed? No, not if the Bohr idea is correct. Aside from the complications of the hydrogen molecule, absorptions corresponding to the Balmer series must all start from the $n = 2$ level, but at ordinary temperatures there would be no H atoms excited to this state and capable of absorbing the light that is present.

A gas of hydrogen atoms could absorb only the much higher energy photons of the Lyman series. These are the only transitions that start from the lowest energy state of the atom.

Rydberg Atoms

You can imagine using a mixture of ordinary visible light and ultraviolet. The ultraviolet might excite an atom to its $n = 2$ energy level, and then a visible photon could excite a Balmer transition from this state to some higher state. This is not practical using ordinary light sources because they are too weak: The atom will return to its ground state long before a visible photon arrives to induce a Balmer transition upwards. With lasers, however, it is possible to induce double absorption. Lasers can produce enormous numbers of photons with very well defined frequency. The frequency of one laser can be adjusted to exactly the right energy to excite atoms from their ground states to some particular excited state. Then if these excited atoms are illuminated with a large number of photons from a second laser set to a wavelength to induce a transition upward from the excited state, it becomes likely for such a transition to occur before the atoms de-excite to the ground state.

By such two-step excitation it is possible to make atoms in states with very high values of n. Any atom, not just hydrogen, in a high-n energy state obeys the Rydberg formula, Eq. 3. Consequently, such highly excited atoms are called "Rydberg" atoms, and the energy states with high values of n are called "Rydberg states." It is as though the excited electron is so far from the nucleus that the negative charge of the other electrons neutralizes all but one of the positive charges of the nucleus. As a result, the atom's nucleus and innermost electrons look to the excited electron like a hydrogen nucleus, and the Bohr model describes it quite well. Atoms have been prepared with n as large as 400!

▓ EXERCISES

> **14.** Find the Bohr radius of a hydrogen atom with $n = 400$.

If you have done the calculation of the previous problem correctly, you found the dimensions of the $n = 400$ atom to be as large as objects that can be seen with a high-powered microscope. However, you could not "see" such a Rydberg atom. An atom in a high-n state is very fragile; any photon of visible light would ionize it.

▓ EXERCISES

> **15.** Confirm the previous statement by finding the energy of the $n = 400$ level and comparing the ionization energy to that of the lowest-energy photon that is visible (wavelength around 700 nm).
>
> **16.** Suppose you irradiated hydrogen atoms with light consisting of photons with a continuous distribution of energies from 0 to 14 eV. What spectral series would be observed in absorption, assuming that you have the right equipment and that double-photon absorption is negligible.

Note that we have discussed only hydrogen *atoms* even though the gas comes as diatomic molecules; molecules are beyond the scope of the Bohr theory. However, the usual way of exciting hydrogen is to pass current at high voltages through a sample of gas. The ions created have enough energy to dissociate the molecules, and we get enough single atoms to produce the atomic spectrum. Usually, radiation from excited molecules is also present, and we just ignore it when studying the atomic spectral series.

The Franck–Hertz Experiment

You can also add energy to atoms by bombarding them with energetic particles. Electrons accelerated through an electric potential difference of a few volts will excite atoms when they collide with them. Just as photons do, colliding electrons lose energy in discrete amounts. That is, only the exact energy difference between two levels is absorbed, because there is no way for an atom to exist at any energy in between. There is an important difference in the way photons and electrons lose energy to atoms.

For photons it is all or nothing. They either lose all their energy and disappear, or they lose nothing and continue on. Only when the photon energy exactly matches the transition energy is the photon absorbed. *Free electrons, on the other hand, can lose part of their energy to the atom and retain the rest.* They can lose an amount equal to the energy of the transition that is induced in the atom, and continue on with whatever energy is left over. Free electrons with any energy over the minimum necessary to induce a transition can and will induce transitions.

A year after the Bohr model was published, J. Franck and G. Hertz bombarded mercury atoms with energetic electrons to exhibit directly the existence of discrete atomic energy levels. They accelerated electrons through a vapor of mercury and showed that the electrons lost energy to the mercury atoms in discrete amounts. Figure 17.3 shows how this was done. Electrons were accelerated from the filament to the accelerating electrode. The accelerating electrode was a mesh, so that the accelerated electrons passed through it. They were then decelerated by the retarding voltage V_{ret} of 2–3 V between the accelerating electrode and the collecting electrode.

As the accelerating voltage V_{acc} was increased from 0 to about 6 V, the number of electrons reaching the collecting electrode, measured by the current meter, increased, but instead of rising steadily as V_{acc} was increased further, the number dropped to a minimum at $V_{\mathrm{acc}} \approx 8\,\mathrm{V}$. As Fig. 17.4 shows, the number of electrons reaching the collecting electrode varied regularly as the accelerating voltage was increased towards 40 V. The collector current fell to a minimum every time the accelerating voltage changed by 4.9 V.

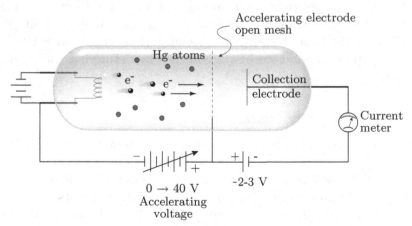

FIGURE 17.3 Schematic of a version of the Franck–Hertz experiment. The arrangement of applied voltages with typical values is shown.

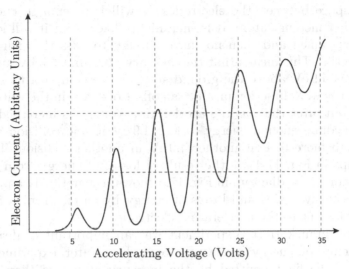

FIGURE 17.4 Typical current vs. accelerating voltage data in a Franck–Hertz experiment on mercury vapor.

The regular fluctuations in the collector current are just what is expected if the free electrons give up their energy to mercury atoms in quanta of 4.9 eV. An electron accelerated through 6 V would have a kinetic energy of 6 eV. For example, this would be enough to permit it to pass through a retarding potential of 3 V and reach the collecting electrode.

■ EXERCISES

17. What would be its kinetic energy when it reaches the collecting electrode?

But if a 6 eV electron loses 4.9 eV of energy in a collision with a mercury atom somewhere between the filament and the accelerating electrode, it will reach the accelerating electrode with only 1.1 eV of kinetic energy. Then it will not have enough kinetic energy left to overcome the retarding potential and reach the collecting electrode, and it will not contribute to the current measured by the meter in Fig. 17.3. This absorption of energy by the Hg atoms will cause the electron current reaching the collecting electrode to decrease.

The most striking feature of the graph in Fig. 17.4 is the sharp drop in the current at periodic intervals of the accelerating voltage. The analysis above suggests that every time an electron gains another 4.9 eV or so in the space between the electrodes, it will have enough energy to excite another mercury atom. It is then highly likely that it will lose the energy it has gained and then not have enough to pass through the retarding potential. This means that the mercury atom must have an excited state that is 4.9 eV above the ground state. Some complications having to do with the work function produce an offset of ≈ 2 V in the voltage of the first dip, but after that, as the accelerating voltage is increased, the electron will acquire enough energy to lose 4.9 eV in one collision and then gain enough more to lose another 4.9 eV in another collision. The succession of dips in Fig. 17.4 are the result of losses of energy from such multiple collisions. The diagram in Fig. 17.5 shows schematically a possible version of successive gains and losses of energy by an electron as it passes from the filament to the accelerating electrode.

The Franck–Hertz experiment was an important confirmation of the existence of energy levels in atoms. In a later experiment Hertz observed the light emitted by the mercury atoms as they decayed back to the ground state. As you would expect, these photons had 4.9 eV of energy.

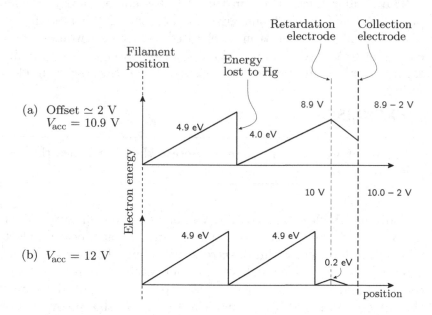

FIGURE 17.5 Diagram showing the mechanism of multiple collisions in a Franck–Hertz experiment. Because the collection and screening electrodes have work functions different from that of the filament, the effective acceleration voltage is 2 V less than the applied voltage.

EXERCISES

18. For defining the voltage differences it is more reliable to pick the voltages of the highest currents just before the start of the dips. Find the voltages of peak currents in the graph above and plot a graph of V_p versus n, the number of collisions. Take the slope to obtain the best value for the voltage difference.

Hydrogen-like Ions

The Bohr model only works well for simple two-body systems like hydrogen atoms or Rydberg atoms. When the atom has more electrons than the hydrogen atom, the potential energy of each electron becomes very complicated, arising now from the constantly changing distribution of the other electrons as well as the nuclear charge. However, there are some other simple two-body combinations of charged particles besides the hydrogen atom. For example, if the helium atom, which has a doubly charged nucleus, is singly ionized, it will look like a hydrogen atom with two units of nuclear charge. It is easy to include the effect of a different nuclear charge in Eqs. 5–11.

Recall that in these equations the product $q_1 q_2$ was written as $-e^2$ because the charge of the hydrogen nucleus is just e. If there were a larger charge on the nucleus, it would have to be written as Ze, where Z is the atomic number of the atom. When the factor of e^2 is replaced with Ze^2, Eqs. 5, 7, 10, and 11 become, respectively,

$$\frac{mv^2}{r} = \frac{Zk_c e^2}{r^2}, \tag{12}$$

$$K = \frac{Zk_c e^2}{2r}, \tag{13}$$

$$r = \frac{n^2 \hbar^2}{mk_c Ze^2}, \tag{14}$$

$$hf = \frac{mZ^2 k_c^2 e^4}{2\hbar^2}\left[\frac{1}{n'^2} - \frac{1}{n^2}\right] = 13.6Z^2\left[\frac{1}{n'^2} - \frac{1}{n^2}\right] \text{ eV.} \tag{15}$$

Equation 15 applies to any low-mass atom that has had all but one of its electrons removed. For example, it describes quite accurately the lines of the spectrum of He^+ and of Li^{++}.

▪ EXERCISES

19. For singly ionized helium, calculate the energies of the first three lines of the series equivalent to the Lyman and Balmer series.

20. Show that some of the "Balmer-series" lines of singly ionized helium are the same as lines of the hydrogen Lyman series. Which ones are they, in general? This correspondence was a source of confusion when spectral lines of He ions were first observed in the solar spectrum; some people speculated that they were seeing a new form of hydrogen not found on Earth.

21. Find the energies of the first lines of the series in doubly ionized lithium that are equivalent to the Lyman and Balmer series.

17.5   HOW ATOMS GOT THEIR (ATOMIC) NUMBERS

Introduction

Mendeleev's periodic table of the elements was a significant advance in chemistry. It made explicit the empirical observation that similarities in the chemical properties of the elements recur periodically as atomic mass increases. For over 40 years it was a useful guide for scientists, but it provided no explanation for the periodicity of properties of the elements. Then, in the early decades of the twentieth century dramatic advances in physics revealed the structure of atoms and uncovered the physical basis of the periodic table. This section tells you about one of these great advances: how physicists learned that the atomic number—the number that specifies the position of an atom in the periodic table—is the number of positive charges in the atomic nucleus.

How many Elements can there be?

In his 1869 table Mendeleev exhibited the periodic recurrence of the chemical properties of the elements by putting elements with similar chemical behavior in rows and ordering them in columns according to their atomic weights. Thus, in one row he put lithium ($A = 7$), sodium ($A = 23$), potassium ($A = 39$), rubidium $A = 85.4$), and cesium ($A = 133$). These alkali metals show similar chemical behavior; for example, they all form similar compounds with oxygen: Li_2O, Na_2O, Rb_2O, and Cs_2O.

Modern periodic tables like the one on p. 642 put elements with similar properties in columns and with increasing mass along the rows. Modern periodic tables are also complete and correctly ordered, while Mendeleev's table had empty spaces corresponding to undiscovered elements, and it had some elements in wrong places.

In the 1890s physicists and chemists discovered the noble gases—helium, neon, argon, krypton, and, later, radon—and added an entire new column to the table. Given that discovery, you might ask: Is today's periodic table complete? Might there be other elements that have been overlooked? The fact that the atomic number is the number of charges in the nucleus Z assures us that the answer to this last question is "No."

Immediately after Rutherford discovered the atom's nuclear core (Chap. 16), Bohr showed that the nuclear charge Ze determines the scale of the energy states of an atom (p. 535). In 1913 H. G. J. Moseley measured the wavelengths of x-rays emitted by many different kinds of atoms and showed that each chemical element is uniquely identified by its nuclear charge Ze, that there is a one-to-one correspondence between Z and a chemical element. In other words, the nuclear charge number Z specifies the position of an element in the periodic table and is, therefore, the same as the atomic number—the serial number of the element in the periodic table. Up to the time Moseley did his experiments the atomic number could only be determined empirically. Moseley found its physical basis: It is the number of positive charges in the nucleus.

From this fact it follows that there can only be as many elements as there are integers Z; once you have found 83 elements with values of Z from 1 to 83, you have found all the elements from hydrogen to bismuth. The only other possible elements must have $Z > 83$. Some of these exist in nature, and it is possible to make others by adding charges to the nucleus.[1]

The properties of x-ray line spectra were the basis of Moseley's discovery that Z is the atomic number, and you need to learn about them to understand his experiment.

X-Ray Line Spectra

In 1905, a decade after Roentgen discovered x-rays, the British physicist Charles Barkla found that a target struck by a beam of high energy x-rays emitted x-rays distinctly different in behavior from those in the incident beam. He called the incident x-rays "primary," and the different outgoing

---

[1] Nuclear physicists and nuclear chemists do this. They have made elements up to $Z = 118$, but these high Z elements are difficult to make and are radioactively very unstable. For example, experimentalists have observed only two or three atoms of $Z = 118$, and these lived only a few milliseconds. It takes a while before these are given official names. In 2010 element $Z = 112$ was officially named copernicium and given the symbol Cn.

x-rays "secondary." By measuring the absorption of the secondary x-rays in sheets of material placed between the emitting target and the x-ray detector, Barkla showed that the energy (frequency) of the secondary x-rays was characteristic of the target (anode) material. For example, secondary x-rays from an iron target were more energetic than those from an aluminum target. He discovered that the secondary x-rays emitted by a target are unique to the chemical element the target is made of, so he called them "characteristic x-rays," and pointed out that they could be used to identify the target material. Barkla had discovered a new means of chemical analysis.[2]

From his measurements of the absorption of x-rays Barkla found that an anode emits two distinctly different types of characteristic x-rays— a more penetrating type (shorter wavelengths, higher energy) that he called K radiation or K x-rays, and a more easily absorbed type (longer wavelengths, lower energy) that he called L radiation.

Later, after x-rays were found to be waves and the x-ray spectrometer was developed, it became clear that Barkla's K and L radiations were x-ray line spectra (see Fig. 17.6). These are x-rays with well defined frequencies that show up as high intensity peaks on the background of continuum radiation discussed in Chap. 14. When the energy of electrons bombarding the anode is increased, the intensity of the emitted x-ray lines increases relative to the background, but their wavelengths remain unchanged. These emissions are called x-ray lines because they are analogous to the spectral lines in the visible light spectra of atoms. And like

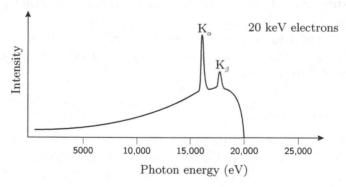

FIGURE 17.6 An x-ray spectrum showing continuum and characteristic x-ray lines from element of atomic number 40 when bombarded with 20 KeV electrons. The subscripts α and β are explained on p. 544 in the text.

---

[2]Charles Barkla was awarded the 1917 Nobel Prize in physics. In addition to discovering characteristic x-rays, he confirmed that x-rays are electromagnetic radiation by showing experimentally that they can be polarized. His doctoral advisor was J. J. Thomson.

visible spectral lines, x-ray lines are a unique fingerprint of the emitting atom. Moseley studied x-ray line spectra, (characteristic x-rays) and discovered a simple relationship that allowed him to predict the frequencies (energies) of x-rays from any element and to see that the charge of the atomic nucleus is the property that gives an atom its identity.

Moseley's Experiment

Moseley[3] knew from Rutherford's work that an atom has a nucleus with a nuclear charge roughly $A/2$ times the elementary charge e, where A is the atomic mass. (And from Barkla's studies of x-ray absorption, he knew that the number of electrons in an atom was also roughly $A/2$ as necessary to balance the nuclear charge and produce an electrically neutral atom.) Consequently, Mendeleev's ordering of the elements by mass number A was roughly the same as ordering them by nuclear charge number Z. Now Moseley showed that Z gave an exact and unambiguous ordering; Z, not A, was an element's serial number in the periodic table.

Moseley designed and constructed an ingenious spectrometer that allowed him to quickly and accurately measure the wavelengths of the characteristic x-rays emitted by various elements. His spectrometer (Fig. 17.7) had two novel features. First, it used photographic film rather than an electrometer to record x-ray intensities. Second, it had multiple targets (anodes) mounted on a carriage inside the x-ray tube, so that he could change the element he was studying without losing the tube's vacuum. Working mostly on his own, in just over one year of intense effort Moseley measured the line spectra of 38 elements. Some of his early results[4] are summarized in Table 17.1 and in Fig. 17.8.

Moseley found that when he plotted the square root of the frequency f (or equivalently, as we plot it in Fig. 17.8, the square root of the photon

---

[3]From 1910 to 1913 Moseley was at the right place at the right time. He was a graduate student in Rutherford's laboratory at the University of Manchester as Rutherford established the nuclear model of the atom (1911) (Chap. 16). Bohr spent four months at Manchester in 1912, and then went home to Denmark and conceived and published in 1913 his revolutionary model of the hydrogen atom. In 1912 and 1913, von Laue, Friedrich, and Knipping in Germany established the wave nature of x-rays, and the Braggs, at the University of Leeds—36 miles from Manchester—devised the first x-ray spectrometer (Chap. 14). Responding to these advances, Moseley, with Rutherford's support and advice from the Braggs, began studying x-rays. He built an x-ray spectrometer and studied x-ray diffraction, and finished his thesis. Then he built a new spectrometer and, in November, measured K x-rays of some 10 chemical elements. That month he moved to Oxford, set up a new laboratory, rebuilt his apparatus, and measured K and L x-rays from some 20 more elements. This work was published in May. A year later he was shot dead by a sniper in World War I.

[4] Moseley, "The high-frequency spectra of the elements," *Phil. Mag.* **26**, 1024–1034 (1913); http://web.mit.edu/8.13/www/pdf_files/moseley-1913-high-freq-spectra-elements-part2.pdf.

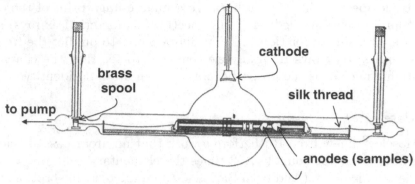

FIGURE 17.7 Moseley's apparatus for measuring K and L x-rays from a succession of samples. The shafts with the brass spools attached can be rotated without breaking the vacuum. When the spool rotates it takes up the silk thread and pulls a carriage along on its wheels. The samples are mounted on the carriage, and when properly positioned a sample becomes the anode struck by electrons from the cathode.

TABLE 17.1   X-ray data for ten elements and Moseley's law

Z	Element	λ (nm)	hf (eV)	Q	A	Q+1
20	Ca	0.3357	3694	19.01	40.09	20.0
21	Sc				44.1	
22	Ti	0.2766	4483	20.97	48.1	22.0
23	V	0.2521	4919	21.97	51.06	23.0
24	Cr	0.2295	5403	23.02	52.0	24.0
25	Mn	0.2117	5854	23.96	54.93	25.0
26	Fe	0.1945	6375	25.00	55.85	26.0
27	Co	0.1796	6904	26.02	58.97	27.0
28	Ni	0.1664	7451	27.03	58.68	28.0
29	Cu	0.1548	8010	28.02	63.57	29.0
30	Zn	0.1446	8575	28.99	65.37	30.0

energy hf) of the K radiation against the atomic number of the emitting substance, a straight line fit the data remarkably well. He saw that he could define a dimensionless constant Q,

$$Q = \sqrt{\frac{hf}{\frac{3}{4}E_0}}, \tag{16}$$

that had the property that it increased in steps of 1 from one element to the next if E_0 is taken to be 13.6 eV. This behavior of Q is apparent

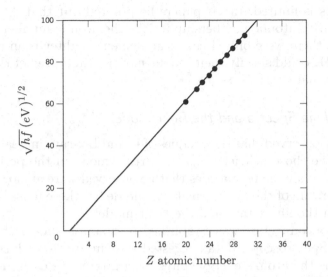

FIGURE 17.8 The data points are the square roots of Moseley's measured x-ray energies vs. atomic number Z. The straight line is the graph Eq. 16.

in Table 17.1, and he observed that $Q + 1$ corresponded to the atomic number of the anode's metal. As he wrote:

> We have here a proof that there is in the atom a fundamental quantity, which increases by regular steps as we pass from one element to the next. This quantity can only be the charge on the central positive nucleus, of the existence of which we already have definite proof [from Rutherford].

▮ EXERCISES

> **22.** Find the wavelength, energy, and Q for the element scandium (Sc), missing from Table 17.1.

Notice from Table 17.1 that if Mendeleev had followed his rule of ordering the elements by their masses, he would have put nickel before cobalt. He broke his rule because of their chemical properties, but Moseley's technique unambiguously established, independently of their chemical properties or their masses, that cobalt's atomic number is one unit less than nickel's.

By establishing that the atomic number is the number of positive charges in the nucleus, Moseley made it clear that the number of possible elements is limited. In his papers he pointed out that 75 elements were known with atomic numbers up to 79—the atomic number of gold—so as of 1913 there were only 4 chemical elements lighter than gold still to be found. He could specify their atomic numbers and predict the wavelengths of their characteristic x-rays.[5]

X-Ray Line Spectra and the Bohr Model

Moseley observed that the energies of K and L x-rays increased in a regular way as he chose heavier elements further along in the periodic table. He could see that the regularities that he observed were analogous to those in the spectrum of the hydrogen atom, and he felt there must be a connection between the line x-rays and the Bohr model.

Such a connection was established by a simple model of multi-electron atoms. Spectroscopic data suggested that in atoms with several or more electrons the atom's energy states form groups. In each group the states are close together in energy, but between the groups there is a considerable separation in energy. These well defined groups of energy states are called "shells" of energy, and there is one shell for each value of the principal quantum number $n = 1, 2, \ldots$. It also turns out that there can not be more than $2n^2$ electrons in a shell with principal quantum number n. This rule is called the "Pauli exclusion principle"[6] or, for short, the "exclusion principle."

Here is how shell structure and the exclusion principle explain the generation of characteristic x-rays. Because of the exclusion principle, electrons in higher-energy shells cannot make transitions to lower-energy shells when these shells are filled with electrons. But when electrons accelerated in an x-ray tube strike atoms of the anode, some of the collisions will knock an electron out of the lowest-energy shell. In such an atom there will then be a vacancy in the $n = 1$ shell that can be filled by the transition of an electron from a higher-energy shell to the lower-energy one. When such a transition occurs, a photon is emitted with an energy equal to the difference between the energies of the two shells as illustrated by Fig. 17.9. This photon is one of the characteristic x-rays.

The Bohr model can predict with useful accuracy the energy of a characteristic x-ray photon. As Eq. 15 shows, the Bohr model predicts this will scale as Z^2. For example, consider the energy of an electron in the $n = 1$ shell of an element with nuclear charge $Z = 40$ (zirconium). According to Eq. 15, its energy will be about $-40^2 \times 13.6 = -22\,000\,\text{eV}$.

---

[5]They have all been discovered—hafnium ($_{72}$Hf), rhenium ($_{75}$Rh), technetium ($_{43}$Tc), and promethium ($_{61}$Pm). These last two are radioactive and must be produced artificially.
[6]Quantum mechanics explains the Pauli exclusion principle, but for now you can be like physicists in the early 1920s and just accept it as an empirically established rule.

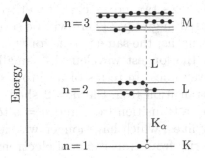

FIGURE 17.9 Schematic representation of energy shells of an atom, illustrating how x-rays can be produced by transitions from a higher-energy shell to a hole in the lower-energy shell.

■ EXERCISES

> **23.** Show that the energy of an electron in the $n = 2$ shell of a $Z = 40$ nucleus would be on the order of $-5500\,\text{eV}$.

Then if an electron in the $n = 2$ shell changes from its energy state of $-5500\,\text{eV}$ to the available $-22\,000\,\text{eV}$ energy state in the $n = 1$ shell, it will lose $-5500 - (-22\,000) = 16\,500\,\text{eV}$. This energy comes out as an x-ray photon.

Because of the presence of other electrons, it is unlikely that the electron making the transition will see the full nuclear charge Z. Then maybe the scale factor will be $(Z - s)^2$ instead of Z^2, where s corrects for the effects of the other electrons. If you put these ideas together in a single formula, you get

$$hf = (Z - s)^2 E_0 \left(\frac{1}{1^2} - \frac{1}{2^2} \right) = \frac{3}{4} E_0 (Z - 1)^2. \tag{17}$$

With $s = 1$ and $E_0 = 13.6\,\text{eV}$ this is exactly the equation that Moseley inferred from his measurements of K_α x-rays, i.e., it is the same as Eq. 16 on p. 540. For zirconium ($Z = 40$) Eq. 17 predicts

$$hf = (40 - 1)^2 13.6 \frac{3}{4} = 15\,500\,\text{eV},$$

which agrees well[7] with the energy of the K_α peak in Fig. 17.6.

---

[7]The model accurately predicts values for $\Delta E = E_n - E_{n'}$, the differences between energy levels, but it inaccurately predicts the energy E_n of any given level: For Zr the Bohr model predicts $E_1 = -22\,\text{keV}$, but the observed value is $-18\,\text{keV}$

Figure 17.6 shows two peaks close together. In general, characteristic x-rays are emitted in groups. There will be several K lines with somewhat different wavelengths; the same is true for the L lines. For the K lines the two lines with the longest wavelengths (smallest frequencies) are called K_α and K_β x-ray lines. In terms of an atom's energy shells, the K_α line arises from a transition from the $n = 2$ shell to the $n = 1$ shell, and the K_β results from a transition from and $n = 3$ to $n = 1$.

The L x-ray lines, which have longer wavelengths (lower energy) than the K x-rays, arise from transitions of electrons from higher-energy shells into a vacancy in the $n = 2$ shell. In particular, the L_α x-rays correspond to transitions from $n = 3$ to $n = 2$, and L_β x-rays from $n = 4$ to $n = 2$.

▪ EXERCISES

24. Calculate the energy of the K_β line from $Z = 40$. Compare your answer to the value of the K_β peak in Fig. 17.6.

17.6 SUMMARY

The Bohr Model

Bohr explained the spectral series of the hydrogen atom. He hypothesized that the energy of the photons emitted by hydrogen was equal to the difference in energies of specific, well-defined energy states. The allowed states were those for which the angular momentum of the electron was an integer multiple of $\hbar$. When in these states, the electron interacted with the nucleus by means of the Coulomb force and satisfied the law of conservation of mechanical energy. These three assumptions justify the following four equations:

$$hf = E_{n'} - E_n,$$
$$rp = n\,\hbar,$$
$$\frac{mv^2}{r} = \frac{k_\mathrm{c} Z e^2}{r^2},$$
$$E = \frac{p^2}{2m} - \frac{k_\mathrm{c} Z e^2}{r},$$

where the symbols are as defined earlier in the chapter. The integer quantity n is the principal quantum number.

From these assumptions Bohr's model predicts that

$$E_n = -\frac{2m\pi^2 k_c^2 Z^2 e^4}{n^2 h^2},$$

$$hf = \frac{k_c^2 Z^2 e^4 m 2\pi^2}{h^2}\left[\frac{1}{n'^2} - \frac{1}{n^2}\right]$$

$$= -13.6 Z^2 \left[\frac{1}{n'^2} - \frac{1}{n^2}\right] \text{ eV},$$

$$r_n = \frac{n^2 h^2}{4\pi^2 m k_c Z e^2} = n^2 a_0,$$

$$a_0 = 0.0528\,\text{nm}.$$

Limitations of the Bohr Model

Bohr's major innovation was the idea of well-defined, discrete energy levels. His model worked well for single-electron atoms, but he as well as others recognized that it was an unsatisfactory theory. It met with very little success in atoms with more than one electron. It was completely unable to predict how rapidly transitions between states occur, nor could it explain the relative intensities of the various spectral lines. Moreover, it was inconsistent in its approach. It used classical derivations for a very nonclassical situation. Having electrons in orbits to which classical mechanics was applied, but then forbidding the electrons to radiate, was definitely an uncomfortable way of producing a theory. Using the wave nature of electrons to help suggest why the orbiting electrons do not radiate is no more satisfactory. To talk of an electron with a specific *orbit* but then consider it to be a wave mixes two different kinds of descriptions, and the good result does not justify the bad logic.

A complete, consistent theory of the atom came with the development of quantum mechanics. This theory, which used several of Bohr's fundamental ideas, superseded the Bohr model.

X-Ray Line Spectra

The existence of x-ray line spectra, sharp peaks ("lines") in the x-ray spectrum, is explained by an extension of Bohr's model. Characteristic x-ray lines suggest the existence of shells of energy in the atom. The exclusion principle says that there can not be more than $2n^2$ electrons in shell n. This means that there cannot be transitions into a full shell. However, when holes are created in a lower-energy shell, for example by bombardment with energetic particles, transitions become possible, and they result in the emission of characteristic x-rays. Transitions to the $n = 1$ shell give rise to K x-rays; transitions to the $n = 2$ shell yield L x-rays.

Moseley's Law, the Atomic Number, and the Periodic Table

Moseley's measurement of K and L x-ray line spectra from many different elements showed that the square root of the energy of these x-ray photons increases linearly with the nuclear charge number Z, i.e., $\sqrt{hf} \propto Z$. His work established that this number is the atomic number—the parameter that orders the elements in the periodic table.

Moseley's work made x-rays a practical, reliable way to identify the chemical composition of a substance.

Moseley's research significantly accelerated the acceptance of Rutherford's idea that every atom has a compact nucleus. His work also gave more credibility to Bohr's idea that atoms have definite energy states and emit photons when they change from a state of higher energy to one of lower energy. The two ideas were so radical that most physicists viewed them with great skepticism. Moseley's results convinced many that these ideas had to be taken seriously.

PROBLEMS

1. Bohr showed that the possible energy states of a hydrogen atom are accurately given by the expression $E_n = -\frac{13.6}{n^2}$ eV, where n is any integer $1 \leq n < \infty$.

   a. Draw to scale a level diagram of hydrogen showing the first 5 or 6 energy states. Label the energies of the lowest 4 states.

   b. Show on your diagram the transitions that give rise to the three lowest energy lines of the Balmer series.

   c. What is the energy of the photon emitted when a hydrogen atom goes from its first excited state to its ground state?

   d. What is the wavelength of that photon?

2. Atoms of the never-discovered element fictitium (Fi) have energy states as shown in Fig. 17.10.

   a. What would be the energies of photons emitted after a vapor of Fi is bombarded with 3.7 eV electrons?

   b. What would be the energies of photons emitted after a vapor of Fi is bombarded with 3.7 eV photons?

   c. Assume that Fi is in its ground state. What is the longest-wavelength photon that this atom can absorb?

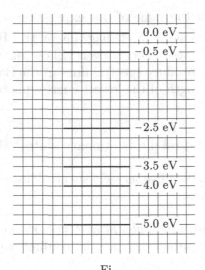

Fi

FIGURE 17.10   Energy levels of a fictitious element (Problem 2).

   d. Fi has a nucleus with a radius of 4 fm. Use the Heisenberg uncertainty principle to estimate the kinetic energy of a proton confined within this region.

3. Bohr derived the energy levels and the corresponding radius r of the hydrogen atom from the following three relations:

$$pr = n\hbar,$$

$$E_{\text{tot}} = \frac{-k_c e^2}{2r},$$

$$\frac{p^2}{2m} = \frac{k_c e^2}{2r},$$

where p is the momentum of an electron of mass m and charge e orbiting a distance r from the center of the positively charged nucleus; n is any integer 1, 2, ...; k_c is the Coulomb force constant; and $\hbar$ is Planck's constant divided by 2π.

   a. Obtain an expression for the radius of the lowest Bohr energy state of an atom in terms of fundamental constants. Evaluate your results.

   b. Use the above relations to show that the possible energy states of an H atom are

$$E_n = -\frac{m k_c^2 e^4}{2\hbar^2 n^2}.$$

c. Show that the value of this expression $= \frac{-13.6}{n^2}$ eV.

d. Draw to scale a level diagram of the H atom showing the four lowest of these energy levels. Show on your diagram the transitions that produce spectral lines in the Balmer series, and calculate the wavelength of the lowest-energy photon emitted in the Balmer series.

4. What is meant by the "stationary states" that Bohr postulated?

5. Give the names of five series of spectral lines that appear in the spectra of hydrogen atoms.

6. What are two fundamental assumptions necessary for Bohr's model of the hydrogen atom that are in conflict with the ideas of classical mechanics?

7. A 50 kV accelerating voltage is applied to an x-ray tube as shown in Fig. 17.11. When electrons from the cathode crash into the target, K_α radiation of wavelength $\lambda_{K_\alpha} = 0.0723$ nm is emitted.

a. What is the atomic number Z of the target material? Hint: Recall that the energy levels of an electron orbiting a nucleus of charge $+Ze$ are given by

$$E_n = -13.6\frac{Z^2}{n^2} \text{ eV}.$$

b. The K_α photons strike a carbon block and undergo Compton scattering, as shown in Fig. 17.11. If the photon is scattered through an angle $\phi = 180°$, the scattered photon has a wavelength $\lambda = 0.0771$ nm.

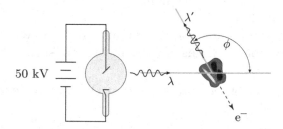

FIGURE 17.11 X-ray generation and Compton scattering geometries for Problem 7.

FIGURE 17.12   Energy levels of the hydrogen atom (Problem 10).

    c. Find the kinetic energy and the momentum imparted to the electron in the scattering event. Use appropriate units. In what direction is p?

8. Who was Henry Gwyn Jeffreys Moseley and when, where, and how did he die?

9. How did Moseley's work relate to Bohr's model of the atom?

10. Figure 17.12 is a diagram representing the possible energy states of a hydrogen atom. The lowest energy state ($n = 1$) has an energy of $-13.6\,\text{eV}$. The next lowest state ($n = 2$) has an energy equal to $-13.6/2^2 = -3.40\,\text{eV}$, etc.

    a. What are the energy and wavelength of the photon emitted when a hydrogen atom changes from its $n = 2$ to its $n = 1$ state?

    b. Identify the transitions that produce the H_α, H_β, and H_γ lines of the Balmer spectrum. These are three visible lines easily observed in laboratory. What is the color of each of the three lines?

11. A certain atom is stripped of all but one of its electrons. The lowest seven energy levels of the remaining electron are shown in Fig. 17.13.

    a. If the electron is initially in the $n = 3$ state, what is the minimum energy that must be added to the ion to remove the electron?

    b. Visible light ($400 < \lambda < 700\,\text{nm}$) is emitted when the electron makes a transition from an initial state n_i to a final state n_f. What is the smallest value of n_f that would result in the emission of visible light?

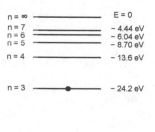

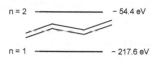

FIGURE 17.13 The energy states of an atom with all but one of its electrons removed (Problem 11).

    c. What is the atomic number Z (i.e., the nuclear charge Ze) of the ion?

    d. In the Bohr model of a one-electron atom or ion, does the electron's kinetic energy increase, decrease, or stay the same when n increases? Provide a mathematical explanation of your answer.

12. What is meant by the binding energy of a system? Give a numerical example of the binding energy of an atom. Of a nucleus.

13. Consider an x-ray tube with an anode made of a Ni–Cr alloy and to which $20\,\mathrm{kV}$ has been applied.

    a. If $Z_{\mathrm{Cr}} = 24$ and $Z_{\mathrm{Ni}} = 28$, what are the energies of the K_α lines ($n_i = 2$ to $n_f = 1$) emitted by each element?

    b. Make a sketch of the x-ray spectrum (I vs. λ) you would expect to see emitted by the Ni–Cr anode. Your wavelength scale should be accurate. Explain the main features of the spectrum.

    c. If you analyze the spectrum by sending the x-rays to a crystal with $d_{100} = 0.3\,\mathrm{nm}$, at what angle ($2\theta$) will $0.28\,\mathrm{nm}$ x-rays be diffracted by the (100) plane?

    d. Find the momentum in eV/c of electrons with a de Broglie wavelength of $0.28\,\mathrm{nm}$.

    e. Sketch an apparatus that could be used to generate electrons with the momentum of part (d). If there are hydrogen atoms in the apparatus, could they be ionized by those electrons? Explain.

TABLE 17.2 L_α X-ray lines measured by Moseley

Element	Z	λ (nm)	Element	Z	λ (nm)
Zr	40	0.6091	Sm	62	0.2208
Nb	41	0.5749	Eu	63	0.2130
Mo	42	0.5423	Gd	64	0.2057
Ru	44	0.4861	Dy	66	0.1914
Rh	45	0.4622	Er	68	0.1790
Ag	47	0.4170	Ta	73	0.1525
Sn	50	0.3619	W	74	0.1486
Sb	51	0.3458	Os	76	0.1397
La	57	0.2676	Ir	77	0.1354
Ce	58	0.2567	Pt	78	0.1316
Pr	59	(0.2471)	Au	79	0.1287

14. Moseley measured the L_α x-ray lines of 24 elements from zirconium to gold. Some of his data are given in Table 17.2.

a. Show that plotted as Moseley would have plotted them, these data lie on a straight line. From your graph determine which elements would produce L_α x-rays with wavelengths of 0.2382 nm and 0.4385 nm.

b. Determine the slope and the screening correction that satisfy

$$hf = A(Z - s)^2.$$

c. Compare your value of A with what you would expect from the Bohr-model explanation of L x-rays.

d. Discuss the significance of the value of s that you obtain.

18.

The Heisenberg Uncertainty Principle

18.1 INTRODUCTION

The photoelectric effect showed that waves behave like particles. A wave with a frequency f has a minimum packet, or quantum, of energy $E = hf$, where h is Planck's constant. Compton showed that when hf is comparable to the rest mass energy mc^2 of an electron, the scattering of electromagnetic radiation from electrons behaves like the scattering of one compact object from another. The particle-like behavior of light seems so prominent in these cases that the quantum of light has been given the particle-like name of "photon." Individual photons can be detected with a photomultiplier tube; such detection also suggests a degree of localization in space that is characteristic of particles rather than waves.

The fact remains, however, that photons—even a single photon!—can behave like a wave. Photons give rise to double-slit interference: When a beam of photons passes through two slits, the characteristic pattern of fringes appears on the screen behind the slits. Depending upon the experimental setup, light can exhibit wave properties or particle properties. Niels Bohr called this combination of properties "wave-particle duality." But isn't such a dual nature self contradictory? A particle is an object with a definite location; a wave can extend over large distances. How can an object be both localized and spread out? How can an entity behave like a particle in some circumstances and like a wave in others?

Moreover, wave-particle duality is not just a quirk of photons. It is a general feature of all quantum systems. In Chaps. 13, 14, and 15, you saw that things you think are particles, such as electrons, can sometimes behave as waves, and things that you think are waves, such as light, can sometimes behave as particles. In Chap. 15 you saw that electrons, and

C.H. Holbrow et al., *Modern Introductory Physics, Second Edition*,
DOI 10.1007/978-0-387-79080-0_18, © Springer Science+Business Media, LLC 1999, 2010

all other objects that you customarily think of as particles, can exhibit wavelike interference and diffraction. Wave–particle duality is a feature of all atomic and subatomic matter. It is, therefore, a profound aspect of nature, and theories of matter and energy must include it.

In the 1920s, physicists created a theory of the behavior of microscopic matter that incorporated wave–particle duality. This theory is called "quantum mechanics." Quantum mechanics revolutionized physics and changed physicists' ideas about causality and measurement and what it means to know something about the physical world. Even now, more than eighty years later, physicists argue heatedly about the meaning of quantum mechanics and the inferences one may draw from quantum theories.

This chapter introduces you to the Heisenberg uncertainty principle. It states a fundamental feature of quantum theory that, among other things, specifies the sizes and energies of atomic and subatomic systems and resolves the apparent contradiction of wave-particle duality.

18.2   BEING IN TWO PLACES AT ONCE

Why can't something be both a particle and a wave? Perhaps you don't think there is much of a problem here. If so, consider the following hypothetical experiment that could certainly be done in principle.

A beam of electrons is allowed to strike two narrow slits close enough together to produce a nice interference pattern on a photographic plate. Imagine that the intensity of the electrons is turned down so that only 1 electron arrives at the photographic plate each hour. It would be tedious and expensive, but imagine that you exposed a plate and developed it after 1 h; and then exposed a second plate for 2 h and then developed it; and then exposed a third plate for 3 h and developed it; and so on. What would you see on the succession of plates?

The answer is what you would expect. On the first plate there would be one exposed grain of silver halide somewhere on the plates from the arrival of 1 electron during the 1 h of exposure. On the second plate there would be two exposed grains of silver halide from the arrival of 2 electrons during the 2 h of exposure; on the third there would be 3 exposed grains; on the fourth 4; and so on.

What would be the pattern of these exposed grains? At first there would be no apparent pattern. The number of exposed grains would be too few to show a pattern. But as the number of exposed grains reached into the hundreds and thousands, a distinct pattern would appear. It would be the pattern of interference fringes you have come to know and love. Although the electrons arrive one at a time an hour apart, they still form an interference pattern. Electrons preferentially arrive at the places corresponding to maxima in the fringe pattern; they do not arrive at places corresponding to minima in the pattern.

The point is that interference occurs for the individual electrons; it occurs electron by electron. The same story is true for photons. Illuminate a pair of closely spaced slits with a laser beam that has been made so weak that only 1 photon goes through the slits each hour. We could detect these with the cumbersome succession of photographic plates used for the electrons, but it speeds things up if instead we imagine an array of photomultiplier tubes in place of the screen. To each of these we attach a counter so we can keep track of how many photons arrive at each photomultiplier tube. About once an hour one of these counters will go "click" and record the arrival of a photon. After many hours a pattern will emerge. The counters near the maxima of the interference pattern will have many counts; the counters near the minima will have few or none. Single photons preferentially arrive at places corresponding to maxima in the fringe pattern; they will not arrive at places corresponding to the minima. This process is illustrated schematically in Fig. 18.1.

Interference occurs for individual photons; it occurs photon by photon.

Does this bother you? If it doesn't, consider the following. Suppose you block one of the slits and do the above two experiments again. What will you see? This time you will see a single-slit diffraction pattern build up slowly one photographic grain or one click at a time. If the slits are narrow enough so that the first single-slit diffraction minimum is off the edge of the screen, then you will see only a smooth, nearly uniform distribution of exposed photographic grains or counter clicks. Electrons (or photons) *will now go to places where they did not go when there were two slits.*

Now think about this. When there are two slits, how does a single photon or electron "know" that it may not go to the places where the minima occur? When there is only one slit, how does a single electron or photon "know" that it can go to the previously forbidden places? If an electron passes through the bottom slit, how does it "know" whether or not the top slit is open?

For one-particle-at-a-time interference to occur, each particle must in some sense pass through both slits. Only then can you have a theory that is both internally consistent and in agreement with experimental observation. But this is a claim that the particle can be in two different places at once! The claim flies in the face of all our experience and the physics of Newton, Maxwell, and Einstein. How can such a curious idea be true?

The idea should make you deeply uneasy because an electron—or any particle—usually has a well defined position. For example, it is detected

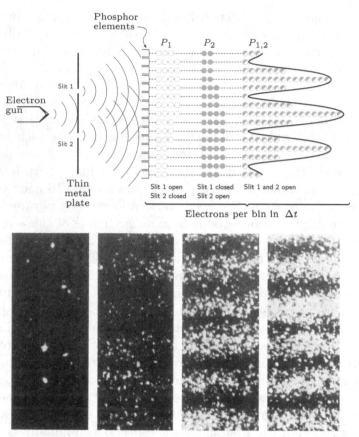

FIGURE 18.1 Hypothetical one-particle-at-a-time interference experiment. The open circles labeled P_1 are counts recorded when only slit 1 is open; the shaded circles labeled P_2 are counts recorded when only slit 2 is open; the half-shaded circles $P_{1,2}$ are the counts recorded when both slits 1 and 2 are open. The set of images in the bottom half of the figure show the gradual emergence of the double-slit interference fringes as particles are recorded over an ever longer time interval. *Taken with permission from J.G. Hey and Patrick Walters, The Quantum Universe, Cambridge University Press, 1987.*

in a well-defined, localized region of space—the photomultiplier tube or the photograph grain. That spatial localization makes apparent that the electron or the photon is a particle.

How can something so localized go through two separate slits at the same time? Quantum theory connects localization to measurement. The theory says that a particle can be prepared experimentally to be in a range of positions and that the act of measurement selects one of these. This means that an electron is in a wide range of positions as it approaches the photographic plate; then interaction of the electron with a grain of silver halide in the emulsion selects one of these positions. When there are two slits, the range of possible positions of the electron or photon does not include those positions that correspond to the minima of the interference pattern.

The urge to deny this picture is strong. "Surely," you will say, "I can measure through which slit the particle passes." You are then demanding that it pass definitely through one or the other slit as you would expect a classical particle to do. You can make such a measurement in any of several ways. For example, you might shine light on electrons as they come through one of the slits and use the resulting Compton scattering to signal the electron's passage. In other words, an electron passing through the illuminated slit will scatter light that can be detected, and the detection of a flash of light tells you that this electron passed through the illuminated slit. This possibility is consistent with our general principle: The particle is to be thought of as being in different places at the same time, but when you measure it you select out one of them. If you measure the next electron, you will find that it went through one slit or the other. The result of each measurement will be perfectly definite, but it will be unpredictable. You might find that 50% of your measurements find the electron passing through one slit and 50% through the other.

But if you can tell through which slit the particle came, then how can it interfere with itself? How can it contribute to producing the double-slit interference pattern? Well, it cannot. And here something quite interesting occurs. *If you do an experiment, any experiment, that determines through which slit the electrons pass, the double-slit interference pattern does not occur.*

This should not be so surprising, because in doing the experiment you have an effect on the electron. If no experiment is done, the electrons that come through the slits have some distribution of momenta in the y direction. But obviously, they do not have all values of p_y. For example, they do not have those values of p_y that would cause them to arrive at the minima of the interference pattern. However, in order to find through which slit the particle came, it is necessary to interact with the particle, for example, by shining light on it. The resulting Compton scattering changes the distribution of components of momentum in the y direction; it introduces components that permit the electron to arrive at places on the screen that were forbidden to it before its position was measured. These new components wash out the double-slit pattern, leaving only the broader single-slit pattern.

Here can be seen the importance of the quantum. According to classical physics we can reduce the energy of an electromagnetic wave of frequency f to an arbitrarily small amount. If you really could do this, then you could determine through which slit the electron came with light that had its energy reduced to the point that its interaction with the electron was negligibly small. But you cannot; light of a given frequency comes with an energy of at least hf, no less, and so it has a momentum of hf/c. This is enough to add the missing components of momentum to the electron and so wipe out the interference pattern.

Of course, you could reduce the frequency f. This works. At a sufficiently low frequency the interference pattern is not affected by the interaction of the photons with the electrons. However, nothing is gained. As you reduce f, you are increasing the wavelength λ. As λ increases, the precision with which you localize the particle gets worse, because waves of wavelength λ will show you where something is to only roughly $\pm\lambda/2$. In fact, just when the photon energy is low enough so that it does not destroy the interference pattern, the wavelength has become so large that you can no longer tell through which slit the electron passed. The experiment no longer selects out a well-defined location.

18.3   HEISENBERG'S UNCERTAINTY PRINCIPLE

There seems to be a conspiracy here. Experiments that exhibit the particle-like behavior of something destroy its wavelike behavior. Experiments that show the wavelike behavior of something destroy its particle-like behavior. As a result, contradictory wave and particle behavior can never occur together. Nature is constructed in such a way that no experiment can ever exhibit the contradictory properties at the same time. Therefore, the apparent contradiction can never arise.

The fact that Nature *always* behaves so that particle and wave properties cannot be exhibited simultaneously is equivalent to the following fundamental principle. Any physical situation that forces an electron into a narrow range of positions Δx will at the same time impart to the electron a wide range of momenta in the x direction Δp_x. The converse is true. Any physical situation that forces the electron into a narrow range of momenta Δp_x gives it a broad range of positions Δx. Werner Heisenberg showed that in general, the product of Δx and Δp_x is never smaller than $h/(4\pi)$. It can be larger, but for most physical systems of interest the product is of the order of h or $h/(2\pi)$. In its exact form the Heisenberg uncertainty principle is written[1]

$$\Delta x \, \Delta p_x \geq \frac{h}{4\pi} = \frac{\hbar}{2}. \qquad \text{Heisenberg uncertainty principle} \qquad (1)$$

---

[1] The quantity $h/(2\pi)$ is used so often that it gets its own symbol, an h with a line through it, which is written $\hbar$ and is called "h-bar." Just as it is useful to know that $hc = 1240\,\text{eV\,nm}$, it is useful to know that $\hbar c = 197\,\text{eV\,nm}$. For quick calculation and easy recollection many physicists take $\hbar c$ to equal $200\,\text{eV\,nm}$.

The uncertainty principle is at the heart of quantum mechanics. It applies to all particles and things built from them. It applies to many pairs of physical quantities other than position and momentum. For example, there is an uncertainty relationship between energy and time: $\Delta E \Delta t \geq \frac{\hbar}{2}$.

You may feel there is a certain vagueness about the principle. What do Δx and Δp_x mean? Is their product greater than $h/(2\pi)$ or is it equal? When is it one or when is it the other? Should the right-hand side of Eq. 1 be $h/(4\pi)$ or $h/(2\pi)$ or even, as it is sometimes written, just h? Vagueness may be appropriate for something called the "uncertainty principle," but it is not as bad as it sounds. Precise definitions of Δx and Δp_x are given below, but you will not need them. More important for you are order-of-magnitude arguments with the uncertainty principle, where approximate values of Δx and Δp_x can be inferred from basic geometry and physical considerations. Also, for such order-of-magnitude calculations, the particular choice of $h/(4\pi)$, $h/(2\pi)$, or h is not very important, and for many purposes it is convenient to use the simplified forms

$$\Delta x \, \Delta p_x \geq \hbar \quad \text{or} \quad \Delta E \Delta t \geq \hbar. \tag{2}$$

rather than Eq. 1.

▪ EXERCISES

1. Show that $\hbar c$ equals $197 \, \text{eV nm}$.

2. Show that $\Delta E \, \Delta t \geq \hbar$ is dimensionally correct.

Δx and Δp_x are defined to be standard deviations of distributions. The idea of a distribution was developed in the appendix of Chap. 5 which shows how gas pressure and temperature connect to molecular velocity. The argument was that the molecules of a gas have various different velocities that span some range. Then you could imagine dividing this range of velocities into small intervals of velocity of width dv. Such intervals are often called "bins." If you label the velocity of the ith bin to be v_i, and then count the number of molecules with velocities in the small range v_i to $v_i + dv$, you will get n_i, the number of molecules in the ith such bin. The set of values n_i represents the distribution of velocities. From this distribution you can compute an average value of the square of the velocity $\langle v^2 \rangle$ using the relationship

$$\langle v^2 \rangle = \frac{\sum n_i v_i^2}{\sum n_i}.$$

Given a distribution of the positions x, you could find $\langle x \rangle$ and $\langle x^2 \rangle$; if you had a distribution of the momenta p_x, you could find $\langle p_x \rangle$ and $\langle p_x^2 \rangle$. The formal definitions of Δx and Δp_x are

$$\Delta x = |\langle x \rangle^2 - \langle x^2 \rangle|^{1/2},$$
$$\Delta p_x = |\langle p_x \rangle^2 - \langle p_x^2 \rangle|^{1/2}.$$

You may recognize that each of these is a quantity called "the standard deviation" of a distribution. It is a measure of the spread of values within the distribution.

The mathematics of distributions and quantities like Δx are the same for quantum mechanics as for calculating Δv for gas molecules. The meaning of the spread of values, however, is quite different for quantum mechanics than for the kinetic theory of gases. Kinetic theory assigns a definite value of velocity to each atom, and the spread of the distribution arises because there are many molecules with different velocities. Quantum mechanics ascribes all the velocities of the distribution to each particle and says that a particle will have a definite velocity only when you measure it.

EXAMPLES

1. To see how the definition of Δx works, consider a uniform beam of electrons directed at a slit of width a. What is Δx for electrons passing through such a slit? Suppose you divide the slit into 10 intervals of width $0.1a$. To find Δx, you first need to find $\langle x \rangle$ and $\langle x^2 \rangle$.

To calculate these quantities you need to place a coordinate system on the aperture. It is convenient to place the origin at the center of the slit. Calculation of $\langle x \rangle$ is then particularly easy. The symmetry of the problem shows that $\langle x \rangle$ will be 0, because there will be as much negative x below this origin as there is positive x above it. A similar symmetry argument is often used to calculate $\langle p_x \rangle$.

If you assume that passage through the slit is as likely at any place within the aperture as any other, calculation of $\langle x^2 \rangle$ is straightforward. Each bin is $0.1a$ wide; the values of x run from $-0.5a$ to $+0.5a$, so the values of x^2 run from $0.25a^2$ down to 0 and then back up to $0.25a^2$. Therefore, the sum runs over 5 intervals from $0.025a^2$ down to 0 and back up another 5 intervals to $0.025a^2$. The process is shown in Fig 18.2. You can write the sum as $0.1(0.1a)^2 2[1 + 4 + 9 + 16 + 25] = 0.11a^2$, so $\Delta x = 0.33a$.

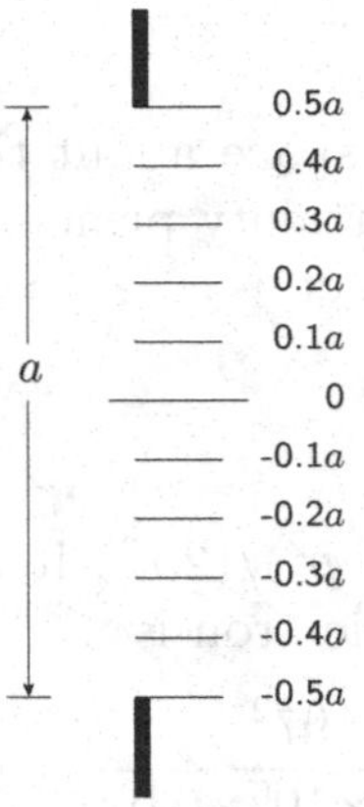

$$\langle x \rangle = \frac{1 \times .1 + 1 \times .2 + \cdots - 1 \times .1 - 1 \times .2 - \cdots - 1 \times .5}{10} a = 0$$

$$\langle x \rangle = 0$$

$$\langle x^2 \rangle = \frac{1 \times .01 + 1 \times .04 + \cdots + 1 \times .01 + 1 \times .04 + \cdots + 1 \times .25}{10} a^2$$

$$\langle x^2 \rangle = .110 a^2$$

$$\Delta x = \sqrt{|\langle x \rangle^2 - \langle x^2 \rangle|} = .3a$$

FIGURE 18.2 Calculation of $\langle x^2 \rangle$ for a slit of width a (viewed from above).

EXERCISES

3. The above example used the largest value of x^2 occurring in each bin. What would be $\langle x^2 \rangle$ and Δx if you chose the value of x^2 at the midpoint of each interval?

For many purposes, however, you can skip all such calculation and observe from the physical setup that $\Delta x \approx a$.

18.4   ATOM SIZES AND ENERGIES FROM THE UNCERTAINTY PRINCIPLE

The uncertainty principle relates average kinetic energy $\langle K \rangle$ to spatial confinement, and you can use this fact to predict important features of systems built up from particles. First, you use the uncertainty principle to find values for $\langle p_x \rangle$ and $\langle p_x^2 \rangle$. If the system has symmetry, $\langle p_x \rangle$ will be 0. If the particle is bound, say if it is an electron going around a proton, then $\langle p_x \rangle$ will also be 0, because on average a bound particle has no net motion in any direction. For systems with $\langle p_x \rangle = 0$, you have the very useful result that $(\Delta p_x)^2 = \langle p_x^2 \rangle = 2m \langle K \rangle$.

EXAMPLES

2. Consider an electron confined to a region of space about the size of an atom, so that $\Delta x \approx 0.1\,\mathrm{nm}$. Then the uncertainty principle tells you that $\Delta p_x \approx \hbar/\Delta x$. Squaring both sides, you get

$$(\Delta p_x)^2 = \langle p_x^2 \rangle \approx \frac{\hbar^2}{(\Delta x)^2}.$$

Because $\langle p_x^2 \rangle = \langle p_y^2 \rangle = \langle p_z^2 \rangle$ and $K \approx (p_x^2 + p_y^2 + p_z^2)/(2m)$, it follows that the average kinetic energy K of the bound electron is

$$K = 3\frac{\langle p_x^2 \rangle}{2m} = \frac{3\hbar^2}{2m(\Delta x)^2} = \frac{3(\hbar c)^2}{2mc^2(\Delta x)^2} = \frac{3\times 197^2}{2\times 0.511\times 10^6\times 0.01} = 11.4\,\mathrm{cV}.$$

Although such calculations are approximate, the result is very informative. It says that simply by virtue of its confinement within the space of an atom, an electron will necessarily have an average kinetic energy of the order of $10\,\mathrm{eV}$. As a general rule, the more closely confined a particle is, the greater is its average kinetic energy.

This result also tells you something about the force holding the electron in confinement. For the electron to be bound to the atom, its total energy, kinetic plus potential, must be negative. This means that in attracting the electron to the atom, the force must reduce the total energy of the electron by more than enough to offset the kinetic energy.

Does it? The hydrogen atom is formed by the electrical attraction between the positively charged proton and the negatively charged electron. These behave like point charges, so you can calculate their energy of interaction from the formula for the electrical potential energy between two point charges. For an atom $\approx 0.1\,\mathrm{nm}$ in diameter, the two charges are separated by a distance of $0.05\,\mathrm{nm}$, and their potential energy is

$$U = \frac{-k_c e^2}{r} = \frac{-1.44}{0.05} = -28.8\ \mathrm{eV}.$$

The total energy would then be $11.4 - 28.8 = -17.4\,\mathrm{eV}$. The total energy is negative, and the fact that the electron and the proton form a bound system about $0.1\,\mathrm{nm}$ in diameter is consistent with the Heisenberg uncertainty principle.

The example above assumed knowledge of the size of an atom and deduced the average kinetic energy of an electron held inside the atom. If instead of knowing the size, you know the forces between the interacting parts of the atom, you can use the uncertainty principle to estimate the atom's size. Like any physical system, the atom will configure itself to achieve the lowest possible total energy. You might think that this would occur when the distance between the two charges goes to zero, because then the electron's potential energy goes to negative infinity. However, the uncertainty principle warns you that confinement of an electron to a smaller and smaller volume of space will lead to an unbounded increase in the electron's kinetic energy. The atom takes on a size that minimizes the sum of these two effects. You can find that size by expressing the total energy in terms of size and then finding the radius that gives a minimum total energy.

▼ EXAMPLES

3. The total energy E of an electron of mass m and charge $-e$ interacting with a proton of charge e separated by an unknown distance r is

$$E = \frac{p^2}{2m} - \frac{k_c e^2}{r}.$$

As you have already seen, the uncertainty principle connects the kinetic energy of the electron to the size of the atom. According to the uncertainty principle, confinement to a region $\Delta x \approx r$, the radius of the atom, means that the electron must have $\Delta p = \sqrt{\langle p^2 \rangle} \geq \hbar/r$. Consequently, its kinetic energy will be

$$\frac{p^2}{2m} \approx \frac{\hbar^2}{2mr^2}.$$

Therefore, the total energy of the atom can be written as

$$E \approx \frac{\hbar^2}{2mr^2} - \frac{k_c e^2}{r}.$$

To find the value of r that minimizes E, differentiate E with respect to r; set the result equal to zero; solve for r.

$$\frac{dE}{dr} = -\frac{2\hbar^2}{2mr^3} + \frac{k_c e^2}{r^2} = 0.$$

solving this for r you get,

$$r = \frac{\hbar^2}{mk_c e^2} = \frac{(\hbar c)^2}{mc^2 k_c e^2} = \frac{197^2}{0.511 \times 10^6 \times 1.44} = 0.0527 \,\text{nm}.$$

This result is very satisfactory, because it is the size of the radius a real atom. This value is also precisely the radius of the hydrogen atom obtained from the Bohr model. Obtaining the right order of magnitude for the radius of atoms using the uncertainty principle is convincing evidence that the principle is fundamental.[2]

EXERCISES

4. Use the expression just obtained for the radius of the hydrogen atom and derive an expression for the total energy of the hydrogen atom in terms only of fundamental constants such as e, $\hbar$, c, and the mass of the electron m.

5. (a) Evaluate the expression you obtained in the previous problem to give a numerical value for the total energy. (b) Your answer should be negative. Why? What does a total negative energy mean?

The arguments relating the energy of an electron in an atom to the size of the atom also apply to the atomic nucleus.

EXAMPLES

4. For example, you can use the uncertainty principle to show that the kinetic energy of a neutron in a nucleus of diameter 10 fm is of the order of 25 MeV. Think of the nucleus as a box 10 fm on a side. Confinement in the x-direction means

$$\Delta p_x\, c \approx \frac{hc}{\Delta x} = \frac{1240[\text{ MeV fm}]}{10[\text{ fm}]} = 124 \text{ MeV}$$

and because $\Delta p_x\, c = \sqrt{\langle p_x^2 c^2 \rangle} a$ it follows that the average kinetic energy associated with the x-motion of a neutron confined to 10 fm is

$$\frac{\langle p_x^2 c^2 \rangle}{2m_0 c^2} = \frac{124^2}{2 \times 939} \approx 10 \text{ MeV}.$$

Because the neutron moves in the y and z directions as well as x, this result implies that the kinetic energy of the neutron in the nucleus will be of the order of $K = 3 \times 10 \approx 30$ MeV.

Notice how this result highlights a fundamental question about the nucleus. If the neutron has a kinetic energy of ~ 30 MeV, what holds

---

[2]The precise agreement with the Bohr model occurs because we knew to choose the form of the uncertainty principle that would produce this agreement.

it in the nucleus? The answer had to be a new force, a force that had not been observed before, a force between nucleons that is immensely strong at short range but weak when the nucleons get farther apart than ~ 10 fm.

EXERCISES

6. Estimate the minimum potential energy required for a neutron to remain bound inside an atomic nucleus.

7. Why didn't Rutherford use the uncertainty principle to estimate the energy of a proton in a nucleus?

18.5   GENERAL FEATURES OF THE UNCERTAINTY PRINCIPLE

The uncertainty principle is a basic law of nature. It embodies the fact that a particle can never be made to exhibit at the same time wavelike properties of interference and particle-like properties such as billiard-ball scattering. No contradiction between wave and particle behavior can occur.

The fundamental nature of the uncertainty principle is also apparent in the fact that you can use it to estimate the size, internal kinetic energy, and binding energy of atomic and nuclear systems. From the uncertainty principle it follows that the smaller, more compact a system of particles is, the higher their kinetic energy is and the more strongly they must be bound to offset the high kinetic energy. The smaller a system of particles is, the greater must be the forces holding the particles together.

In general a quantum system exists in a superposition of many different states, e.g. different locations, but when a measurement is made on the system, the system takes on some particular, definite value from among the many possible. Which one of the possible values will be obtained in any given measurement is entirely unpredictable. Quantum theory can only predict the probability of each outcome, *i.e.,* how many times a given value will be obtained when a large number of measurements are made on identical systems.

Notice that the uncertainty principle also implies that the act of measurement of one quantity will cause the particle to take on a range of new values of some complementary quantity. For example, consider the Heisenberg relations for position and momentum, $\Delta x \, \Delta p_x \approx \hbar$. Any useful measurement of position puts the system into a narrow range of locations Δx, but, according to the uncertainty principle, this will impart to the system a broad range of momenta Δp_x. The result will always be that the more precisely you measure position, the less precisely you will know momentum, and conversely. The uncertainty principle places fundamental, general limits on what can be measured and what can be learned about nature from measurement.

PROBLEMS

1. 650 nm light passes through two very narrow slits separated by 10 μm and then strikes a screen 1 m distant.
   a. Describe what appears on the screen and where.
   b. Suppose the screen were replaced with an array of photon counters, and the intensity of the incident light were reduced to 1 photon per minute. What would be detected on the arrays of counters?
   c. If one of the two slits were blocked, what would appear upon the screen? Assume that the width of the single slit is 1 μm.
   d. What is the puzzle implicit in your answers to (b) and (c), and how does the Heisenberg uncertainty principle resolve this puzzle?

2. How does the Heisenberg uncertainty principle resolve the apparent paradox that light and electrons each exhibit wave properties in some circumstances and particle properties in other circumstances?

3. State clearly the Heisenberg uncertainty principle. Explain what the symbols mean. Explain what the uncertainty principle is saying about the behavior of physical systems.

4. Use the uncertainty principle to estimate the kinetic energy of an electron confined to the region of an atom. (You should know a reasonable dimension for an atom.)

5. Use the uncertainty principle to estimate the kinetic energy of a neutron confined to a nucleus.

6. If the average kinetic energy of an electron in the outer regions of an atom is 2 eV, what will be its average kinetic energy if it is confined to a nucleus?

7.

  a. Explain how the uncertainty principle sets the scale of energies of small, bound systems.

  b. Suppose you had reason to believe that there is a particle called the "quark" that is confined to a region of about 1 fm. Estimate the momentum of the quark. What do you need to know before you can estimate its kinetic energy?

8. Estimate the minimum possible kinetic energy of an alpha particle confined to a nucleus of diameter 7 fm.

9. Can an electron be confined to the volume of a nucleus by the Coulomb attraction between the electron and the nuclear charge? To answer this, find the approximate average kinetic and potential energies of the electron.

  a. Take the nuclear charge to be $Z = 100$ and estimate the average distance between electron and nuclear charge to be ≈ 1 fm. (This approximation concentrates all the nuclear charge into a tiny sphere of radius less than 1 fm; this is not a good approximation but it won't change the answer.) Calculate the average potential energy of the electron.

  b. Estimating the uncertainty Δx to be ≈ 1 fm, evaluate the average kinetic energy.

  c. Use your answers to (a) and (b) to answer the question.

C H A P T E R

19.

Atoms, Photons, and Quantum Mechanics

19.1   INTRODUCTION

Quantum mechanics was the outcome of physicists' twenty-five year struggle to understand the behavior of matter and light at the atomic level. This struggle began in 1900 when Max Planck explained the spectrum of light from a hot body by an ad hoc assumption that atoms absorb and emit light in bundles of energy. In 1905 Einstein argued convincingly that light is itself quantized in bundles of energy and used the idea to explain the photoelectric effect (Chap. 13). Rutherford and Moseley showed (Chaps. 16 and 17) that the atom is made of discrete elements, and Bohr showed (Chap. 17) that atoms take on definite, or as we say today, quantized states of energy.

The need to resolve the paradox of wave-particle duality became acute when G. P. Thomson and Davisson and Germer confirmed de Broglie's idea that particles can behave like waves (Chap. 15). The resolution occurred in 1925 when Heisenberg, Born, Kramers, Schrödinger, and Dirac formulated a coherent theory with well defined rules for calculating properties and behavior of light and atoms—the theory of quantum mechanics.

You have already met some quantum mechanics in the Heisenberg uncertainty principle (Chap. 18). It embodies a basic feature of quantum mechanics: A physical entity can be in a superposition of distinct measurable states subject to a fundamental limitation. The more states of one kind the entity possesses, the fewer states it has of another kind, e. g., it can have many positions but few momenta, or vice versa. The principle is general; it applies not just to momentum and position, but also to other pairs of physical quantities.

C.H. Holbrow et al., *Modern Introductory Physics, Second Edition*,
DOI 10.1007/978-0-387-79080-0_19, © Springer Science+Business Media, LLC 1999, 2010

It may be fundamental and general, but the uncertainty principle does not predict details. It does not tell you what energy states an atom will have or what their lifetimes will be. To make quantitative predictions of outcomes of experiments you need detailed rules. This chapter presents some of these and shows you how quantum mechanics combines an expanded conception of superposition with some basic ideas of probability to accurately describe the interference of individual photons in a Mach-Zehnder interferometer.

19.2   BASIC IDEAS OF QUANTUM THEORY

Superposition and the Uncertainty Principle

Superposition is a fundamental principle of quantum physics. The superposition of waves was introduced in Chap. 10, but for quantum mechanics the idea must be enlarged, reinterpreted, and made to satisfy restrictions imposed by "indistinguishability," and by the uncertainty principle. Indistinguishability and the uncertainty principle each specify limits on superposition that result in a consistent description of the atomic and subatomic world.

For example, to understand what it means for an entity to be "spread out" requires an enlarged meaning of superposition. You must think of the entity as simultaneously occupying many different positions, i.e., being in a superposition of different locations, being in more than one place at once. The quantity Δx is a measure of how many locations are occupied by the entity, and Δp_x is a measure of the range of the entity's momenta.

Nature eliminates the self-contradiction of wave-particle duality by placing an important restriction on superposition. This restriction is described by the Heisenberg uncertainty principle which says that an entity can not have a narrow range of positions and a narrow range of momenta at the same time. The equation $\Delta x \, \Delta p_x \geq \hbar$ tells you that objects can either behave as particles (Δx is small) or as waves (Δp_x is small and therefore $\Delta \lambda$ is small), but not as both simultaneously.

You have seen how double-slit interference exhibits this limitation. If you set up your experimental apparatus to localize a particle enough to know which slit it passes through, it loses its wave nature and there is no interference pattern. If you arrange your apparatus to make Δx large enough so that you can't tell through which slit the particle passes, then Δp_x becomes small, and the particle has a wavelength well enough defined to result in an interference pattern. Although wave and particle properties are contradictory, Nature avoids the contradiction by behaving only in ways that do not manifest both properties at the same time.

The uncertainty principle and superposition are fundamental to understanding atoms and their behavior. Both ideas apply far beyond the particular case of a single entity occupying a range of positions or possessing a range of momenta. Many other pairs of physical properties obey an uncertainty principle, e. g. energy and time, and angular momentum and angular position. All aspects of quantum behavior exhibit superposition, e. g. a hydrogen atom can be in a superposition of energy states.

Random chance is a basic property of quantum theory. Every quantum event is fundamentally random. It is pure chance whether a photon reflects from or passes through a beam splitter. It is pure chance whether a radioactive atom decays or does not decay. Quantum theory uses "probability amplitudes" and their associated "probability" to deal with such randomness.

The following sections explain these ideas and illustrate them by using them to analyze experiments that try to answer the following questions about how photons behave.

- When a single photon passes through an interferometer, how does it "know" not to go to the places on a screen where interference minima occur? That is, does interference really occur "one photon at a time"?

- Suppose you add to your interferometer a special gadget. When the gadget is turned on, you can know which path the photon took through the interferometer, but when it is turned off, you can not know. What happens to interference when you turn the gadget on? When you turn it off?

19.3 DOWN CONVERSION, BEAM SPLITTING, COINCIDENCE COUNTING

Photon experiments often use down converters, beam splitters, and coincidence counters, so you need to understand what these devices do. Begin by considering a simplified, idealized experiment that tests what happens to a photon incident on a reflecting surface. The next section (p. 575) will show you how actual experiments use these devices to exhibit the surprising behavior of single photons.

When light strikes a glass surface, some light reflects from the glass and some passes through. Ordinary window glass reflects about 4% of the incident light and transmits the rest. By coating the glass with a thin layer of reflecting metal, such as aluminum, you can increase how much light is reflected. Any glass surface used to divide an incident beam of light is called a "beam splitter." If it has been coated to reflect 50% and to transmit 50% of the light, it is said to be a "50–50 beam splitter."

The splitting of a beam is easily explained when light is viewed as a wave. Upon striking the beam splitter, the wave divides into two waves of smaller amplitude but unchanged frequency; the reflected part comes back from the surface, and the transmitted part passes through it. But if light consists of photons, i.e. quanta or chunks, what happens to a photon when it strikes the glass? Does the photon separate into two pieces? Separation seems unlikely because the energy of the quantum would then be divided and the outgoing pieces would have frequencies (and, of course, wavelengths) much different from those of the incident photon.

There have been two obstacles to experiments to answer this question. First, there had to be advances in technology before it became possible to be sure that only one photon was in the apparatus. Second, detectors detect only a fraction of the photons incident on them. This fraction is called the "efficiency" of the detector. For example, the maximum detection efficiency of photomultiplier tubes is about 30% and only for a narrow range of wavelengths. Again, technological advances have helped, but it is still the case that no experiment can tell you what happens to every photon.

The first obstacle is overcome by an ingenious device called a "down converter." A down converter converts one short wavelength photon into two longer wavelength photons. This is a relatively rare occurrence; for example, when a 350 nm photon enters a carefully prepared beta barium borate crystal, roughly one time out of 10^{11} the incident photon is absorbed and two 700 nm photons emerge in slightly different directions.

A down converter makes it possible to be sure you are recording the behavior of a single photon and not more. To do this you detect coincidences between the pair of photons that come out of the down converter. If you detect one of the pair, you know that the other one exists, and if you detect the second one within some short time interval of your detection of the first, you can be quite sure that it is a single photon that has passed through your apparatus.

Figure 19.1 shows a diagram of a setup to detect a photon coming out of a beam splitter. One of the photons produced by down-conversion is directed onto the beam splitter BS, which has outputs viewed by detectors D_1 and D_2; the other photon is sent to detector D_0. You arrange the experiment so that detectors D_1 or D_2 record photons only if you also record a photon at D_0. This method of detection is called "coincidence counting"; you record a photon in D_1 or D_2 only when you detect it within a narrow interval of time—say a few nanoseconds—of the time that you

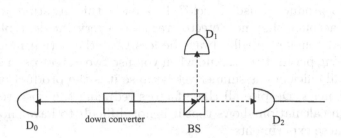

FIGURE 19.1 A schematic diagram of an apparatus to measure the outputs from a beam splitter BS after a photon passes through it. One photon from a pair produced in a down converter is detected in detector D_0; the other photon is detected in D_1 or D_2. The recorded data are coincidences of D_1 or D_2 with D_0.

detect a photon in D_0. Having detected a photon at D_0, you know there is another photon in your apparatus, and when you detect an output at D_1 or D_2 in coincidence with the detection at D_0 you can be quite sure the photon at D_1 or D_2 is the other down-conversion photon.

With 100% efficient detectors (unrealistic idealization) you could test directly whether a photon splits at BS. You could count $N(D_0)$, the number of photons detected by D_0, and also record *all* the coincidences between D_0 and D_1, i.e., $N(D_0, D_1)$, and *all* the coincidences between D_0 and D_2, i.e., $N(D_0, D_2)$. If you found that

$$N(D_0) = N(D_0, D_1) + N(D_0, D_2),$$

you could conclude that every photon entering the apparatus went either to D_1 or to D_2 and that no photon divided at the beam splitter. However, because efficiencies are of the order of 80% or less, there are always photons that pass undetected through the apparatus; maybe one of these split.

You could also look for coincidences between D_1 and D_2, thinking that if a photon split, its pieces would arrive in each of these detectors at the same time. In principle this is o.k., but in practice the pieces of the "split photon" would have energies below the threshold of the detectors, and there would be no detection even if they were present. All experiments that have actually been done are consistent with the conclusion that a photon does not split at the beam splitter.

Given that only about one out of 10^{11} photons incident on a down-converter crystal converts into two photons of equal lower frequencies, you might ask: Are there enough photons in the beam entering the crystal to produce a useful yield? To answer this question you need to take into account that no detector registers every incident photon; your observed count rate will always be less than the actual number of photons arriving per unit time. And when you use two detectors in coincidence, the overall efficiency is smaller yet because it is the product of the individual efficiencies. Despite all these factors reducing the observed count rate, a rough calculation shows that it is not difficult to have enough photons to do these experiments.

▼ EXAMPLES

1. To estimate how many photons might be incident on the down-converter crystal, assume for convenience of calculation that you have $\lambda = 310$ nm light from a 1.6 mW laser entering the down converter. (This is modest laser power; a typical red laser-pointer has a 5 mW beam.) A 310 nm photon has an energy of $hf = \frac{1240}{310} = 4$ eV. A beam power of 1.6 mW is $1.6 \times 10^{-3}/1.6 \times 10^{-19} = 10^{16}$ eV s^{-1}. Therefore, this 1.6 mW beam delivers 2.5×10^{15} photons s^{-1}, and you might reasonably expect

$$\frac{2.5 \times 10^{15}}{10^{11}} = 25\,000 \text{ down converted pairs per second.}$$

This means there will be 50 000 photons in all, 25 000 for each detector.

2. However, because of geometry, detectors intercept only a small fraction of these photons. Moreover, as noted above, no detector detects every photon incident on it. Suppose your detector can collect only 10% of the photons emitted from the crystal, and suppose it can register about 10% of the photons that it collects. The percentage of successful detections is called the "detector efficiency." (The efficiency is the same thing as the probability of registering a detection.) Thus, if 25 000 photons s^{-1} come out of the crystal, and 2500 of them reach a detector and the detector is 10% efficient, you will observe a count rate of $2500 \times 0.10 = 250$ counts s^{-1}.

3. But notice what a 10% efficiency does to the probability of coincidence detection. If each detector has an efficiency of $\epsilon = 0.10$, i.e., 10%, the probability that you will register counts in both detectors at the same time is 1%, the *product* of the two efficiencies, i.e., for this example the overall efficiency is $0.10 \times 0.10 = 0.01$, and you will see a count rate of about 25 s^{-1}. Count rates are often give in units of hertz; so for this example you can expect the rate of coincidences to be 25 Hz.

19.4   INTERFERENCE OF QUANTA

If a photon does not split at a beam splitter, what determines whether a photon is reflected or transmitted by the beam splitter? The disturbing answer is that the result is pure chance. There is nothing about the incident photon or about the splitter that determines which way the photon will go from the beam splitter. If you are not disturbed by this statement, think about it until you are. Einstein never was able to accept that the result was in principle unpredictable. As Feynman has pointed out,[1] this behavior contradicts the belief that it is an essential feature of science that if you do two experiments with identical set ups, identical initial conditions, identical particles, etc., you must get identical results. *Experiments with beam splitters show this is not so.* Experiments identical in every respect can have different outcomes, and the outcome for any particular photon can not be predicted.

A second profound question has already been touched upon. If photons don't divide, how do they interfere? How can the passage of individual photons (or electrons) through an interferometer produce an interference pattern? As already mentioned, the answer is that in some sense an individual particle must be in more than one place at the same time. You can make this seem less weird if you say instead that the principle of superposition needs to be reinterpreted, but the reinterpretation requires serious revision of your intuitions about the nature of reality.

To see why such reinterpretation is necessary, consider the interference of photons in a Mach-Zehnder interferometer—a type of interferometer that is particularly convenient for working with laser beams. As shown schematically in Fig. 19.2, a Mach-Zehnder interferometer consists of a pair of beam splitters and a pair of mirrors. The two photons from the down-conversion source are directed so that one (called the idler) goes to detector F; and the other (called the signal) enters the interferometer and comes to the beam splitter BS_1. The beam splitter offers the signal photon a choice of paths of length ℓ_1 or ℓ_2; the mirrors M_1 and M_2 direct the photon to the second beam splitter BS_2, so that a photon can end up either at D_1 or at D_2. Notice that there is no way to tell whether the signal photon arrived at D_1 (or D_2) by path ℓ_1 or ℓ_2; the two paths are indistinguishable.

If $\ell_1 \neq \ell_2$, then for light of wavelength λ there is a phase difference $\phi = 2\pi \left(\frac{\ell_2 - \ell_1}{\lambda} \right)$ associated with the difference in path lengths. The detector

---

[1] Richard Feynman, *The Character of Physical Law*, MIT Press, Cambridge (1965), p. 147.

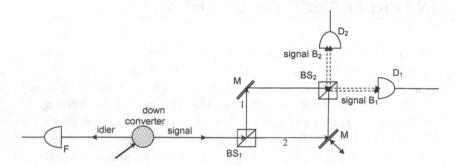

FIGURE 19.2 A photon from a laser (not shown) enters a down converter and two photons emerge—the idler photon and the signal photon. The signal photon enters the Mach-Zehnder interferometer. By detecting only signal photons that arrive at D_1 or D_2 in coincidence with idler photons arriving at F, you can be sure that the resulting interference pattern arises from just one photon at a time passing through the interferometer.

at D_1 will register photons when the phase difference is a multiple of 2π radians; no photons will be detected when the phase difference is an odd multiple of π radians. This is interference.

If, while counting photons in the detector at D_1, you vary the length of one path, say ℓ_2, you get variations in the number of counts as shown in Fig. 19.3. By analogy with visible interference patterns, this pattern of cyclic variation in the number of counts as you change the difference between the path lengths is called a fringe pattern.

To be sure you are looking at the behavior of only one photon in the interferometer, you use the same trick as in Sect. 19.3: You use a down converter as a source of pairs of photons. You send one of a pair directly into detector F; you send the other through the interferometer to detector D_1. By detecting them in coincidence, you see a single photon's contribution to the interference pattern. The data shown in Fig. 19.3 are counts of coincidences. These data show that the interference pattern builds up one photon at a time.

You can explain the observed interference by treating the light entering the interferometer as a wave, but that evades the fact that at D_1 the light passed through the interferometer and was detected as single photons.[2] How can you reconcile the granularity of light with its wavelike behavior?

---

[2]You can attach an audio speaker to the detector, so that it produces a single distinct click when a photon arrives. This is what physicists are thinking of when sometimes they speak of detection as occurring "one click at a time."

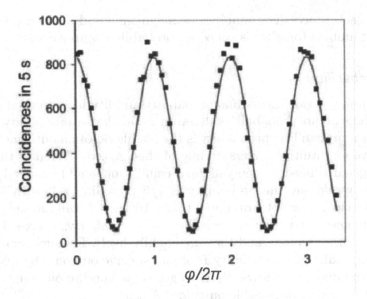

FIGURE 19.3 Photon counts recorded by a detector at D_1 in Fig. 19.2. The solid curve is what quantum mechanics predicts.

19.5 PROBABILITY AMPLITUDES AND PROBABILITIES

Introduction

You can get a consistent description of wave-like and particle-like behavior of an object by shifting emphasis from the object itself to its processes. Stop worrying about the nature of a photon at a beam splitter and focus on what can happen to it; it can undergo reflection or transmission. Make a theory that predicts the probability of each possible outcome. For the 50–50 beam splitter, the probability should be 1/2 for reflection and 1/2 for transmission.

However, in order for quantum theory to account for interference and other effects of superposition, the theory must associate with each possible process or outcome a phase of some sort as well as some measure of how probable the outcome will be. Such an association is achieved by assigning to each process two numbers, a magnitude and a phase angle. This pair of numbers together is called a "probability amplitude."[3]

Probability

Before you can understand probability amplitudes, you need to know a few things about probabilities. Keep in mind that a probability amplitude is not a probability; probability is the likelihood of an outcome. You already have an intuitive understanding of this. Asked, "What is the probability of getting 'heads' when you flip a coin?" you would answer 1/2. Similarly, you would say the probability is 1/6 of rolling a 3 with a cubical die. Consciously or not, you noted that a two-sided coin has only two possible outcomes, and they are equally likely. Similarly, a six-sided die has 6 possible outcomes, and they are equally likely. For such simple cases, you can predict the probability P of any specific outcome by taking the ratio of the number of ways $N(A)$ to get some specific outcome A among the total number of possible outcomes N_{total}:

$$P = \frac{N(A)}{N_{\text{total}}}.$$

What if you ask a more complicated question like "What is the probability of getting three heads in a row when you flip a coin?" You could work this out by writing out all the possibilities: HHH, HHT, HTH, HTT, THH, THT, TTH, TTT. There are eight possibilities, $N_{\text{total}} = 8$, and only one way to get HHH, $N(\text{HHH}) = 1$ so $P(\text{HHH}) = 1/8$. But if the probability of each outcome is independent of the others, their joint probability is the product of the individual probabilities. For the example here, the individual probability of getting each H is each 1/2 so the probability

$$P(\text{HHH}) = P(\text{H})\,P(\text{H})\,P(\text{H}) = 1/8.$$

In general *multiply the probabilities of independent events to get their joint probability,* i.e., if the probability of outcome 1 is $P(1)$ and the probability of outcome 2 is $P(2)$, the probability of obtaining both outcomes 1 and 2 is

$$P(12) = P(1)\,P(2). \qquad \text{probability of joint events} \qquad (1)$$

It is also useful to know that the probability of an outcome consisting of a combination of several mutually exclusive possibilities is the sum of the probabilities of obtaining each individual outcome. For example,

---

[3] Sometimes people call the magnitude itself an "amplitude." The use of the word "amplitude" to designate one part of something that is also called an "amplitude" is unfortunate, but deal with it as best you can.

suppose you ask "What is the probability of getting at least one tails when you flip a coin twice?" The possibilities are HH, HT, TH, and TT. Each has a probability of 1/4. The probability of getting at least one tails is 1/4+1/4+1/4 = 3/4. In general *the probability of an outcome that is a set of mutually exclusive events is the sum of the probabilities of the occurrences of the individual events.*

$$P(\text{T}) = \sum_{i=1}^{n} P(\text{x}_i\text{T}) \qquad \text{probability of mutually exclusive events} \qquad (2)$$

where $P(\text{x}_i\text{T})$ represents the probability of getting anything and at least one T.

■ EXERCISES

> **1.** When you roll a die, what is the probability that you get an even number? Which of Eqs. 1 and 2 applies here?
>
> **2.** When you roll a die, what is the probability that you will roll either and even number or a 3? Which of Eqs. 1 and 2 applies here?
>
> **3.** What is the probability you will get an even number on the first roll and the number six on the second roll? Which of Eqs. 1 and 2 applies here?

As you know, predictions using probability are not exact. When you perform N measurements, looking each time for an outcome that has a probability P, you do not expect to get exactly NP outcomes; i.e., if you flip a coin 1000 times, it is quite unlikely that you will get exactly 500 heads. But the more times you flip the coin, the closer to 1/2 will become the ratio $N_{\text{heads}}/N_{\text{tosses}}$. In one experiment 100 tosses gave 54 heads; 1000 tosses gave 480 heads; and 10 000 tosses gave 5018 heads. You can see that the ratio approaches closer and closer to 1/2, i.e., .54 → .480 → .5018. This effect is called the "law of large numbers."

According to Newtonian mechanics (often called classical mechanics), there is actually nothing probabilistic about rolling a die or flipping a coin. If you know the initial position and velocity of a die and the impulse you impart to it as you roll it, you can calculate the outcome by the laws of mechanics. You resort to probability because it is impractical to find the initial conditions with precision sufficient to predict an outcome, and it is easier to assume that these conditions vary randomly from throw to throw; you do not assume that the conditions are in principle unknowable. But quantum systems always possess a superposition of different initial positions and velocities, and, therefore, it is in principle impossible to know the initial conditions and predict the outcome of any single measurement. Where it is a convenient tool in classical physics, probability is fundamental to quantum physics.

Probability Amplitudes

For correct description of the interference phenomena that occur in quantum systems, a phase angle needs to be associated with each possible outcome. This is done by assigning to each possible quantum process a "probability amplitude," a mathematical entity that has both a magnitude and a phase angle. Thus in Fig. 19.1 (p. 573), the probability amplitude for a photon to pass through a 50–50 beam splitter and arrive at detector D_2 is $p_1 = (\frac{1}{\sqrt{2}}, 0 \text{ rad})$.

The probability of an outcome is the square of the magnitude of its probability amplitude. This is important; it is the connection between probability, something you can measure, and probability amplitude, something theory can predict. The square of the magnitude of p_1 is 1/2, so the probability of a photon being transmitted by the beam splitter is 1/2.

■ EXERCISES

---

4. In Fig. 19.1 the probability amplitude is $p_2 = (\frac{1}{\sqrt{2}}, \frac{\pi}{2} \text{ rad})$ for a photon to arrive at detector D_1 by reflection from the beam splitter. What is the probability of this outcome? Remember, the probability is the square of the magnitude of the probability amplitude.

Given that there are only two possible outcomes, you are not surprised at the answer. In fact, you probably wonder, "Why bother with $1/\sqrt{2}$? Why not just give the probability as 1/2? Why complicate things with a probability amplitude?" The answer is that with the probability amplitude and a few simple rules you can construct the probability amplitude of a complicated process from simpler probability amplitudes of different parts of the process.

Product Rule for Probability Amplitudes

Product Rule: When two (or more) processes occur in succession, the probability amplitude for the outcome is a special kind of product of the individual probability amplitudes. The probability amplitude for a succession of processes has *a magnitude which is the product of the magnitudes of the individual processes*, and it has *a phase which is the sum of the phases of the individual processes.* As a specific example, suppose a photon reflects first from one beam splitter and then from another. The probability amplitude for the first event is $p_1 = (\frac{1}{\sqrt{2}}, \frac{\pi}{2} \text{ rad})$ and for the second event it is $p_2 = (\frac{1}{\sqrt{2}}, \frac{\pi}{2} \text{ rad})$. The probability amplitude of the combination of the two events has a magnitude of $\frac{1}{\sqrt{2}} \frac{1}{\sqrt{2}} = \frac{1}{2}$ and a phase of $\frac{\pi}{2} + \frac{\pi}{2} = \pi$.

Addition Rule for Probability Amplitudes

Addition Rule: When an outcome can occur by two or more processes that are mutually exclusive and indistinguishable, their probability amplitudes add, and—this is important—they add like vectors. Consequently, it is possible for one amplitude to cancel some or all of another. Then the probability, which is the square of the magnitude of the combined amplitudes will be larger or smaller depending on the relative phase of the amplitudes; the combined amplitude can even be zero. Such variations in probability can show up as an interference pattern in your observations. The general idea is applied in the following specific example.

▼ EXAMPLES

4. Here's how to use the rules to predict the probability that a photon passing through the Mach-Zehnder interferometer shown in Fig. 19.2 (p. 576) will be detected by D_1. First, use the Product Rule to find the probability amplitude for each path. For path 1 it is $p_1 = (\frac{1}{2}, \frac{\pi}{2})$; for path 2 it is $p_2 = (\frac{1}{2}, \frac{\pi}{2} + \phi)$ where ϕ is some additional phase change that you introduce by adjusting the apparatus. Second, use the Addition Rule to find the probability amplitude for the overall outcome: add p_1 and p_2 vectorially as in Fig. 19.4. The probability P_{B_1} that a photon will arrive at detector D_1 is the square of the magnitude of the combination of p_1 and p_2:

$$P_{B_1} = p_1^2 + p_2^2 + 2p_1p_2 \cos\phi = \frac{1}{2}(1 + \cos\phi). \tag{3}$$

Do you see how the difference in phases of p_1 and p_2 leads to interference in the probability P_{B_1}? This equation produces the smooth curve that fits the data in Fig. 19.3.

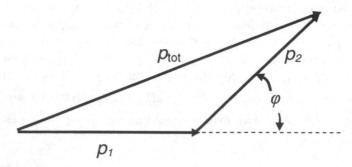

FIGURE 19.4 Probability amplitudes p_1 and p_2 add like vectors.

■ EXERCISES

5. Calculate P_{B_2}, the probability a photon will be registered by detector D_2 in Fig. 19.2.

6. What should be the value of $P_{B_2} + P_{B_1}$? Is it?

Indistinguishability

To apply the Addition Rule correctly you need to understand what is meant by "indistinguishable." A couple of examples may help. Consider the two paths through the Mach-Zehnder interferometer shown in Fig. 19.2. A photon can arrive at detector D_1 by either path: it can reflect from BS_1 and pass through BS_2, or it can pass through BS_1 and reflect from BS_2. If the two path lengths are the same, i.e. $\ell_1 = \ell_2$, nothing in the apparatus or the outcome can tell you which path the photon took. The two paths are indistinguishable.

Or suppose, as shown in Fig. 19.5, two identical photons, γ_1 and γ_2, come from a down converter and arrive at a beam splitter at the same time. There are four possible outcomes, each with its own probability amplitude: photons γ_1 and γ_2 arrive together in detector D_1 (call the probability amplitude $p_{D_1D_1}$); the two photons arrive together in detector D_2 (for which the probability amplitude is $p_{D_2D_2}$); γ_1 arrives at D_1 and γ_2 arrives at D_2 (for which the probability amplitude is $p_{D_1D_2}$); γ_1 arrives at D_2 and γ_2 arrives at D_1 (for which outcome the probability amplitude is $p_{D_2D_1}$). Notice that nothing in the apparatus or the outcome can distinguish between the last two possibilities. Consequently, the probability amplitude for registering a photon in D_1 and D_2 at the same time

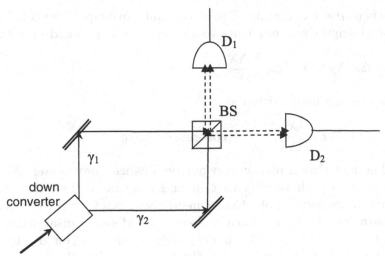

FIGURE 19.5 Two photons, γ_1 and γ_2, enter a beam splitter BS at the same time, one at each entry port.

(coincidence) is the (vector) sum of the two probability amplitudes $p_{D_1 D_2}$ and $p_{D_2 D_1}$. (As you will see in problem 5, for this particular case the sum turns out to be zero.)

What might make different paths to an outcome be distinguishable? In the first example, if one path were longer than the other, say $\ell_1 > \ell_2$, then a photon would take longer to travel path 1 than to travel path 2, and the paths could be distinguished by the photon travel times. In the second case, suppose one photon had a measurably different frequency from the other. Then it would in principle be possible to distinguish between $p_{D_1 D_2}$ and $p_{D_2 D_1}$, and the Addition Rule would not be applicable.

The Uncertainty Principle, Coherence Length, and Indistinguishability

Given that interference occurs only between indistinguishable processes, you might think that as soon as two path lengths in an interferometer differ by any amount there would be no interference. This is not the case. The uncertainty principle shows that there is always a range of differences in the path lengths for which the paths are indistinguishable. You can see that this is true in two different ways.

First, any set of photons will have some spread of wavelengths $\Delta\lambda$ however small. As a result, they will have a spread of momenta. This is because $p = \frac{h}{\lambda}$ and differentiation shows you that

$$\Delta p = \frac{h\,\Delta\lambda}{\lambda^2}$$

where p is momentum. The uncertainty principle connects this spread of wavelengths and momenta to a corresponding spread in space Δx:

$$\Delta x \, \Delta p \approx h = \Delta x \frac{h \, \Delta \lambda}{\lambda^2}$$

which can be rewritten as

$$\ell_c = \Delta x = \frac{\lambda^2}{\Delta \lambda}. \qquad \text{coherence length} \qquad (4)$$

The fact that a photon can not be localized any better than Δx means that two path lengths that differ by no more than Δx are equivalent and indistinguishable. Consequently, you can vary the difference between path lengths by as much ℓ_c before interference disappears. This quantity ℓ_c is important for understanding the behavior of any interference phenomenon, and it is called the "coherence length."[4]

Second, the time-energy version of the uncertainty principle limits how distinguishable one path is from another by differences in a photon's travel time along the different paths. There is always some finite interval of time Δt for which the two possibilities will be indistinguishable as long as the travel times differ by less than Δt. If a photon has a spread of frequencies Δf, then it has a corresponding spread of energies $\Delta E = h \Delta f$. This spread of energy means, according to the uncertainty principle, that you will always be unsure by an amount Δt of the instant t_0 at which the photon was created. As a result, travel times $(t_{\text{detection}} - t_0)$ are always uncertain by the same Δt and

$$\tau_c = \Delta t = \frac{h}{\Delta E} = \frac{1}{\Delta f},$$

where τ_c is called the "coherence time."

Notice that the coherence time and coherence length of photons are directly related by the speed of light c, i.e., $\ell_c = c \tau_c$.

■ EXERCISES

7. Show that Δf and $\Delta \lambda$ are connected by the relation $\Delta f = c \frac{\Delta \lambda}{\lambda^2}$.

8. Then show that if $\ell_c = c \tau_c$, $\ell_c = \frac{\lambda^2}{\Delta \lambda}$.

---

[4]Using h rather than $\hbar/2$ in the uncertainty principle gives a quantum definition of coherence length that is the same as the definition used in classical wave theory.

> **9.** The spread in frequencies Δf is often referred to as "bandwidth." What are the coherence length and coherence time of a helium-neon laser ($\lambda = 632.8\,\text{nm}$) that has a bandwidth of $1.5\,\text{GHz}$?
>
> **10.** Using this He-Ne laser, by approximately how much can you lengthen one arm of an interferometer compared to its other before you will cease to observe interference?

There is more to say about indistinguishability and about coherence. For example, there are partially distinguishable processes that result in partial interference. You can look forward to learning about such things in some other book.

19.6   RULES OF QUANTUM MECHANICS

The properties of probability amplitudes and the idea of indistinguishability make it possible to reduce quantum theory to a few simple rules.[5]

1. Outcomes of measurements are described in terms of probabilities P. Values of P are predicted from *probability amplitudes*. A probability amplitude has both a magnitude p and a phase ϕ, and the probability P of an outcome is the square of the magnitude of the probability amplitude for that outcome, i.e.,

$$P = p^2. \tag{5}$$

2. When an event consists of several sub-events *in sequence*, the probability amplitudes for each sub-event multiply together according to the Product Rule, illustrated on p. 581. The magnitude of the total probability amplitude is the *product* of the magnitudes of the individual probability amplitudes; and the phase of the total probability amplitude is the *sum* of their phases.

$$p = p_1 p_2 \qquad \text{and} \qquad \phi_{\text{total}} = \phi_1 + \phi_2. \tag{6}$$

---

[5]To see how elegantly the great twentieth-century physicist Richard Feynman has done this reduction, read his popular lectures *QED: The strange theory of light and matter*, (Princeton University Press, 1985). For a more detailed treatment read R.P. Feynman, R.B. Leighton and M. Sands, *The Feynman Lectures on Physics* (Addison-Wesley, Reading, 1965) Vol. 3, Chap. 3.

3. When there are alternate ways by which an event can occur and these ways are *indistinguishable*, the total probability amplitude is the (vector) sum of the individual probability amplitudes as illustrated in Example 4, and the probability P is the square of the magnitude of this combination. Thus, for the case of two probability amplitudes p_1 and p_2 differing in phase by ϕ, the corresponding probability P is

$$P = p_1^2 + p_2^2 + 2p_1p_2 \cos \phi. \tag{7}$$

4. Now comes the big difference between distinguishable and indistinguishable. When alternate ways by which an event can occur are *distinguishable*, the total probability of the event is the sum of the *probabilities* of the individual alternatives. This means

$$P = P_1 + P_2 = p_1^2 + p_2^2, \tag{8}$$

and there is no term with phase dependence; there is no interference. Interference only occurs between probability amplitudes of *indistinguishable* processes.

5. There is also a rule that tells you the phase of a probability amplitude of a photon after it has traveled some distance. When a photon with a wavelength λ travels a distance ℓ, its probability amplitude acquires a phase $2\pi\ell/\lambda$. If there is only one path, you don't need to worry about the phase, but when there are two paths of lengths differing by less than the coherence length, i.e., ℓ_1 and ℓ_2 such that $|\ell_2 - \ell_1| \leq \ell_c$, the interference is governed by the phase difference ϕ where

$$\phi = \frac{2\pi}{\lambda}(\ell_1 - \ell_2). \tag{9}$$

Does Interference Occur One Photon at a Time?

As you have seen, the answer is "Yes." The data in Fig. 19.3 show that the observed interference pattern builds up one photon at a time. Moreover, the rules of quantum theory predict that the interference pattern should vary as the solid curve in Fig. 19.3. This curve is a graph of Eq. 3. You can see that to within the usual statistical fluctuations of such counts, the data confirm the theory.[6]

---

[6]You can find more detail on this experiment in "Interference with Correlated Photons: Five Quantum Mechanics Experiments for Undergraduates," by E.J. Galvez, C.H. Holbrow, M.J. Pysher, J.W. Martin, N. Courtemanche, L. Heilig, and J. Spencer, in *American Journal of Physics* V. 73, p. 127 (2005).

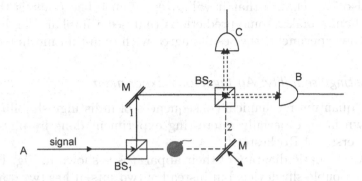

FIGURE 19.6 Diagram of a Mach-Zehnder interferometer with a bomb in one of its arms.

Spookiness of Superposition

The "bomb" experiment is a hypothetical illustration of how strange the consequences of superposition in quantum mechanics can be. Consider a Mach-Zehnder interferometer set so the paths are indistinguishable and $\phi = \pi$. Thus the detector at B will detect no photons because $P = 0$ (see Eq. 3). However, if the paths are made distinguishable by blocking one of the arms then the probability of detecting a photon at B becomes $P = 1/4$. Half of the photons will be blocked; 1/4 of the photons will go to port B; and 1/4 will go to port C.

Now suppose you have a bomb that gets triggered when a single photon hits it. If you place this hair-triggered bomb in arm 2 as shown in Fig. 19.6, it makes the paths distinguishable, because, if a photon gets through without setting off the bomb, you know that it took the other path; and if the bomb goes off (and you survive) then you know the photon took that path. Now send one photon into the interferometer. Because the bomb makes the paths distinguishable, there is 25% probability that the photon will reach B. The fact that a photon reaches B means that the photon that reaches it "knows" that there is a bomb in arm 2 without going through it! (Remember: if the bomb were not there, the two paths would be indistinguishable and because of the π phase difference, no photons would reach B.) The photon detects the bomb without touching it! (Of course, this is not a failsafe way to detect bombs: there is also 50% chance that it will go kaboom.) The point is that quantum mechanics makes some predictions that seem implausible, but when you do the experiment, the results agree with quantum mechanics.

Indistinguishability: An Ingenious Experiment

The quantum mechanical consequences of indistinguishability are vividly shown in an especially interesting experiment done by physicists at the University of Rochester.[7]

A schematic diagram of their apparatus is shown in Fig. 19.7. It works like a double-slit device, but instead of two slits, it has two down-converter crystals of lithium iodate, labeled NL1 and NL2 in the figure. Into each comes a beam of 351.1 nm light brought from an argon laser by way of the beam splitter BS_P. From each crystal come two beams of photons; one is called the "idler" beam and is labeled i_1 or i_2; the other is called the "signal" beam and is labeled s_1 or s_2. The exact values of wavelengths of the idler and signal photons depend upon the geometry of the crystal and the angle of incidence of the entering photon. In this experiment the signal photons have wavelengths of 632.8 nm and the idler photons have wavelengths of 788.7 nm.

▓ EXERCISES

---

11. Show that the frequencies of the outgoing beams add up to the frequency of the incoming beam. Do this without converting the wavelengths to frequencies.

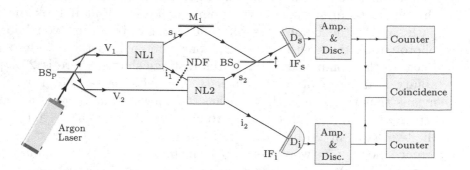

FIGURE 19.7 Schematic diagram of two-crystal analogue of double-slit interference apparatus. *Taken with permission from X.Y. Zou, L.J. Wang, and L. Mandel, Phys. Rev. Lett. 67, 318–321 (1991) ©1991 The American Physical Society.*

---

[7]X.Y. Zou, J.L. Wang, and L. Mandel, "Induced Coherence and Indistinguishability in Optical Interference," *Phys. Rev. Lett.* **67**, 318–321 (1991).

The signal beams are brought together by the mirror M_1 and recombined at the output beam splitter BS_O. As Fig. 19.7 shows, there are two different paths, either $NL1 \rightarrow M1 \rightarrow BS_O \rightarrow D_s$ or $NL2 \rightarrow BS_O \rightarrow D_s$, by which signal light can reach the detector D_s, and differences in the distances along these paths will give rise to phase differences that result in interference.

After photons pass through BS_O, they pass through a filter IF_s, which eliminates any idler light so that only signal photons reach the detector D_s. When the phase of the probability amplitude of the signal photons is varied by shifting the output beam splitter BS_O a little, the count rate in the detector varies as shown in curve A of Fig. 19.8. It is an interference pattern.

You know that double-slit interference occurs only as long as the apparatus is arranged so that the two different paths by which the photon can reach the output detector are indistinguishable. In an exactly analogous way, an interference pattern occurs in the Rochester apparatus only as long as the apparatus is arranged so that there is no way to tell from which crystal the signal photon came to BS_O; that is, as long as the two paths are indistinguishable.

This is true even though the experimenters can tell from which crystal a photon has come without disturbing the s_1 or s_2 photons. They use the i_1 and i_2 photons from the down converters to identify from which crystal a signal photon comes. No s_2 photon is emitted without an i_2

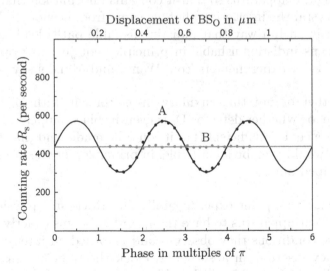

FIGURE 19.8 Measured photon counting rate as function of displacement of BS_O. The distinction between curves A and B is explained in the text. *Taken with permission from X.Y. Zou, L.J. Wang, and L. Mandel,* Phys. Rev. Lett. *67, 318–321 (1991)* ©*1991 The American Physical Society.*

photon. Therefore, if a detector D_i, set up as shown in Fig. 19.7, records the presence of an i_2 photon at the same time D_s detects a photon, you know that the photon at D_s must have come from NL_2. This is just the same as knowing from which slit a photon came in a double-slit setup, although it is accomplished without any direct interaction with an s_1 or an s_2 photon. What is the result? It is curve B of Fig. 19.8; there is no interference.

Well, then, how do they get interference in the first place? To get interference the apparatus must be set up so there is no possible way to distinguish which path a photon took in reaching D_s. In an apparatus in which it is *in principle* impossible to distinguish which path the photon takes, interference occurs. This is like all the previously considered examples of interference, where a particle—an electron, a photon, a neutron, or a buckyball—contributed to the build-up of an interference pattern by, in some sense, passing through two slits at once.

The experimenters arranged their apparatus to make the two paths indistinguishable by carefully aligning it so that i_1 photons passed through the crystal NL_2 and emerged exactly along the path that i_2 photons followed as shown in the diagram. Then the detector D_i cannot tell whether a photon arriving at D_s came from NL_2 or NL_1. Under these circumstances the interference pattern of curve A was obtained.

It is important to understand that it is not the presence of D_i or the actual detection of i_2 photons that eliminates the interference pattern. Just arranging the apparatus so that it contains information that distinguishes which crystal the interfering photons came from is enough. Nature is constructed in such a way that indistinguishable paths lead to interference. This means indistinguishable in principle, not just in practice. Perhaps the Rochester experimenters Zou, Wang, and Mandel say it better:

> Whether or not this auxiliary measurement with D_i is actually made, or whether detector D_i is even in place, appears to make no difference. It is sufficient that it could be made, and that the photon path would then be identifiable, in principle, for the interference to be wiped out.

To make this point experimentally the Rochester physicists carefully aligned their apparatus to have the i_1 and i_2 beams exactly coincide. Under these conditions they observed and recorded an interference pattern. Then they inserted an absorber to block the i_1 beam, as shown by the dashed line in Fig. 19.7. With i_1 blocked, any photon detected at D_i signals that a photon detected at the same time with D_s is from NL_2. Under these new conditions, with an absorber blocking i_1, the interference pattern disappeared, and the observers measured curve B in Fig. 19.8. Notice that D_i has nothing to do with this result; it does not have to be present at all. Blocking i_1 makes it possible in principle to distinguish the s_2 photons from the s_1 photons, and that is enough to guarantee that there is no longer any interference.

▉ EXERCISES

12. Suppose the experimenters misaligned their apparatus just enough so that the counter at D_i could distinguish i_1 photons from i_2 photons by the slightly different directions from which they were coming. Would the interference pattern be present? Explain your answer.

19.7 SUMMARY

This chapter emphasized photons rather than atoms because we wanted you to see experimentally observed quantum behavior, and it is easier to perform experiments on quantum behavior with photons than with atoms. Nevertheless, atoms can be made to show the same quantum behaviors as photons. Atoms can exhibit interference and interesting superpositions of their properties. For example, physicists have made an atom laser in which atoms act collectively and coherently just as photons do in an optical laser. Of course, like photons, atoms can also behave as discrete particles. Whatever the behavior of atoms, it is well described with probability amplitudes and the same quantum rules as describe the behavior of photons.

Like a photon, an atom can interfere with itself, and to explain interference you need the strange idea that the atom occupies a superposition of different positions all at the same time. Just as in the case of a photon, superpositions of an atom obey the uncertainty principle: An atom with a narrow range of locations always has a broad range of momenta; an atom with a narrow range of momenta has a broad range of locations.

In general, a quantum system exists in a superposition of many different states, e.g. different locations, but when a measurement is made on the system, the system takes on some particular, definite value from among the many possible. Which one of the possible values will be obtained in any given measurement is entirely unpredictable. Quantum theory can only predict the probabilities of outcomes, *i.e.*, how many times a given value will be obtained when a large number of measurements are made on identical systems.

The concept of indistinguishability was also introduced in this chapter. Only indistinguishable processes can lead to interference. The ideas of coherence length and coherence time connect indistinguishability and the uncertainty principle. It provides the latitude that permits processes to be indistinguishable over some range of differences of their properties. Indistinguishability and its consequences are very important in quantum theory, and the next chapter has more to say about these.

PROBLEMS

1. When a laser pointer is projected on a screen we see a red ($\lambda = 670$ nm) spot with an area of ~ 1 mm^2. The intensity of the light reaching the screen is 3 mW.

    a. Find the number of laser photons reaching the screen per unit time.

    b. If you think of these photons as uniformly distributed through the volume of the laser beam, how far apart are they? (Hint: How fast are they going?)

2. Consider the Mach-Zehnder interferometer shown in Fig. 19.2 (p. 576). Each beam-splitter reflects half and transmits half of the intensity of the light that is incident on it. The interferometer has detectors D_1 and D_2 at its output ports.

    a. The intensity of the incident light is I_0. If you block one of the arms, what is the intensity of the light exiting each of the interferometer ports (in terms of I_0)?

    b. Both arms are now unblocked. The intensity of the light exiting to D_1 is $I_1 = (I_0/2)[1 + \cos \phi]$, where ϕ is the phase difference between the beams from the two arms due to the difference between the lengths of the two arms. Find an expression in terms of I_0 and ϕ for the intensity of photons exiting to D_2. Hint: energy must be conserved.

    c. The wavelength of the light is 900 nm. Find I_1 when the difference between the lengths of the two arms (i.e., $\ell_1 - \ell_2$) is 2 μm.

    d. Consider a signal photon entering the interferometer. The lengths of the two arms are now the same (i.e., $\ell_1 = \ell_2$).

        i. If the paths are indistinguishable, what is the probability that the photon will be detected at D_1?

    ii. If the paths are indistinguishable, what is the probability that the photon will be detected at D_2?

   iii. If the paths are distinguishable, what is the probability that the photon will be detected at D_1?

   iv. If the paths are distinguishable, what is the probability that the photon will be detected at D_2?

3. Consider the Mach-Zehnder interferometer shown in Fig. 19.9. It has 50–50 beam splitters, which reflect 50% of the intensity of the light incident on them and transmit the other 50%.

  a. Find the probability of transmission through a 50–50 beam splitter.

  b. Find the magnitude of the probability amplitude of going through the beam splitter.

  c. The probability amplitude of going through the two beam splitters is the product of the individual probability amplitudes. What is the magnitude of the probability amplitude for going from A to B when arm 1 is blocked?

  d. What is the probability of a photon to go from A to C when arm 2 is blocked?

  e. If the two paths of the interferometer are **distinguishable**,

    i. What is the probability of going from A to B when neither arm is blocked?

   ii. What is the probability of going from A to C when neither arm is blocked?

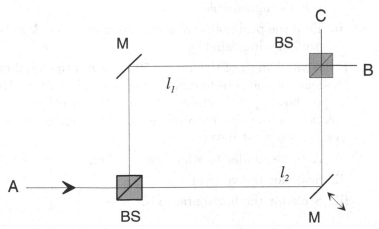

FIGURE 19.9 Diagram of a Mach-Zehnder interferometer.

f. If the paths through the interferometer are **indistinguishable**,

    i. What is the probability for a photon to go from A to B when $\phi = 3\pi$?

    ii. If 500 photons per second are incident to the interferometer from A, how many photons reach B in 1 s when $\phi = 3\pi/2$?

    iii. If 500 photons per second are incident to the interferometer from A, how many photons reach B when $\phi = 3\pi$?

    iv. Based on the previous question, how many photons reach C in 1 s when $\phi = 3\pi$?

    v. If 500 photons per second are incident to the interferometer from A, what is the value of ϕ if 250 photons reach B in 1 s?

4. Consider the Mach-Zehnder interferometer shown in Fig. 19.9. Assume it has 20-80 beam splitters so that 20% of the light is reflected through each beam splitter while 80% of the light is transmitted.

  a. Find:

    i. The probability amplitude for a photon to be reflected by the beam-splitter.

    ii. The probability amplitude for a photon to be transmitted by the beam-splitter.

    iii. The probability for a photon to go from A to B through arm 1 if arm 2 is blocked.

  b. If the arms of the interferometer have the exact same length and both arms are unblocked:

    i. Find the probability of a photon going from A to B if the arms are distinguishable.

    ii. Find the probability of a photon going from A to B if the arms are indistinguishable.

  c. If the wavelength of the light is 500 nm and the number of photons incident on the interferometer per second is $N = 1000$, make a graph showing the number of photons detected at B in 1 s as a function of the difference between the lengths of the two arms. Your graph must have:

    i. A curve similar to what you would get in the lab.

    ii. Scale for the vertical axis.

    iii. Scale for the horizontal axis.

5. Figure 19.5 on p. 583 shows an apparatus set to deliver simultaneously a photon to each entry port of a beam splitter. Given that the probability

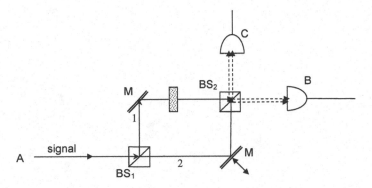

FIGURE 19.10 Diagram of a Mach-Zehnder interferometer with a neutral density filter in arm 1. (Problem 6).

amplitude is $(\frac{1}{\sqrt{2}}, \frac{\pi}{2})$ to reflect from a beam splitter and $(\frac{1}{\sqrt{2}}, 0)$ to pass through it, show that the probability of detecting coincidences between D_1 and D_2 is zero if the two photons are initially in phase. In other words, show that as long as the arms are of equal length, the two photons both go to D_1 or they both go to D_2; they never go one to D_1 and the other to D_2.

6. Consider the interferometer in Fig. 19.10. It has a neutral density filter with a transmission **amplitude** of 1/2 in arm 1.

    a. What is the magnitude of the probability amplitude of going from A to B via arm 1?

    b. What is the probability of going from A to B when $\phi = 4\pi$?

7. Figure 19.7 on p. 588 shows a box marked "coincidence." It represents a device that the experimenters used to tell them when photons arrived at the two counters D_s and D_i at nearly the same time, i.e., in coincidence with each other. Explain why the experimenters needed this device. Would they have observed an interference pattern without it?

Entanglement and Non-Locality

20.1   INTRODUCTION

Before you finish this book you need to learn about experiments that support the belief that quantum indeterminacy is a fundamental feature of atoms, indeed, of the entire physical world. To understand these experiments you need to know about a remarkable and important feature of quantum superposition called "entanglement." As you will see, entanglement not only helps to establish that indeterminacy is a basic feature of reality, it also reveals surprising, non-local connections between quantum systems far apart from each other. It shows that quantum mechanics is a non-local theory. Non-locality is as strange as fundamental indeterminacy. This chapter discusses both.

An analogy may help you appreciate better how superposition makes the characteristics of a quantum object indefinite. Suppose you own many quantum socks that are either green or orange when you observe them. As a skilled quantum mechanic, you can prepare your socks in states that are superpositions of the two colors. Then when you measure a sequence of your socks (i.e., look at them one after another), sometimes the measured sock will be green and sometimes orange. Quantum theory can not predict which color you will see; it can only predict what fraction of many observed socks will be green (or orange).[1] It seems that the socks have no definite color until you observe them, that a quantum system has no definite property until you measure it. Nature is not locally deterministic.

---

[1] It is possible to prepare quantum systems to be in a state with only one possible outcome; then you can predict it.

C.H. Holbrow et al., *Modern Introductory Physics, Second Edition*,
DOI 10.1007/978-0-387-79080-0_20, © Springer Science+Business Media, LLC 1999, 2010

This idea contradicts deeply engrained intuitions and leads to hot arguments about what is reality. Some physicists have argued that quantum mechanics is correct as far as it goes; it just does not go far enough; it is an incomplete theory. They argue that there are variables that determine the outcome of each measurement, e. g., whether a sock will be green or orange, but these variables are hidden from quantum theory, and a more complete theory will take these hidden variables into account and assign properties unambiguously to each object: a green sock will be green before you measure it.

For many years no one could think of a way to test whether there might be valid hidden-variable theories. Then in 1964 Irish physicist John Bell provided a breakthrough insight that hidden variable theories—theories in which objects have definite properties whether you measure them or not—must obey a set of constraints now called "Bell's inequalities." Experiments show that Nature violates Bell's constraints just as quantum theory predicts; a quantum object does not have definite properties until you measure them.

Indeterminacy and non-locality are fundamental aspects of Nature; they can be made to show up in any quantum system. It is particularly convenient, however, to use entangled, polarized photons to test these ideas, and some experiments that do this are described below. But first you need to learn about the polarization of light.

20.2   POLARIZATION

Photons—and electrons, many kinds of atoms, nuclei, and other quantum systems—can be polarized. This means that the individual particles have something about them that specifies an orientation in space. It is as though a photon carries an arrow that points in either of two directions—in the direction in which the photon is traveling or opposite to that direction. A stream of photons with their arrows all pointing in the same direction is said to be "polarized."[2]

The Wave Picture of Polarization

The classical wave picture provides another way to visualize polarization. Chap. 10 told you that light is a transverse wave. This means that the

---

[2]These photon are actually *circularly* polarized. It as though each photon is like a spinning top; the photon's arrow points along the axis of the top—along the direction of travel for right-handed spin; opposite the direction of travel for left-handed.

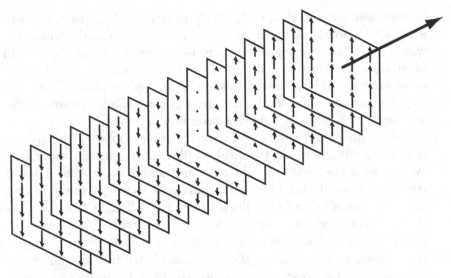

FIGURE 20.1 A representation of an electromagnetic plane wave traveling in the direction of the large arrow. The planes are $\lambda/24$ apart and extend out to a large distance (many, many wavelengths). The electric field is indicated by the small arrows and is the same at every point in any given plane.

wave oscillations are perpendicular to the wave's direction of travel. Light waves (and other kinds of electromagnetic waves) are transverse oscillations of electric and magnetic fields. The electric field of the wave oscillates perpendicular to the direction in which the light is traveling, and the magnetic field oscillates perpendicular to both the direction of travel and to the direction of the electric field. The magnitude of the wave's magnetic field B is related to the magnitude of its electric field E by $B = E/c$.

You can produce light waves in which the electric field everywhere in space oscillates back and forth along the same direction. Figure 20.1 is a representation of such a wave at a single instant in time (a snapshot). The wave's direction of propagation is shown by the large arrow; the small arrows show the direction and magnitude of the electric field. Although the arrows representing the electric field are drawn only on finite planes, you should imagine that the wave fills all space (an idealization). The electric field has the same magnitude and points in the same direction on any plane surface perpendicular to the direction of travel. Because the field is oscillating in time and space, at a given instant of time the magnitude of the field is different on different plane surfaces, but regardless of the variation of magnitude, the fields point either negatively or positively in the same direction everywhere in space.

Such waves are said to be "linearly polarized" because the electric field everywhere points along the same line. They are also described as "plane waves" because, everywhere on any plane perpendicular to the direction of motion, the electric field has the same magnitude and direction. Not all electromagnetic waves are linearly polarized or plane waves, but these are the simplest kind, and they are a good approximation to the waves produced by a laser or a down-conversion crystal.[3]

Using the wave picture, you describe linear polarization by the direction along which the electric field oscillates. Imagine an x-y-z coordinate system with the beam of light traveling along the z-axis. Then its electric field is oscillating along some direction lying in the x-y plane. The direction you ascribe to the linearly polarized light is the direction of the electric field, e. g., the angle θ that it forms with the x-axis. Thus, 60° linearly polarized light is a wave with an E field oscillating everywhere in space along a line at 60° to the x-axis as shown in Fig. 20.2.

It is common to specify the electric field by its components in the coordinate frame you have chosen. For this example, the components are $(E_0 \cos\theta, E_0 \sin\theta)$ where E_0 is the value of the wave's electric field at its maximum. When the electric field oscillates parallel to the horizontal direction ($\theta = 0$ in Fig. 20.2), the light is said to be horizontally polarized. Similarly, when the electric field oscillates parallel to the vertical axis ($\theta = \pi/2$) the light is said to be vertically polarized.

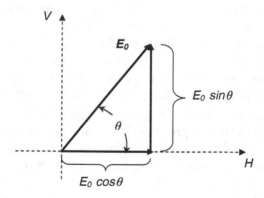

FIGURE 20.2 Electric field vector E_0 decomposed into components aligned with orthogonal polarization axes; $E_0 \cos\theta$ will be transmitted (or absorbed) as horizontally polarized (H) light; $E_0 \sin\theta$ will be transmitted (or absorbed) as vertically polarized (V) light.

---

[3]The linearly polarized photons that are the quanta of these linearly polarized waves are actually in superpositions of their two fundamental states of circular polarization.

EXERCISES

---

1. What are the components of E_1 and E_2 in Fig. 20.3?

Sheet Polarizers

Light from the Sun or from a lamp is unpolarized. That is, the probability amplitudes of the arriving photons have randomly different phases. Or, in terms of the wave picture, the incoming light is made up of a mix of waves with polarizations in randomly different orientations.

You can produce linearly polarized light by passing unpolarized light through a "sheet polarizer." This is a sheet of transparent plastic containing long-chain molecules oriented parallel to each other. These molecules absorb light polarized parallel to them, and this direction is the polarizer's "extinction axis." The direction at right angles to the extinction axis is the polarizer's "transmission axis" because light polarized parallel to this direction is almost all transmitted by the polarizer.[4]

The lenses of polarizing sunglasses are made of sheet polarizers oriented to block oscillations parallel to the horizon. When light reflects from asphalt on the surface of a road or from water on a lake or pond, the oscillations parallel to the horizon are reflected more efficiently than the others and increase glare. By blocking this part of the reflected light, polarizing sunglasses significantly reduce glare. Polarizing sheets are also used as lenses in glasses for viewing 3D movies. One viewing system

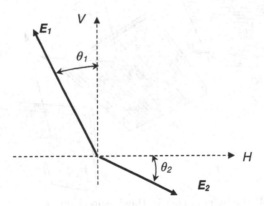

FIGURE 20.3 Figure for Exercise 1.

---

[4]Polaroid sheet is a commonly available kind of sheet polarizer. Notice that the so-called "axis" is not really an axis; it is a line of direction that is the same everywhere on the sheet.

projects two images polarized at right angles to each other; the polarizers in front of your eyes are oriented perpendicular to each other, so the left eye detects one image and the right eye the other, and you see them as a single 3D image.

A simple experiment demonstrates polarization and the effects of polarizing sheets. Take two polarizing sheets and put them with their transmission axes at right angles from each other. The first sheet absorbs all of one component of an incident oscillating electric field; the second one absorbs the rest. Nothing is transmitted, so this combination of two polarizing sheets is essentially opaque. Each polarizer absorbs half the original intensity. This fact leads to a handy rule:

- When unpolarized light of intensity I_0 is incident on a polarizing sheet, half of the intensity is transmitted: $I_T = I_0/2$.

Consider a light wave polarized in the V direction and perpendicularly incident on a polarizing sheet. Figure 20.4 shows the possible cases. The polarizing sheet is represented by the square, and the parallel lines denote its transmission axis. In Fig. 20.4a the polarization of the light is parallel to the transmission axis of the polarizer, and all of the light is transmitted. If the incident electric field is E_0 then the transmitted electric field is $E_T = E_0$ and its intensity is $I_T = I_0$.

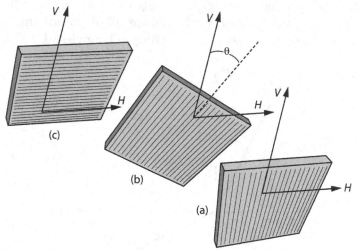

FIGURE 20.4 Light polarized in the V direction (a) enters a polarizing sheet with its transmission axis parallel to the V direction, (b) enters a polarizing sheet with its transmission axis oriented at an angle θ relative to the V direction of the incident light's polarization, or (c) enters a polarizing sheet with its transmission axis oriented perpendicular to the V direction.

If the V-polarized incident electric field is perpendicular to the transmission axis (Fig. 20.4c), then there is no component of electric field parallel to the transmission axis; all of the polarized light is parallel to the extinction axis, and it is all absorbed. The transmitted field and intensity are $E_T = 0$ and $I_T = 0$.

What happens in the intermediate case shown in Fig. 20.4b? Here the transmission axis forms an angle θ with the V-direction of the incident polarization. Resolve the incident V-polarized electric field into two components, one parallel to the transmission axis ($E_0 \cos \theta$), and the other parallel to the extinction axis ($E_0 \sin \theta$). The former component is transmitted and the latter is absorbed. This is a general case, and you need to remember the following.

1. The transmitted electric field is $E_T = E_0 \cos \theta$.

2. The transmitted intensity is $I_T = I_0 \cos^2 \theta$ (in optics this is known as Malus's law.).

3. The light emerging from the polarizing sheet is polarized along the transmission axis of the polarizer.

Light Through Polarizers: The Quantum Picture

For the situations described above, the quantum predictions of how much light will emerge from a polarizer must be the same as the classical predictions, because experiments show that the classical predictions are correct. With the correct probability amplitudes, you can analyze the behavior of polarized photons using the rules of quantum mechanics given in Sect. 19.6 (p. 585).

Suppose a polarizing sheet lies in the x-y plane of an x-y-z coordinate system and is oriented with its transmission axis parallel to the x-axis. Imagine that light, linearly polarized in the x-y plane at an angle θ with respect to the x-axis, is incident along the z-axis as shown in Fig. 20.2. Using the wave picture, you can think of the linearly polarized wave as made of two parts, the one polarized parallel to the x-axis is transmitted, and the other polarized parallel to the y-axis is absorbed.

But you should think of a photon in the quantum way. A photon does not split and have some part pass through the polarizer while the rest is absorbed. Instead, a photon that is linearly polarized at some angle θ relative to the transmission axis of the polarizer has two probability amplitudes: $(\cos \theta, \phi)$ to pass through the polarizer and emerge polarized parallel to the transmission axis and $(\sin \theta, \phi)$ to be polarized parallel to the extinction axis and be absorbed.

These probability amplitudes give the same result as the classical wave picture. For example, if $\theta = 60°$ then the magnitude of the probability amplitude of a photon to emerge polarized parallel to the horizontal direction is $\frac{1}{2}$. This corresponds to a projection of the electric field vector of $E_0 \cos 60° = E_0 \frac{1}{2}$ along the horizontal direction; a similar correspondence holds for the photon's probability amplitude and the electric field's component along the vertical axis, i.e., $E_0 \sin 60° = E_0 \frac{\sqrt{3}}{2}$. For either the photon picture or the wave picture, when the polarizer's transmission axis is horizontal, the transmitted intensity is $I_0 \frac{1}{4}$ (i.e., proportional to the square of the field); when the polarizer's transmission is vertical, the transmitted intensity is $I_0 \frac{3}{4}$.

■ EXERCISES

2. What is the square of the probability amplitude for transmission of linearly polarized photons incident on a polarizer at an angle $\theta = 60°$ relative to its transmission axis? What is the probability that a photon will pass through the polarizer?

3. What is the probability that a photon will pass through if you rotate the polarizer so that its transmission axis rotates by an additional 90°?

4. Which of the quantum rules in Sect. 19.6 (p. 585) did you use to answer the previous questions?

This example shows that you can think of the state of a photon as a superposition of other states. You can think of the photon as in a pure state of linear polarization at 60° to the x-axis, but you can also think of it as a superposition of two states, one parallel to the y-axis and the other parallel to the x-axis. Just as there is an infinite number of ways to resolve a vector into components, there is an infinite number of possible superpositions that correspond to any given polarization state. You can use whatever superposition is most convenient.

To sum up: If a photon, linearly polarized at some angle θ to the transmission axis of a polarizer, is incident on the polarizer:

- The magnitude of the probability amplitude for the photon to be transmitted through the polarizer is $\cos \theta$.

- The probability that the photon will be transmitted is $\cos^2 \theta$.

- Any photon that is transmitted will have *a new polarization*; it will be polarized parallel to the transmission axis of the polarizer. The act of measurement puts the photon into a new state.

A Polarizer Changes a Photon's State

This last statement is fundamental. The polarizer *projects* (to use quantum jargon) the transmitted photon into a new polarization state. This is not surprising. Both from the Heisenberg uncertainty principle and from experiments with interferometers, you know that determining the path of a photon changes it from its initial state that is a superposition of a broad range of locations to a different state that is a superposition of a narrow range of positions. Measurement projects a photon from its initial state into a quite different final state. Similarly, when you measure the polarization of a photon by passing it through a polarizer, you project the photon from its initial state of polarization into a state of polarization parallel to the polarizer's transmission axis.

The following experiment with three polarizers shows you that polarizers actually change the state of transmitted photons.

Consider unpolarized photons incident on a polarizer $\mathcal{P}_1$ with transmission axis oriented vertically as shown in Fig. 20.5. Half of the incident photons are transmitted, and they emerge vertically polarized.

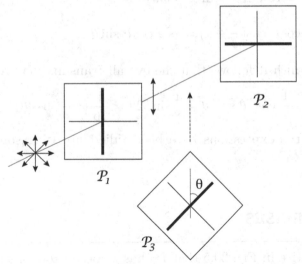

FIGURE 20.5 A polarizer experiment. Unpolarized light is incident on a polarizer $\mathcal{P}_1$ with transmission axis (thick line) oriented vertically followed by a polarizer $\mathcal{P}_2$ oriented horizontally. No light comes out. Then a third polarizer $\mathcal{P}_3$ oriented at an angle θ is inserted between $\mathcal{P}_1$ and $\mathcal{P}_2$. Now light comes out of $\mathcal{P}_2$.

Next, pass these transmitted photons into a second polarizer $\mathcal{P}_2$ with its transmission axis oriented horizontally. No photons emerge from the second polarizer.

Now, place between the vertical and horizontal polarizers a third polarizer $\mathcal{P}_3$ oriented at an angle θ with respect to the vertical as shown schematically in Fig. 20.5. After you do this, photons emerge from polarizer $\mathcal{P}_2$ although previously they did not.

Are you surprised that adding polarizer $\mathcal{P}_3$ between polarizers $\mathcal{P}_1$ and $\mathcal{P}_2$ causes light to come out? By orienting $\mathcal{P}_3$ with its transmission axis at an angle θ relative to the transmission axis of $\mathcal{P}_1$, you cause the photons from $\mathcal{P}_1$ to be projected into a state of linear polarization at an angle $(\pi/2 - \theta)$ relative to the transmission axis of polarizer $\mathcal{P}_2$. Consequently, $\mathcal{P}_2$ now receives photons in a state that it can project into transmitted photons.

▼ EXAMPLES

1. You can use the product rule to do this analysis. Note that the magnitude of a photon's probability amplitude is $\frac{1}{\sqrt{2}}$ to pass through $\mathcal{P}_1$, $\cos\theta$ to pass through $\mathcal{P}_3$, and $\cos(\frac{\pi}{2} - \theta)$ to pass through $\mathcal{P}_2$. Apply the product rule to find that the magnitude of the probability amplitude to pass through all three polarizers is

$$\frac{1}{\sqrt{2}} \cos\theta \, \cos(\frac{\pi}{2} - \theta) = \frac{1}{\sqrt{2}} \cos\theta \sin\theta$$

from which it follows that the overall transmission probability P_T is

$$P_T = \frac{1}{2} \cos^2\theta \sin^2\theta = \frac{1}{8} \sin^2 2\theta = \frac{1}{16}(1 - \cos 4\theta)$$

where the expressions have been tidied up using some trigonometric identities.

EXERCISES

5. Notice in Fig. 20.5 that P_T has zeroes at $\theta = 0$, $\frac{\pi}{2}$, π, and $\frac{3\pi}{2}$. This means that the intensity of the transmitted beam will be zero when $\mathcal{P}_3$ is set at these angles. Can you visualize rotating $\mathcal{P}_3$ and confirm that there will be no transmission at these angles?

6. A vertically polarized photon is incident on a pair of polarizers. The first one has its transmission axis oriented at an angle θ with the vertical. The second's transmission axis is oriented horizontally.

  a. What is the probability for a photon to be transmitted through the two polarizers when $\theta = 30°$?

  b. What is the final polarization orientation of the photon?

Indistinguishability and the Quantum Eraser

The "quantum eraser" is an interesting experiment that uses polarization to distinguish which path a photon takes while passing through an interferometer. To set up this experiment put two polarizers in each interferometer arm as shown in Fig. 20.6 where each BS is a 50–50 beam splitter. Orient the transmission axis of the first polarizer at 45° from the vertical; set the transmission axis of the second one to be vertical. Arrange the light entering the interferometer to be polarized vertically.

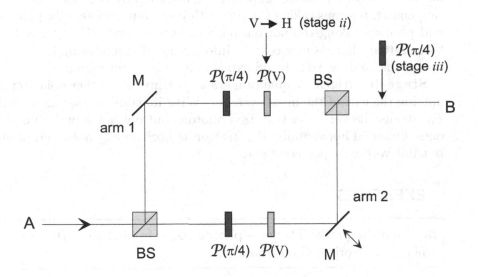

FIGURE 20.6 Experimental layout for the quantum eraser experiment; the light entering from A is polarized perpendicular to the plane of the diagram; each BS is a 50–50 beam splitter.

EXERCISES

7. What is the magnitude of the probability amplitude to go from A to B through arm 1? (You will first have to calculate the probability amplitude of going through the two polarizers.)

8. What do you predict will happen to the interference pattern when the polarizers are set so that you can tell which path a photon took to reach the detector at the output of the interferometer?

Consider the experiment in three stages.

Stage i: Begin with the pairs of polarizer set as in Exercise 7. Then the pairs of polarizers in each arm absorb half of the photons, but the other half come out still vertically polarized. Nothing about a transmitted photon tells you which path it took to go from A to B. The two paths are indistinguishable, and there is interference.

Stage ii: Next rotate the second polarizer in arm 1 from vertical to horizontal. The magnitude of the probability amplitude for passing through both polarizers is the same as before, but now photons coming out of the second polarizer in arm 1 are horizontally polarized. In this arrangement, photons taking the top path come out horizontally polarized, and photons taking the bottom path come out vertically polarized. The transmitted photons now contain information distinguishing between the two paths, and, as you should expect, there is no interference.

Stage iii: To create a quantum eraser you install another polarizer $\mathcal{P}_{out}$ outside the exit of the interferometer. With its transmission axis vertical $\mathcal{P}_{out}$ transmits vertically polarized photons but not horizontally polarized ones. Oriented horizontally, $\mathcal{P}_{out}$ transmits horizontally polarized photons but not vertically polarized ones.

EXERCISES

9. Find the probability for a photon to go from A to B (both arms) when $\mathcal{P}_{out}$ is oriented horizontally.

Now perform some quantum erasure. Rotate $\mathcal{P}_{out}$ so that its transmission axis is at an angle of 45° with the vertical axis. Then half of the horizontally polarized photons and half of the vertically polarized photons will go through $\mathcal{P}_{out}$, and—this is important—all the transmitted photons will have their polarizations oriented at 45°. Now that they have the same polarization, the transmitted photons no longer carry information distinguishing which path they took to get through the interferometer. The paths have been made indistinguishable! And ...(sound of trumpets)... there is once again interference. Rotating the output polarizer $\mathcal{P}_{\text{out}}$ to 45° *erases* the distinguishing information.

■ EXERCISES

10. For stage iii of the experiment
   a. Find the magnitude of the probability amplitude for going from A to B through arm 1.
   b. What is the orientation of the polarization reaching B coming from arm 1?
   c. Find the magnitude of the probability amplitude for going from A to B through arm 2.
   d. What is the orientation of the polarization reaching B coming from arm 2?
   e. Find an expression for the probability for going from A to B in stage iii as a function of ϕ, the phase difference created by moving the mirror in arm 2.

20.3 ENTANGLED QUANTUM STATES

Entanglement provides a vivid example of non-locality, and it raises fundamental questions about the nature of reality. It also makes possible experiments that show conclusively that quantum mechanics is a non-local theory.

Entanglement by Analogy

To help you understand entanglement, recall your orange and green socks. Suppose you have a pair of them, one green and one orange, and you don't know which sock is on which foot because you put them on in the dark. Later you look and see that the sock on your left foot is orange. What color is the sock on your right foot? You don't need to look; you know it's green. Physicists say the socks are correlated. Measurement of the properties of one sock determines the properties of the other.

■ EXERCISES

11. Suppose the sock on your left foot is green. Without looking, what is the color of the sock on your right foot?

12. If the socks were known to be both the same color, would it still be correct to say they are correlated?

Now if each sock was a quantum object, it could be in a superposition of orange and green. (This is the same idea as being in two different positions at the same time.) Then the sock on your foot has no definite color until you make a measurement, i.e., until you look at it. (That's the creepiness of quantum behavior.) But when you look, you will observe a definite color. It will be green, or it will be orange. That's what Chapters 18 and 19 told you about quantum measurements. They also told you that there is no way to know in advance which color you will get. If you have 100 of these quantum socks, and you look at them one after the other, sometimes you will get orange and sometimes you will get green.

Now comes entanglement. It is possible to prepare the two quantum socks in a combined state such that if you measure one and find it to be green, the other will surely be orange, but if you measure the first sock and find it to be orange, then the second is surely green.

If you have many pairs of socks prepared in this peculiar state, you can't predict what color you will find when you measure the first sock. Sometimes it will be green and sometimes it will be orange, but once you measure it, you know that you will always get the other color when you measure the second one.

Physicists now know how to prepare photons, electrons, atoms, and molecules to behave like this. Quantum objects in such a state are said to be "entangled," and their combined state is called an "entangled state."

Non-Locality of Entanglement

Neither entangled quantum sock is green or orange until it is measured. Then, purely randomly, it becomes either green or orange—that's ordinary quantum behavior. What's new here is that measuring the first sock puts the second sock into a definite state even though there is no mechanism for one to affect the other. When Erwin Schödinger, discovered that quantum mechanics predicted this behavior, he said "I don't like it, and I'm sorry I ever had anything to do with it."

Einstein realized that such behavior was independent of how far apart the socks are. If you put on only one quantum sock in the dark, travel to the Moon, and then look at your foot, as soon as you see a green sock, you know that anyone on Earth who looks at the sock you left behind will find it to be orange. He thought this was implausible and scornfully referred to it as *spukhafte Fernwirkungen*—spooky action at a distance.

Entanglement raises in a new way the recurring question: Does an unmeasured object have definite properties that are present before you measure them? In the case of the entangled orange and green socks, after you measure one sock and find it green, you know that the other sock will be orange whenever you look at. Surely, the second sock had the definite property of being orange before you saw its green partner. This view is called "local realism." Most people are by experience and intuition convinced local realists.

Einstein believed an object possessed a property whether you measured it or not. He argued that entanglement confirmed his view. If you can arrange by measuring one object to know for certain that a measurement on a second object will give a certain value for a property, that property must belong to the object. If he had been considering the quantum socks, he would have said the second sock was orange all along. As he (and Boris Podolsky and Nathan Rosen) put it, the fact that quantum mechanics can not predict a result in advance of a measurement does not mean the theory is wrong; it means that the theory is incomplete. Physicists should look for a deeper more complete theory.

Indeterminacy and non-locality are basic to quantum mechanics. Quantum theory has built into it the feature that properties are really unattached to any particular object until they are measured, and the long-distance correlations of non-locality are a necessary consequence of superposition as it is used in quantum theory. Of course, the real question is not "What did Einstein think?", but "How does Nature behave?" Quantum theory certainly asserts indeterminacy and non-locality, but what about Nature? Is it local or non-local? Do hidden variables underlie the indeterminacy? Does experiment support one or the other view?

For many years most physicists believed that physics could not answer such questions. Einstein's concerns were dismissed by Wolfgang Pauli when he wrote to Max Born (one founder of quantum mechanics to another), "One should no more rack one's brain about the problem of whether something one cannot know anything about exists all the same, than about the ancient question of how many angels are able to sit on the point of a needle."

Then, in 1964, the Irish physicist John S. Bell showed that, at least in principle, there were experiments that could tell whether Nature is realistic. He showed that predictions made by any hidden-variable theory satisfy a certain set of inequalities, while predictions made by quantum theory do not. It took twenty years before actual experiments could be performed, but since then experiments of ever better quality have shown unambiguously that Nature violates Bell's inequalities. The creepy indeterminacy of quantum mechanics is correct.

20.4   BELL'S INEQUALITY

There is nothing mysterious about Bell's inequalities. Bell realized that if you have a set of objects possessing (or not possessing) three different properties A, B, and C, and if you group the objects according to their distinct, well defined properties, the numbers of the objects in each group *must* satisfy certain inequalities. The following example shows you what Bell had in mind.

Take a handful of coins from your pocket and spread them out on a table (get together with a friend if you are short of money). Sort the coins into groups according to the properties given in the following table.[5] The table lists both the property, e. g., for these coins A means copper colored, and its negation $\overline{A}$ (read as "not A") means silver colored.

state	meaning	inverse state	meaning
A	copper (penny)	$\overline{A}$	silver (quarter, dime, nickel)
B	date before 1995	$\overline{B}$	date 1995 or later
C	heads up on table	$\overline{C}$	heads down on table

---

[5]In case you are not familiar with U.S. currency: quarters, dimes and nickels are worth 25, 10 and 5 cents, respectively, and they are all made of silver-colored metal alloys. Pennies are worth 1 cent and are made of a copper-colored metal alloy. Other currencies may have more interesting options.

Suppose that your handful of coins lays out as follows:

Coin	Year	Orientation	State
penny	1993	heads up	$A\ B\ C$
nickel	1989	heads up	$\overline{A}\ B\ C$
quarter	2001	heads up	$\overline{A}\ \overline{B}\ C$
dime	1993	heads up	$\overline{A}\ B\ C$
nickel	1995	heads up	$\overline{A}\ \overline{B}\ C$
penny	1997	tails up	$A\ \overline{B}\ \overline{C}$
quarter	2000	tails up	$\overline{A}\ \overline{B}\ \overline{C}$
penny	1984	tails up	$A\ B\ \overline{C}$
quarter	1991	tails up	$\overline{A}\ B\ \overline{C}$
nickel	1985	heads up	$\overline{A}\ B\ C$
penny	1975	tails up	$A\ B\ \overline{C}$
penny	1985	heads up	$A\ B\ C$
quarter	2002	tails up	$\overline{A}\ B\ \overline{C}$
nickel	2004	tails up	$\overline{A}\ \overline{B}\ \overline{C}$

Find the number of coins with property A and property $\overline{B}$; call this number $N(A,\overline{B})$. There is only one such coin, the 1997 penny, so $N(A,\overline{B}) = 1$. There are three coins older than 1995 with their heads up, so $N(B,\overline{C}) = 3$. Can you see that $N(A,\overline{C}) = 3$?

Bell pointed out that such data will *always* satisfy the inequality

$$N(A,\overline{B}) + N(B,\overline{C}) \geq N(A,\overline{C}). \qquad \text{a Bell inequality} \qquad (1)$$

This is certainly true for the present example: $3 + 1 \geq 3$.

The inequality is general. It applies no matter how you label the properties. (See Problem 5.) The diagram in Fig. 20.7 shows the general validity of Eq. 1; by inspection you can see that the combined areas of $N(A,\overline{B})$ and $N(B,\overline{C})$ will always be greater than or equal to $N(A,\overline{C})$. A more formal proof of the general validity of Eq. 1 is given in the appendix at the end of this chapter (p. 621).

The proof that the inequality in Eq. 1 is always true requires only one assumption—that an object possesses one property or the other, but never some strange superposition of them. In Einstein's world real objects possess each property as unique and distinct, i.e., the properties A and $\overline{A}$ are distinct and mutually exclusive. The same is true for B and $\overline{B}$, and for C, and $\overline{C}$. If you can't believe that a coin can exist as a superposition

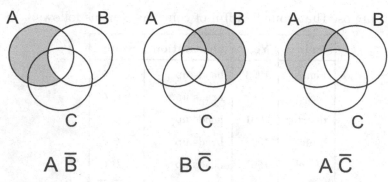

FIGURE 20.7   Venn diagrams of a Bell inequality. The shaded part of the left-hand diagram shows everything included in A but not included in B; the shaded part of the middle diagram is everything included in B but not in C; the shaded part of the right-hand diagram is everything included in A but not in C.

of being a penny and a nickel, how can you believe that an atom or a photon can exist as a superposition of two (or more) states? One answer: practice, practice, practice ... until you do.

20.5   VIOLATING BELL'S INEQUALITY

Here is a possible experiment to test if reality is non-local. Prepare entangled photons in one or another of three different polarizations, A, B, and C, chosen because quantum theory predicts they will result in the violation of Bell's inequality in Eq. 1. Measure these particular photon states, and observe that, as quantum theory predicts, your results violate Bell's inequality. Quantum theory is correct.

Beam-Splitting Polarizer

For these experiments you need "two-beam polarizers." Unlike sheet polarizers, which absorb one polarization state and transmit the other, two-beam polarizers transmit both polarization states, but as separated beams. If you send a beam of unpolarized light into a two-beam polarizer, the photons will emerge in two separated beams. The photons in one beam will be polarized at right angles to those in the other.

A two-beam polarizer has a transmission axis. When the photons incident on a two-beam polarizer are all polarized parallel to this transmission axis, only one beam comes out. If now you rotate the transmission axis (by rotating the polarizer), a second beam will appear. When the transmission axis reaches 45°, the two beams will be of equal

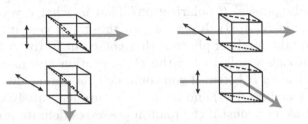

Transmission axis vertical   Transmission axis horizontal

FIGURE 20.8 Polarizing beam splitters are different than sheet polarizers. They transmit and reflect instead of transmit and absorb. The transmission axis is in the plane perpendicular to the plane of the splitting surface.

intensity. When the transmission axis reaches 90°, the first beam will have disappeared and all the photons will be in the second (reflected) beam.

Because a two-beam polarizer has one beam of input and two beams of output, it acts like the beam splitters used in interferometers. For this reason, it is often called a "beam-splitting polarizer," or a "polarizing beam splitter." Photons polarized perpendicular to the beam splitter's reflection plane are reflected; photons polarized parallel to its reflection plane are transmitted. Thus, the polarizing beam splitter has its transmission axis in the plane perpendicular to the surface that splits the light. This is illustrated in Fig. 20.8 for splitting angles of 90°.

Measuring the Photon's State of Polarization

When photons are incident on a two-beam polarizer at some angle θ to its transmission axis, the photons that would have been absorbed by a sheet polarizer are now transmitted in a second beam. If a photon linearly polarized at an angle θ to the transmission axis enters the polarizer, the magnitude of its probability amplitude to emerge polarized parallel to the transmission axis is, as before, $\cos\theta$. Its probability amplitude to emerge in a polarization state perpendicular to the transmission axis is $\sin\theta$. (For a photon entering a sheet polarizer, $\sin\theta$ was the probability amplitude that it would be absorbed.)

Now imagine you set up a polarizer in a photon's path in one of the three orientations—A, B, or C. When the polarizer is in the A orientation, some photons will come through in one beam—call it the A beam—and some will come through in the other beam—call it the $\overline{A}$ beam. A similar description applies to the B and C orientations and their outgoing beams.

What does a local realist say when a photon comes out in the A beam? The local realist says the photon possessed A polarization when it entered; that's why it came out in the A beam. If it came out in the $\overline{A}$ beam, that's because it did not possess any A polarization when it entered.

What if you want to know how many of the photons possessing A polarization also possess $\overline{B}$ polarization? That is, what if you want to measure the value of $N(A, \overline{B})$. It's wrong to think you can answer this question by taking the $N(A)$ of photons that come out in the A beam and passing them through a polarizer in the B orientation and measuring how many come out in its $\overline{B}$ beam. If you could do this, you could measure $N(A, \overline{B})$, $N(B, \overline{C})$, and $N(A, \overline{C})$ and see if these numbers satisfy the Bell inequality in Eq. 1 as they must if the photon possesses definite polarizations before you measure them.

The major obstacle to this approach is that passage of the photon through the A polarizer changes the photon's state. Although all the photons coming out in the A beam are A polarized, you have no idea what passage through the A polarizer did to the B or $\overline{B}$ polarizations of the photon. You have no right to assume after the first measurement that the photon possesses all the properties that it had before you made your measurement. You have no reason to expect your second measurement with the B polarizer to give you the value of $\overline{B}$ polarization that the photon would have had if it had gone through B first. The photon coming out of your measurement apparatus is quite different from the one that went in; you can not measure the same photon twice.

Entanglement offers a way around this difficulty. You can arrange your down-converter so that each idler and signal photon pair emerge with their polarizations entangled. Then the two photons have correlated polarizations. For example, your apparatus can be adjusted so that an idler photon and the accompanying signal photon are always polarized parallel to each other.[6] Then if you measure the polarization of an idler photon and find it in the A state, you know without making a measurement on it that the signal photon is also in the A state.

Notice that this arrangement does not mean that the outgoing photon beam is polarized. If you measure the polarizations of a succession of idler photons, you will find they vary randomly; the beam of idler photons is unpolarized. Nevertheless, the signal photon's polarization is always parallel to whatever polarization the idler photon has. To check that this is true, you could set up polarizing beam splitters to detect both the idler and the signal photon. If the polarizing beam splitters are oriented with both transmission axes horizontal, you will observe, as shown in Fig. 20.9, that when one photon is transmitted through one polarizer, its partner is *always* transmitted by the other polarizer, and when one photon is reflected the other one is *always* reflected.

---

[6] You can have other correlations; for instance, the down-converter can be set to produce idler and the signal photons that are always polarized perpendicular to each other.

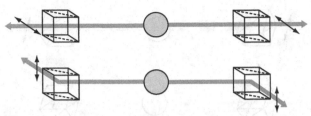

FIGURE 20.9 Possible outcomes when photons with entangled parallel polarization are incident on polarizing beam splitters that are both set to reflect the light in the horizontal plane.

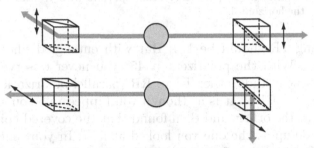

FIGURE 20.10 Diagram of the behavior of photons with entangled parallel polarization incident on polarizing beam splitters with their transmission axes perpendicular to each other.

Or you could orient a polarizer horizontally in front of one beam and another polarizer vertically polarized in front of the other beam. As shown in Fig. 20.10, with the polarizers perpendicular to each other, when one photon is transmitted the other is reflected. Figures 20.9 and 20.10 illustrate that the polarizations of these entangled photons are always parallel. Note that the polarizers will give the same results as in Fig. 20.9 if they are oriented with their transmission axes at 45° to horizontal (shown in Fig. 20.11) or *any* other angle, as long as they are parallel to each other.

Perhaps this sort of correlation does not bother you. After all, a measurement of one photon tells you about the other. If you pick up a left shoe out of a pair, you know for sure the other one is a right shoe. A measurement of a property of one shoe tells you about a property of the other.

And yet, when you think harder, you can see this is strange. These photons are somehow connected. Suppose they were not. Suppose they were simply independent parallel photons with each photon in a definite state of polarization—say horizontal—when it arrived at a polarizer. Then if each photon is incident on a polarizer at 45° to vertical, each will have an equal chance of being transmitted or reflected, and there will be four possible outcomes: TT, RR, TR, and RT—both transmitted, both reflected, left transmitted and right reflected, and *vice versa*. The probability for

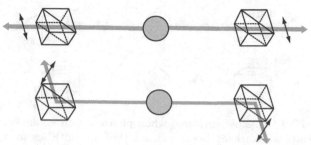

FIGURE 20.11 Diagram of the possible behavior of photons with entangled parallel polarization incident on polarizing beam splitters with transmission axes oriented at 45° from the horizontal.

observing TR should be 1/4. But with entangled photons this is never the case. With the polarizers at 45° you never observe TR or RT. Any measurement finds either TT or RR (parallel polarizations)—each with a probability of 1/2. It is as though you flipped two coins, covered one up, looked at the other, and then found that the covered coin always had the same side up as the one you looked at first. In your entanglement apparatus the sequence of polarization states of a single photon is random like the flip of a coin, but when you compare measurements of the polarization state of the first photon with measurements of the polarization state of the second, you find they are correlated.

Stranger still, suppose you put each analyzer (polarizer and detector) far away from the other and have each operated by someone who randomly changes the polarizer's orientation, even after the photons are already in flight. Suppose that the detectors are so far apart that there is not time enough for a signal to reach one detector from the other while the photons are in flight. Then neither polarizer can "know" the setting of the other. Nevertheless, if the operators get together afterwards and compare notes, they will find that whenever the polarizers happened to have the same orientation, either both photons were transmitted or both were reflected; and whenever the polarizers happened to be oriented perpendicularly, one photon was transmitted and the other photon was reflected (see Problem 11).

A local realist might argue that each photon has a definite polarization and carries with it a set of instructions for what value it should take for any polarizer orientation. As long as the photons carry the same instruction sets, you can explain the behavior of entangled photons. This argument attributes "hidden variables" to the photon. The photon has a definite polarization, and the results of measurement are not actually random. "But," says the local realist, "the instruction set is hidden and can only be known by a theory that is more complete than quantum mechanics." The advocate of quantum theory replies, "Alright, if each photon has a set of definite properties and instructions that tell how these properties are to show up in a polarizer at any orientation, then the photons must satisfy Bell's inequalities. Let's make some measurements and see if they do."

20.6 TESTING BELL'S INEQUALITY: THEORY AND EXPERIMENT

What does quantum mechanics predict? Notice that the ratio of $N(A, \overline{B})$ to N_T, the total number of photons passing through the apparatus, will be the probability $P(A, \overline{B})$ that a photon possesses the properties A and $\overline{B}$. If $N(A, \overline{B})$, etc., obey the Bell inequality in Eq. 1, so will the probabilities, i.e.,

$$P(A, \overline{B}) + P(B, \overline{C}) \geq P(A, \overline{C}). \tag{2}$$

Equation 2 is equivalent to Eq. 1.

Quantum mechanics predicts definite values of $P(A, \overline{B})$, $P(B, \overline{C})$, and $P(A, \overline{C})$. Suppose the transmission axes of the idler and signal polarizers differ by an angle θ. For example, the left one might be vertical, and the right one might be at an angle θ from the vertical. The unpolarized idler photon has a 50% chance of passing through the vertical polarizer. But, once it does, it declares itself to be a vertically-polarized photon, so its partner *must* also be vertically polarized. The angle between this vertically-polarized signal photon and the axis of the second polarizer is θ, so the probability the signal photon will be transmitted is $\cos^2 \theta$. The probability that it is reflected (not transmitted) is $1 - \cos^2 \theta = \sin^2 \theta$. Then the overall probability that the idler photon will be transmitted through the vertical polarizer, and, therefore, that the signal photon is in the A state, and that the signal photon will not be transmitted but will be reflected (i.e., be in the $\overline{B}$ state) through the B polarizer oriented to θ is from the product rule $(1/2) \sin^2 \theta$.

Now look at the probabilities for three specific cases. Let outcome A be the transmission of vertically polarized photons; let B be the transmission of photons by a polarizer at $30°$ to vertical; and let C be the transmission of photons by a polarizer at $30°$ to B, or $60°$ to the vertical, as shown in Fig. 20.12.

The joint probabilities are:

$$P(A, \overline{B}) = \frac{1}{2} \sin^2 30° = \frac{1}{8}$$

$$P(B, \overline{C}) = \frac{1}{2} \sin^2 30° = \frac{1}{8}$$

$$P(A, \overline{C}) = \frac{1}{2} \sin^2 60° = \frac{3}{8} \tag{3}$$

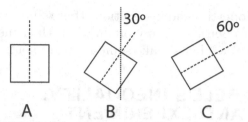

FIGURE 20.12   Three orientations of transmission axes of polarizers used to exhibit violation of a Bell inequality. The dashed lines represent the transmission axes of the polarizers.

This quantum mechanics prediction violates the Bell inequality Eq. 1. From the results in Eq. 3

$$P(A, \overline{B}) + P(B, \overline{C}) = 2/8 \ngeq P(A, \overline{C}) = 3/8.$$

How might you measure $P(A, \overline{B})$? You could direct the idler photons into the A polarizer set to transmit vertically polarized photons and send the signal photons into the B polarizer set to reflect $\overline{B}$ photons. Then you would count the coincidences between idler photons in the A state and signal photons shown by the B polarizer to be in the $\overline{B}$ state. Because the idler photon was found to be in the A state you know that the photon entering the B polarizer was also in the A state. Photons reflected from the B beam splitter must be also be in the $\overline{B}$ state. The number of counts of coincidences of idler and signal photons is $N(A, \overline{B})$.

What do experiments show? Since 1982 ever better experiments have shown complete agreement with the predictions of quantum mechanics. These experiments show that Bell's inequalities are violated. Consequently, polarization-entangled photons have no instruction sets; there are no hidden variables. When the polarization is measured, the results confirm that quantum mechanics is right. The world is non-local.

At the 1927 Solvay conference, Einstein repeatedly confronted Bohr with arguments to show that quantum mechanics must be wrong. The universe must be deterministic, he believed, because "God does not play dice." "Einstein," said Bohr after several days of refuting subtle and penetrating arguments, " stop telling God how to run the world."[7]

---

[7]This is a quotation in the tradition of Thucydides, i.e. largely made up. There are many versions of it. For an elaborate version see Richard Rhodes, *The Making of the Atomic Bomb* (Simon and Schuster, New York, 1986), pp. 133.

Appendix: Formal Proof of a Bell Inequality

For three properties A, B, and C, there are only eight possible outcomes:

$$N(A,B,C), N(\overline{A},B,C), N(A,\overline{B},C), N(\overline{A},\overline{B},C),$$
$$N(A,B,\overline{C}), N(\overline{A},B,\overline{C}), N(A,\overline{B},\overline{C}), N(\overline{A},\overline{B},\overline{C}).$$

Notice that

$$N(A,\overline{B}) = N(A,\overline{B},C) + N(A,\overline{B},\overline{C}). \tag{4}$$

Similarly,

$$N(B,\overline{C}) = N(A,B,\overline{C}) + N(\overline{A},B,\overline{C}). \tag{5}$$

Add Eqs. 4 and 5 to get:

$$N(A,\overline{B}) + N(B,\overline{C}) = N(A,\overline{B},C) + N(A,\overline{B},\overline{C}) + N(A,B,\overline{C}) + N(\overline{A},B,\overline{C}). \tag{6}$$

Note that on the right-hand side

$$N(A,B,\overline{C}) + N(A,\overline{B},\overline{C}) = N(A,\overline{C}). \tag{7}$$

Therefore,

$$N(A,\overline{B}) + N(B,\overline{C}) = N(A,\overline{C}) + N(A,\overline{B},C) + N(\overline{A},B,\overline{C}), \tag{8}$$

from which it follows that

$$N(A,\overline{B}) + N(B,\overline{C}) \geq N(A,\overline{C}). \qquad \text{A Bell inequality (p. 613)}$$

PROBLEMS

1. A vertically polarized photon is incident on a pair of polarizers. The first one, $\mathcal{P}_1$, can be rotated to have its transmission axis oriented at any angle θ with the vertical. The second one, $\mathcal{P}_2$, has its transmission axis oriented horizontally.

    a. How will the intensity of the beam of photons emerging from $\mathcal{P}_2$ vary as a function of θ as $\mathcal{P}_1$ is rotated?

    b. At what angles of $\mathcal{P}_1$ will the transmission through $\mathcal{P}_2$ be a maximum?

    c. How will your answers change if the incident beam of photons is unpolarized?

2. A vertically polarized photon is incident onto a polarizer with a transmission axis oriented an angle of $80°$ counter-clockwise from the horizontal.

a. What is the probability amplitude of the photon to go through the polarizer?

b. Now after the first polarizer put a second polarizer that has its transmission axis oriented 10° clockwise from the transmission axis of the previous polarizer.

    i. What is the probability that an incident photon is transmitted through the two polarizers?

    ii. If you add a third polarizer after the second one, what should be the orientation of its transmission axis so that no photon is transmitted.

3. In doing a new quantum eraser experiment we rotate the polarization of the light going through one of the arms of the Mach-Zehnder interferometer using three polarizers. The photons that go into the interferometer are vertically polarized. See Fig. 20.13.

a. In the polarization-rotating arm the transmission axis (TA) of the first polarizer is oriented 30° with the vertical, the TA of the second polarizer is oriented 60° relative to the vertical, and the TA of the third polarizer is oriented horizontally. What is the probability that the photon will go through all three polarizers?

b. In the other arm of the interferometer we want to keep the orientation of the polarization vertical but we want to provide a probability amplitude for transmission through the two polarizers that is the same as the one in the other arm. Explain how we can do this with two polarizers, indicating the angle that each polarizer makes with the vertical. Make a diagram.

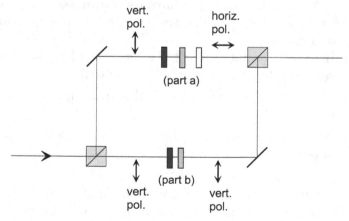

FIGURE 20.13 Diagram of a Mach-Zehnder interferometer with polarizers for Problem 3.

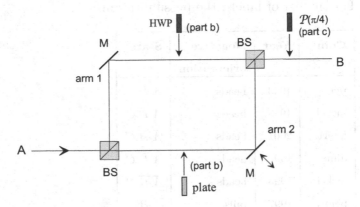

FIGURE 20.14   Diagram of a Mach-Zehnder interferometer for Problem 4.

4.   A Mach Zehnder interferometer has beam-splitters that have an uneven ratio of reflection to transmission. The reflection probability is $1/3$ and the transmission probability is $2/3$. The 900-nm incident light is vertically polarized (see Fig. 20.14).

    a. Find the probability amplitude for a photon to go from A to B via arm 1.

    b. We now put a "half wave plate" (HWP) in arm 1 that rotates the polarization by 90°. Arm 2 has compensating plate that creates the same phase shift but does not rotate the polarization. Both components transmit all of the light that reaches them. Find the probability that the photon reaches B when the lengths of the two arms are the same.

    c. We now put a polarizer after the interferometer oriented at 45° with the horizontal. Find the probability for the photon going from A to B when the lengths of the two arms are the same.

    d. If we increase the length of one of the arms by 1350 nm, what is the probability of going from A to B?

5.   Bell's inequalities are independent of how you label the categories. For the example on p. 613 you could just as well have labeled the categories as

State	Meaning	Negative state	Meaning
A	Date before 1995	$\overline{A}$	Date 1995 or later
B	copper	$\overline{B}$	silver
C	Heads up on table	$\overline{C}$	Heads down on table

For this set of labels, the possibilities are

Coin	Year	Negative orientation	State
penny	1993	heads	$A\ B\ C$
nickel	1989	heads	$A\ \overline{B}\ C$
quarter	2001	heads	$\overline{A}\ \overline{B}\ C$
dime	1993	heads	$A\ \overline{B}\ C$
nickel	1995	heads	$\overline{A}\ \overline{B}\ C$
penny	1997	tails	$\overline{A}\ B\ \overline{C}$
quarter	2000	tails	$\overline{A}\ B\ \overline{C}$
penny	1984	tails	$A\ B\ \overline{C}$
quarter	1991	tails	$A\ \overline{B}\ \overline{C}$
nickel	1985	heads	$A\ \overline{B}\ C$
penny	1975	tails	$A\ B\ \overline{C}$
penny	1985	heads	$A\ B\ C$
quarter	2002	tails	$\overline{A}\ B\ \overline{C}$
nickel	2004	tails	$\overline{A}\ B\ \overline{C}$

Show that when the items are categorized as above, the values of N satisfy the inequality: $N(A, \overline{B}) + N(B, \overline{C}) \geq N(A, \overline{C})$.

6. A caterer makes up a large order of sandwiches on rye bread or white, with mustard or without, with ham or pastrami.

    a. Why do you know that the number of rye bread sandwiches without mustard plus the number of pastrami sandwiches with mustard is always greater than or equal to the number of pastrami sandwiches on rye?

    b. Is the above result true if the caterer makes 12 ham-on-white with mustard, 4 ham-on-white without mustard, 6 ham-on-rye with mustard, 7 pastrami-on-rye with mustard, 2 pastrami-on-rye without mustard? Show that these numbers are consistent with Eq. 1.

    c. What if you invert the order to 12 pastrami-on-rye without mustard, 4 pastrami-on-rye with mustard, 6 pastrami-on-white without mustard, 7 ham-on-white without mustard, and 2 ham-on-white with mustard? How do you know this order of sandwiches will satisfy Eq. 1?

7. Is it really true that the number of male A-students plus the number of non-A varsity athletes is greater than or equal to the number of male varsity athletes?

8. For what circumstances will Eq. 1 be an equality? Use Fig. 20.7 to explain your answer.

9. On checking Bell's inequality Eq. 1 with mutually parallel polarization-entangled photons,

    a. Is it violated using polarizers A vertical, B at 22.5°, and C at 22.5° from B and 45° from A?

    b. Is it violated for A vertical, B at 50°, and C at 50° from B and 100° from A?

    c. If A is vertical, B forms an angle θ with A, and C forms an angle θ with B and 2θ with A, find the values of θ for which the inequality is violated. The inequalities are *not always* violated by an indeterministic quantum description, but they are *never* violated by a deterministic description.

10. Consider the experiment of Fig. 20.15.

We use a different type of entangled state, one where the photons are mutually perpendicular. That is, regardless of where a first polarizer is oriented, once a first photon is detected with a given polarization the other is found to have a polarization perpendicular to the first one. We send the mutually perpendicular polarization-entangled photons to a pair of polarizing beam splitters. The left polarizer is vertical and the right polarizer forms an angle θ with the vertical.

    a. Show that the probability that the left photon gets transmitted and the right photon gets reflected is

$$P_{V,\theta} = \frac{1}{2} \cos^2 \theta. \tag{9}$$

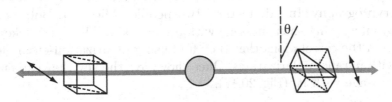

FIGURE 20.15 Figure for Problem 10.

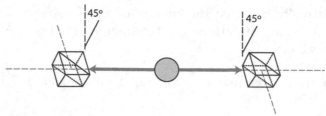

FIGURE 20.16   Figure for Problem 11.

    b. What is the probability that the right photon gets transmitted and the left photon gets reflected?

    c. What is the probability that the both photons get transmitted?

11. Consider the arrangement of Fig. 20.16, where both polarizers are oriented at an angle of 45°.

    a. If we use a source of mutually parallel entangled photons,

        i. What is the probability that both photons get transmitted?

        ii. What is the probability that the left photon gets transmitted and that the right photon gets reflected?

    b. If both photons are vertically polarized,

        i. What is the probability that both photons get transmitted?

        ii. What is the probability that the left photon gets transmitted and that the right photon gets reflected?

    c. If half the time the two photons are both horizontally polarized and half of the time the photons are vertically polarized.

        i. What is the probability that both photons get transmitted?

        ii. What is the probability that the left photon gets transmitted and that the right photon gets reflected?

    d. Can experiments distinguish between cases (a), (b), and (c)?

12. It is possible to communicate securely using quantum mechanics. This method is now commercially available, and it is called quantum cryptography. In this method two people, Alice and Bob, communicate by encrypting their message with a secret key. Their secret key is derived from the correlations that they find using polarization-entangled photons with parallel polarizations. Once they have the key they can encrypt their message with it. (Fig. 20.17).

    Consider the scheme of Fig. 20.17. Alice has a source of mutually parallel polarization-entangled photons. She sends one photon through a polarizer and a detector. She sends the other photon to Bob, who is far away. Bob also has a polarizer and a detector.

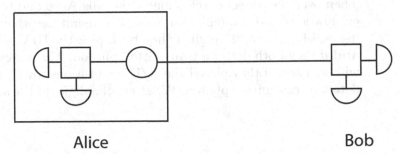

FIGURE 20.17   Figure for Problem 12.

a. If Alice detects a photon when her polarizer is horizontal (H), what is the probability that Bob will detect the partner photon when his polarizer is oriented:

   i. horizontal (H)

   ii. vertical (V)

   iii. +45° to the horizontal (diagonal i.e., D)

   iv. −45° to the horizontal (antidiagonal i.e, A)

b. Alice and Bob decide to detect eight photons pairs by randomly setting their beam-splitting polarizers in one of two possible transmission-axis orientations: vertical (V) or diagonal (D). The polarizer orientations and results that each got are given below where T and R are, respectively, a transmitted or a reflected photon:

Pair	Alice's polarizer	Alice's result	Bob's polarizer	Bob's result
1	V	R	V	T
2	V	T	V	T
3	V	T	D	T
4	D	T	D	R
5	D	R	D	R
6	V	T	V	T
7	D	T	D	R
8	D	T	V	R

Then over the unsecure telephone Bob calls Alice and tells her his orientations and she tells him hers. The secret key that they will use will be have a "1" digit if they both pick the H/V orientations **and** if they both detect a transmitted photon. The secret key digit will be a zero if they picked the A/D orientations **and** if they both detect a transmitted photon. What are the digits of the secret key?

C H A P T E R **21.**

Epilogue

This book has introduced you to the physicist's special way of looking at and trying to understand nature. Out of the many ways to make such an introduction, we chose to present and develop important evidence, ideas, and reasoning that have led to our present-day conception of the atom. We chose this approach partly because we think it is interesting physics and partly because the idea of the atom is so important. It is the basis of all our modern technologies, from computers to gene manipulation, from pharmacology to agriculture, mining, manufacturing, transportation, communications, and management of the environment. The atom as it has been elaborated in the past two centuries is fundamental to physics, chemistry, biology, geology—to all the natural sciences. It is arguably one of the most important ideas in human history.

Despite all the details, calculations, exercises, and explanations offered to familiarize you with the atom, we have left an immense amount unsaid. But if all has gone as we hope, you now know enough to be able to explore the richness of the atomic idea further on your own.

For example, there is within the atom more and deeper structure than we have begun to describe. There are other fields than the familiar electric and magnetic fields. There are particles within particles. Physicists have studied the protons and neutrons that lie within the atomic nucleus and found that they, too, have parts, which have been given the unlikely names of quarks and gluons. There is a successful theory of these entities called quantum chromodynamics. Some aspects of this realm of matter are nicely described in the closing chapters of Sheldon Glashow's *From Alchemy to Quarks*, Brooks/Cole Publishing Co., 1993. Steven Weinberg connects this deeper structure to cosmology and the structure of the universe in *The First Three Minutes*, Basic Books, 1988. Some of what physicists are

C.H. Holbrow et al., *Modern Introductory Physics, Second Edition,*
DOI 10.1007/978-0-387-79080-0_21, © Springer Science+Business Media, LLC 1999, 2010

thinking about the very large as well as the very small can be found in *The New Physics*, edited by Paul Davies, Cambridge University Press, 1989.

But even without going deeper into the atom, there is much more to be said. For example, atoms are tiny magnets. Their magnetic properties have remarkable consequences. Most of the knowledge of the world is now stored on magnetic tapes and disks. These would not exist without our understanding of the magnetic behavior of atoms. Great advances in medical imaging have followed from our understanding of the magnetism of atoms and their nuclei. It is from our understanding of the magnetism of atoms that we understand such wonderful objects as pulsars. To learn more, read James D. Livingston's *Driving Force: The Natural Magic of Magnets*, Harvard University Press, Cambridge, MA, 1996.

Our understanding of atoms is being advanced by a revolution in experimental control and manipulation. We can now hold a single atom in a trap made of electromagnetic fields and light waves and then use lasers to prod and probe it with extreme precision. We can map the interaction of individual atoms from moment to moment as they combine chemically. We can watch quantum jumps in a single atom. Using the technology of trapping we can create a large-scale, directly observable quantum state called Bose–Einstein condensation. Quantum interference between beams of atoms has been observed, and we can make a laser of atoms (instead of light).

One of the most striking advances in single-atom manipulation is a device called the "scanning tunneling microscope," or STM. It can be used to image and manipulate single atoms. When a sharp tungsten point, as sharp as a single atom at its tip, is brought near atoms sitting on a surface, an electric current flows. The quantum properties of the flow of electrons restrict the current to such a small region of space that as the tip is moved across the surface, the variation in the current can outline the presence of single atoms. Figure 21.1 shows a pair of STM scans. The first shows two conical mounds; each is a molecule of O_2 sitting on a flat surface of platinum atoms. Before the second scan was made, the tip of the STM was brought down to within $0.6\,\text{nm}$ of one of the mounds and a small voltage pulse was applied. The second scan shows that the effect of the voltage pulse was to divide the mound into two smaller mounds—single oxygen atoms. The STM has revealed the dissociation of a single molecule into its constituent atoms. Other remarkable examples of this kind of manipulation are shown at http://www.almaden.ibm.com/vis/stm/gallery.html, IBM's gallery of STM images.

The strange mysteries and ambiguities of the quantum nature of the atom promise further remarkable changes in our technology and society. You have seen the evolution of our picture of the atom from the tiny, hard, featureless ball that explains the gas laws to a complicated assembly

FIGURE 21.1 Upper left: Two O_2 molecules are revealed by the scan of an STM; lower right: Scan of the same two molecules after a voltage pulse has been delivered to one. The scan shows that the O_2 molecule has been separated into two O atoms. *Picture courtesy of Wilson Ho, Department of Physics and Astronomy, University of California, Irvine.*

of electrons and a nucleus made up of protons and neutrons. You have seen that as we learned more about the atom, its inner parts got fuzzy. The particle-like behavior of light and the wavelike behavior of particles blurred the insides of the atom. Bohr's model was correct in its idea of well defined internal states of energy that can be represented by a level diagram, but it was wrong in its simple planetary images. There are no well defined orbits. In Tom Stoppard's play *Hapgood* the physicist Kerner says

> So now make a fist, and if your fist is as big as the nucleus of one atom then the atom is as big as St. Paul's, and if it happens to be a hydrogen atom then it has a single electron flitting about like a moth in the empty cathedral, now by the dome, now by

the altar.... Every atom is a cathedral. I cannot stand the pictures of atoms they put in schoolbooks, like a little solar system: Bohr's atom. Forget it. . . . an electron does not go round like a planet, it is like a moth which was there a moment ago, it gains or loses a quantum of energy and it jumps, and at the moment of quantum jump it is like *two* moths, one to be here and one to stop being there; an electron is like twins, each one unique, a unique twin.

The atom, which began as a hard, featureless ball, is now a moth-filled cathedral that can be depicted as a level diagram.

The fuzziness, the flittering uncertainty, the property of being in more than one state at the same time, are all integral parts of our contemporary understanding of the atom, and they may be the bases of some surprising practical uses. Objects that can be in several different states at the same time may make possible quantum cryptography with unbreakable codes that will warn their users when someone tries to listen in. There is a prospect of designing computers that use these multiple-state systems to achieve massively parallel computation with speeds and capabilities that are impossible in principle with the kinds of computers we now use. Such possibilities are described in *Schrödinger's Machines: The Quantum Technology Reshaping Everyday Life* by Gerard J. Milburn, W. H. Freeman and Co., New York, 1997.

With traps and cooling and scanning probe microscopes, with ever more refined lasers, with a deeper appreciation that quantum mechanics means what it says, our understanding of the atom improves day by day. This understanding has already had extraordinary consequences for human society; the future promises unimaginably more. We hope that our book has helped to prepare you to understand that future.

Of course, there is more to physics than atoms. One of the powerful attractions of physics for physicists is its universal applicability. Whether you are studying the collision of quarks or galaxies, the flow of electrons or the sliding of sand piles, a plasma in a star or a vortex in superfluid helium, the bending of beams in a building or the folding of proteins, the laws of physics apply. We invite you to participate further in the exciting enterprise of revealing and savoring this universality. We invite you to study more physics.

A P P E N D I X

Useful Information

Just as you need to know your name and address and telephone number and e-mail address to locate yourself in the world, so must you know some basic information to locate yourself in physics. Like competent professionals in any field, a practicing physicist carries a large amount of factual baggage. Starting out in physics you will need only the small backpack of facts presented in Tables A.1 and A.3.

Then there is information that you need occasionally. Some of that is collected here for your convenience. If you don't find what you want here or in the text, try the library. Ask a reference librarian to help you find what you want to know, or look in the *Handbook of Chemistry and Physics*. You can also use the World Wide Web to find constants:

> http://physics.nist.gov/cuu/Constants/index.html

will supply you the very latest, most precise values from NIST (National Institute of Standards and Technology).

A.1 SI PREFIXES

You need to know the SI prefixes. They tell you the order of magnitude of the units of whatever physical quantity they are attached to. It is absolutely essential that you know them.

They are widely used, and when you are wrong about them, you make mistakes of factors of thousands! Maybe you can absorb them by osmosis as you use them; maybe you need to get them by heart by the purest of rote learning; maybe you can come up with a clever mnemonic;

TABLE A.1   SI prefixes

Factor	Prefix	Symbol	Factor	Prefix	Symbol
10^{18}	exa	E	10^{-1}	deci	d
10^{15}	peta	P	10^{-2}	centi	c
10^{12}	tera	T	10^{-3}	milli	m
10^{9}	giga	G	10^{-6}	micro	μ
10^{6}	mega	M	10^{-9}	nano	n
10^{3}	kilo	k	10^{-12}	pico	p
10^{2}	hecto	h	10^{-15}	femto	f
10^{1}	deka	da	10^{-18}	atto	a

maybe (and this would be best) you can learn them attached to particular physical situations and quantities, as suggested in Chap. 2. However you do it, learn them! They are listed in Table A.1.

A.2   BASIC PHYSICAL CONSTANTS

You need to know some basic physical constants. These set the scale of the phenomena of the physical world.

Which ones are most important depends on the physical situation under consideration. In this book, with its emphasis on atoms and their parts, the elementary charge; the masses of the electron, the proton, and the neutron; and the values of the Planck and Boltzmann constants are very important. When you deal with macroscopic quantities of atoms in the laboratory, Avogadro's number and Earth's gravity are important. For convenient reference Table A.2 lists the official values of these constants in SI units. Table A.3 lists the ones you need to know in the units in which you need to know them.

A.3   CONSTANTS THAT YOU MUST KNOW

You need to be able to calculate quickly and easily with these constants. For this purpose, you need the constants expressed as much as possible in terms of units chosen to match the natural scale of atoms. Electron volts (eV) and nanometers (nm) are convenient for atoms, while megaelectron volts (MeV) and femtometers (fm) are a good choice for nuclei. It is also often simpler to work with masses in units of eV/c^2.

TABLE A.2   Basic physical constants

Name of constant	Symbol	Value
Atomic mass unit	m_u or u	1.661×10^{-27} kg
Avogadro constant	N_A	6.022×10^{23} mol^{-1}
Bohr radius	a_0	5.292×10^{-11} m
Boltzmann constant	k_B	1.381×10^{-23} J·K^{-1}
		8.617×10^{-5} eV·K^{-1}
Charge-to-mass ratio of electron	e/m	-1.759×10^{11} C·kg^{-1}
Coulomb constant	k_c or $\frac{1}{4\pi\epsilon_0}$	8.988×10^9 N·m^2·C^{-2}
Electron mass	m_e	9.109×10^{-31} kg
Elementary charge	e	1.602×10^{-19} C
Faraday constant	F	96485 C·mol^{-1}
Intensity of Earth's gravitational field	g	9.82 N·kg^{-1} (m·s^{-2})
Molar gas constant	R	8.314 J·mol^{-1}·K^{-1}
Neutron mass	m_n	1.675×10^{-27} kg
Planck constant	h	6.626×10^{-34} J·s
		4.136×10^{-15} eV·s
	$\hbar = \frac{h}{2\pi}$	1.055×10^{-34} J·s
		6.582×10^{-16} eV·s
Proton mass	m_p	1.673×10^{-27} kg
Rydberg constant	R_∞	1.09737×10^7 m^{-1}
Speed of light	c	2.99792458×10^8 m·s^{-1}

Table A.3 gives constants, combinations of constants, and masses in terms of these more convenient units. The combinations simplify calculations of energies, wavelengths, and frequencies that are frequently made in this course. The Remark column tells you when the constant is one that you absolutely need to know. No kidding!

These constants are of fundamental importance. One goal of this book is to show how the constants interrelate and how they specify the scale of observed effects and phenomena. They specify the scales and magnitudes

TABLE A.3   Constants in convenient energy units

Name	Symbol	Value	Remark
Planck constant	h	4.14×10^{-15} eV·s	
	hc	1240 eV·nm	know this one!
Reduced Planck constant: $\frac{h}{2\pi}$	$\hbar$	6.58×10^{-16} eV·s	
	$\hbar c$	197 eV·nm	know as ≈ 200 eV nm
	$\hbar c$	197 MeV·fm	know as ≈ 200 MeV fm
Coulomb force numerator	$k_c e^2$	1.44 eV·nm	know this
Thermal energy at $T = 300$ K	$k_B T$	0.0259 eV	remember as $\approx 1/40$ eV
Bohr radius	$a_0 = \frac{\hbar^2}{ke^2 m_e}$	0.0529 nm	
Fine structure constant	$\alpha = \frac{ke^2}{\hbar c}$	1/137.036	no units
Rydberg energy	hcR_∞	13.61 eV	know this one
Electron mass	$m_e c^2$	511 keV	know this
Proton mass	$m_p c^2$	938.3 MeV	know ≈ 938 MeV
Neutron mass	$m_n c^2$	939.6 MeV	know m_n is 1.29 MeV $> m_p$
Atomic mass unit	u	931.50 MeV/c^2	remember 1 u $\approx m_p$
Speed of light	c	3×10^8 m·s^{-1}	know this
Elementary charge	e	1.6×10^{-19} C	know this

of the quantities with which physicists have built a consistent and informative picture of the microphysical world and its connection to the macrophysical world where we live and do physics.

A.4   MISCELLANEOUS

Table A.4 contains some constants used occasionally in this course, including constants having to do with Earth, Moon, and Sun.

Table A.5 gives conversion factors between some especially common English units and their metric equivalents.

TABLE A.4 Miscellaneous occasionally used constants

Name	Symbol	Value	Units	Remarks
Earth's mass	$M_\oplus$	6×10^{24}	kg	10 moles of kilograms
Earth–Sun distance	R_{ES}	1.5×10^{11}	m	1 A.U.
Earth radius	$R_\oplus$	6.366×10^6	m	$2\pi R_\oplus = 40\,\text{Mm}$
Earth–Moon distance	R_{EM}	3.82×10^8	m	$60\,R_\oplus$
Moon's mass	$M_)$	$0.01234\,M_\oplus$		$M_\oplus/81$
Sun's mass	$M_\odot$	2×10^{30}	kg	$333\,000\,M_\oplus$
Viscosity of air	η	18.3	μPa·s	at 20°C
Speed of sound in air	v_s	343	$\text{m}\,\text{s}^{-1}$	at 20°C

TABLE A.5 Some conversion factors between english and metric units

English	English	metric
1 in		2.54 cm
1 ft	12 in	30.48 cm
1 mile	5280 ft	1609.3 m
3.28 ft		1 m
0.396 in		1 cm
1 mph	1.467 ft/s	0.447 m/s
0.621 mph	0.911 ft/s	1 km/hr
2.24 mph	3.28 ft/s	1 m/s
1 lb	16 oz	453.5 g
1 oz		28.3 g
2.205 lb		1 kg

A.5 NAMES OF SOME SI DERIVED UNITS

Table A.6 lists some names of composite SI units. There are also a number of non-SI units that are still in use because because they are deeply embedded in engineering practice or every day life (because many people are unwilling to change their habits of thought). Table A.6 lists some of these non-SI units along with their abbreviations and their SI equivalents. The entries in the table are in alphabetical order according to their abbreviations.

TABLE A.6   Commonly used units and abbreviations

quantity	Name	Abbrev.	SI units
current	ampere	A	A
length	Angstrom	Å	10^{-10} m
pressure	atmosphere	atm	101.3 kPa
area	barn	b	10^{-24} m^2
pressure	bar	bar	100 kPa
energy	calorie	cal	4.1858 J
electric charge	coulomb	C	A·s
viscosity	centipoise	cp	10^{-3} Pa·s
energy	electron volt	eV	1.602×10^{-19} J
magnetic field	gauss	G	10^{-4} T
frequency	hertz	Hz	s^{-1}
energy	joule	J	kg·m^2·s^{-2} = N·m
temperature	kelvin	K	K
mass	kilogram	kg	kg
volume	liter	L	10^{-3} m^3
length	meter	m	m
pressure	millimeters of mercury	mm Hg	133.32 Pa
volume	cubic meter	m^3	m^3
amount	mole	mol	mol
force	newton	N	kg·m·s^{-2}
electric field	newton per coulomb	N·C^{-1}	N·C^{-1}
pressure	pascal	Pa	N·m^{-2}
viscosity	pascal seconds	Pa·s	
angle	radian	rad	rad
time	second	s	s
magnetic field	tesla	T	kg·s^{-1}·C^{-1}
pressure	torr	torr	133.32 Pa
mass	atomic mass unit	u	1.6605×10^{-27} kg
electric potential	volt	V	J·C^{-1}
electric field	volts per meter	V·m^{-1}	N·C^{-1}
power	watt	W	J·s^{-1}
angle	degree	°	1.7453×10^{-2} rad

TABLE A.7 SI base units

Name	Symbol	Definition
meter	m	The meter is the length of path traveled by light in vacuum during a time interval of 1/299 792 458 of a second.
mass	kg	The kilogram is the unit of mass. It is equal to the mass of the international prototype of the kilogram. (The international prototype is a platinum–iridium cylinder kept at the BIPM in Sèvres (Paris) France.)
second	s	The second is the duration of 9 192 631 770 periods of the radiation corresponding to the transition between the two hyperfine levels of the ground state of the cesium-133 atom.
ampere	A	The ampere is that constant current that if maintained in two straight parallel conductors of infinite length, of negligible circular cross section, and placed 1 meter apart in vacuum, would produce between these conductors a force equal to 2×10^{-7} newton per meter of length.
kelvin	K	The kelvin is the unit of thermodynamic temperature. It is the fraction 1/273.16 of the thermodynamic temperature of the triple point of water. (The Celsius temperature scale is defined by the equation $t = T - T_0$, where T is the thermodynamic temperature in kelvins and $T_0 = 273.15\,\text{K}$.)
mole	mol	The mole is the amount of substance of a system that contains as many elementary entities as there are atoms in 0.012 kg of carbon-12.
candela	cd	The candela is the luminous intensity, in a given direction, of a source that emits monochromatic radiation of frequency 540×10^{12} hertz and that has a radiant intensity in that direction of 1/683 watt per steradian.

A.6 SI BASE UNITS

There are seven units that form the basis of the SI. In this book we use six of them. Table A.7, which gives their names, symbols, and definitions, is provided here just for your general information. You will find it more useful and informative to remember the looser definitions that are given in the chapters where they are introduced.

TABLE A.8   Some chemical atomic masses

Element	Symbol	Z	Mass (u)	Phase	Density $(g\ cm^{-3})$
hydrogen	H	1	1.00797	gas H_2	
helium	He	2	4.0026	gas He	
lithium	Li	3	6.939	solid	0.534
beryllium	Be	4	9.0122	solid Be	1.848
boron	B	5	10.811	crystalline B	2.34
carbon	C	6	12.01115	amorphous C	≈ 2.0
nitrogen	N	7	14.0067	gas N_2	
oxygen	O	8	15.9994	gas O_2	
fluorine	F	9	18.9984	gas F_2	
aluminum	Al	13	26.981538	solid	2.6989
silicon	Si	14	28.0855	solid	2.33
iron	Fe	26	55.844	solid	7.874
cobalt	Co	27	58.93320	solid	8.9
nickel	Ni	28	58.69	solid	8.902
copper	Cu	29	63.546	solid	8.96
zinc	Zn	30	65.40	solid	7.133
tantalum	Ta	73	180.9479	solid	16.6
silver	Ag	47	107.8681	solid	10.5
gold	Au	79	196.96654	solid	19.32
lead	Pb	82	207.2	solid	11.35
uranium	U	92	232.0289	solid	18.95

A.7   ATOMIC MASSES

Table A.8 lists some useful chemical atomic masses and densities of elements that are solids at room temperature. If you need to know the density of any gaseous element, you can calculate it. There is a periodic table on p. 643.

TABLE A.9   Masses of some Nuclides

Name of Nuclide	Symbol	Z	Nuclide Mass (u)	% Natural Abundance	Half-life
hydrogen	^{1}H	1	1.007825	99.985	
deuterium	^{2}H or D	1	2.01410	0.015	
tritium	^{3}H or T	1	3.016050		12.26 y
helium-3	^{3}He	2	3.016030	0.00013	
helium-4	^{4}He	2	4.002603	100.0	
lithium-6	^{6}Li	3	6.015125	7.42	
lithium-7	^{7}Li	3	7.016004	92.58	
beryllium-9	^{9}Be	4	9.012186	100.	
boron-10	^{10}B	5	10.012939	19.78	
boron-11	^{11}B	5	11.009305	80.22	
carbon-12	^{12}C	6	12.000000	98.89	
carbon-13	^{13}C	6	13.003354	1.11	
carbon-14	^{14}C	6	14.003242		5730 y
nitrogen-14	^{14}N	7	14.003074	99.63	
nitrogen-15	^{15}N	7	15.010599	0.37	
oxygen-16	^{16}O	8	15.994915	99.759	
oxygen-17	^{17}O	8	16.999133	0.037	
oxygen-18	^{18}O	8	17.999160	0.204	
fluorine-19	^{19}F	9	18.998405	100.0	

A.8   MASSES OF NUCLIDES

For determining how nuclei will behave, the difference between masses of atoms may be important. When this is the case, you need to know the individual atomic masses quite precisely. Table A.9 lists some of the more important elements and their nuclides and their masses.

A.9 PERIODIC TABLE OF THE CHEMICAL ELEMENTS

The periodic table of the elements is the basic map of the material world. Since Moseley's work made clear the structure of the table, chemists and physicists have used it as a guide for searching for new elements. The table succinctly shows which elements are likely to have analogous chemical properties, and it provides practical help for making and understanding new chemical compounds.

PERIODIC TABLE OF THE CHEMICAL ELEMENTS

1																	18
hydrogen 1 H 1.0079																	helium 2 He 4.0026
lithium 3 Li 6.941	beryllium 4 Be 9.012											boron 5 B 10.811	carbon 6 C 12.081	nitrogen 7 N 14.007	oxygen 8 O 15.999	fluorine 9 F 18.998	neon 10 Ne 20.180
sodium 11 Na 22.990	magnesium 12 Mg 24.305											aluminium 13 Al 26.982	silicon 14 Si 28.086	phosphorus 15 P 30.974	sulfur 16 S 32.065	chlorine 17 Cl 35.453	argon 18 Ar 39.948
potassium 19 K 39.098	calcium 20 Ca 40.078	scandium 21 Sc 44.956	titanium 22 Ti 47.867	vanadium 23 V 50.942	chromium 24 Cr 51.996	manganese 25 Mn 54.938	iron 26 Fe 55.845	cobalt 27 Co 58.933	nickel 28 Ni 58.693	copper 29 Cu 63.546	zinc 30 Zn 65.38	gallium 31 Ga 69.723	germanium 32 Ge 72.64	arsenic 33 As 74.921	selenium 34 Se 78.96	bromine 35 Br 79.904	krypton 36 Kr 83.798
rubidium 37 Rb 85.468	strontium 38 Sr 87.62	yttrium 39 Y 88.906	zirconium 40 Zr 91.224	niobium 41 Nb 92.906	molybdenum 42 Mo 95.96	technetium 43 Tc [98]	ruthenium 44 Ru 101.07	rhodium 45 Rh 102.906	palladium 46 Pd 106.42	silver 47 Ag 107.868	cadmium 48 Cd 112.411	indium 49 In 114.818	tin 50 Sn 118.710	antimony 51 Sb 121.760	tellurium 52 Te 127.60	iodine 53 I 126.904	xenon 54 Xe 131.293
cesium 55 Cs 132.905	barium 56 Ba 137.327	57–70 *	lutetium 71 Lu 174.967	hafnium 72 Hf 178.49	tantalum 73 Ta 180.948	tungsten 74 W 183.84	rhenium 75 Re 186.207	osmium 76 Os 190.23	iridium 77 Ir 192.217	platinum 78 Pt 195.08	gold 79 Au 196.967	mercury 80 Hg 200.59	thallium 81 Tl 204.383	lead 82 Pb 207.2	bismuth 83 Bi 208.980	polonium 84 Po [209]	astatine 85 At [210]	radon 86 Rn [222]
francium 87 Fr [223]	radium 88 Ra [226]	89–102 **	lawrencium 103 Lr [262]	rutherfordium 104 Rf [261]	dubnium 105 Db [262]	seaborgium 106 Sg [266]	bohrium 107 Bh [264]	hassium 108 Hs [277]	meitnerium 109 Mt [268]	darmstadtium 110 Ds [281]	roentgenium 111 Rg [272]	copernicium 112 Cn [285]	ununtrium 113 Uut [286]	ununquadium 114 Uuq [289]	ununpentium 115 Uup [288]	ununhexium 116 Uuh [293]	ununseptium 117 Uus [294]	ununoctium 118 Uuo [297]

*Lanthanide series

lanthanum 57 La 138.905	cerium 58 Ce 140.116	praseodymium 59 Pr 140.908	neodymium 60 Nd 144.242	promethium 61 Pm [145]	samarium 62 Sm 150.36	europium 63 Eu 151.964	gadolinium 64 Gd 157.25	terbium 65 Tb 158.925	dysprosium 66 Dy 162.500	holmium 67 Ho 164.930	erbium 68 Er 167.259	thulium 69 Tm 168.934	ytterbium 70 Yb 173.054

**Actinide series

actinium 89 Ac [227]	thorium 90 Th 232.038	protactinium 91 Pa 231.036	uranium 92 U 238.029	neptunium 93 Np [237]	plutonium 94 Pu [244]	americium 95 Am [243]	curium 96 Cm [247]	berkelium 97 Bk [247]	californium 98 Cf [251]	einsteinium 99 Es [252]	fermium 100 Fm [257]	mendelevium 101 Md [258]	nobelium 102 No [259]

Periodic table of the chemical elements as of 2010. Data are from Michael E. Wieser and Michael Berglund, *Pure Appl. Chem.*, 81(11), 21312156 (2009) and from http://physics.nist.gov/cgi-bin/Compositions/ stand_alone.pl?ele=&ascii=html&isotype=some. From top to bottom in each box are the element name, its atomic number, its atomic symbol, and its atomic weight. For radioactive elements that do not occur in nature the atomic mass of the longest lived isotope is given in square brackets.

Index

Printed in the United States
By Bookmasters